Oil and Wine Presses in Israel from the Hellenistic, Roman and Byzantine Periods

Edited by

Etan Ayalon
Rafael Frankel
Amos Kloner

BAR International Series 1972
2009

Published in 2016 by
BAR Publishing, Oxford

BAR International Series 1972

Oil and Wine Presses in Israel from the Hellenistic, Roman and Byzantine Periods

ISBN 978 1 4073 0505 9

BAR Publishing is the trading name of British Archaeological Reports (Oxford) Ltd. British Archaeological Reports was first incorporated in 1974 to publish the BAR Series, International and British. In 1992 Hadrian Books Ltd became part of the BAR group. This volume was originally published by Archaeopress in conjunction with British Archaeological Reports (Oxford) Ltd / Hadrian Books Ltd, the Series principal publisher, in 2009. This present volume is published by BAR Publishing, 2016.

Printed in England

BAR
PUBLISHING

BAR titles are available from:

BAR Publishing
122 Banbury Rd, Oxford, OX2 7BP, UK
EMAIL info@barpublishing.com
PHONE +44 (0)1865 310431
FAX +44 (0)1865 316916
www.barpublishing.com

This volume is dedicated to Dr. Zeev Meshel,

a colleague, friend and teacher,

on his seventy-fifth birthday

Amos Kloner, Rafael Frankel and Etan Ayalon

Preface

In recent years large numbers of Hellenistic, Roman and Byzantine period oil and wine presses have been excavated in Israel. They are of particular interest because of the remarkable variety of installations found, and because, in spite of the small size of the country, they evince great regional diversity. As a result, several books have been devoted to the various aspects of the subject (Frankel, Avitsur and Ayalon 1994; Frankel 1999), but in the present book for the first time the reader will find in one volume a corpus of short reports on almost all the types of these installations.

The book opens with a short introduction which summarizes what is known about the typology, chronology and regional diversity of these installations, and also discusses relevant references in ancient written sources. The reports are presented and numbered from north to south and east to west, and deal with both oil and wine presses in the same order. Many of the installations have been published previously, but all the reports have been prepared especially for this volume, some by the editors but in most cases by the excavators themselves. The opinions expressed are those of the excavators, but an attempt has been made to use a uniform terminology. In some cases, where the editors' views differed from those of the excavators, editors' notes have been added. A bibliography is attached to each report, and a general bibliography can be found at the end of the volume. Some articles on related subjects also appear at the end of the book.

The preparation of this volume lasted more than ten years, because of our wish to make sure the latest up-to-date discoveries and new data could be included. The reports were received from many contributors in both Hebrew and English, and in varying degrees of detail, that required translating and considerable editing.

We wish to thank the people and institutions that have helped us in preparing this volume: the late Jay C. Jacobson, who translated many of the reports into English; Nili Graicer who prepared the illustrations for publication; the Krauthammer Chair for Archaeology at Bar-Ilan University that gave much financial help; the Israel Antiquities Authority for the permission to include as yet unpublished sites in this book; and last but not least – the contributors themselves.

Etan Ayalon, Rafael Frankel and Amos Kloner

November 2008

Contents

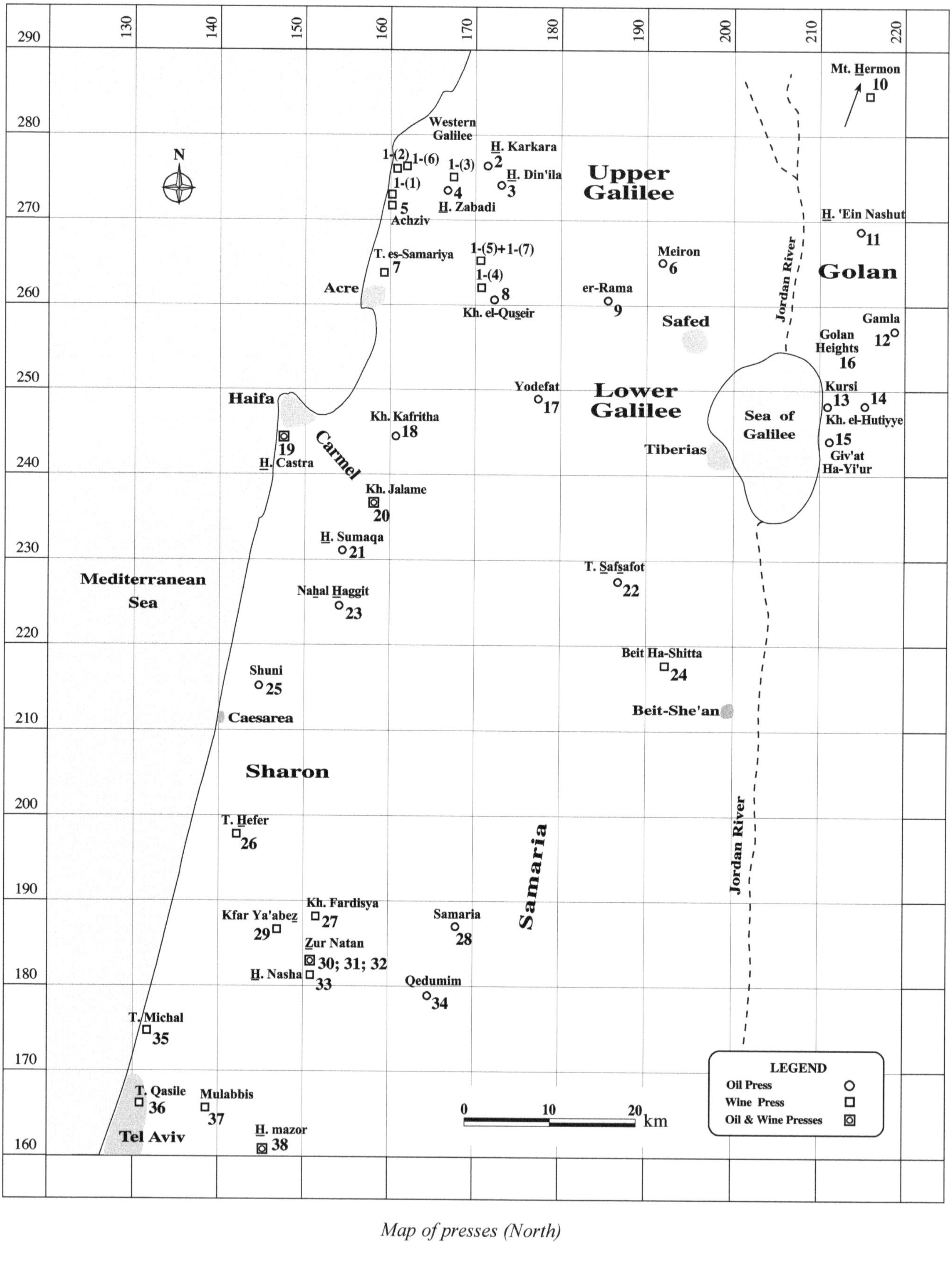

Map of presses (North)

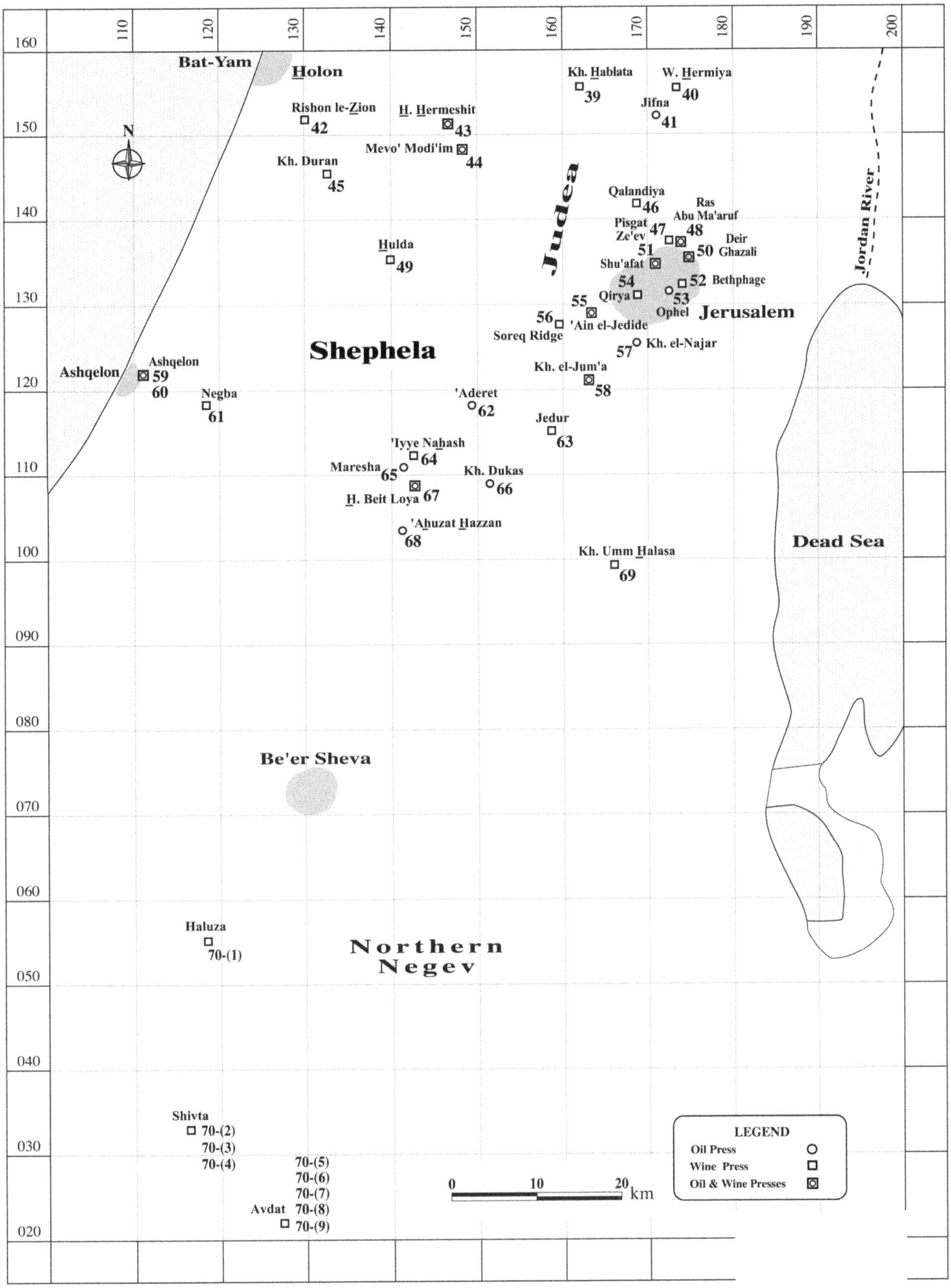

Map of presses (South)

Introduction

Rafael Frankel

In this book we wish to present the reader with a selection of installations for the production of wine and oil from Israel of the Hellenistic, Roman and Byzantine periods.* Many such installations have been found in Israel from earlier periods also but the peak in their development, in the number of installations found, in the technology used and in their variety is towards the end of the Byzantine period. Several factors combined to create this situation. The first is that, particularly in the case of agricultural installations, old techniques were rarely completely abandoned when new methods were introduced and as a result the old and the new are found in use side by side. Another factor is the great regional diversity in types of agricultural installations used, these regional differences often continuing over long periods.

Our knowledge as regards the methods used to produce wine and oil is derived from four main sources. The first of these are the ancient written sources, especially the works of the Roman writers Pliny the Elder, Vitruvius, Cato, Varro, Columella and Palladius, those of Hero of Alexandria, written in Greek but surviving in Arabic, and the Geoponica, also in Greek but of a later date. Of particular importance in our case are the Hebrew and Aramaic Biblical and Talmudic sources as these reflect the situation in ancient Israel itself. The second source consists of the ancient pictorial depictions ranging from Egyptian wall paintings and Athenian Vases to Roman reliefs, Byzantine mosaics and mediaeval manuscripts. The third source is the pre-industrial evidence, installations using traditional methods, many of which were still in use throughout the Mediterranean basin at the beginning of the 20th cent. In remote places some are still operating today thus making it possible not only to describe the installation but also to record the procedures used. Dalman (1928–1942) and Avitsur (1976, 1994) have amassed much data in this field in Israel. The final source is that described in this book, the archaeological remains. As the wooden and other perishable parts of the installations from archaeological contexts usually do not survive attempts at reconstruction must inevitably be based on the evidence from the other three sources and preferably on those, geographically and chronologically close to the installation concerned. In other aspects also only by integrating the knowledge derived from all four sources can the full chronological, technical and cultural significance of the evidence be assessed and understood.

The production of olive oil and wine has much in common. Both involve extracting a liquid from a fruit. For this reason the installations are in some cases very similar and therefore it is not always clear whether installations from archaeological contexts were for producing wine, oil or another product. This is so particularly in the case of simple installations or when only parts of an installation are found.

In both oil and wine production there were three basic stages, the first and second stages in both processes being very similar. Olives were first crushed to produce a pulp, then the pulp was pressed to extract the liquid and finally the pure oil was separated from the black watery lees. In the case of wine the grapes were also first crushed almost always by treading with bare feet. Then the grape skins and stalks, the 'rape', were pressed to extract the remaining must and finally the must was fermented. Although in both processes the fruit was first crushed and then pressed, the difference was that in the case of the olive nearly all the liquid was extracted in the second stage using a press while in that of the grape nearly all the must was extracted already in the first stage by treading. In both cases the substance to be pressed, the olive pulp or the rape was either placed in special containers, sacks or frails (Hebrew/Aramaic – *'eqel*, plural – *'aqalim*; Latin – *fiscus, fiscellus, fiscina*) or enclosed in wooden frames or a thick rope was wound around them spirally to form a cylinder.

In more sophisticated techniques the processes were more complicated. In the case of the olive often a first oil was collected from the olives after crushing and before pressing. This was the "first oil" of Talmudic literature, the finest oil for use in the temple rite (Mishnah *Men.* 8:4; Babylonian Talmud *Men.* 86A), known in Latin as *lixivium* (Columella XII 52 11).[1] Then the olive pulp was often crushed and pressed more than once, the oil being of poorer quality in each consecutive pressing. In the case of the grape the first must (*prototropum* – Greek [Pliny XIV 11 85], *mustum lixivium* – Latin [Columella XII 41; Cato XXIII 2]) that seeped from the grapes before they were trodden was specially prized. After treading and pressing the rape, the remaining skins and stalks were often broken up and pressed again. The must from the pressing of the rape was known in Latin as *mustum tortivum* (Columella XII 36) and that from the second pressing as *circumsicium* (Varro I 54 3). Additional poor quality wines were made

* This introduction is based primarily on Frankel 1999 and there will be found the data on which statements made here are based. However, many of the installations described in this book do not appear there and in certain cases views expressed there have been revised or modified.

by soaking the remaining skins and stalks in water and fermenting the liquid produced. This was known in Hebrew as *temed*, in Greek as *deuterius* (Pliny XIV 12 86) and in Latin as *lora* (Cato XXV; Varro 54 3). Lees wine (*vinum faecatum* – Cato CLIII) was made by soaking the dregs of the wine in water and fermenting them. It was also usual to add a wide variety of substances to the wine including pepper, honey, salt water, and many herbs. Also common was to boil the must down. In Hebrew boiled down must was known as "cooked wine", *yayin mebushal*, but in Latin the must of different concentrations had different names: the most concentrated was called *sapa*, that slightly less so *defrutum* and the least concentrated *carenum* (Pliny XIV 11 80; Columella XII 20 2; Palladius XI 18). These products were used as a sweetener and as an ingredient in cooking but were also added to other must to aid fermentation.

Simple Installations

The simplest installations consist of flat treading floor, slightly sloping, connected by a channel or pipe to a collecting vat. Rock-cut versions are found in large numbers in the countryside throughout Israel particularly in those areas where the rock exposed is "*nari*" that is particularly easy to cut. Built examples, however, are also known as well as examples in which the treading floor and sometimes also the vat were cut into a single moveable slab of stone. This simple installation is without doubt the Biblical *gat* or *yeqeb*, the term *gat* laying the stress on the treading floor and *yeqeb* on the vat. Both are somewhat misleadingly translated as wine press. Here the term winery will be used in order to distinguish these installations from true presses but in the descriptions in this book the term wine press is usually used. There is evidence for the existence of this installation already in the Chalcolithic period. As far as is known at present the vine was extremely rare in the southern Levant in the Chalcolithic period so that these early simple treading installations apparently served for the production of olive oil. The olives were probably crushed on the treading surface using a stone roller, the mash was then perhaps put in a sack and pressed by the weight of stones on the same surface, the liquid flowing into the vat and the oil separated there. Although the grape was apparently grown on a large scale in the region by the Early Bronze Age, in Israel throughout the Bronze Ages there is not always a clear differentiation between installations that were for producing wine and those for oil although one type has been definitely identified as an oil press (Frankel 2006). In the Iron Age there were marked technical developments and wine and oil producing installations from this period are clearly distinguishable. These changes are not as yet reflected in Biblical Hebrew, however, in which the terms *gat* and *yeqeb* still refer both to wine and oil.

The differentiation between the installations for wine and oil production are clearly reflected in Talmudic literature where *gat* refers only to wine production (in Talmudic literature the vat is no longer called *yeqeb* but instead the word *bor* is used) and there is a completely different term for an oil press – *bet bad* ("the house of the beam"). In spite of the fact that the simple treading installations were apparently sometimes also used for oil production particularly in the early period, these treading installations are clearly nearly always for wine production. The fact that wineries are found nearly always in the countryside and not in the settlements themselves shows that they were in the midst of the vineyards as also reflected in Biblical texts (Isa 5: 1–2; Matt 21: 33; Mark 12: 1). This type of winery, consisting of treading floor and comparatively large collecting vat that was in the open and in which the first fermentation was nearly always carried out in the vat, is common in the Levant, Egypt and North Africa. In Egypt a type is found in which the vat instead of being sunken in the ground as in the Levant, is completely above ground that has the advantage that the vat can be emptied through an outlet at its base. In Europe, however, wineries with large collecting vats are rare. There the must was usually collected directly in pottery vessels. The difference is probably because south of the Mediterranean it very rarely rains in the vintage season whereas in the north rain at that time is common.

Developed Wineries

There are very large wineries consisting only of treading floor and collecting vat from later periods (e.g. Achziv [Site 5], 4th cent. CE) but usually in more sophisticated wineries, in addition to these two basic elements, other components and devices are added. These parts are placed in relation to each other in different ways to form various plans some of which are typical of certain regions.

The intermediate vats placed between the treading floor and collecting vat served to clarify the must. In some the outlet was at the top. In these cases the intermediate vats were usually large and apparently acted as settling tanks. Others were small and with outlets at the bottom. In these the must was cleared by sieving.

Other additional components are compartments placed around the treading floor. These vary in form and size and probably served different purposes. Some are large and connected directly to the treading floor and the grapes were almost certainly placed in them to stand for some time before treading, a practice recommended by Columella (XII 27) to produce sweet wine. Others are similar but are connected to the treading floor by a pipe, in some cases having also small intermediate vats between the compartments and the main treading floor. These were probably to allow for separate collection of the first must that seeped from the grapes before treading. Two interesting and unusual examples are those from Tel Hefer (Site 26) and H. Mazor (installation M4; Site 38) in which the small collecting vats of the subsidiary treading floors face away from the main treading floor. In these cases the first must could be collected without entering the main floor and interfering with the treading of a previous batch of grapes, and after the first must was collected the grapes could then be easily transferred from the ancillary surfaces to the main treading floor. Another exceptional winery is that from Mulabbis (Site 37) where the auxiliary floors are built above compartments with vaulted roofs and the first must flowed through openings in the floors to the compartments below. A third type of compartment is much smaller and often roofed with miniature half cupolas (e.g. Tell Qasile [Site 36]; and round examples at H. Nasha [Site 33] and Zur Natan [Site 31]). These were probably used to contain the various substances that were added to the must.

Other additional components are the presses. The discussion of these, however, will be postponed and will appear below together with that on the oil presses.

A component that has led to considerable controversy is the large slotted roller. This device is found almost only at sites in the Mt. Carmel region and is associated with installations that are considered by some scholars to be wineries, who explain that they were used to press the grapes either after or instead of treading. Others, however, consider these rollers as so unusual that they suggest that the installations with which they are associated are not wineries and served some other unspecified purpose (e.g. Dar 1999: 92–94). It must be pointed out that rollers without slots are also found in wine presses particularly, but not only, in eastern Lower Galilee (Frankel 1999: 146–147, Map 38; see also, for instance, the installation at Beit Dagan [Peilstöcker and Kapitaikin 2000: 59*-60*]; and paper by Amit and Baruch in this book: 429–440).

Certain clearly defined plans of wineries can be distinguished. The simplest plan is the Linear Plan in which the treading floor, intermediate vat and collecting vat are all on one axis. This is the main type in Galilee (e.g. Western Galilee [Site 1]). In the Galilean wineries the intermediate vat is usually a small sieving vat or even only a small sieving cavity on the edge of the vat. The plan found mainly in the central regions of the country is the Four Rectangle Plan in which the intermediate vat, usually large, and the collecting vat are next to each other, the treading floor and the two vats fitting neatly into a fourth rectangle (e.g. Tel Michal [Site 35] from the Hellenistic period and from later periods Shu'afat [Site 51]; Kh. Fardisya [Site 27]; Zur Natan [Site 31]; H. Hermeshit 401–402–407 [Site 43]; Tell Qasile [Site 36]; H. Nasha [Site 33]).

Another type found mainly in the Negev highlands (Site 70) has a Complex Plan consisting of large compartments connected directly to the treading floor, two round collecting vats with an intermediate vat in common and an unusual type of screw base in the centre of the treading floor connected to the vats by pipes or channels. In these wineries three consecutive batches of grapes could be processed simultaneously, one in the compartments, one being trodden into one vat and one fermenting in the other vat. Wineries that are extremely similar are found further north also, near Ashqelon (Site 59), at Negba (Site 61), at Kh. Duran-Rehovot (Site 45) and at H. Castra at the foot of Mt. Carmel (Site 19). They also have two collecting vats with an intermediate vat in common and large compartments around the treading floor. There are, however, also differences. The collecting vats are not round as in the Negev wineries. In the Negba winery and winery C at Ashqelon the vats are octagonal and those in wine press E at Ashqelon and in that from Kh. Duran are square. The main difference between the northern examples and those in the Negev is that in those from the north the compartments were connected to the main floor by pipes allowing for the separate collection of first must. The famous Gaza and Ashqelon wines known from Byzantine documents and clearly connected to the well known Gaza and Ashqelon amphorae were almost certainly produced in these wineries (Mayerson 1985, 1992, 1993). The difference between those from the Negev and those further north was that the Negev wineries could produce one type of wine only, probably for export, whereas those further north, nearer population centres could also produce quality products probably for the local market. Around Jerusalem there were also several developed wineries not uniform in plan and lacking the two round vats with common intermediate vat of the Negev wineries but like the northern examples of the latter having the capability to produce various products for the local market (Ras abu Ma'aruf [Site 48]; Bethphage [Site 52]).

The floors of built wineries of the Byzantine period are usually of mosaic as are some from the Late Roman period while those from that period and earlier are usually plastered. A remarkable exception is Tell es-Samariya (Site 7) where wineries with mosaic floors have been dated to the Hellenistic period.

One interesting question is that of the production of red wine. Although red wine is mentioned frequently both in classical and Talmudic literature the technology of the producing of this type of wine is apparently not referred to in either. In modern practice the main difference between the method of production of white wine and of red wine is that in the latter case the grape skins are left in the must during the fermentation, this imparting the colour to the wine. This was presumably true in antiquity also. As regards the types of wineries discussed above, the existence of intermediate vats would clearly impede the transference of the grape skins from the treading floor to the collecting vat, as would the connection between the floor and vat being a pipe. Therefore the explanation for the fact that a very large and beautifully made winery at Achziv (Site 5), for example, lacks intermediate vats could be that it was intended to produce red wine.[2]

Oil Production

Crushing the Olives

The simplest ways of crushing olives were either by using a stone roller or a mortar and pestle and there is archaeological evidence for the use of both methods from the Iron Age. The use of rollers in wine presses in later periods has already been mentioned and they were sometimes also used for crushing olives in later periods, as shown by pre-industrial examples. See, for example, the photograph taken in Jerusalem in the early 20th cent. of two women using this method (Frankel 1994 Fig. 19).

From the Hellenistic period, however (see, for instance, Yeivin 1966; Magen 1993a, 2000), and perhaps from slightly earlier the main method has been to use the rotary olive crusher. In this device a round vertical crushing stone (an edge runner) rotates both on its own axis and also around a circular crushing basin. The axle on which the crushing stone rotated was also the handle that was operated either manually or by an animal (see Fig. 8.7). The axle usually turned on or was inserted in a pivot fixed in a socket in the centre of the basin. This, however, was not always the case. Sometimes there is no socket and then there must have been two crushing stones on one axle that kept each other in position (H. Karkara, H. Zabadi [Sites 2, 4]). In crushers with sockets, however, there was clearly often or perhaps usually only one crushing stone. In most cases the crushing surface is flat and the crushing stones wheel-shaped but in others the crushing surface is concave and the crushing stones convex (Maresha [Site 65]). The basins of the crushers with flat surfaces vary. Some have sunken sockets (Zur Natan [Site 30]), others raised sockets (Golan [Site 16]) while others have neither sockets nor central protrusions (H. Beit Loya [Site 67]). There are also crushers with convex crushing surface but without a socket, some with central protrusions (H. Karkara, H. Zabadi [Sites 2, 4]) and some without (H. Din'ila [Site 3]). The origin and development of the rotary olive crusher and the historical relationship between the various types are not clear and await the results of further research.

There is some controversy as to both the Latin and Hebrew terms used to describe the rotary olive crusher and crushing stone. The type found in the Campanian *villae rusticae* around Pompeii was that with convex crushing surface which led scholars to identify it with the *trapetum* referred to by Cato (XX-XXII) and other Latin writers on farming matters. As here Brøndsted (1928: 111–112) suggested that the crusher with flat crushing surface, that he thought to be later was the *mola olearia* mentioned by Columella (XII 52 6) who lived some two hundred years later than Cato. In the southern Levant, however, both these types appear side by side already in the Hellenistic period, and Brøndsted almost certainly misinterpreted the Latin texts. The Latin term for both types was probably *trapetum* while the *mola olearia* was apparently a type of olive crusher found in North Africa and Spain that was very similar to the *mola*, the Donkey Mill used for grinding wheat (Frankel 1993a).

As for the Hebrew terms, it has long been accepted that the crushing basin was the *yam* and the crushing stone the *memel*, terms appearing in a Mishnaic legal dicussion. In this case also, however, the text has almost certainly been misunderstood and the terms used were *rehaym 'elyonot* – upper millstone and *rehaym tahtonot* – lower millstone (Frankel 1999: 188–189).

Oil and Wine Presses

The use of a press is a basic part of the process of oil production but was usually also part of that of wine production. One type of press, that of the single fixed screw, usually found in the centre of the treading floor of wineries, was apparently used only in wine production and in some other cases also the presses used in wine production are slightly different from those used for oil. Essentially, however, the types of presses used in both processes were the same and they will here be discussed together.

The simplest method of pressing was by the direct pressure of weights placed on a board laid directly on the material to be pressed. This method is attested for in pre-industrial procedures and in Talmudic literature, but does not leave any obvious archaeological remains and will not concern us here. Another method attested for in Egyptian wall paintings from the Pharaonic periods and also from pre-industrial examples is the sack press that was operated by torsion, by twisting the sack. It is possible that the two cup marks often found on either side of the treading floor of simple wineries served to support the frame of such a sack press.

Lever Presses

The first type of press that exploited the principles of mechanical advantage was the lever press in which a beam acting as a lever was anchored at one end (Fulcrum) and force exerted on the free end of the beam (Effort) brought pressure to bear on the material to be pressed that was placed between these two points below the beam (Load). The closer the material to be pressed was to the fulcrum the greater the mechanical advantage achieved. A great variety of lever (beam) presses developed varying at all three points (the anchoring point – Fulcrum; the pressing point – the frails – Load; and the pressure point – weights etc. – Effort). They are usually classified, however, by the method the force (Effort) was applied. The main types were 1) The Lever and Weights press; 2) The Lever and Drum Press; 3) The Lever and Screw Press.

Lever and Weights Press; Lever and Drum Press

The earliest clear evidence for lever and weights presses (oil presses) is from the last stages of the Late Bronze Age from Cyprus (Hadjisavvas 1992: 21–27) and Syria (Callot 1987). The earliest evidence from Israel is from Tel Dan in Iron Age I (Biran 1980; Stager and Wolff 1981; Borowski 1982).

In Iron Age II different types of oil presses can already be distinguished in different regions and there is also evidence for the use of lever and weights presses in wineries. In ancient Israel in the Iron Age the main distinguishing feature in the oil presses was the manner in which the oil was collected. In the north in nearly all cases the oil was collected laterally. The olive mash was pressed on a press-bed and the liquid extracted flowed to a lateral vat.

This was the usual method throughout the ancient world. In the south, however, most of the presses were equipped with central vats that combined the functions of the press-bed and the lateral collecting vat. The frails of olive mash were pressed on crisscross laths or perhaps on a perforated board placed over the opening of the central vat, the liquid flowing directly down into the vat (Fig. 1: a). As we will see this unusual method of oil collection continued into later periods but remained a local technique and hardly spread out of a limited area in the southern Levant. The weights used in the simple presses of the Iron Age varied slightly but were all with one bore only.

The lever and weights presses used in wineries were usually very simple, the evidence for their use being only a small hole in the wall of the treading floor. The grape skins and stalks remaining after treading were apparently pressed directly on the treading floor and the weights probably hung over the vat (Bet Ha-Shitta [Site 24]; Ras Abu Maʻaruf – first stage [Site 48]). One unusual type found is the open topped niche with small holes on either side to insert a rod to secure the beam. This device is found mainly in the Jerusalem area (e.g. Ras et-Tawil [Site 47] and Soreq Ridge [Site 56]) but has also been found in other areas (e.g. H. ʻAqav on Mt. Carmel [Hirschfeld 2000: 32–36] and Zur Natan in western Samaria [Site 31]).

In the Hellenistic, Roman and Byzantine periods several different types of lever and weights presses can be distinguished. The typical southern press is the Maresha press (Maresha [Site 65]), twenty two examples of which have been found at the site of Maresha alone and many more in other parts of Judea. It is very similar to the lever and weights press found in the region in the previous period (Fig. 2: a). The beam is anchored in a hole in the wall and the liquid collection is central. There were three main innovations. The first was the pair of plain piers placed on either side of the central collecting vat that served to support the frails. The second innovation was the type of weight that was with a bore in the form of a reversed T, and the third was the fact that the weights stood in a special pit. Another type of press found in the north-western regions of the country is the Zabadi press (H. Zabadi [Site 4]; H. Karkara [Site 2]; Fig. 2: b), an installation that was different from the Maresha press in almost every way. The collection was not central but lateral. There were also two piers but in this case they were slotted piers and placed not at the pressing point but at the fulcrum and their purpose was to anchor the beam. There were two collecting vats connected by a small bore at the top thus allowing only the oil floating on the watery lees to flow into the second vat. The weights as in the south were in a special pit. They were, however, of a different type having a small depression filled with lead that clearly served to secure a hook or ring. There is pre-industrial evidence for the use of both types of weight in the region and in both cases the weights were raised using a winch, a wooden rod with a rope wound round it, attached to the beam, either by hanging below it or by resting on top of the forked end of the beam (Dalman 1928–1942, IV: Pls. 56–57; note that the press from et-Tafile [Pl. 56] also has central oil collection). It is almost certain that this was also the method used in the presses of the southern Levant in antiquity.

In the other areas of the country between the regions where these two main types were dominant other types of presses are found, these often combining elements from both types. In Samaria many of the presses have lateral collection with plain piers (e.g. Qedumim [Site 34]). There is also a local variety of collecting vat with a stone lid having two openings, through one of which the liquid flowed into the vat while from the other the oil was taken out (e.g. Samaria [Site 28]; Qedumim [Site 34]). Around Mt. Carmel there are examples of the opposite, presses lacking plain piers but with central collection (e.g. Kh. Kafrita [Site 18] and H. Castra [Site 19] presses E and F).

One element found in some presses the use of which is illustrated by the written sources is a central collecting vat unconnected to any pressing equipment. On these vats frails of crushed olives were left standing to allow the first oil (*lixivium*) to seep into the vat. This oil was considered to be of particularly high quality. The remaining oil was extracted later in an ordinary press (e.g. Kh. el-Quseir [Site 8] collecting vat A2, H. Castra [Site 19] oil presses C, E, and Qedumim [Site 34] area C).

There are two types of beam presses that are common in other parts of the Mediterranean basin but are either absent or rare in Israel. The first is that with the Semana weight, a rectangular weight with dovetail mortices at either end. This weight is found in the Aegean, southern France, Spain and North Africa. It operated using a winch attached to the weight. Weights somewhat similar in form and apparently working on the same principle were found at Ashqelon (Site 60) and at nearby sites.

Another type is the beam and drum press described by Cato (XVIII oil press ; XIX wine press) and by Pliny the Elder (XVIII 74 317) and found both in archaeological and pre-industrial installations in Italy, former Yugoslavia and France (Fig. 1: b). In this press there is no weight and the beam is lowered by a rope and drum attached to the ground. This can be regarded as the typical Roman press and probably developed from the Semana Press (Frankel 2001). There is no evidence for the use of this type of press in ancient Israel or in other parts of the southern Levant.

Lever and Screw Press

The lever and weights press continued in use after the introduction of the more sophisticated screw presses until modern times. Of particular interest is the fact that at H. Castra (Site 19) there was a return to the use of the lever and weights presses after the use of screw presses. In fact in one case a screw base served in secondary use as a beam weight.

From ancient written sources (see below) we learn that the screw was already in use in the 1st cent. BCE and in

Israel a small number of screw presses have been found that can be dated to the Roman period. The archaeological evidence, however, suggests that screw presses only came into general use in the Byzantine period.

In this type of press a screw attached to the free end of the beam lowers the beam and exerts the pressure (Fig. 1: c). In western Europe there are pre-industrial lever and screw presses in which the lower end of the screw is attached to the press frame or to the ground. As the wooden components survive only in very exceptional circumstances, such presses would leave no recognizable archaeological remains. In all the pre-industrial examples from the eastern Mediterranean, however, the lower end of the screw was attached to a stone screw weight and a great variety of screw weights have been found in archaeological contexts throughout the Mediterranean basin providing clear evidence for the use of this type of press (*contra* the reconstruction in Magen 2008: 317–319. Magen's dating of all these presses to the Early Moslem period also contradicts the archaeological finds).

The ancient screw weights found in Israel are usually cylindrical, the average height and diameter being approximately one metre (Fig. 3). Typologically they can be divided into two groups. The first group consists of those with a round socket in the centre of the upper surface and with additional mortices that vary in form thus distinguishing the specific types. The second group has no socket but instead a mortice in the centre of the upper surface, in this case the form of mortice distinguishing the various types. The two groups clearly worked on different principles.

The weights with sockets are common in pre-industrial contexts (although they are smaller than the ancient weights) and these demonstrate how they functioned. The bottom end of the screw was usually in the form of a small sphere, connected to the screw by a narrow neck. In some cases outside Israel a metal pin with a spherical end was attached to the bottom end of the screw. The wooden sphere or metal spherical pin end was inserted into the socket of the weight and was held by a board in which there was a round aperture, or by two boards in each of which there was a semicircular aperture, that fitted round the neck (see Avitsur 1994 Fig. 118). These boards were then attached to the weight with the aid of tenons that were fixed in the mortices on the weight. Thus the screw was secured to the weight in a way that allowed it to revolve in the socket but not to work free from the weight. Screwing the screw into the nut that was fixed to the free end of the beam first lowered the beam and when the beam could descend no further raised the weight from the ground.

There are apparently no exact pre-industrial parallels for the second type, the weights without sockets. Clearly, however, in this case the screw was fixed to the weight in the mortice and did not rotate. There are pre-industrial lever and screw presses in which the screw is fixed and the nut is turned (e.g. the Fenis press – Drachman 1932: 54, Fig. 13), and this is apparently the way these weights functioned. If this is so, when the weights were raised in the air theoretically the nut, screw and weight would all turn together. There are, however, several ways in which the screw and weight could be held to allow the nut to be turned and this is almost certainly the way these presses were operated although there is no archaeological evidence to show exactly how this was actually done.

Six main types of screw weight are found in Israel – three of the first group with sockets and three of the second without sockets. Those with sockets are 1) the Samaria weight, 2) The Kasfa weight and 3) the Dinʿila weight.

1. The Samaria weight (Fig. 3: 1) has two external mortices usually dovetailed but sometimes in the form of a reversed T and sometimes there are four mortices instead of two. The Samaria weight is common in the centre of the country (e.g. Samaria [Site 28]; Shuni [Site 25]; Zur Natan [Site 32]; Ras Abu Maʿaruf [Site 48]) and also in other parts of the Mediterranean basin. This type was probably introduced from the western Mediterranean.

2. The Kasfa weight has two internal mortices, closed dovetail mortices connected directly to the socket (Fig. 3: 2). This type is found in small numbers in the centre of the country (e.g. Shuʿafat [Site 51]; H. Mazor [Site 38];) but is common in North Syria and was probably introduced from there. A very similar type, however, is also found in southern France.

3. The Dinʿila weight has a dovetailed channel, two open dovetail mortices connected to the socket (Fig. 3: 3). This type is found only in Upper Galilee and on Mt. Hermon (e.g. H. Dinʿila [Site 3]; H. Karkara [Site 2]) and it is probable that it developed in this region or in its vicinity. It should be pointed out, however, that pre-industrial screw weights that are very similar to these are found in Portugal and also in Sicily.

The three types of weights without sockets are 4) the Miʿilya weight, 5) the Bet Ha-ʿEmeq weight and 6) the Luvim weight.

4. The Miʿilya weight is the simplest of the screw weights found in Israel, having a closed dovetail channel (Fig. 3: 4). The few examples so far identified are also found in Upper Galilee and it also probably developed in this region (e.g. Kh. el-Quseir Press A [Site 8]).

5. The Bet Ha-ʿEmeq weight has a central dovetail mortice in the centre of its upper surface (Fig. 3:5, 3a). This type of mortice consists of an approach channel with vertical sides connected to a dovetail mortice with sloping sides. The object to be secured, in this case the screw, terminates in a dovetailed tenon that is first inserted into the straight-sided approach channel from above, then pushed sideways into the mortice and there wedged into position. This type of weight is typical of Upper Galilee as is also this type of mortice (Kh. el-Quseir Presses B & C [Site 8]).

6. The Luvim weight has a square mortice widening on two adjacent sides (Fig. 3: 6). In this case a square ended screw projecting in two directions was inserted into the mortice. This type of screw weight has not yet been reported *in situ* in a complete press but several examples have been recorded in surveys, up to now mainly on the Carmel and in the Sharon coastal plain. Square mortices of various types are found in other types of installations in the central and southern regions of the country.

There are other types of screw weights of which individual examples have been identified. For two of these clear parallels have been found in other countries suggesting that it was from there that they were introduced (H. Beit Loya [Site 67]; 'Ain el-Jedide [Site 55]) (Frankel, Patrich and Tsafrir 1990).

The complete oil presses shown by the presence of screw weights to have been lever and screw presses are in other respects often identical to the lever and weights presses found in the same regions (e.g. H. Karkara – H. Zabadi [Sites 2, 4]; Maresha – H. Beit Loya [Sites 65, 67]). Some elements, however, are associated specifically with lever and screw presses. One of these are perforated piers that like the slotted piers served to anchor the press beam (e.g. H. Din'ila [Site 3]; Kh. el-Quseir Presses B, C [Site 8]). The difference between the two types is that the slotted piers allowed for adjustment of the height of the fixed end of the beam, whereas in presses equipped with perforated piers the beam end was at a constant height. Apparently the advantages and greater efficiency accruing from the change to the use of the screw led to pressing routines that no longer entailed readjusting the height of the beam end during pressing.

Another element connected specifically to lever and screw presses are two holes on either side of the beam. Those found on either side of the middle section of the beam were for posts that kept it in position (H. Beit Loya [Site 67]; Samaria [Site 28]). In the three oil presses at Kh. el-Quseir (Site 8) there were two holes on either side of the free end of the press beam. The screw weights of the Quseir presses show that in these three presses it was the nut that was turned and not the screw. In presses in which the screw is turned and the nut fixed turning the screw in the reverse direction raises the beam. However, in presses in which the nut is turned, turning the nut the other way will not have this effect so that in the case of the Kh. el-Quseir presses the posts secured in the holes were probably to hold tackle that raised the free end of the beam after pressing.

Direct Pressure Frame Presses – Single Screw, Double Screw and Wedge Presses

From pre-industrial and ancient pictorial evidence we know of several types of presses that exert direct pressure without the aid of a beam\lever. These include various types of single and double screw presses and wedge presses (Fig. 1: d). The stone bases of all of these presses except one are similar. They consist of a press-bed or central vat on either side of which are mortices the purpose of which was either to secure the press frame of a single screw press or of a wedge press or to secure two screws of a double screw press. All that usually remain of ancient presses are the stone elements. Therefore as opposed to pre-industrial examples in which we know the exact form of the press, in archaeological contexts when we find a press-bed with two mortices, we cannot be sure of what type the press was. Typological discussion of the archaeological remains of direct pressure presses must therefore be based on different criteria from those used when discussing pre-industrial examples. One exception is the single fixed screw press, the archaeological remains of which are different from those of the other presses consisting of a single mortice only. This type will therefore be discussed separately.

There are, however, two types of press, in which the super-structure is also of stone so that in these cases we know the exact form of the press. The first of these is the Grooved Pier Press ("the Judean Pier Press") that consists of two piers standing on either side of a central vat, each with vertical grooves on the posterior and anterior faces and on the lateral faces square mortices connected to the top of the stone by short channels and to each other by a bore. Dalman (1928–1942, IV: 226–227) showed that these piers were to hold a single rotating screw press; the square mortices secured tenons that held the threaded board in which the screw rotated that was placed above the piers. B. Frankel (1981a, b) demonstrated that the posterior and anterior grooves served to hold rods that helped to withstand lateral pressure and keep the piers in position (see article by A. Kloner on this type of press in present volume and reconstruction Fig. 66.1).

Recently Yehoshua Dray has suggested that in this press during pressing the two piers were raised into the air and in effect acted as a pair of weights. Dray has actually built a full scale model that works in this manner. It is almost certain that sometimes the piers were raised from the ground just as in some cases the slotted piers of beam presses were probably momentarily raised from the ground, but it is very doubtful that this was the way the Judean grooved pier press was intentionally and consistently operated (see Dray's article in present volume).[3]

The diffusion of this standard type is very clearly defined and is limited to the southern parts of the country – Judea and the Shephelah. There are, however, also variants of the standard type and these are found in other parts of the country including Galilee (e.g. Tel Safsafot [Site 22]).

The second type of press with stone super-structure is the Cross Press that is cut into vertical rock faces, usually inside caves (Fig. 4). The threaded board was in the arms of the cross and the screw in the vertical section (Ophel, Jerusalem; Jifna [Site 41]; Kh. Marah el Jum'a [Site 58]).

In the basalt areas of the Golan and of eastern Galilee the 'Ein Nashut press base is found consisting of a basalt slab, rectangular with rounded corners, with round press-bed in the centre and two open mortices on either side. A sub-type found in the centre of this region around Sea of Galilee has a central collecting vat with radial grooves, and two examples also from this area have central openings, probably to allow the expressed liquid to flow directly to a removable vessel placed below them. The Zippori press in which the screw base is rectangular with square mortices is found mainly in the centre of the country and is usually of limestone.

Another variant is the Rama press in which there are central dovetail mortices on either side of the press-bed. This type is found in small numbers in northern Lower and eastern Upper Galilee. A somewhat similar press is the Mishkana press in which the mortices are square but with an approach channel. Only three examples of this press have been recorded, all in south-eastern Lower Galilee.

Single Fixed Screw Press

In Israel in antiquity the types of presses described so far served in most cases as oil presses although presses of similar types described in classical literature and similar pre-industrial examples from Europe were very often wine presses. In rare cases presses of these types were also found in archaeological contexts in Israel associated with wineries and were clearly also wine presses. There is, however, one type of installation that was apparently used only as a wine press and most of the large Byzantine wine presses in the southern Levant were of this type.

The archaeological evidence for this press is the presence of a single mortice usually located in the centre of the treading floor of a winery (Figs. 1: e; 5). The mortice was either cut into a stone that was then emplaced there or cut directly into the bed-rock floor. This mortice secured a fixed screw. The grape skins and stalks remaining after treading were piled up around the screw and a pressing board placed on them, the screw protruding through a hole in the centre of the board. A nut that was turned on the screw exerted pressure directly on the pressing board and thus on the rape, pressing out the remaining must.

In the more sophisticated wineries the mortice was connected directly to the vats by covered channels or by pottery or lead pipes (Tell Qasile [Site 36]; Bethphage [Site 52]).

These press mortices varied in several ways: in the form of the press-bed, in the form of the mortice and in whether or not the mortice was connected directly to the collecting vat (or intermediate vat) by a pipe or channel.

The press-beds were usually round (Kfar Ya'abez [Site 29]) or rectangular (Tell Qasile [Site 36]; Bethphage [Site 52]; Shu'afat [Site 51]), the explanation being that on the round press-bed the material to be pressed was held in place by a wound rope while on the rectangular bed it was enclosed within a wooden frame.

As in the screw weights there is a marked regional variation in the types of mortices used. In Upper Galilee the only type found was the central dovetail mortice – the Hanita Press (see Fig. 3a), while in other parts of the country most of the mortices were square – the 'Ayalon Press (Kfar Ya'abez [Site 29]; Shu'afat [Site 51]; Ras Abu Ma'aruf [Site 48]), although some were dovetailed as in the north (Wadi Hermiya [Site 40]; Bethphage [Site 52]).

There is also some variation in the form of the square 'Ayalon Mortices. Some widen at the bottom on one side, some on two adjacent sides and in some the mortice is open from the bottom. In the latter the screw was probably inserted from below before the stone was put in position. In the square mortices of the Negev wineries (Site 70) there is no obvious way to anchor the screw and as they are connected to the collecting vats by channels Mazor has suggested that these are not screw bases but sumps to allow the must to flow directly to the vats. In this case, however, the screw bases were apparently of wood and the screw end protruding below was emplaced in the square mortice without being fixed in position.

It is of great significance that while this type of press is not mentioned in classical literature nor is there archaeological evidence for its presence in Europe in antiquity, in the southern Levant in addition to the many archaeological examples it also appears in depictions in mosaic pavements both in Jordan and Lebanon and is hinted at in Talmudic literature (Frankel 1999: 192–193).

Oil Separation

The final stage in the production of olive oil is the separation of the oil from the watery lees (by weight olives consist of c. 20% oil, 40% solid waste and 40% watery lees). All methods of separation are based on the fact that the oil is lighter than the watery lees and tends to float on them. The simplest method of separation is to skim the floating oil off using a shell or a ladle. In Latin the man who ladled the oil was called *capulator* and the expressed liquid was often poured from one pan to another so that the oil should rise faster (Cato LXVI; Columella XII 52 11 12). There were also more sophisticated methods of separation. In overflow decantation the floating oil flowed through an outlet at the top of the vat or jar into a second receptacle. In underflow decantation first the lees and then the oil flowed out through an outlet at the bottom. In the western Mediterranean there was a device that combined both methods allowing the oil and the lees to flow out through different outlets (Frankel 1999: 174–176). In Israel presses with lateral oil collection such as those at H. Zabadi and H. Karkara often had two connected collecting vats and the oil was separated by overflow decantation. In presses with central collection there are often additional vats in the vicinity and these were probably to separate the oil by skimming.

Conclusions - Technical Development and Regional Diversity

Almost all the different types of installations described in the previous paragraphs functioned side by side at the end of the Byzantine period, creating a rich and variegated mosaic of regional types. This was the product of creative technical development within the framework of several different parallel regional cultures.

This phenomenon is manifested to the greatest extent in the history of the press. A basic technical sequence of press types can be presumed that is not only technologically logical but is reflected in classical texts (Pliny; Hero of Alexandria). 1) Lever and Weights Press; 2) Lever and Drum Press; 3) Lever and Screw Press; 4) Direct Pressure Presses: Screw and Wedge Frame Presses and Single Fixed Screw Press. In this development the two most important changes were the introduction of the lever and the introduction of the screw. The earliest clear evidence for the use of the lever in pressing is from Cyprus and the northern Levant (Ugarit) in the Late Bronze Age. From the written sources (e.g. Pliny XVIII 74 317; Vitruvius VI 6) we learn that the screw was already in use in the 1st cent. BCE but amongst the archaeological remains in the southern Levant there are few examples from the Roman period and nearly all the screw presses are from the Byzantine period or later. Fine examples of Lever and Weights Presses are still found, however, in the Byzantine period and even later, including pre-industrial examples.

Not all the types mentioned above appear in all the regions, however. In Israel, for instance, the Lever and Drum Press is not found; in antiquity the Single Fixed Screw Press does not appear outside the southern Levant and in Italy there is apparently no evidence for the Lever and Weights Press, there the main press being the Lever and Drum Press.

In Israel two integrated primary local technical cultures can be discerned, that of Judea in the south and that of Phoenicia in the north-west, while other subsidiary cultures existed in Samaria, Lower Galilee and the Golan, etc. In the Jerusalem area the technical culture was largely eclectic in character.

The first difference between the Judean and the Phoenician culture is that in the former the oil collection was central while in the latter it was lateral. At a later stage pairs of piers were added to the presses: in the south the Judean Plain Piers, and in the north, the Phoenician Slotted Piers. These two types of piers were completely different in form, function and in their position in the press.

With the introduction of the screw to the southern Levant in each region it was exploited in a different way, in each case in that best suited to the installations that already existed in the region previously. In Judea there are very few examples of the Lever and Screw Press and the main type was the Grooved Pier Screw Press. This was clearly adapted from the Maresha Lever and Weights Press in which there were also two piers on either side of a central vat. All that was needed was to add grooves and mortices to the piers and add the threaded board, tenons and screw. Another reason that this type was preferred could have been a paucity of tall trees to supply press beams as compared to regions further north.

In Western Galilee the dominant press was the Lever and Screw Press, again enabling the exploitation of press elements from previous periods, in this case the almost monumental slotted piers of the Zabadi press.

Another marked difference between the two regions is the preference of square mortices in the south and of dovetail mortices in the north, especially the central dovetail mortice.

A particularly interesting aspect of the subject of ancient presses is the regional diversity of the types of mortices used in single fixed screw wine presses and in screw weights. The development of these two types of devices was clearly interconnected.

The single fixed screw wine presses (the Hanita and 'Ayalon presses) are closely related to those lever and screw presses in which the screw is fixed and the nut turned (those using the Bet Ha-'Emeq, Luvim and Mi'ilya screw weights). Both work on the same principle, the screw is fixed and the nut turned; both use the same regional mortice types, the square mortice in the south and the dovetail and central dovetail mortice in the north and perhaps most important these installations and devices are known only from the southern Levant and are unknown in ancient Europe. They appear neither in the Classical texts nor in European archaeological contexts while in the Jerusalem Talmud there is apparently an indirect reference to the fixed screw wine press (see below). This all suggests that these installations developed in the region.

Of the other screw weights, those in which the screw was turned and the nut fixed, the Samaria weight clearly originated in the western Mediterranean, but the Din'ila weight is found primarily in Upper Galilee and was based on the dovetail channel as were the other two northern weights: the Bet Ha-'Emeq and Mi'ilya weights and probably derived from them. It was suggested above that the Kasfa weight probably reached Israel from northern Syria or from a region connected to it but technically it can be regarded as an improved version of the Din'ila weight that developed somewhere in Phoenecia or Syria (Frankel 1993b; 1997).

All these devices were in use side by side in the Byzantine period but as yet there is no archaeological evidence that can aid in dating the introduction of each type. Hero of Alexandria, however, who wrote at the latest in the 1st cent. CE describes a screw weight which conforms exactly to the Din'ila weight. Therefore if the model proposed here is correct, that the Din'ila weight developed from the other dovetail weights then the various devices using a fixed screw were already in use even earlier.

The installations presented in this volume reveal the remarkable variety of techniques and devices found in one small section of the complicated mosaic of local technical cultures that were spread throughout the Mediterranean basin, each developing separately but influenced by and influencing the others. Even techniques such as the use of the screw developed in different ways in different regions. The extent and borders of these technical cultures are in many cases closely related to those of political entities changing in extent and character together with these.

Thus the study of these ancient crafts not only reveals important aspects of ancient technology, economics and day to day life but mapping the variegated regional technical cultures contributes a new and independent delineation of ancient human geography.

Some References to Oil and Wine Presses, Oil Mills and Wineries in Ancient Literature

Classical Literature--Wine Production

Treading

Although there are many references to the treading of grapes, the written sources usually take this part of the process for granted and the only ancient detailed description of treading as such is apparently that in the Geoponica (VI 11) that dates to the Byzantine period. There it is stressed that the workers who did the treading should be fully dressed, wearing girdles and should wash their feet and that both before putting the grapes in the winery (*lenos*) and during the treading, they should remove leaves, sour grapes and dry bunches. After treading once, they were to move the kernels so that the liquid flowed into the vat (*upolenion*), then to tread them again and finally to press them. The type of press was not specified, the instructions being to place the grape skins under a pressing board.

The only actual description of a winery/treading installation in classical literature is apparently that of Palladius (I 18), one of the later Roman agronomists who probably wrote in the 4th cent. CE. The raised treading floor (*calcatorium*, a very rare term) was between two collecting vats (*lacus*) from which channels or earthenware pipes took the must to large storage jars (*dolia*) and to wooden(?) casks (*cupae*). In this installation the *lacus* was in effect an intermediary vat and not a collecting vat like those of the Levant.

Presses

Cato the Elder. The earliest detailed description of a press is that of Cato the Censor (the Elder), the Roman statesman (234–149 BCE). It appears in his handbook *De Agricultura*, the earliest surviving agricultural textbook. He describes how to build a pressing room in which there were four olive presses (XVIII) and four olive crushers (XX, XXI, XXII) and also describes a wine press that was very similar to the oil presses (XIX). The presses were lever and drum presses. The beam (*prelum* – 7.4 m long) was anchored between two piers (*arbores*) with slots (*forminae*). These slotted piers were secured in a pier base (*forum* – Brun 1986: 242, n. 34; *lapis pedicinus* – Drachman 1932: 105–106). The cross pieces that were inserted in the slots were perhaps termed *cunei*. On the press-bed (*ara*) was a circular groove (*canalis rotundis*). The pressing board (*orbis olearius*) was constructed with "Phoenician Joints" (*punicanus coagmenta*). The free end of the beam was lowered by a drum (*suculum*) that was secured between two wooden piers (*stipites*) and turned by special levers/hand spakes (*vectes*).

Vitruvius. Vitruvius (c. 25 BCE) in his treatise on Architecture, *De Architectura* (VI 6 3) stipulates: "The pressing room itself if the pressure is exerted by means of levers and a beam and not worked by turning screws should not be less than forty feet long".

Pliny the Elder. In his encyclopedic work *Naturalis Historia* (XVIII 54 317) Pliny (c. 23–79 CE) gives a short survey of the history of the press. The text is apparently corrupted and although it would appear to be referring to four types of wine presses, Jungst and Thielscher (1954: 108–109) have convincingly restored the text showing that in fact there were only three presses, the second and third being one press. The first press that was used by "our forefathers" was a lever and drum press, the second (formerly thought to be two presses) introduced "within the last hundred years" was a lever and screw press that raised a box of stones, and the last press that was "invented" "within the last twenty two years" was a direct pressure single screw press.

Hero of Alexandria. There is some controversy as to when Hero of Alexandria was active, some dating him to the 1st cent. BCE. The date usually accepted is, however, the 1st cent. CE. The writings of Hero of Alexandria concerned with presses appear in his book *Mechanica* that, although written in Greek, has only survived in Arabic.

Hero gives instructions as to how to construct four types of press, all wine presses (*Mechanica* III 13–20). As opposed to Cato and Pliny who were clearly referring to presses that they knew, Hero was almost certainly attempting to invent improved versions of existing devices and there is considerable doubt if they were ever actually made.

Hero's first press was a lever, weight and drum press. A rope attached to a drum that was secured to the ground passed through a pulley attached to the end of the beam, then through a pulley attached to the weight and finally was apparently attached to the beam end. After the weight was raised by turning the drum it was tied directly to the beam. The grape skins and stalks, the rape, were enclosed in a wound rope. In the text the vat is not mentioned but on the diagram in the Leiden manuscript a jar appears.

Hero's second press is a lever and screw press with a screw weight. The weight is in the same form as the Din'ila weight described above but it did not operate in the way we presume that the Din'ila weight operated. Instead of the screw rotating in the weight it is the nut which is in the form of a sleeve with internal female thread that rotates in the weight. The "nut" is attached to the weight in such a way that it can turn but not work free. The upper end of the screw is attached to the beam by a complicated block, the description of which is not clear. The lower end is inserted into the sleeve that when turned "swallows" the screw, lowers the beam and raises the weight. As already explained above Hero's screw weight is of the type defined here as the Din'ila weight.

Hero's third press was a double screw direct pressure press in which the screws are attached to the base in such a way as to allow them to rotate but not to work free. Two female screw threads were cut in the pressing board and turning the screws in these lowered it and brought pressure to bear.

Hero's fourth press is a direct pressure frame press with one rotating screw.

In addition Hero mentions wedge presses, without describing them in detail, but stating that they were used for making perfume (*Mechanica* II 4).

Justinian's Digesta. In the Emperor Justinian's monumental legal corpus, the Digesta (Mommsen and Kreuger 1985) known also as the *Pandectae*, there is reference to an oil press. The section (19 2 19) is quoted from Ulpian (early 3rd cent. CE) who in turn quoted Neratius (early 2nd cent.). The text discusses the obligations of the lessor of a farm to his tenant. The opening part of the discussion is clear. The lessor must provide storage jars (*dolia*), a press (*praelum* – the beam, but used here for the press as a whole) and a crusher (*trapetum*), these equipped with ropes (*instructa funibus*). After stating other legal provisons Neratius stipulates that the tenant supplies the frails (*fiscus*). He then continues that if *regulae* are used instead of frails the lessor supplies them. Pliny (XV 2 5) mentions *regulae* as a recent invention. Drachman (1932: 60–62) has shown that *regulae* were wooden frames that were used instead of frails or ropes to enclose the grape skins or olive mash during pressing and he equated them with the *galeagra*, the constructing of which Hero of Alexandria describes in detail. In the Digesta at this point together with the *regulae* appears a list of all the components of the press that the lessor must supply: *praelum* – beam, *suculum* – drum, *regulas* – pressing frames, *tympanum* – literally drum but used also by Cato for pressing board, and "the screws used to raise the beam" (*cocleas quibus releuatur praelum*).

The last phrase is clearly problematical on at least two counts. In a lever and screw press the primary purpose of the screw is to lower the beam. Turning the screw the other way, however, does raise the beam and someone not completely cognizant with the process could think this to be the main function of the screw. A more serious problem is that both a drum and a screw are mentioned. The lever and drum press, however, has no screw while the lever and screw press has no drum. There are three possible explanations for this confusion. The first is that Neratius included in his list components from different types of press that the lessor had to supply. The second possible explanation is that the final phrase was a later addition although this must have been very early as the screw already appears in Cyrillus' Greek translation (6th cent.) (Scheltema and Holwerda 1957: 1184). Cyrillus, however, omits the explanation that the screw raises the beam. (Another point of interest in Cyrillus' translation is that he translates *praelum* in both its meanings, beam and press as βάδδην is without doubt the Aramaic\Hebrew term בד בדים). The third possibility is that there actually were Drum and Screw Presses. Rossiter and Haldenby (1989) have suggested reconstructing, as a Lever, Drum and Screw Press, a type of press found at Pompeii and at sites nearby, that had previously been reconstructed as a Lever and Drum Press (Brun 1986: Fig. 33; Frankel 1999: 91).

Biblical and Talmudic Literature

There is no distinction between oil and wine production in the Hebrew Bible, and although the term *gat* is usually associated with treading the same is sometimes true for the term *yeqeb* (Isa 16: 10) and both wine and oil flowed into the *yeqeb* (Joel 2: 24). The only time that the production of oil is specifically mentioned the olives are also trodden (Michah 6: 15).

In Talmudic literature there is clear distinction between wine and oil installations and *gat* is only a winery. The treading floor is usually referred to as the *gat 'eliona* – the upper winery although occasionally as simply *gat* and the collecting vat as *gat tahtona* – lower winery or *bor shel gat*, the pit/vat of the winery or simply *bor* – vat. From the context it is clear that the first fermentation took place in the vat itself (e.g. *Tosef. Ter.* 7: 15).

The term used for an oil press was *bet ha-bad* – the "house of the beam" although the term usually used for the beam of a press was not *bad* but *qorah*. A list of the parts of a beam and weights press appears in a legal discussion as to which components the vendor is obliged to sell when he agreed to sell an oil press. His obligation was limited to selling the three fixed parts: *yam* – vat; *memel* – press-bed; and *betulot* – slotted piers, wheras he was only obliged to sell the three removable parts: *'akirim-'akidim* – weights; *galgal* – winch; and *qorah* – beam, if he specifically stated that he would sell the oil press with all that was in it (BB 4, 5; Frankel 1999: 188–190).

A discussion that appears in several versions that vary slightly (two in the Tosefta, two in the Babylonian Talmud and one in the Jerusalem Talmud) is concerned with purifying, cleaning the parts of a press. The press was apparently a screw press and only three parts are mentioned: *lulabin* – screws; *'adashah* – press-bed; and *pasin* – apparently the pressing board. Much discussion is devoted to how to clean the frails (sing. – *'eqel*) made of different materials. It is of great interest that the version in the Jerusalem Talmud is different from the others in several ways. It omits all reference to the oil presses mentioned in the other versions speaking only of a winery (*gat*), mentions the word for frail in the singular instead of the plural and adds two substances to the materials from which the frails were made, wood and rope. The version in the Jerusalem Talmud was clearly referring to the fixed screw wine press (the 'Ayalon and Hanita presses). In this type of press the usual frail could not be used as the material to be pressed was not put under the press as in all the other types of presses but piled up round the screw and then enclosed in a wooden frame or wound rope (see Frankel 1999: 192–193).

Another discussion concerned the Sabbath. "Both [the schools of Shammai and Hillel] agree that they load the olive press beams and the *'igulim* of the wineries" (Mishnah *Shab.* 1: 9). The injunction allowed the setting up of the presses before the Sabbath so that the liquids could flow of their own accord during the Sabbath. The question is what is the meaning of the term *'igulim*. The literal meaning is circles. In the parallel passage in the Tosefta (*Shab.* 1: 21) it is stated that they "hang the *'igulim*". Danby (1933: 101) translates "rollers" but probably the meaning is beam weights. Both the wine and the oil press were apparently lever and weights presses but in the case of the oil press the large beam was the main element that was referred to while in the winery the beam was insignificant and it was the weights that were referred to.

An important source for all that regards wine and oil are the injunctions regarding the wine and oil to be used in the temple rites (Mishnah *Men.* 8, oil: 3–5; wine: 6, and related passages in the Talmuds, Tosefta etc.). The passage includes lists of places from which the finest olives and grapes were brought, types of olive groves and vineyards from which produce was forbidden and types of wine and oil that were unfit for temple use. The olives were divided into three grades according to the way they were treated before pressing and in each case there were three grades of oil according to the order in which they were pressed (compare to the three grades of oil in classical literature, e.g. Pliny XV.4.18, Columella XII.52.11, Diocletian's price list [Lauffer 1971: 102–103]). The first grade of olives was taken straight from the tree to the press, the second was stored on the roof before pressing and the third was stored indoors and then dried on the roof before pressing. The explanation is apparently that the olives were often picked faster than they were pressed and had to be stored, their quality deteriorating during storage. As regards the pressing the Mishnah combines elements from two differing contentions, one that was based on the installations and methods used at the time (the Roman period) and the divergent view of Rabbi Judah who was of the opinion that the oil should be produced in the way it was in the days of Solomon's temple. The argument is presented in the Babylonian Talmud (*Men.* 86.A): first the general opinion was brought "they said--bring it into the olive press and grind it in the olive crusher (*reyhaym*) put it in frails (*salin*) the oil that comes out is the first; he loaded it under the beam the oil that comes out is the second; he repeated (it), dismantled (it) crushed (it) and loaded (it under the beam) that was the third". This is a clear and terse description of the usual procedure (as regards the first oil, see note 1 above). We are then told that "Rabbi Judah says: 'He did not grind it in an olive crusher but crushed it in a mortar, he did not load it under a beam but under stones, he did not put it in baskets but into around the basket". The parallel version in the Tosefta (*Men.* 9: 7) reads "he did not put the loaf [of crushed olives] in the basket but the basket into the loaf". Rabbi Judah was clearly stipulating the simplest way of producing oil, crushing in a mortar and then pressing by putting stones in a basket on the olive mash. Rabbi Judah Ha-Nasi who compiled the Mishnah adopted Rabbi Judah's precept as regards the first oil only which, according to the Mishnah was to be crushed in a mortar. The second and third oils were to be produced according to the usual method but in these cases the Mishnah also brings Rabbi Judah's divergent view. It is of interest that as regards grapes "one does not bring them from the trellissed grapes only from vines on the ground" – apparently again a preference for more primitive, traditional methods.

Notes

1. See reference to central vats that served to produce "first oil" below, p. 5.

2. Recently Y. Dray (2003) has suggested that in the production of red wine in wineries such as that at H. Castra (Site 19) the grapes were trodden and the must fermented in the large compartments placed around the treading floor and that the only activity that took place on the central treading floor, that Dray calls a working area, was pressing out the last drops from the rape in the screw press. Dray carried out an experiment at H. Castra in which he made wine in this manner. He also suggested that in simple wine presses the must was fermented on the treading floor. Dray's suggestions are unacceptable for several reasons. Firstly in many wineries the method that he proposes could not be implemented because it was not possible to block off the compartments from the central treading floor or in other cases to block off the treading floor from the collecting vat(s). This is so in all the wineries in the Negev Highlands, for example, that otherwise are very similar to that at H. Castra and clearly functioned in a very similar manner. Secondly, in those wineries where theoretically Dray's method could be carried out such as that of H. Castra the logical procedure would be to collect the fermented wine directly from the compartments and not to let the precious liquid flow over the large treading floor to the collecting vats. In fact in Dray's method the large treading floors and the enormous collecting vats of many wineries are in effect superfluous. The only way to explain the size of the collecting vats and the fact that there are often two is that the must fermented in the vats. Evidence for this appears in Talmudic literature: "the fermenting collecting vat [בור], and how long does it ferment? Three days" (Tosef. *Ter.* 7.15). Thirdly, the basic elements of all wineries and in early and simple wine presses the only elements are two, the treading floor on which the grapes were trodden and the collecting vat to which the must flowed and where it usually fermented. Clearly the functions of these two elements remained the same also in more complex wineries, additional elements serving additional functions. Finally it must be stressed that although experiments in which ancient installations are actually operated are of great importance their success does not prove the way in which they functioned neither does their failure prove the opposite.

3. Several examples of piers with posterior and anterior grooves but

without lateral mortices were recorded in surveys (Frankel 1999: 127, 129, Map 27, T[ype] 4618). This is very probably an intermediate type between the plain piers of the Maresha Press and the grooved piers of the Judean Grooved Pier Press. Note also the piers with grooves on their inner faces at Kh. el-Jum'a (Site 58), there part of a lever and weights press and at Kh. el-Najar (Site 57), there supporting a screw press, in both cases also perhaps intermediate types.

Bibliography

Avitsur, S. 1976. *Man and his Work: Historical Atlas of Tools and Workshops in the Holy Land.* Jerusalem. (Hebrew).

Avitsur, S. 1994. "Olive Oil Production in the Land of Israel: Traditional to Industrial". In: R. Frankel, S. Avitsur and E. Ayalon. *History and Technology of Olive Oil in the Holy Land.* Arlington and Tel Aviv: 91–157.

Ben David, H. 1998. "Oil Presses and Oil Production in the Golan in the Mishnaic and Talmudic Periods". *'Atiqot* 34: 1–61 (Hebrew).

Biran, A. 1980. "Two Discoveries at Tel Dan". *IEJ* 30: 89–98.

Borowski, O. 1982. "A Note on the Iron Age Cult Installation at Tell Dan". *IEJ* 32: 58.

Brøndsted, J. 1928. "La basilique des cinq martyrs à Kapljuc". In: E. Dyggve and J. Brønsted, *Recherches à Salone I.* Copenhagen: 33–176.

Brun, J.-P. 1986. *L'Oléiculture antique en Provence: Les huileries du department du Var.* Paris.

Callot, O. 1987. "Les huileries du Bronze Récent à Ougarit: Premiers éléments pour une étude". In: M. Yon, *La centre de la ville Ras Shamra, Ougarit III, 38ᵉ-44ᵉ campagnes (1978–1984).* Paris: 197–212.

Cato 1979. "Marcus Porcius Cato: De Agri Cultura". In: *Cato and Varro* (trans. W.D. Hooper). London: 2–157.

Columella 1969–79. *Columella, Lucius Junius Moderatus: De Re Rustica* (trans. H.B. Ash, E.S. Forster and E. Heffner), 3 vols. London.

Dalman, G. 1928–1942. *Arbeit und Sitte in Palästina.* 7 Vols. Gütersloh (Reprinted Hildesheim 1964).

Danby, H. (trans.). 1933. *The Mishnah.* London.

Dar, S. 1999. *Sumaqa, A Roman and Byzantine Jewish Village on Mount Carmel, Israel* (BAR IS 815). Oxford.

Davidson Weinberg, G. (ed.). 1988. *Excavations at Jalame, Site of a Glass Factory in Late Roman Palestine.* Columbia.

Drachman, A.G. 1932. *Ancient Oil Mills and Presses.* Copenhagen.

Dray, Y. 2003. "Wine Production in the Improved Byzantine Wine Press". In: Y. Eshel (ed.). *Judea and Samaria Research Studies* 12: 219–228 (Hebrew).

Frankel, B. 1981a. "Kh. Dukas – excavation report". *Israel Antiquities Authority Archives.*

Frankel, B. 1981b. "Notes and Remarks Regarding Oil Production in Juda". *Teva Va-Aretz* 23: 183 (Hebrew).

Frankel, R. 1993a. "The *Trapetum* and the *Mola Olearia*". In: M.-C. Amouretti & J.-P. Brun (eds.). *Oil and Wine Production in the Mediterranean Area* (BCH Supplement 26). Athens and Paris: 477–481.

Frankel, R. 1993b. "Screw Weights from Israel". In: M.-C. Amouretti and J.-P. Brun (eds.). *Oil and Wine Production in the Mediterranean Area* (BCH Supplement 26). Athens and Paris: 107–118.

Frankel, R. 1994. "Ancient Oil Mills and Presses in the Land of Israel". In: R. Frankel, S. Avitsur and E. Ayalon. *History and Technology of Olive Oil in the Holy Land.* Arlington and Tel Aviv: 19–89.

Frankel, R. 1997. "Presses for Oil and Wine in the Southern Levant in the Byzantine Period". *DOP* 51: 73–84.

Frankel, R. 1999. *Wine and Oil Production in Antiquity in Israel and Other Mediterranean Countries.* Sheffield.

Frankel, R. 2001. "Cato's Press a Reappraisal". In: J.-P. Brun and P. Jockey (eds.). *Techniques et sociétés en Mediterranée, Hommage à Marie-Claire Amouretti.* Paris: 313–325.

Frankel, R. 2006. "Two installations for the production of olive oil". In: I. Finkelstein, D. Ussishkin and B. Halpern (eds.). Megiddo IV, The 1998–2002 Seasons. Tel Aviv: 618–629.

Frankel, R., Patrich, J. and Tsafrir, Y. 1990. "The Oil Press at Horvath Beit Loya". In: G.C. Bottini, L. Di Segni and E. Alliata (eds.). *Christian Archaeology in the Holy Land: New Discoveries* (Studium Biblicum Franciscanun, Collectio Maior 36). Jerusalem: 287–300.

Geoponica 1895. *Geoponica sive cassiani bassi scholastici de resticos ecologae* (ed. H. Beckh). Leipzig.

Hadjisavvas, S. 1992. *Olive oil Production in Cyprus from the Bronze Age to the Byzantine Period.* Nicosia.

Hero of Alexandria – *Herons von Alexandria Mechanik und Katoptrik* (L. Nix and W. Schmidt, trans.). Leipzig 1900.

Hirschfeld, Y. 2000. *Ramat Hanadiv Excavations, Final Report of the 1984–1998 Seasons.* Jerusalem.

Jungst, E. and Thielscher, P. 1954. "Catos Keltern und Kollergange". *Bonner Jahrbücher* 154: 32–93.

Lauffer, S. 1971. *Diokletians Preisedict*. Berlin.

Magen, Y. 1993a. "Mount Gerizim and the Samaritans". In: F. Manns and E. Alliata (eds.). *Early Christianity in Context, Monuments and Documents* (Studium Biblicum Franciscanum, Collectio Maior, 38). Jerusalem: 91–148.

Magen, Y. 2000. "Mt. Gerizim – A Temple City". *Qadmoniot* 33: 74–118 (Hebrew).

Magen, Y. 2008. "Oil Production in the Land of Israel in the Early Islamic Period". In: *Judea and Samaria Researches and Discoveries* (JSP 6). Jerusalem: 217–257.

Magen, Y. 2008. "Oil Production in the Land of Israel in the Early Islamic Period". In: *Judea and Samaria Researches and Discoveries* (JSP 6). Jerusalem: 217-257.

Mayerson, P. 1985. "The Wine and Vineyards of Gaza in the Byzantine Period". *BASOR* 257: 75–80.

Mayerson, P. 1992. "The Gaza Wine Jar (Gazition) and the Lost Ashkelon Jar (Askalonium)". *IEJ* 42: 76–80.

Mayerson, P. 1993. "The Use of Ascalon Wine in the Medical Writers of the Fourth to the Seventh Centuries". *IEJ* 43: 169–173.

Mishnah – see Danby 1933.

Mommsen, T. and Kreuger, P. 1985. *The Digest of Justinian* (trans. A. Watson). Philadelphia, Penn.

Palladius 1807. *The Fourteen Books of Palladius Rutilius Taurus Aemilianus on Agriculture* (trans. T. Owen). London.

Palladius 1898. *Agricultura* (ed. J.C. Schmitt). Leipzig.

Paton, W.R. and Myres, J.L. 1898. "On Some Karian and Hellenic Oil-Presses". *JHS* 18: 209–217.

Peilstöcker, M. and Kapitaikin, A. 2000. "Bet Dagan". *ESI* 20: 59*-60*.

Pliny the Elder, 1958–1963. *Naturalis Historiae*. 10 Vols. (trans. H. Rackham, W.H.S. Jones and D.E. Eichholz). London and Cambridge.

Rossiter, J.J. and Haldenby, E. 1989. "A Wine-Making Plant in Pompeii Insula II 5". *Échoes du Monde Classique / Classical Views* 33 n.s. 8: 229–239.

Scheltema, H.J. and Holwerda, D. 1957. *Basilicorum Libri, LX* (Series B volumen III in Libr. XV-XX). Groningen and The Hague.

Stager, L.E. and Wolff, S.R. 1981. "Production and Commerce in Temple Courtyards: An Olive Press in the Sacred Precinct at Tel Dan". *BASOR* 243: 95–102.

Varro 1979. "Marcus Terentius Varro: Rerum Rusticarum". In: *Cato and Varro* (trans. W.D. Hooper). London: 161–528.

Vitruvius 1962. *De Architectura* (trans. F. Granger), 2 vols. Cambridge.

Yeivin, Z. 1966. "Two Ancient Oil Presses". *'Atiqot* (HS) 3: 52–62 (Hebrew).

Yevin, Z. and Finkielsztejn, G. 1999. *Castra at the Foot of Mount Carmel, the City and its Secrets*. Haifa.

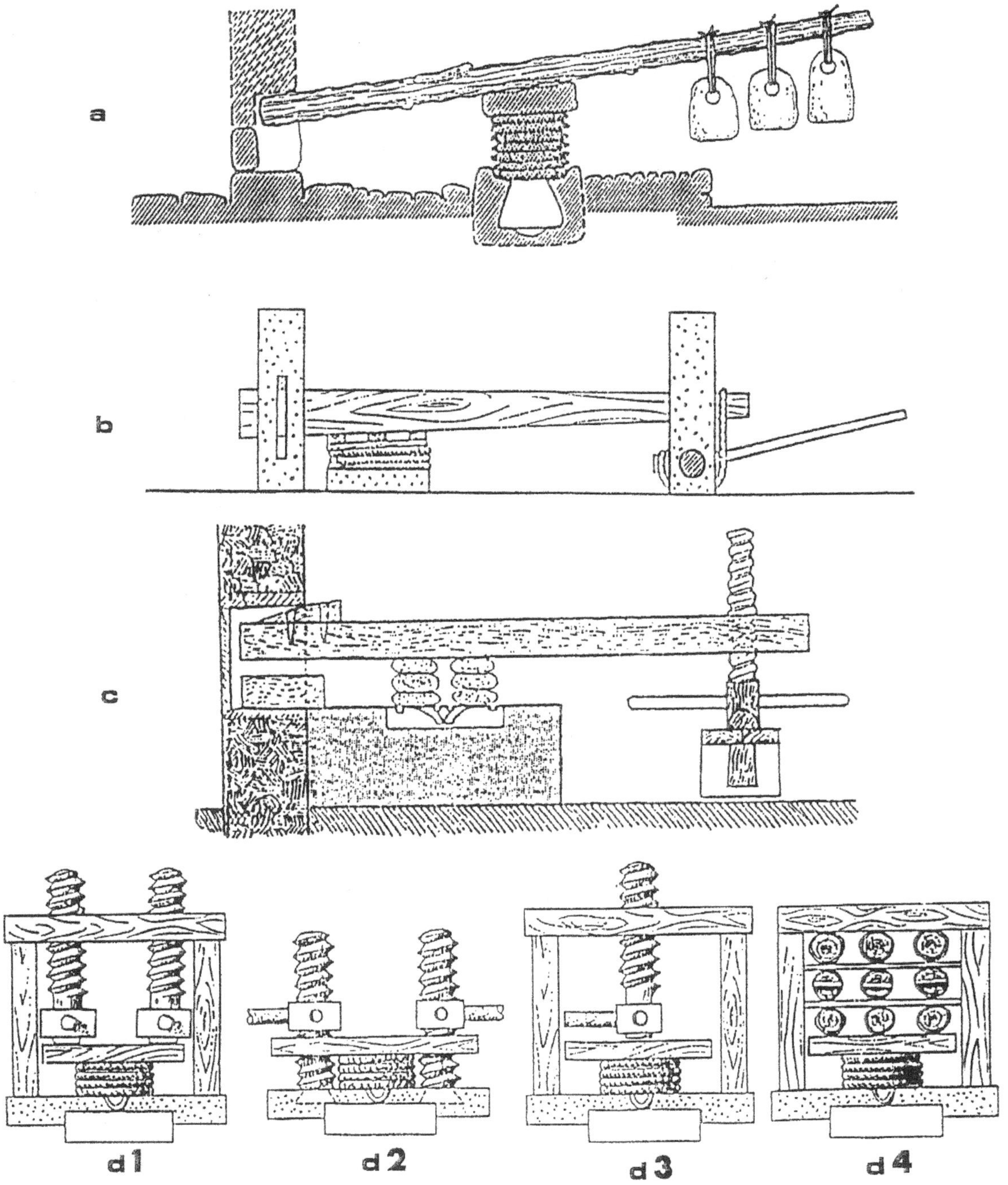

e
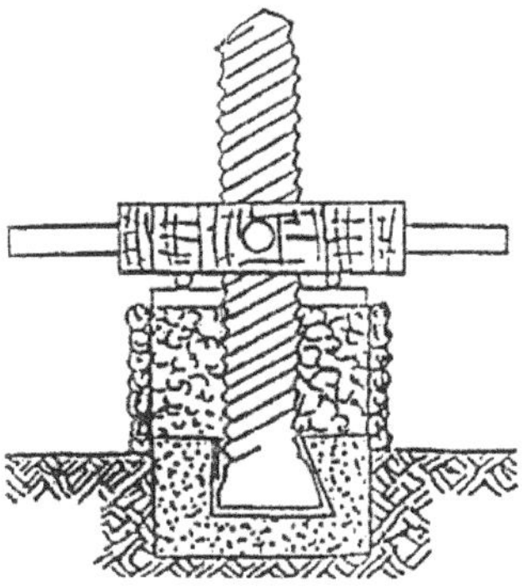

Fig. 1. The main types of presses: a. Lever and weights press; b. Lever and drum press; c. Lever and screw press; d. Rigid-frame direct pressure presses (d1. Double rotary screw press; d2. Double fixed screw press; d3. Single rotary screw press; d4. Wedge press); e. Single fixed screw press (drawings: a, b, d, e after Frankel 1994 Figs. 31B, 44, 63, 8275; c after Paton and Myres 1898: 210).

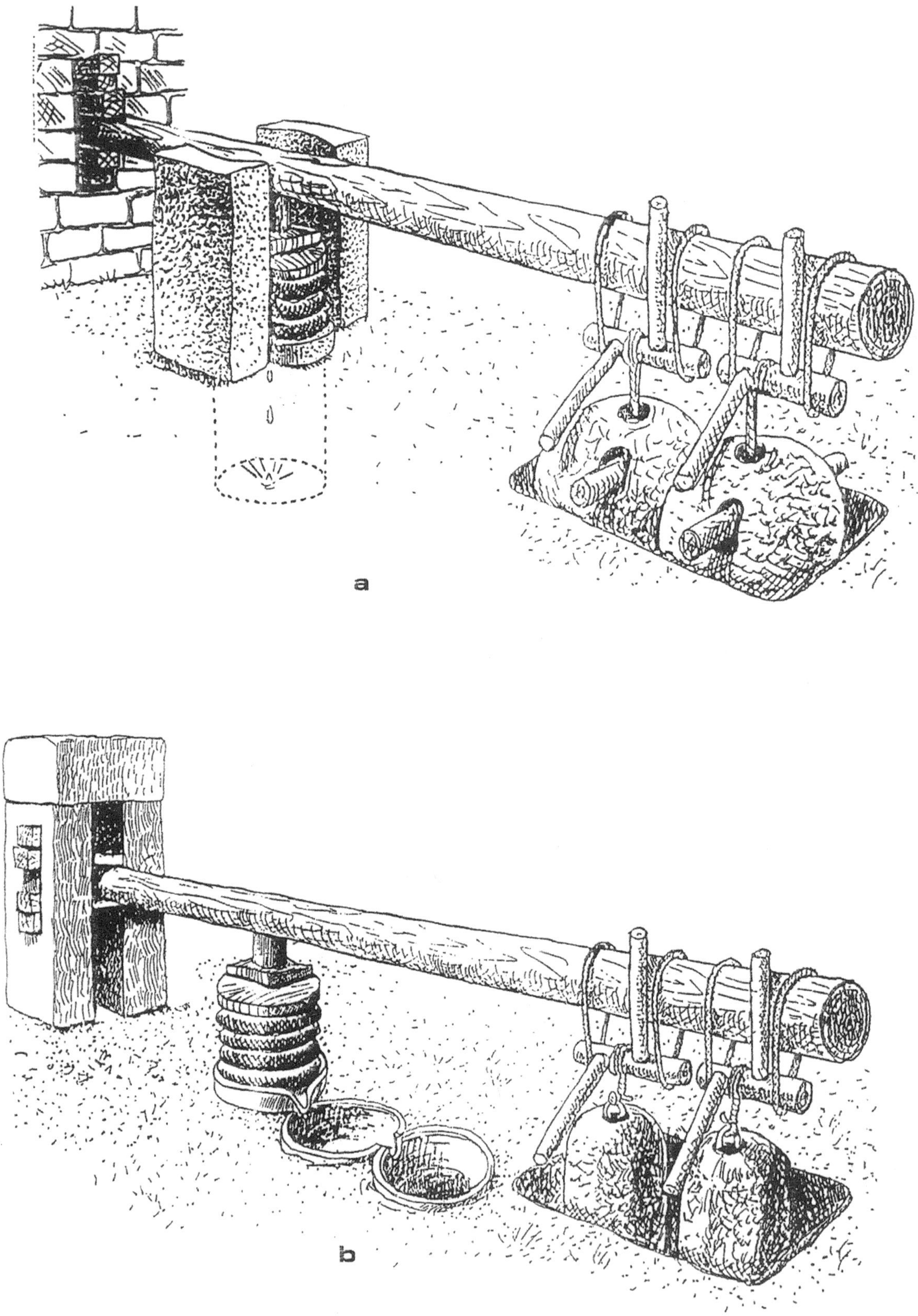

Fig. 2. a. Southern improved lever and weights press, or the Maresha press; b. Northern improved lever and weights press, or the Zabadi press (drawings: after Frankel 1994 Figs. 35, 38).

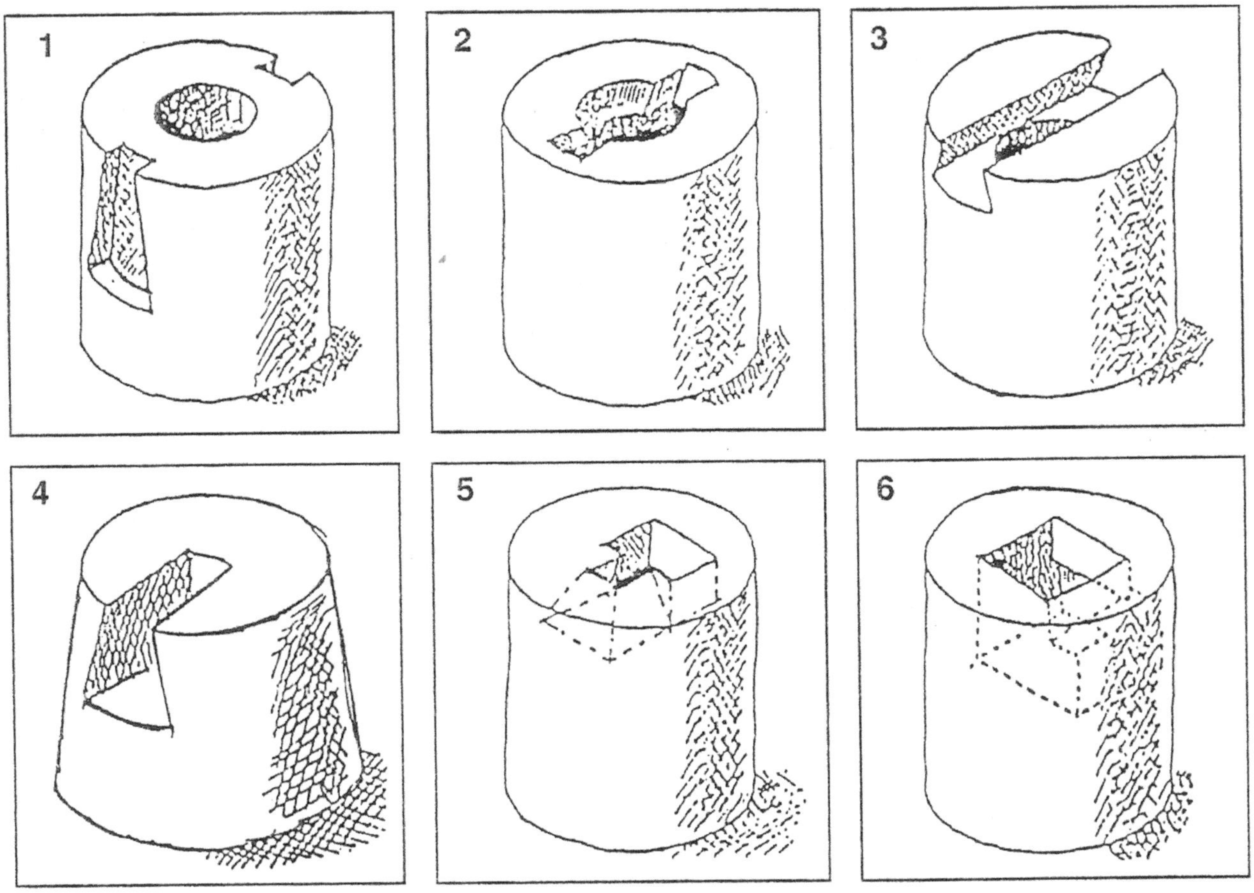

Fig. 3. Six main types of screw weights: 1. Samaria weight; 2. Kasfa weight; 3. Din'ila weight; 4. Mi'ilia weight; 5. Bet Ha-'Emeq weight; 6. Luvim weight (drawings: after Frankel 1994 Fig. 61).

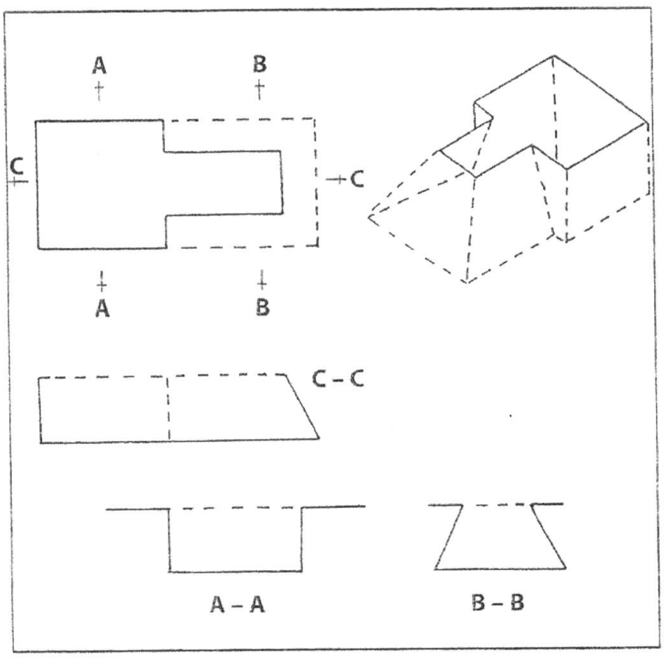

Fig. 3a. Central dovetail mortice, the Phoenician type (drawing by Hagit Tahan)

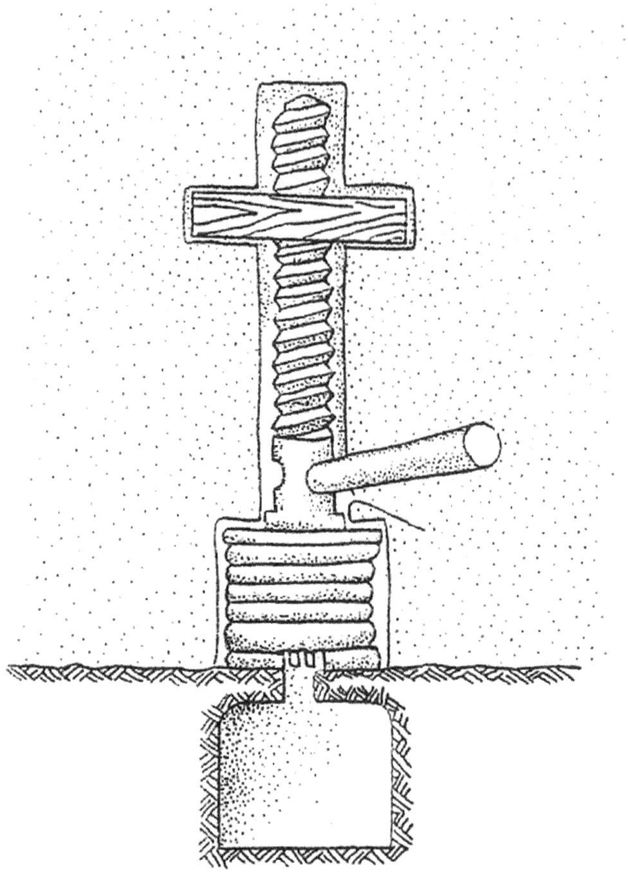

Fig. 4. Cross press – reconstruction (drawing: after Frankel 1994 Fig. 74).

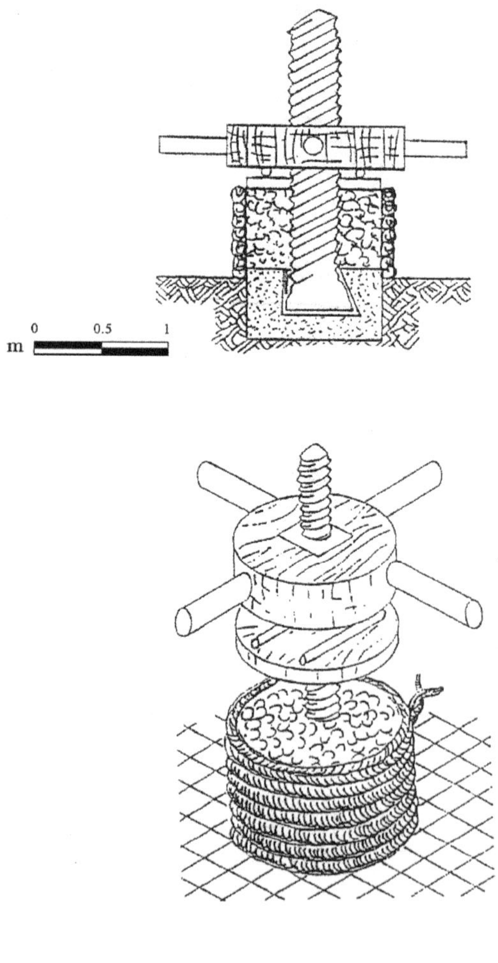

Fig. 5. Fixed screw wine press – section and reconstruction (drawings after Frankel 1994 Fig. 82).

1. Some Wine Presses in Western Galilee

Rafael Frankel

During an archaeological survey of Western Galilee hundreds of wine presses were recorded. The majority were simple rock-cut installations consisting of a sloping treading floor connected to a collecting vat at a lower level. Some were, however, more sophisticated, incorporating other elements and seven of these have been selected to demonstrate the types of wine press found in this region.

As these installations were recorded in a survey in some cases they were partly covered with soil so that some dimensions were not ascertained and are missing from the chart.

The Wine Presses

1) Survey Number 152793/01 (MR 1599/2734)

The wine press is situated on the sea shore 1 km north of the nearest ancient site, Tel Achziv, occupied almost continuously from the Middle Bronze Age onwards. The installation is rock-cut and consists of two square treading floors on either side of a square collecting vat (Fig. 1). There is a ledge around the collecting vat. The floors are connected to the vat by a bore and a cavity cut into the ledge. The southern treading floor is damaged.

2) Survey Number 162706/01 (MR 1604/2765)

The wine press is located about 200 m from the sea shore at the north-eastern corner of a medium sized site (15 dunams) dating to the Persian and Roman periods.

The installation is rock-cut, the treading floor is square and the collecting vat round (Fig. 2). Around the vat is a ledge and the treading floor and vat are connected by a channel and a cavity cut into the ledge.

3) Survey Number 162775/06 (MR 1675/2751)

The wine press is situated on the edge of a wide flat spur. The nearest site is H. Zurit, a small site 750 m away on the hill above dating from the Roman and Byzantine periods. In close proximity to the installation was a large underground water cistern (250 m^3) probably for irrigation.

The wine press is rock-cut, the treading floor and collecting vat are both square (Fig. 3). Around the plastered vat was a ledge. The floor and vat were connected by a bore and a cavity cut into the ledge. In the walls of the vat were small niches that served as steps.

4) Survey Number 172600/03 (MR 1706/2603)

This small wine press is situated 1.5 km from the nearest site, Yirka, where pottery was found from the Iron Age onwards. The installation is rock-cut (Fig. 4). The treading floor is square. The vat is round, narrower at the top and widening lower down. There is a ledge around the vat that is connected to the treading floor by a bore and a channel cut in the ledge.

5) Survey Number 172610/04 (MR 1710/2606)

This small wine press is located about 500 m to the north-east of No. 4. It is rock-cut and consists of a square treading floor, a round vat and an additional triangular floor above the square one (Fig. 5). Around the vat is a ledge and the treading floor is connected to the vat by a bore and channel cut into the ledge.

6) Survey Number 162706/02 (MR 1602/2767)

The wine press is very close to the sea shore next to a small site, H. Misrefot Yam, at which only Byzantine sherds were found.

The installation is rock-cut and, at least partly, paved in mosaic (Fig. 6). The treading floor was probably square. It is connected by a channel to a small intermediate vat. A bore at the bottom of this vat connects it to the collecting vat that is rectangular in shape and steps lead down to it. In the centre of the treading floor is a central dovetail mortice for a fixed screw press. The pressing surface in which it is cut is slightly raised above the floor and is square in shape with rounded corners.

7) Survey Number 172605/03 (MR 1703/2654)

The wine press is at the north-eastern corner of H. Habai, a small Byzantine site (1.7 dunam) on a western spur of the Galilean mountains. The installation was rock-cut but there are also traces of mosaics (Fig. 7). Both treading floor and collecting vat were square. There were also an intermediate vat, steps in the collecting vat and a central dovetail mortice in a free standing rectangular stone. The latter was not *in situ* but was found in the immediate vicinity of the wine press and was clearly connected to it.

Conclusions

Four of the seven wine presses were not connected directly to an ancient site. This is very usual, wine presses being usually in the actual vineyard. Of the remaining three only Nos. 6 and 7 were associated with one period sites, both Byzantine thus dating them to this period. The dates of the others could not be determined with certainty.

The seven wine presses can be divided into three groups each of a different type.

Type 1 (Nos. 1–3). The characteristic these installations have in common is the ledge around the vat and the cavity cut into the ledge. The dimensions of these components are also very similar in the three wine presses. The ledge is almost certainly to accommodate a lid or cover and the cavity to allow the must to flow under the lid. It is probable that the cavity also served to contain sprigs of Thorny Burnet to strain the must.

Type 2 (Nos. 4–5). These two small wine presses are located in the same vicinity and several other wine presses of the same type are found in the same area although less well preserved or less exposed. These installations also have ledges around the vat but in Type 2 a channel replaces the cavity of Type 1.

Type 3 (Nos. 6–7). These two wine presses have an intermediate vat instead of a cavity. Such vats are found in many wine presses and are usually considered, probably correctly, to be settling tanks. The intermediate vats in wine presses 6 and 7 are, however, small and the outlets close to the bottom suggesting that they were filled with sprigs of Thorny Burnet and were used as sieves. In both installations there was also a central dovetail mortice in the centre of the treading floor. These served to hold a single fixed screw to press the must that remained in the grape skins after treading. This type of mortice is typical of the northern regions and probably originated in Phoenicia.

The three types are clearly related. Type 2 was a smaller local variant of Type 1 while Type 3 apparently developed from Type 1, the intermediate vat developing from the cavity of Type 1. The two wine presses of Type 3 are both from the Byzantine period suggesting that Type 1 is perhaps earlier.

Dimensions of Wine Presses

Treading Floor 1	Length	Width	Area
1	4.4	3.7	16.3
2	4.2	4.1	17.2
3	3.9	3.8	14.8
4	2.3	2.3	5.3
5	1.9	1.9	3.6
6	4.5?	4.5?	20.5?
7	4	4	16

Treading Floor 2	Length	Width	Area
1	?	3.7	?
5	1.9	1	1

Collecting Vat No	Length	Width	Depth	Volume
1	2.4	2.2	?	?
2	2.4	2.4	?	?
3	2.5	2.3	1.85	10.6
4	0.75	1.5	1.5	1.28
5	1	0.8	?	
6	3.2	2.6	1.3	10.8
7	3	3	2.0+	18.0+

Int. Vat	Length	Width	Depth	Volume
1	0.35	0.5	0.3	0.05
2	0.3	0.5	0.2	0.03
3	0.35	0.35	0.25	0.03
6	0.65	0.55	0.47	0.17
7	1	0.5	0.3	0.15

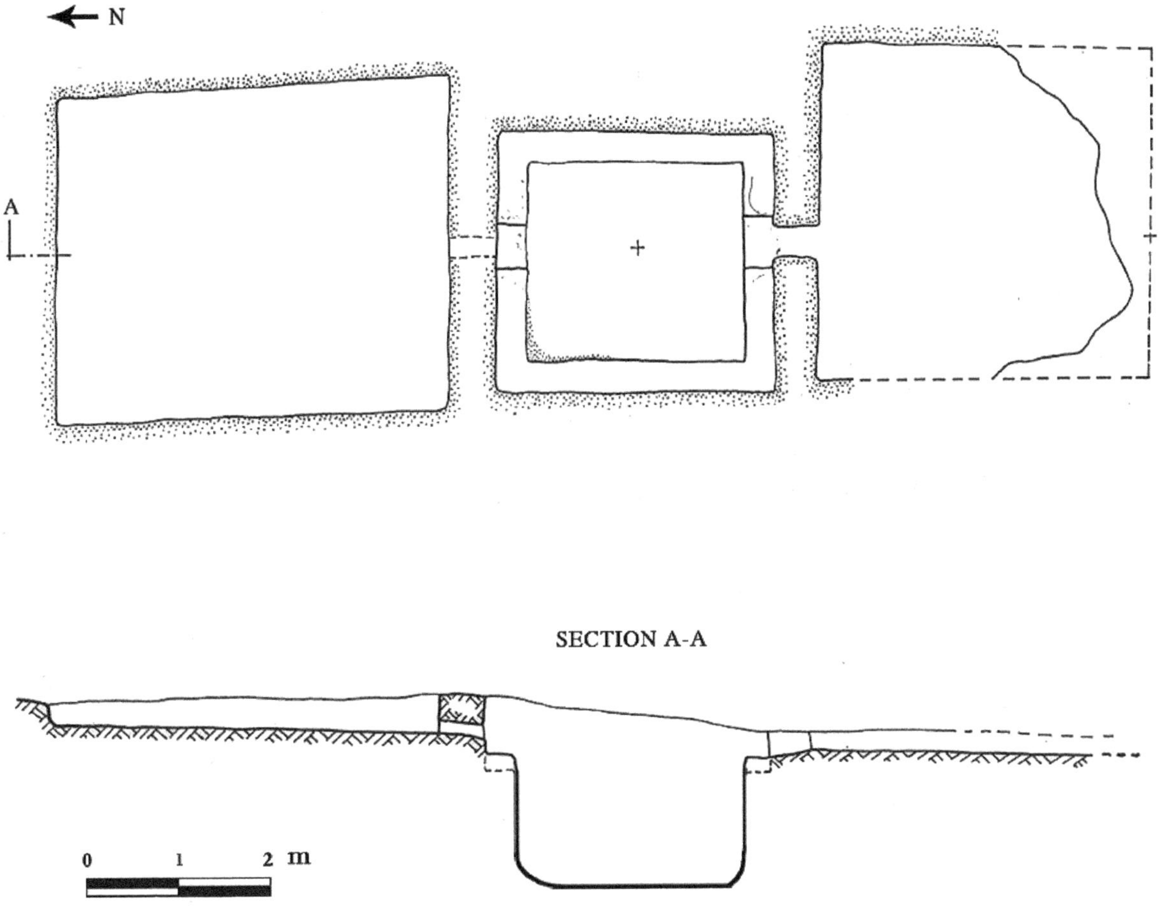

Fig. 1.1. Wine press No. 1, plan and section.

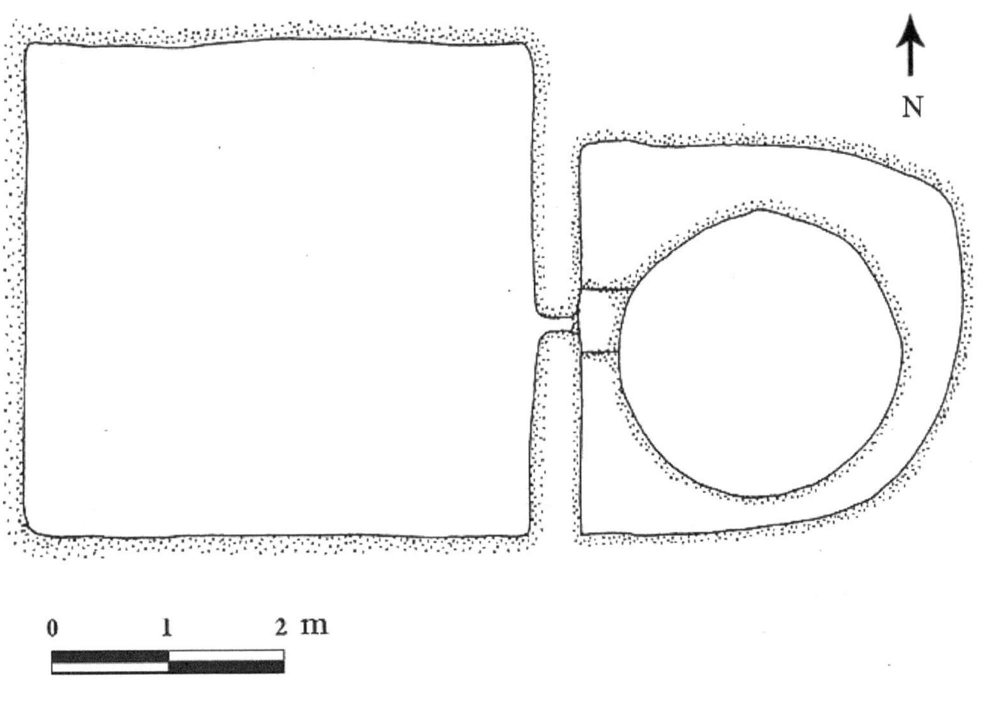

Fig. 1.2. Wine press No. 2, plan.

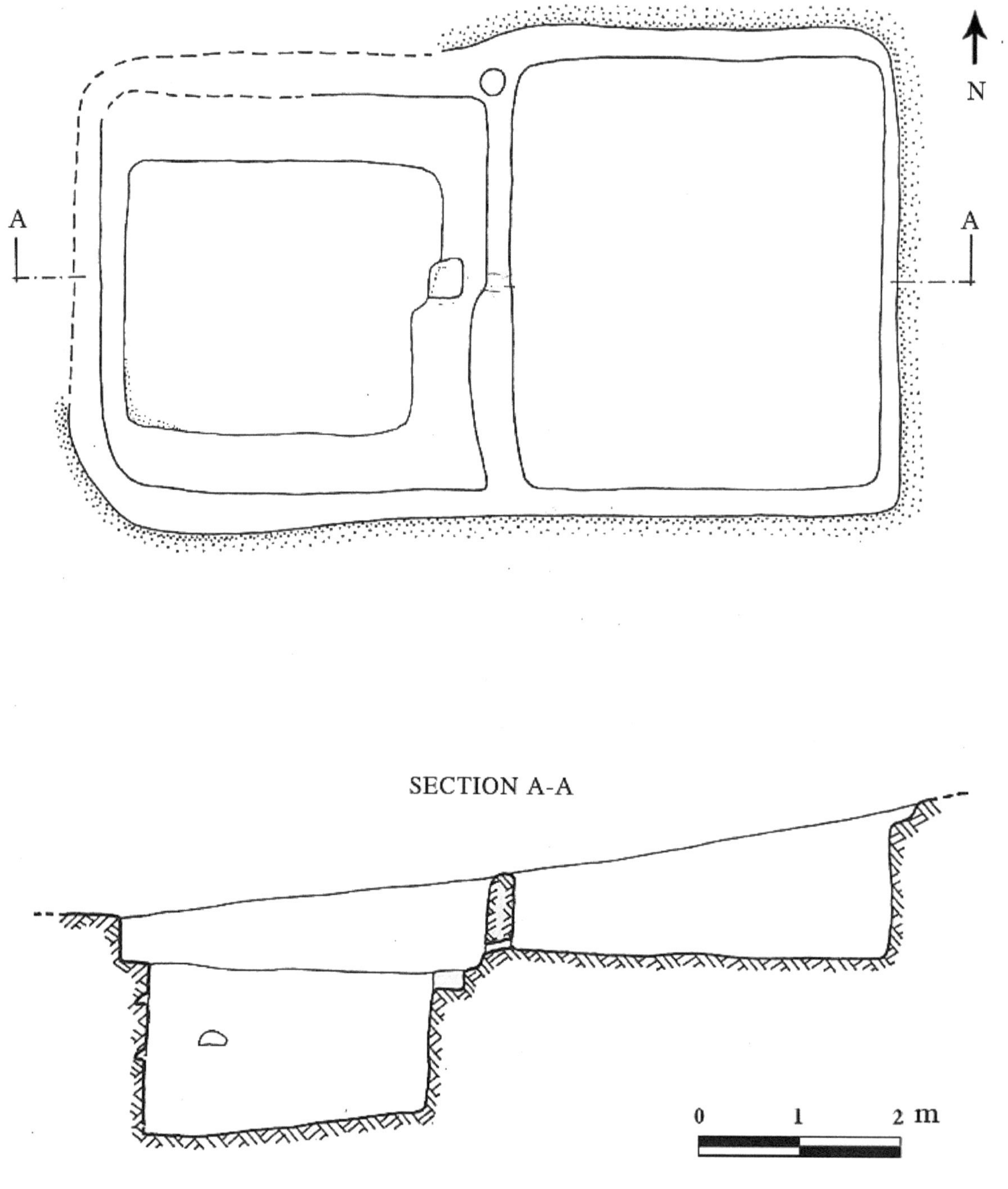

Fig. 1.3. Wine press No. 3, plan and section.

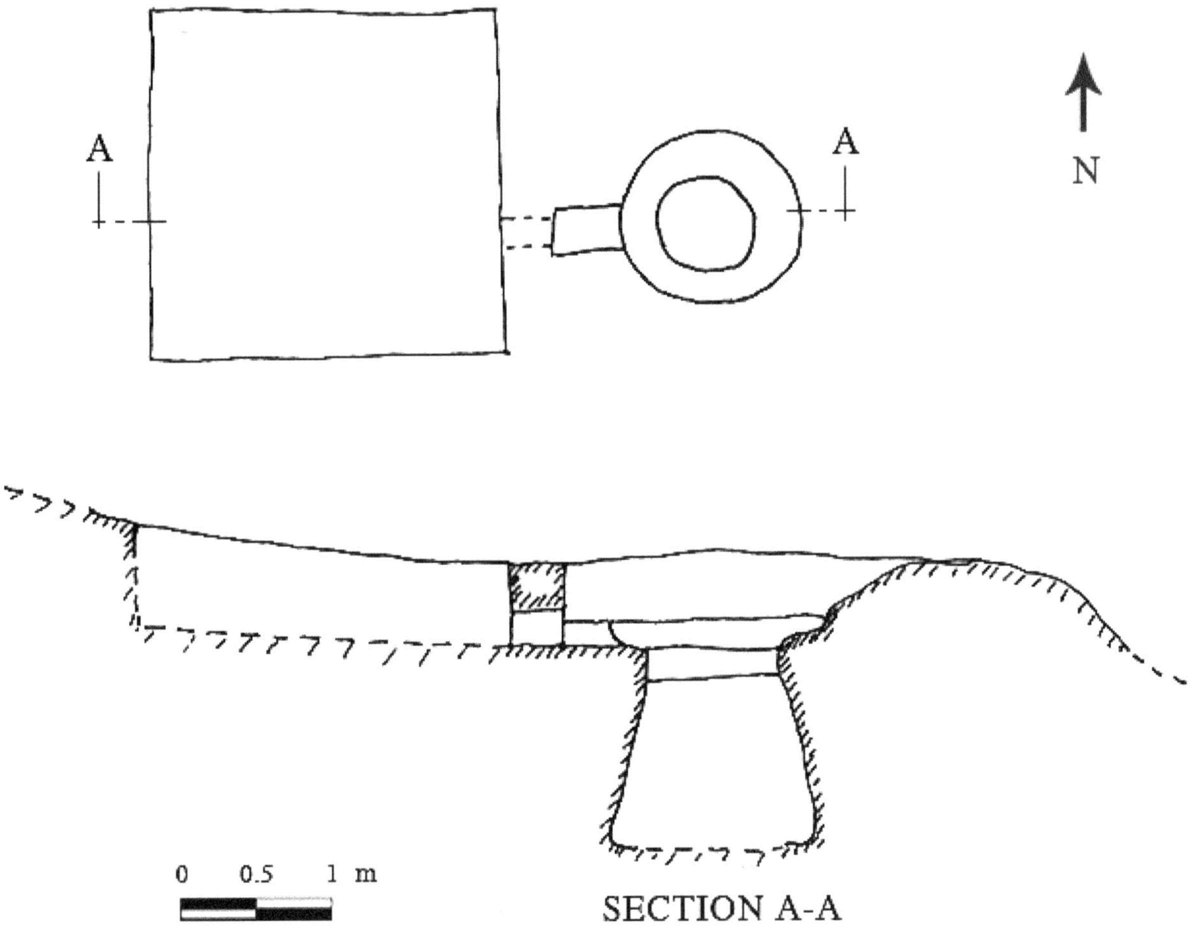

Fig. 1.4. Wine press No. 4, plan and section.

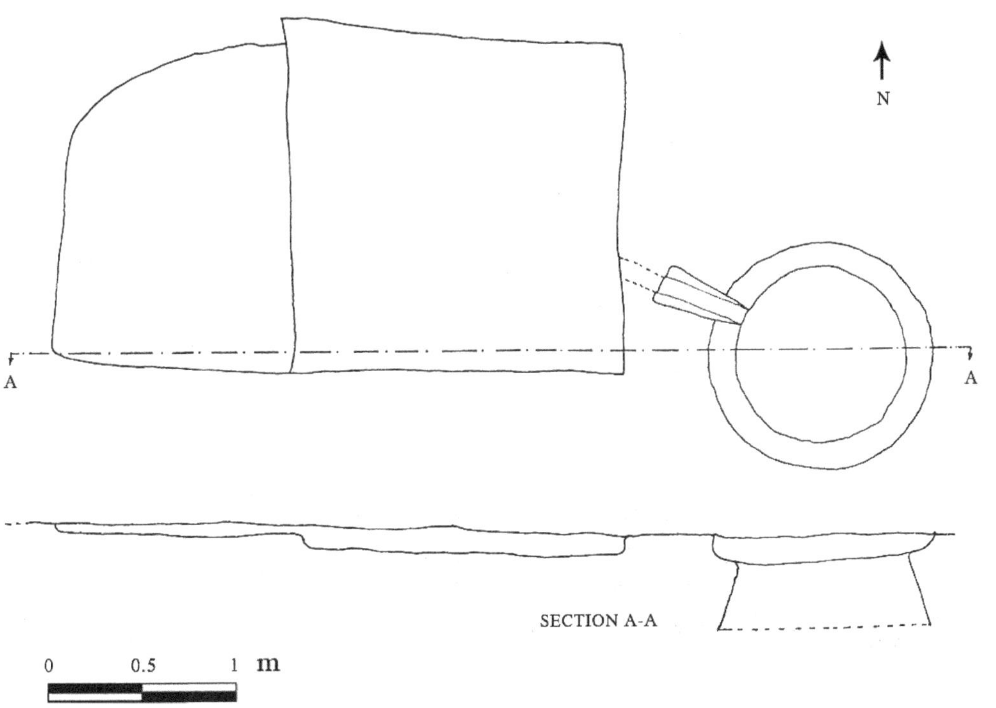

Fig. 1.5. Wine press No. 5, plan and section.

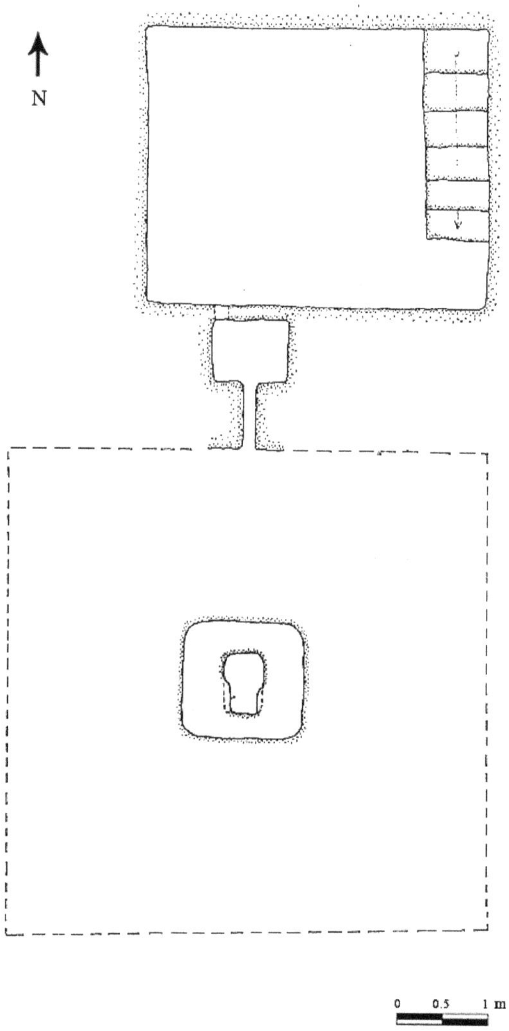

Fig. 1.6. Wine press No. 6, plan.

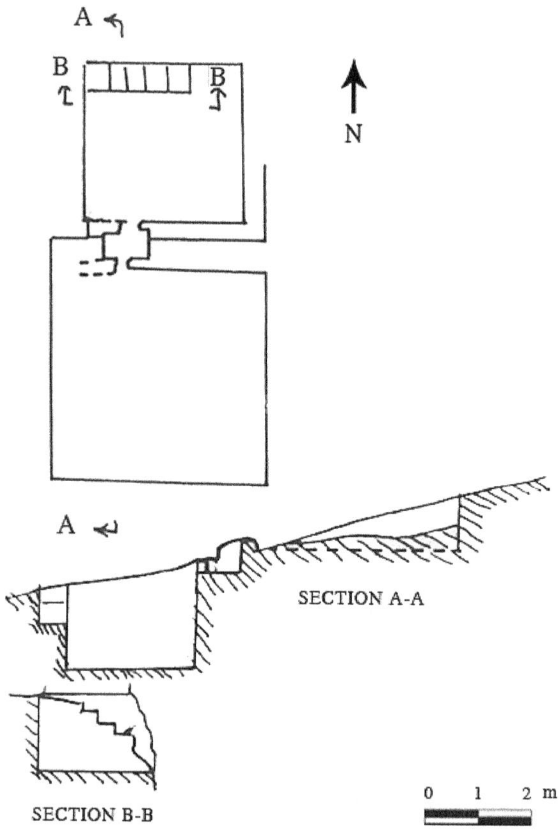

Fig. 1.7. Wine press No. 7, plan and sections.

2. The Oil Presses of H. Karkara

Rafael Frankel

The Site

H. Karkara (MR 1708/2755) is a large site (18 dunams) on the peak of a spur to the south of the Bezet stream. It is located 5 km north-west of H. Zabadi (Site 4). Sherds have been collected at the site from the Early and Middle Bronze Ages and from the Hellenistic period onwards. Sufficient buildings and walls stand to allow a coherent plan of the site to be drawn up without excavation. A church at the site was dated by a mosaic inscription to the last quarter of the 5th cent. CE (Avi-Yonah 1966).

The Oil Presses

Seven oil presses were recorded at the site. The southernmost was excavated in 1981–1982 (Frankel 1992: 46–49). The press is very similar to that of H. Zabadi (Site 4) and therefore the description and discussion are based on comparison to the latter. The H. Karkara installation is a twin lever press with a crushing basin in a separate room (Figs. 1–2). The crushing basin is also concave with central protrusion but lacking a socket. The slotted piers are almost identical to those of Zabadi. The slots in the two central piers are also open-ended while those in the outer two are closed. The only difference is that instead of small L-shaped grooves on the upper part of the piers there are open notches. In each press there is a large press-bed, two connected round collecting vats and two weights. In addition to the slight difference between the slotted piers at Karkara and those at Zabadi the presses vary in two other ways. While the press components at Zabadi are all free standing and were brought to the site, the press at Karkara is largely cut in bed-rock and as a result the rims of the vats are at floor level. This made it necessary to cut a pit for the weights to allow the beam to descend sufficiently and also to cut work pits in order that the workers ladling the oil would have to bend over to a minimum degree. To make room for the work pits the vats were staggered showing that the work pits were planned and not an after-thought. The second difference between the press at Karkara and that at Zabadi is the character of the weights. While one weight in each press is identical to those of Zabadi the second is quite different. The second weight of each press while also being cylindrical in shape has a socket and dovetail channel on its upper surface. These are screw weights of a type known as the Din'ila weight, one of several types that are used in lever and screw presses in which the screw turns and the threaded board is fixed above the beam. The screw is attached to the weight in such a way that it cannot work free. Turning the screw first lowers the beam and finally lifts the weight into the air. The Karkara press, therefore, combined two techniques. It was both a lever and weights press and a lever and screw press. The beam weights were almost certainly only raised after the screw weights were already in the air otherwise they would have interfered with the turning of the screw.

In both weight pits secondary construction could be discerned. The southern weight pit was narrowed by the addition of a line of stones and the northern was filled with stones including broken press components leaving only a round pit in which the screw weight stood. It is very usual that weight pits of screw weights fit them exactly, the reason being that the screw and weight must remain approximately perpendicular to allow the screw to turn easily. It is probable, therefore, that the Karkara press was originally a lever and weights press, that the screw weights were added later and that the weight pits were narrowed at the same time.

Chronology

A complete cooking pot found on the floor of a collecting vat is from the 7th cent. CE providing an exact date for the final use of the press.

Bibliography

Avi-Yonah, M. 1966. "An Addendum to the Episcopal List of Tyre". *IEJ* 16: 209–210.

Frankel, R. 1992. "Some Oil Presses from Western Galilee". *BASOR* 286: 39–71.

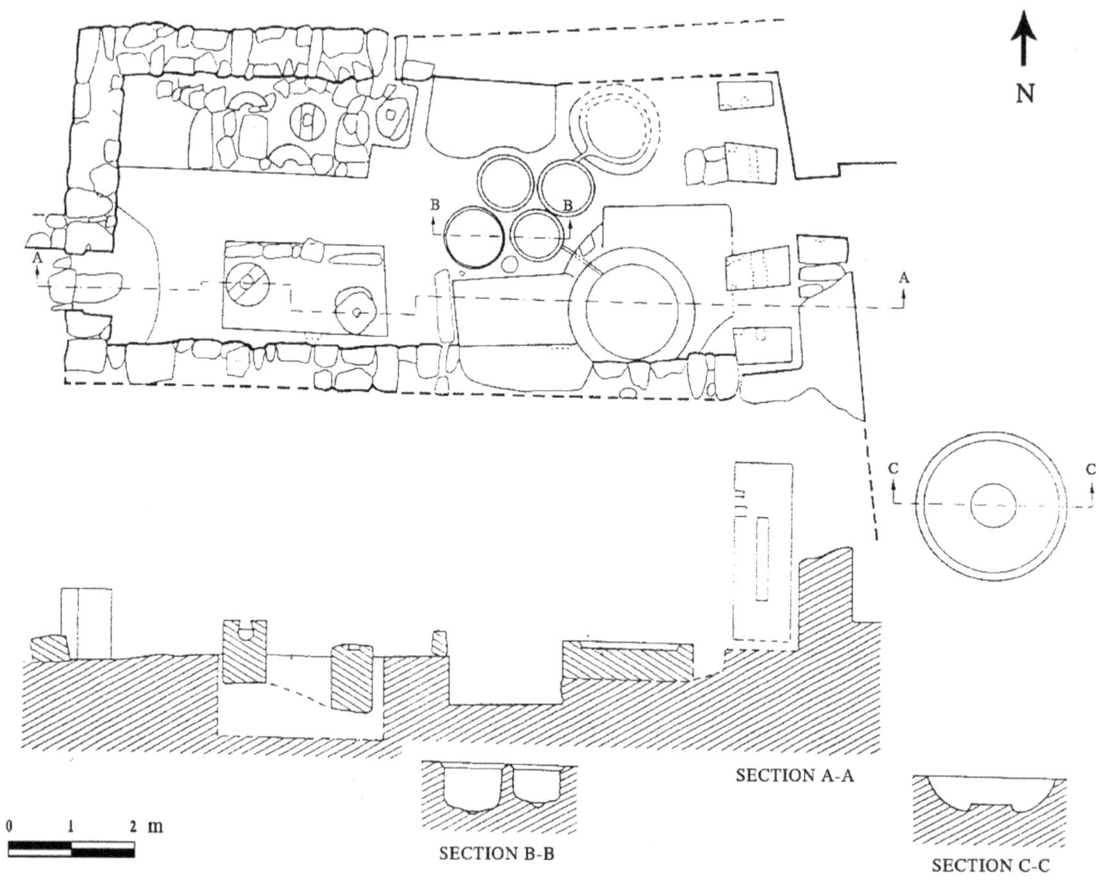

Fig. 2.1. Karkara oil press, plan and sections.

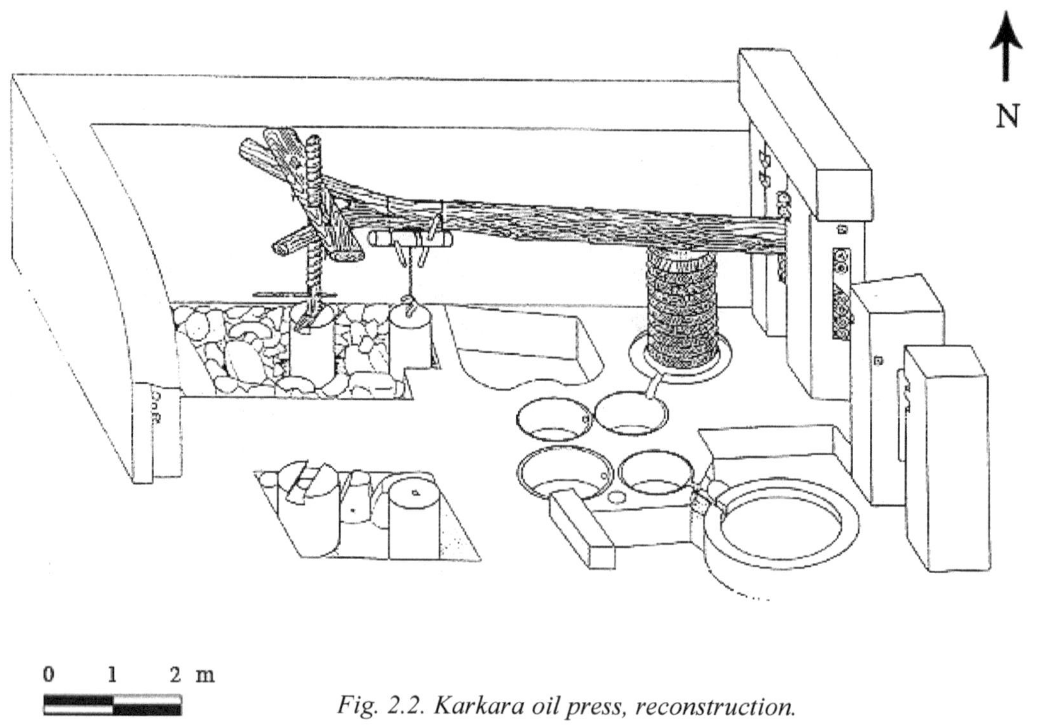

Fig. 2.2. Karkara oil press, reconstruction.

3. The Oil Presses of H. Din‘ila

Rafael Frankel

The Site

H. Din‘ila (MR 1734/2746) had been preserved sufficiently to allow for a detailed plan to be drawn up in the survey that was carried out before excavations began. The plan showed that the site originally consisted of a square courtyard (33×33 m) that later expanded into a small village (c. 80×95 m – 4.5 dunams). Excavations showed that the courtyard core dated from the Roman period and that the expansion took place in the Byzantine period. The site was resettled in the Mameluke period when the settlers often reused the floors of previous periods but also added walls and made other changes.

The Oil Presses

Seven oil presses were recorded in the survey, the visible parts of which were all similar. Three were excavated, A in the northern part of the site in 1984 (Frankel 1985b: 112–113) and B and C in the western part of the site in 1986 (Frankel 1986a). Presses A and B were outside the central courtyard while C was within it. In all three cases the presses no longer functioned in the Mameluke period, the press rooms having been converted to other purposes.

Press A

This was a lever and screw press (Fig. 1). The beam was anchored between two perforated piers, one with a round hole, the other with an angular channel, these serving to secure the rod to which the beam was attached. Oil collection was lateral. The press-bed was round as was the single collecting vat. There was a weight pit and also a shallow work pit. The screw weight had a groove trapezoidal in section with a socket in the centre, the “Din‘ila Weight”. In the room to the north of the press there was a very unusual crushing basin, concave in section, very deep and lacking a central protuberance. The weight was not found *in situ* and in the final stage, the Mameluke period, the region of the weight pit was turned into a water cistern. As Mameluke sherds were found below the press-bed, initially it was suggested that the press was from the Mameluke period (Frankel 1985b: 113), but after the later excavations showed that the Byzantine floors had been reused in the Mameluke period it is now clear that this was also the case in Press A and that this press was also from the Byzantine period.

Press B

This press was very similar to Press A with perforated piers, lateral oil collection, round press-bed, single round collecting vat and Din‘ila screw weight (Figs. 2–3). Only the crushing basin was different. It was of the more usual type, with flat crushing surface and protruding socket. The ceramic evidence shows clearly that this press was from the Byzantine period. The weight was again not found *in situ* but outside the press room and the crushing basin was found between the collecting vat and weight pit where it must have been placed after the press no longer functioned as it would have interfered with the operation of the beam.

Press C

This press had perforated piers similar to those of Presses A and B and a deep crushing basin like that of Press A but otherwise it was different from the other presses (Figs. 2–3). The press-bed was made up of segments laid on the stone floor and had a central opening and around it circular and radial grooves similar to those of central collecting vats (see Kh. el-Quseir, Site 8). However, under the central opening there was no vat and the collection was lateral with two connected vats that allowed the oil that floated on the watery lees in the first vat to flow into the second one. It is as if a device had been introduced from another region and put into use without fully understanding how it functioned. There was a long weight pit in which there was room for three weights but only one was found *in situ* but it was damaged so that it was not possible to determine of what type it was. A Roman lamp was found in the weights pit showing that this press was already functioning in the Roman period, presumably as a lever and weights press. It was not possible to determine which parts of the press were Roman nor what changes were made in the Byzantine period. However, the fact that only one weight was found in the weights pit suggests that in the Byzantine period it was also a lever and screw press. In the Mameluke period a wall was built above the collecting vats dividing the press room into two, the parts of the press-bed were rearranged to serve as paving stones and the weights pit was filled in, the top of the weight probably being broken at the same time.

Conclusions

It is not possible to determine the exact character of Press C in the two periods during which it was in use. All the other presses at the site, however, including those not excavated have perforated piers and Din'ila weights. The Din'ila weight is the most sophisticated of the screw weights to be found in the region and the fact that the simpler Bet Ha'Emeq weight is found mainly in the coastal plain and not in this more remote mountain region is probably to be explained by the fact that the technique of the lever and screw press penetrated this region at a late stage when the more developed screw weight was already in use. The fact that the olive presses were not used in the Mameluke period doubtless bears on the economic history of the region at this period.

Bibliography

Frankel, R. 1985b. "Western Galilee, Oil Presses". *ESI* 4: 110–114.

Frankel, R. 1986a. "Horvat Din'ila". *ESI* 5: 21–23.

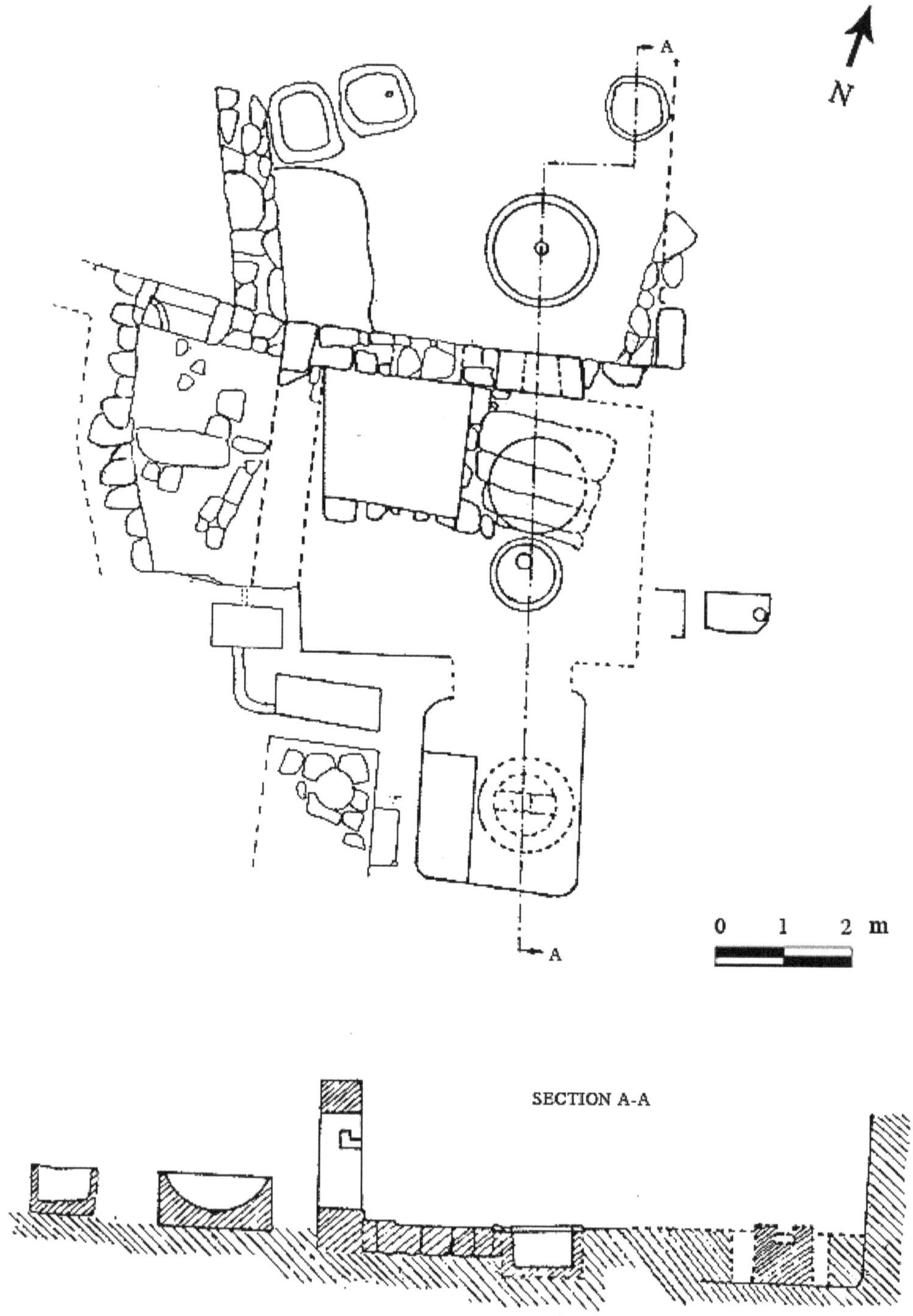

Fig. 3.1. Din'ila, oil press A, plan and section.

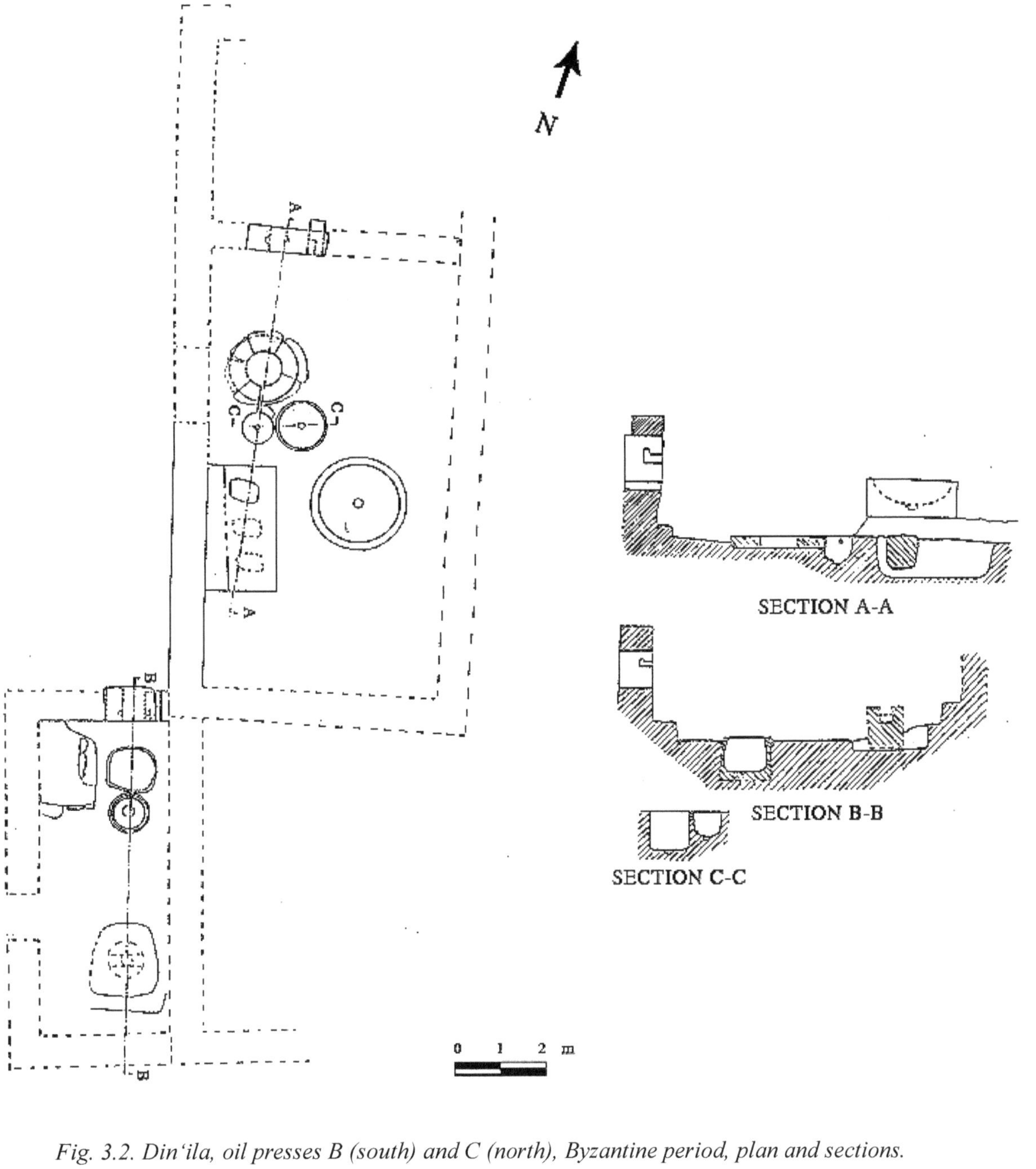

Fig. 3.2. Din'ila, oil presses B (south) and C (north), Byzantine period, plan and sections.

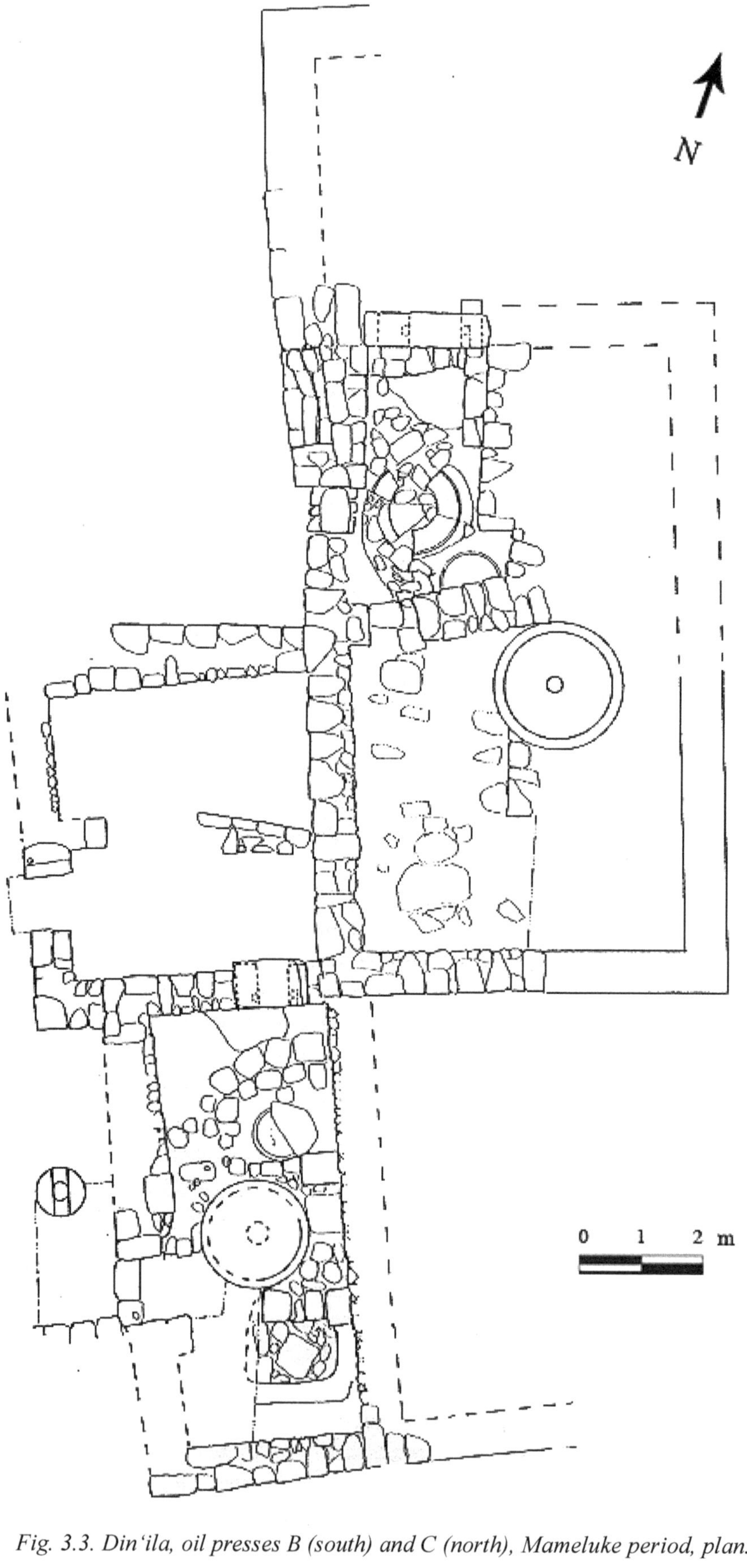

Fig. 3.3. Din'ila, oil presses B (south) and C (north), Mameluke period, plan.

4. The Oil Presses of H. Zabadi

Rafael Frankel

The Site

The site is in the western foothills of Western Galilee (MR 1663/2739), 5 km south-east of H. Karkara (Site 2). It consists of two building complexes, the eastern square, 38×38 m and the second, 30 m to the south-west, rectangular, 28×20 m.

The Oil Presses

There was an oil press in the north-western corner of the western complex and one in the north-eastern corner of the eastern complex. The latter was excavated in 1983 (Frankel 1992: 40–46).[1] It consists of two parallel lever and weights presses in one room and a crushing basin in a neighboring room to the west (Figs. 1–2).

The crushing basin was concave in section with a central protrusion but lacks a central socket (Fig. 3). Two convex millstones apparently rotated in the basin on one shaft that acted both as axle and handle.

In each press the stone components that were all found *in situ* were (Figs. 4–5): two slotted piers that served as the fulcrum of the press and in which the beam was anchored, a round press-bed on which the frails of crushed olive pulp were placed, two round free standing connected collecting vats and two identical beam weights.

The slots in the piers served to hold horizontal cross-pieces that secured the "fixed" end of the beam. The four piers of the two presses were in line. In the outer two the slots were closed, only penetrating the piers partway but on the inner piers (the eastern of the western press and the western of the eastern press) the slots were open ended, cut right through the piers, thus making it possible to insert the cross-pieces into the slots from the passage between these two piers. On the upper part of the inner face of one pier of each press there was also a small square hole and opposite it on the other pier an L-shaped groove. These without doubt served to hold a rod to which the beam end was attached when the cross-pieces in the slots were re-adjusted in order to raise or lower the beam end.

The large round press-bed had a circular groove that connected to an outlet that led to the circular collecting vats. These had a stepped rim to accommodate a lid. A sloping pipe cut below the rim connected the two vats of each press, allowing the pure oil that rose to float on the watery lees in the first vat to flow into the second thus separating the oil from the lees.

The two weights of each press were cylindrical and in the middle of the upper surface of each there was a small depression, in one of which traces of lead were found. This suggests that the weights were hung on ropes that were attached to iron hooks that were secured in the lead.

Chronology

Lamp fragments were found on the floor of the press dating from the 2nd cent. CE, showing that the press was already in use at that time, but a complete cooking pot from the 3rd cent. dates the abandonment of the site.

Notes

1. After the excavation the oil press was dismantled by robbers but the Israel Antiquities Authority succeeded in retrieving the elements of the press and it is now being reconstructed in an Archaeological Park of the Jewish National Fund at Kabri.

Bibliography

Frankel, R. 1992. “Some Oil Presses from Western Galilee”. *BASOR* 286: 39–71.

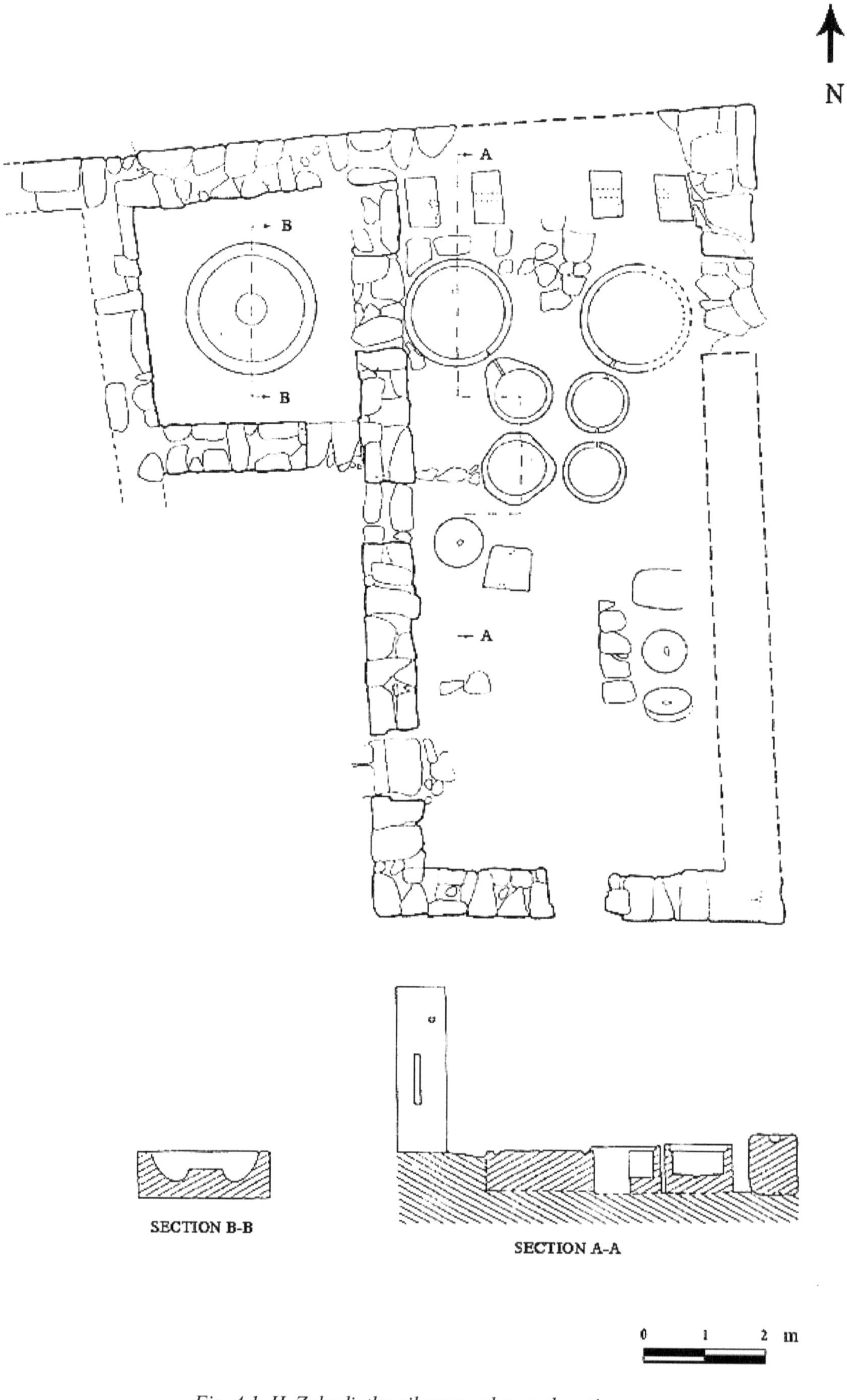

Fig. 4.1. H. Zabadi, the oil press, plan and sections.

Fig. 4.2. H. Zabadi, the oil press from the south.

Fig. 4.3. H. Zabadi, the crushing basin.

Fig. 4.4. H. Zabadi, the oil press from the north. Piers are in the foreground.

Fig. 4.5. H. Zabadi, the oil press from the east showing press-beds, collecting vats, and weights.

5. A Wine Press from Achziv

Danny Syon

The Site

Achziv is situated on the Israeli coast, some 5 km south of the Lebanese border. It comprises an ancient tell, with finds dating from the Middle Bronze Age IIb onwards, up to modern times, and extensive cemeteries to the north, south and east. Its situation, on the ancient coast road, contributed towards its importance as a city and fort. The tell was excavated in 1963–1964 on behalf of the Israel Department of Antiquities and the Rome University Oriental Institute. The cemeteries were intermittently excavated as salvage excavations on behalf of the Israel Department of Antiquities and recently by the Hebrew University Archaeological Institute (Prausnitz 1993; Mazar 1994).

The Wine Press

The site was discovered accidentally by telephone company works in the parking lot to the east of the present National Park, that is situated on the tell (MR 1601/2727). The salvage excavations were carried out on behalf of the Israel Antiquities Authority by the writer in April 1994 (Syon 1998).

The excavated area was 24×9.5 m and elements of a large wine press were discovered (Figs. 1–2). The press consists of two sets of treading floors and collecting vats: The western set consists of three treading floors (24, 13, 14) from which the must flowed to two collecting vats (11, 15; Figs. 3–4) and the eastern set – two collecting vats, to each of which must flowed from two treading floors (25 and 23 to 20; 22 and 17 to 21; Fig. 5).

The must flowed from the treading floors into the vats by way of lead pipes passing through low parapets that separate the floors and the vats, and guiding channels built of mortar. Most pipes have survived *in situ*. All floors and vats are floored with coarse, white mosaic pavements. The vats are surrounded by a narrow ledge of mosaic and their walls are plastered in a gray, watertight plaster. Because of imposed time limitations, only two vats were partially excavated to the bottom. In one of them (No. 20) a series of triangular niches were found straddling the corner, built into the wall. These served as means of getting in and out of the vat, and probably all vats have the same niches. The depth of the excavated collecting vats is 1.23–1.35 m and based on the dimensions of the other vats, their individual and collective volume can be summarized as follows (rounded to the nearest hundred litres):

Vat No.	litres
15	24,200
11	20,000
21	7,900
20	6,900
total	59,000 litres

The press complex (Fig. 6) shows evidence of many repairs and in the trench created by the telephone company works, an older mosaic pavement could be seen some 0.28 m below the top floors, suggesting a major renovation, perhaps along a different plan. The low parapets between the floors show in places up to five successive layers of plaster. Evidence of wall foundations discovered in several places around the perimeter of the complex suggests that it was surrounded by a wall, subsequently robbed. The size of the complex precludes its having been roofed over, but in most cases installations of this type did not have roofs.

Dating

There is very little evidence for dating the press; the finds on the treading floors consist of accumulated rubbish and pottery from biblical times up to the present. Two items, however, help dating the later phases and the demise of the installation. A coin of Constantius II, dated 351–354 CE was found embedded in the plaster of one of the parapets and in another place a rubbish dump, containing large amounts of late 4th cent. CE pottery was found. It lay outside the wine press but on the same level as the treading floors as if it was dumped against the wall of the installation after it went out of use but before the walls were robbed. This dump indicates the date of the end of the use of the press.

Summary

The Achziv wine press is one of the largest installations of its type so far excavated in Israel, and suggests an extensive wine production centre in the western Galilee in the Late Roman period. Its situation at Achziv may imply an exporting centre of local wine through the small but active port of the city[1].

Notes

1. I thank R. Frankel for his insights and helpful suggestions both on site and in the course of the preparation of this summary.

Bibliography

Mazar, E. 1994. "A Burial Ground of the Roman Period at Gesher Haziv". *'Atiqot* 25: 77–93.

Prausnitz, M.W. 1993. "Achzib". *NEAEHL* I: 32–35.

Syon, D. 1998. "A Winepress at Akhziv". *'Atiqot* 34: 85–99 (Hebrew).

Fig. 5.1. Achziv, aerial photograph of the excavation.

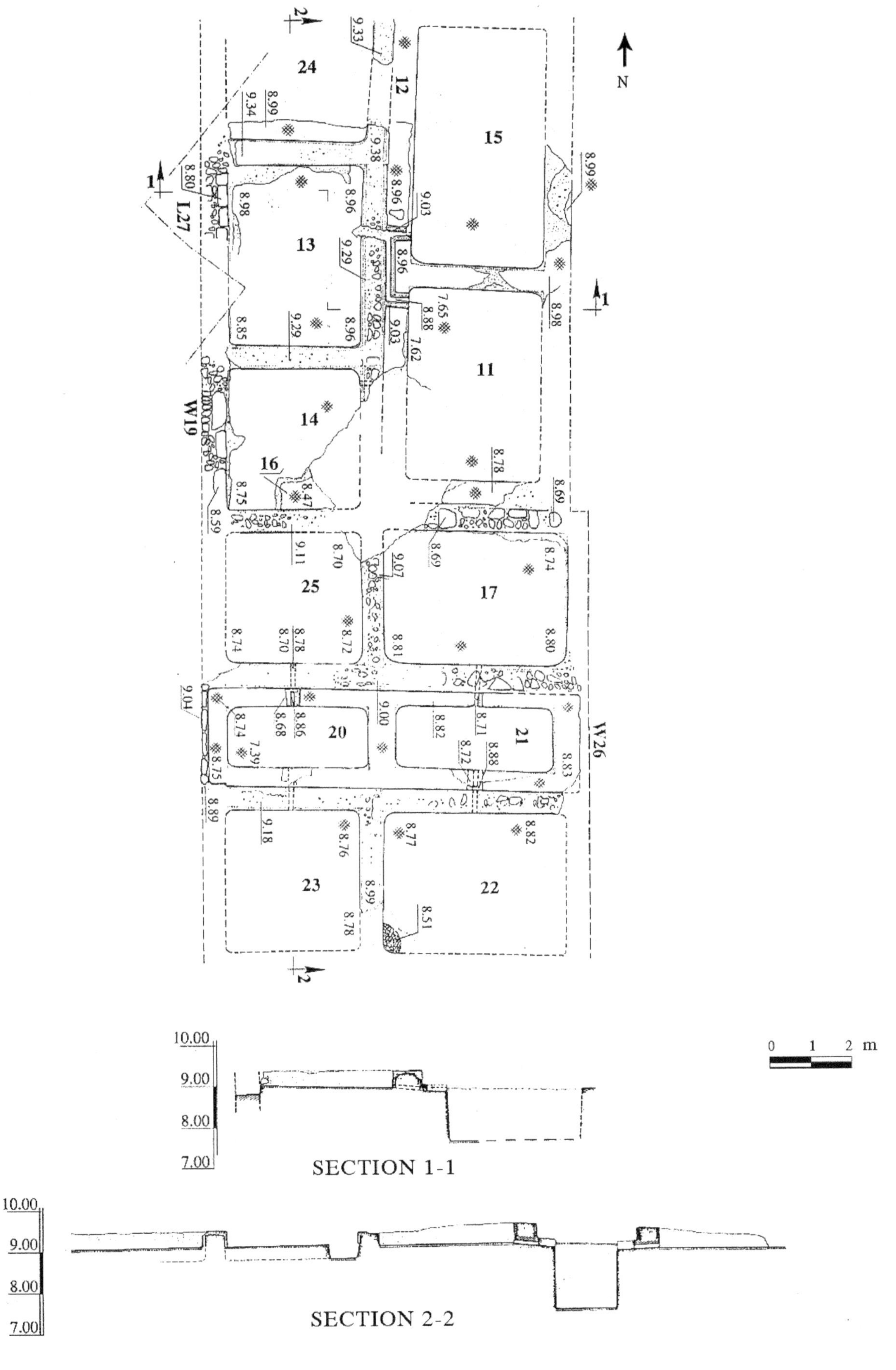

Fig. 5.2. Achziv, plan and sections of the wine press.

Fig. 5.3. Achziv, vats 11 and 15 (in foreground) and at the back treading floor 13. Note parapet around treading floor, the narrow mosaic floor around the vats, the distribution channel between the treading floor and the narrow floor, and the two channels leading from the distribution channel to the vats. View to west.

Fig. 5.4. Achziv, vat 15 with channel leading to it, view to west (see also Fig. 4).

Fig. 5.5. Achziv, vat 20, plastered channel from treading floor 25, view to north.

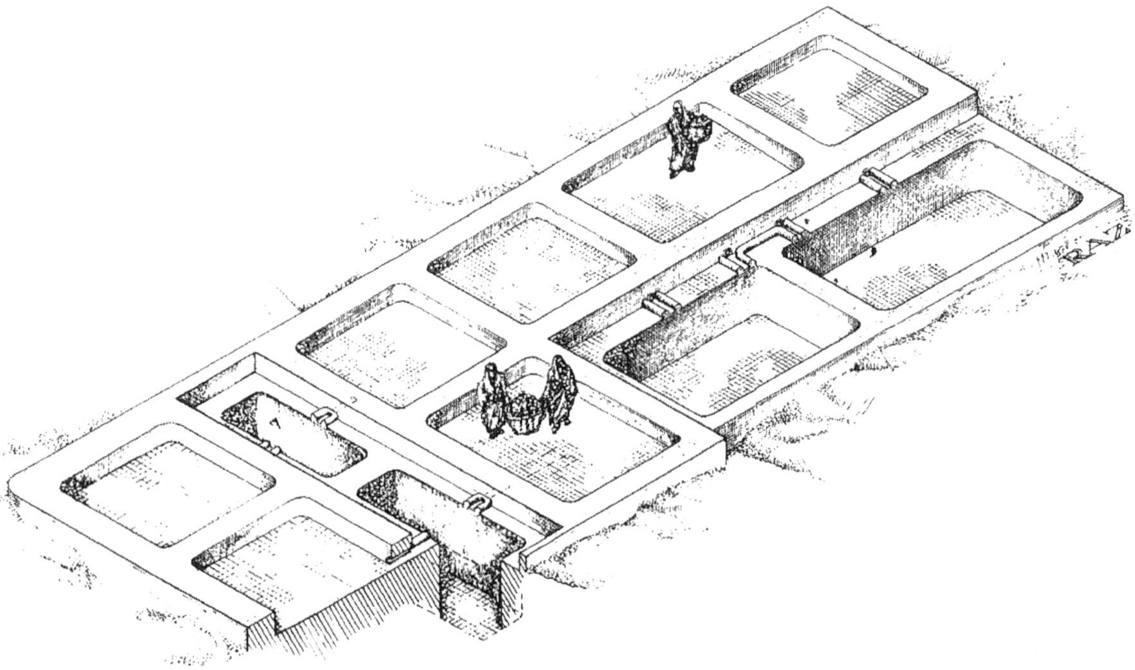

Fig. 5.6. Achziv, reconstruction of the wine press.

6. The Oil Press at Meiron

Nurit Feig

The Site

The excavations at Meiron (MR 1916/2655) were carried out in the winter of 1987 by the author, and were located 150 m south of the synagogue (Feig 2002: 87–114). Meiron is situated north of the Meiron Stream along one of the eastern foothills of Mt. Meiron. It was mentioned by Josephus (Life: 188; War II: 573) and several times in the Mishnah. The site was first surveyed and documented by Kohl and Watzinger (1916: 80–81, Pl. 11), and was excavated by Meyers since 1971. Meyers explored the town of Meiron, its private houses, cisterns and tower, as well as public houses, and the synagogue. The site is dated from the Hellenistic till the Byzantine period, and after a time of abandonment, it was resettled from the mid 8th to the 14th cent. (Meyers, Strange and Meyers 1981: 209).

The Press

In the excavated area some more private buildings were discovered. On the western border of the excavation some installations connected with the oil industry were uncovered (Figs. 1–2). The crushing basin was found *in situ* (L. 808) and 2 m to the west a complete well preserved round crusher was discovered in secondary use (L. 810; Fig. 3). The basin with a central socket was made of hard limestone and was 1.26 m in diameter and 0.25 m. in height. Its dimensions could have accommodated the crusher. West of the crushing basin, 2 cm lower than its base a large storage jar was found sunk upside down in the soil. Half a metre to the east of the crushing basin a large well plastered vat was uncovered (L. 817). It was built of large ashlar stones and measured 1.4×1.4 m and 0.7 m deep. Two small shallow cup marks hewn into the rock are located to the east and north of the basin. A few small plastered vats were found in the northern section of the site on a smooth sloping area (L. 803, 805–806). One of these, L. 803, is a well built plastered vat 20 cm deep, 0.7 m in diameter, built directly on bed-rock.

Two metres to the south of the crushing basin, a press-bed was uncovered *in situ* (L. 826, 836; Fig. 4), classified by Frankel as the 'Ein Nashut Screw Press Base (T732; Frankel 1999: 130–132). At the centre of the press-bed a small channel led to a small round collecting vat, in which some layers of plaster were uncovered. South of the press-bed two conical stone weights with a hole close to their edge were found in secondary use as part of a wall connecting the area of the press-bed to another house (L. 814). A third weight was found in secondary use in a house in the northern part of the excavation that predated the press-bed. The presence of these weights shows that there had previously been another press, probably a beam press with stone weights.

The Processing of the Oil

The crusher and the crushing basin were used for crushing the olives that had been brought in baskets. The next stage in the process is the pressing, that was probably done with a beam and weights press, the weights of which, found in secondary use south of the press-bed, belonged to this phase. It is possible that the weights were found standing in their original position, as there is enough space between the crushing basin and the weights for a beam. However, no evidence remained of how or where the beam was anchored. At a certain stage the beam and weights press went out of use, and another installation, probably more efficient, was installed instead. This system proved to be a direct frame press located to the south of the crushing basin. This type of press-bed can be used with a single rotating screw or with two fixed screws. The frails were piled on the centre of the press-bed and pressed by the screws, the liquid pouring by the small channel to the plastered vat attached to it.

The vat to the east of the crushing basin probably served to separate the oil from the watery lees.

Dating

The majority of the finds found in the houses and installations included pottery and glass vessels, metal objects and coins dated to the Roman period, primarily the 2nd–3rd cent. CE. Three phases were observed at this site according to architectural changes. Phase III is dated to the Hellenistic period, and Phases II-I are Roman. The press-bed belonged to the last phase, Phase I, so the beam and weights press belonged to Phase II. The crushing basin was obviously used in Phase II, but we can not tell about its use in Phase I.

Bibliography

Feig, N. 2002. "Salvage Excavations at Meron". *'Atiqot* 43: 87–114.

Frankel, R. 1999. *Wine and Oil Production in Antiquity in Israel and other Mediterranean Countries*. Sheffield.

Kohl, H. and Watzinger, C. 1916. *Antike Synagogen in Galilea*. Leipzig.

Meyers, E.M., Strange, J.F. and Meyers, C.L. 1981. *Excavations at Ancient Meiron, Upper Galilee, Israel 1971–72, 1974–75, 1977*. Cambridge, Mass.

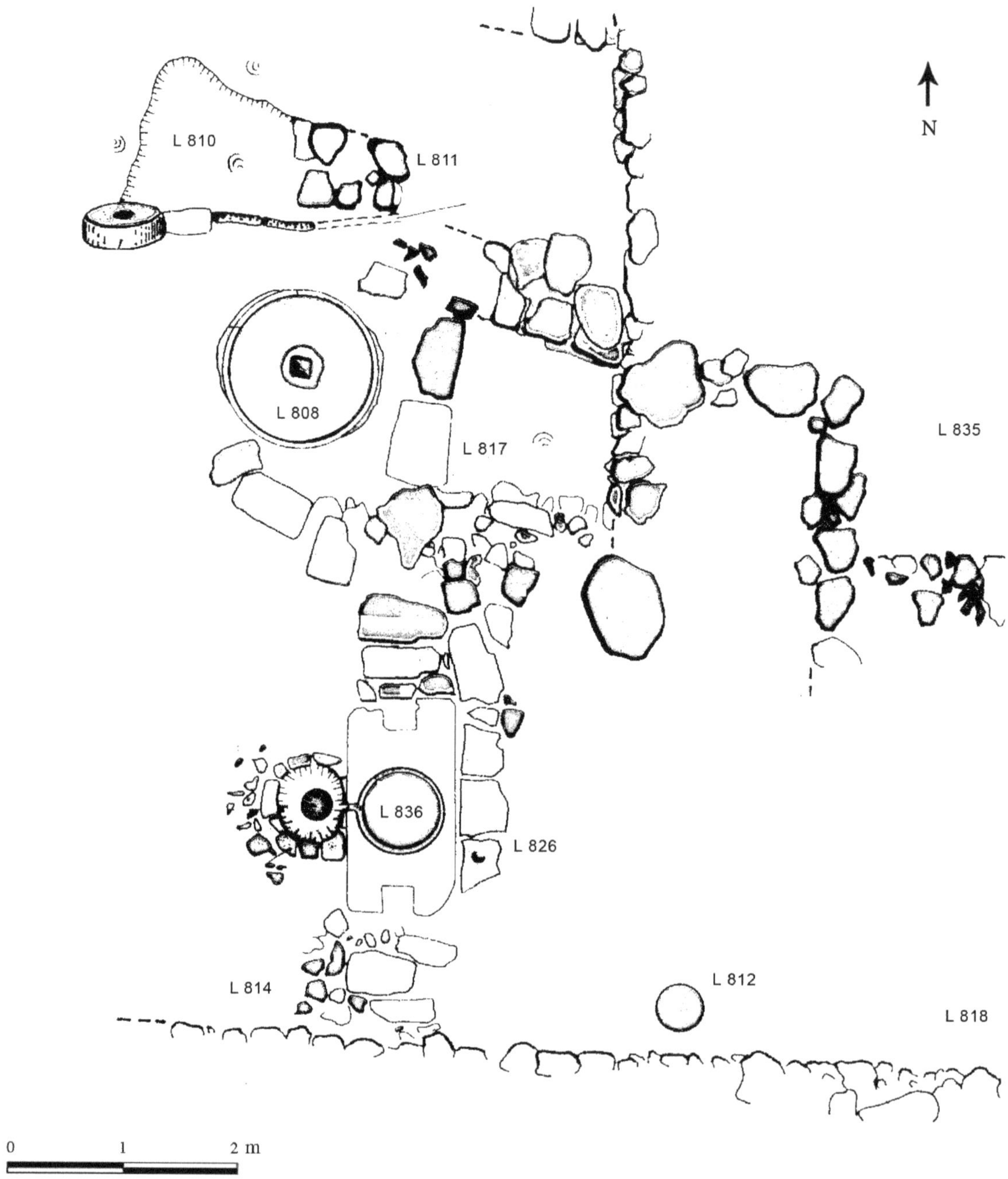

Fig. 6.1. Meiron, plan of the oil press.

Fig. 6.2. Meiron, general view of the oil press, looking south-west: the crushing basin (right) and the press-bed and collecting vat (left).

Fig. 6.3. Meiron, the crushing basin in situ, *looking west. Behind it – the crushing stone in secondary use.*

Fig. 6.4. Meiron, the press-bed and collecting vat, looking west.

7. Hellenistic Wine Presses at Tell es-Samariya

Nurit Feig

The Site

A large group of wine presses (Fig. 1) was discovered beside the Acre-Nahariya road on the eastern edge of Tell es-Samariya (MR 159/262), during a salvage excavation carried out by the author on behalf of the Department of Antiquities during the summer of 1984 (Feig 1990: 73–91).

Excavations carried out by Yogev in 1985 on the summit of the tell and on the eastern slope revealed remains dated to the EB, MBII, Persian, Hellenistic and Mameluke periods (Yogev and Rochman 1985: 103–104). Along the road on the eastern edge of the tell a large cemetery was uncovered that extends over 250 m^2 and comprises EB, LB, Persian, Hellenistic and Roman burials (Ben Yosef 1966: 21; Peleg 1984: 72, 1991: 131–152; Messika 1995: 134–135). 100 m east of this cemetery a burial cave with wall paintings was uncovered, dated to the Byzantine period (Foerster 1986: 416–431). Only a few metres separate the Roman burial ground from the area of the Hellenistic wine presses, indicating that the Roman inhabitants did not know about them.

The Presses

The industrial area can be divided into five clusters, described from north to south:

1. Locus 419 consists of a floor paved with crude white mosaic tesserae (3.5 m north-south and at least 2.5 m east-west). The floor was well preserved, although the western part covered by the road was not excavated and it is probable that the mosaic floor continues westwards. The area was surrounded by a wall 0.5 m wide built of small stones laid side by side that was preserved to a height of 5 cm. On the eastern part of the mosaic floor two recesses were purposely unpaved, possibly for the placing of crushing and pressing installations. Some 0.7 m to the north of the floor was a plastered collecting vat (1.7 m north-south, at least 0.7 m east-west and 0.5 m deep) of which only the eastern portion was excavated. The mosaic surface was probably used as a treading floor to which was attached a movable pressing installation. There is no indication of a channel or gutter connecting the mosaic floor to the vat on the north although it is logical to suppose that such a bond did exist. No remains of weights, beams or screws used in mechanical wine presses were found so we have no evidence for the type of pressing installation used here.

2. Locus 420 is a floor that was plastered a number of times. It is trapezoid in shape (3.3×3.0×4.0 m) and its surrounding wall was only partially preserved. The floor slopes down to a round plastered pit, 0.8 m in diameter and 9 cm deep that was located slightly to the east of the centre of the floor. Compartment 404 to its south could have been used to store grapes before treading. The floor slopes westwards and the first must that seeped out due to the weight of the grapes probably flowed in this direction. From there the grapes were brought for treading to floor 420. No collecting vat was found beside this surface, but it could have been in the unexcavated area to the west, as in complexes 401 and 409. The depression in the floor was possibly used to collect must from the second pressing of the rape, like depression 402 in complex 401 and depression 416 in complex 409.

3. Complex 401 that measures 9×9 m includes two treading floors (Fig. 2). The northern one, L. 412 (3.8×3.8 m) was only partly excavated but it is possible to reconstruct its dimensions. The floor slopes slightly southwards in the direction of the collecting vat (L. 401) though the gutter connecting them was not preserved. The better preserved floor (L. 405) (4.2×4.2 m) was surrounded by a wall of two courses, 0.5 m wide and 0.26 m high. It too sloped westwards, towards the gutter leading to the vat. In the north-eastern corner of the treading floor was a small round plastered basin (L. 402, 0.8 m in diameter, 0.4 m. deep), that also inclined westwards; they are connected by a narrow channel. Three super-imposed layers of plaster are apparent on floor 405. From both treading floors the must flowed to the large, partially preserved collecting vat (L. 401) (2.7×2.6 m, 1.4 m deep). Around the vat is a ledge 0.6 m wide. It was well plastered and slanted to the north-eastern corner in which was found a plastered depression, whose radius is 1 m and depth 0.45 m. On the eastern side of the vat a gutter was uncovered that led from treading floor 405. There was evidence that it was blocked, possibly at a later stage, and then built upon. This gutter was constructed with a strong inclination so that the floor was 0.35 m higher than the bottom of the channel. It is logical to conclude that a similar gutter connected vat 401 with treading floor 412 to its north. Compartments 403 and 404 are part of this complex, although L. 404 could be associated with L. 420. These two compartments were coated with numerous layers of plaster. Both slope slightly to the west.

Compartment 403 was probably used as a storage unit for grapes brought from the vineyard before their transfer to the treading floor. The plastered floor indicates that the juice that seeped from the weight of the stored grapes was collected. From here the grapes were taken for treading by foot on treading floors 405 and 412, the must draining into collecting vat 401 via the gutters. In this vat, that had a capacity of 7,000 litres, the must was left to clear and then transferred to jars for fermentation. The settling pit in the north-eastern corner of the vat was used for settling during the clearing process. It is possible that L. 404 was utilized for the same purpose as L. 403, even though it was some distance away from the treading floor so that it could have been used with L. 420. The purpose of the round pit in the corner of treading floor 405, whose capacity is 200 litres, is not clear. It may have been used in the next stage when the rape was processed a second time to press out any additional liquid, and the must was collected in this pit. The must produced from the second pressing was usually stored separately as it was of a lower quality. A similar phenomenon, of a treading floor with a small pit at its side, is known from Athens and Olynthus, both dated to the Hellenistic period (Dorpfeld 1895: 168–169 Figs. 5–6; Robinson 1938: 343 Pls. 82–84; Wace 1921–1923: 69, Pl. 1–31).

4. Complex 409 also consists of two treading floors, a collecting vat and associated installations (Figs. 3–6). The total area of this complex extends to 9×10 m. The southern treading floor, L. 410, that was in a poor state of preservation, had remains of two layers of plaster, while the upper floor was paved with crude white mosaic tesserae of which only a few stones survived. From the mosaic floor a channel 0.2 m long led through the northern wall to a well preserved gutter. Part of it was built into the wall and part was exposed in collecting vat 409 (Fig. 5). The second treading floor, L. 408, was well preserved but little remained of the channel leading to the gutter. The gutter was incorporated in the wall in an arrangement similar to that in floor 410. The common collecting vat L. 409 remained intact, its floor slanting to the south-east towards a small round depression. The collecting vat is coated with several layers of plaster, especially in the associated depressions. To the south-east another treading floor, L. 418, was uncovered with collecting vat 411 in its corner. East of floors 410 and 418 the rock is raised and bears signs of plaster – perhaps another treading floor. This floor was possibly connected with the round vat L. 415 uncovered to its north.

As in the previous complex here too there are two treading floors, L. 408 and L. 410 from which the must flowed into the collecting vat (6,000 litres capacity). It is to this type of structure that the Tosefta (Ter. 3.7) refers to in the following verse: "two treading floors to one vat". Pit 416 (160 litres capacity) situated in the western part of floor 410 was probably also used for second pressing, like pit 402 (see above). The must settled and cleared in the vat and was then transferred to jars to continue the processing.

Floor 418 and vat 411 with vat 415 in its corner and the worn treading floor nearby were probably used to produce a particular type of wine or to press a different type of grapes. It is possible that they were used to store grapes before pressing, the must seeping from them during storage flowing into the plastered pits. Another possibility is that the pressing of the rape took place here, similar to the process used in treading floor 410.

5. Complex 414. This plastered vat is not close to any treading floor. The rock surface here is high and no remains of plastered areas were discovered on it. The vat was possibly associated with three hewn plastered compartments to its west. There is evidence of a narrow channel that led from the central compartment to vat 414, but it gives no indication as to the process that took place here. Maybe the compartments were used to store the grapes before processing, or some sort of beam press was constructed in them, though no evidence was found for any wood or stone installations.

Conclusions

The finds from the site, unearthed only in vat 414, include dozens of broken jars, bowls and other types of pottery vessels. All are dated to the Hellenistic period (Feig 1990 Figs. 13–17).

The capacity of the various collecting vats is impressive and reaches a total of 13,500 litres. The grape harvesting season lasts 6–8 weeks during which it was possible to carry out 10–12 processing cycles. So the maximum production for the season would be 135,000–160,000 litres. To produce this quantity of wine some 270–320 tons of grapes were needed. These can be harvested from a vineyard of 200–240 dunams. Vineyards usually formed some 10–20% of the cultivated area. This means that the total area under cultivation was 2,000–4,800 dunams. This area extended some 5–8 km north of Acre, between the foothills and the coast, the region under jurisdiction of Acre-Ptolemais in the Hellenistic Period.

Editors' Note

The dating of the wine presses to the Hellenistic period presented in the article is based only on the pottery found in vat 414 that is situated at a considerable distance from the actual installations. This evidence is in our opinion insufficient making the suggested date very doubtful especially as no other wine press with mosaic pavements has as yet been found from such an early period.

Bibliography

Ben Yosef, Y. 1966. "Lohamei Hageta'ot". *HA* 17: 21 (Hebrew).

Dorpfeld, W. 1895. "Die Ausgrabungen am Westabhange der Akropolis II". *Mitteilungen des kaiserlich deutschen Archäologischen Institut* 20: 161–206.

Feig, N. 1990. "Wine Presses of the Hellenistic Period from Tell es-Samariya". *Israel – People and Land.* Yearbook of the Eretz Israel Museum, Tel Aviv 5–6 (23–24): 73–91 (Hebrew).

Foerster, G. 1986. "A Painted Christian Burial Cave near Kibbutz Lohamei Hageta'ot". In: M. Yedaya (ed.) *The Western Galilee Antiquities.* Tel Aviv: 416–431 (Hebrew).

Messika, N. 1995. "Lohame Hageta'ot (Giv'at Yassaf)". *ESI* 14: 134–135.

Peleg, M. 1984. "Lohamei Hagetao't". *ESI* 3: 72.

Peleg, M. 1991. "Persian, Hellenistic and Roman Burials at Lohamei Hageta'ot". *'Atiqot* 20: 131–152.

Robinson, D.M. 1938. *Excavations at Olynthus VII: The Hellenic House.* London.

Wace, A.J.B. 1921–1923. "Excavations at Mycenae". *ABSA* 25.

Yogev, O. and Rochman, A. 1985. "Tell es-Samariya". *ESI* 4: 103–104.

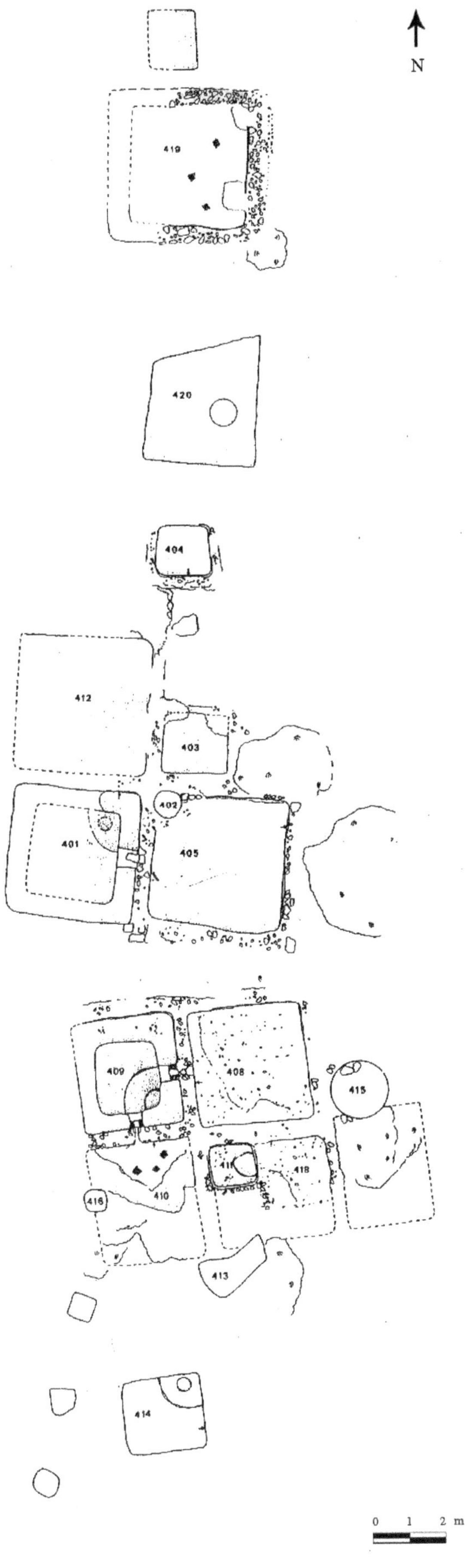

Fig. 7.1. Samariya, general plan of the wine presses.

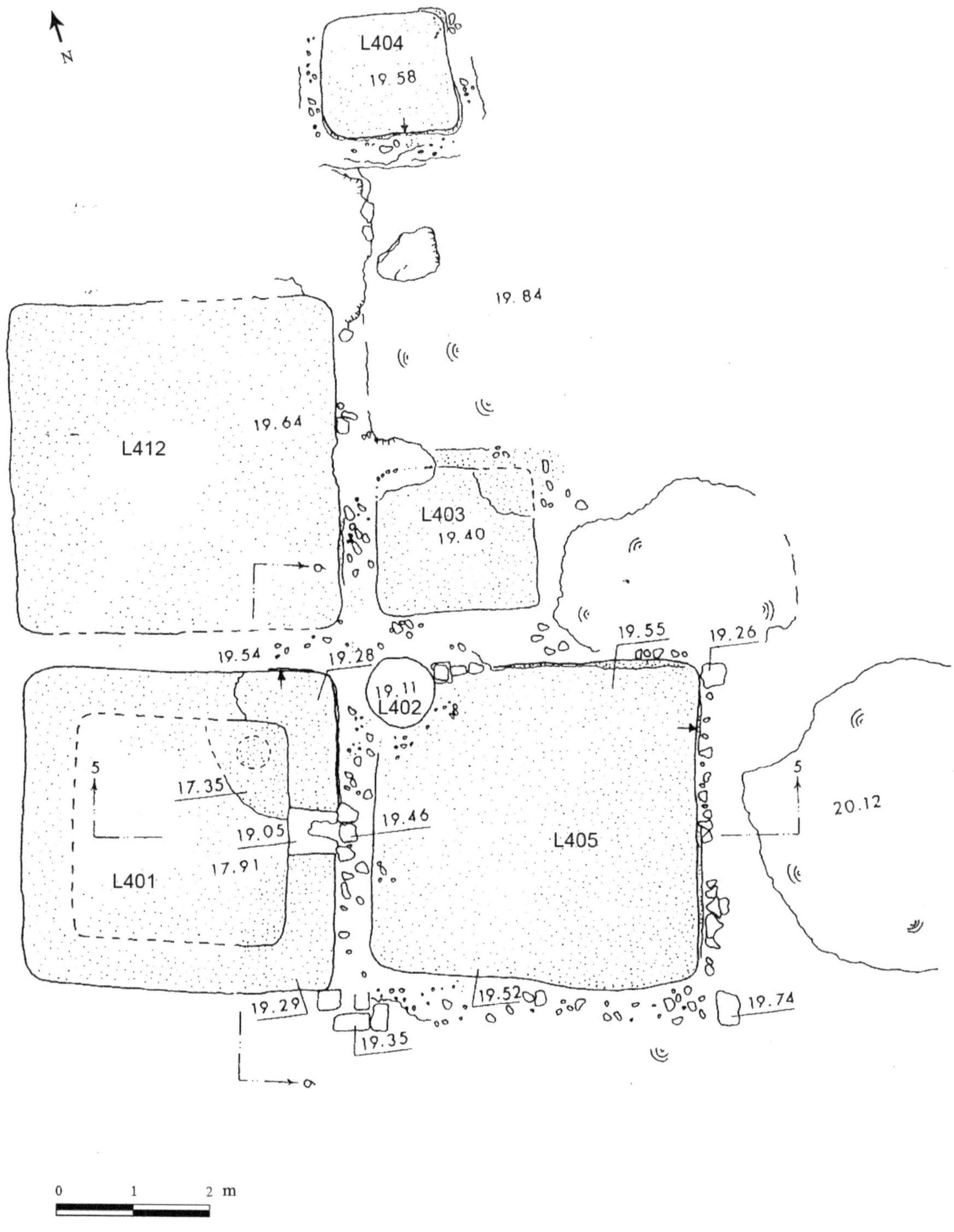

Fig. 7.2. Samariya, plan and sections of Complex 401.

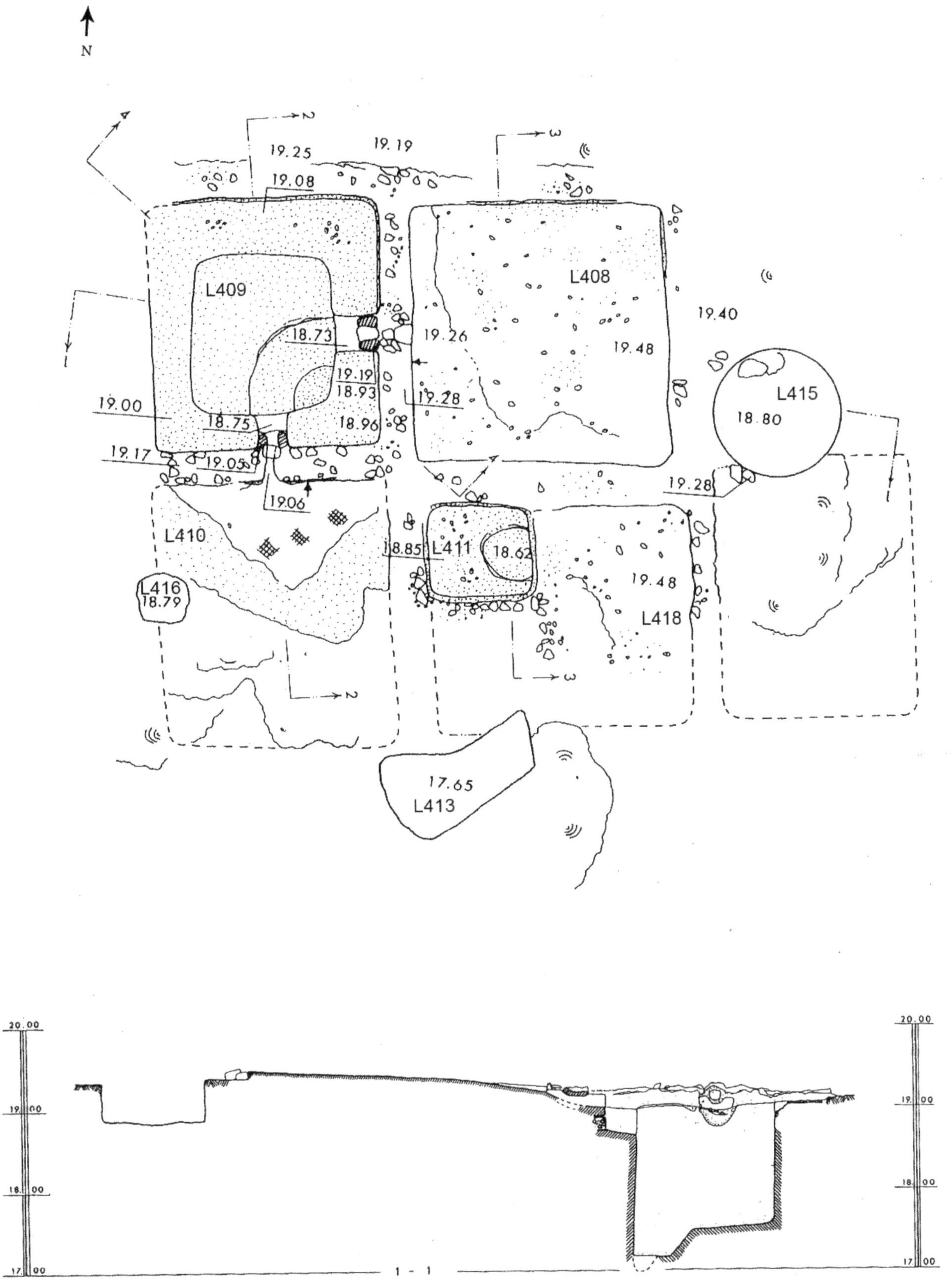

Fig. 7.3. Samariya, plan and sections of complex 409.

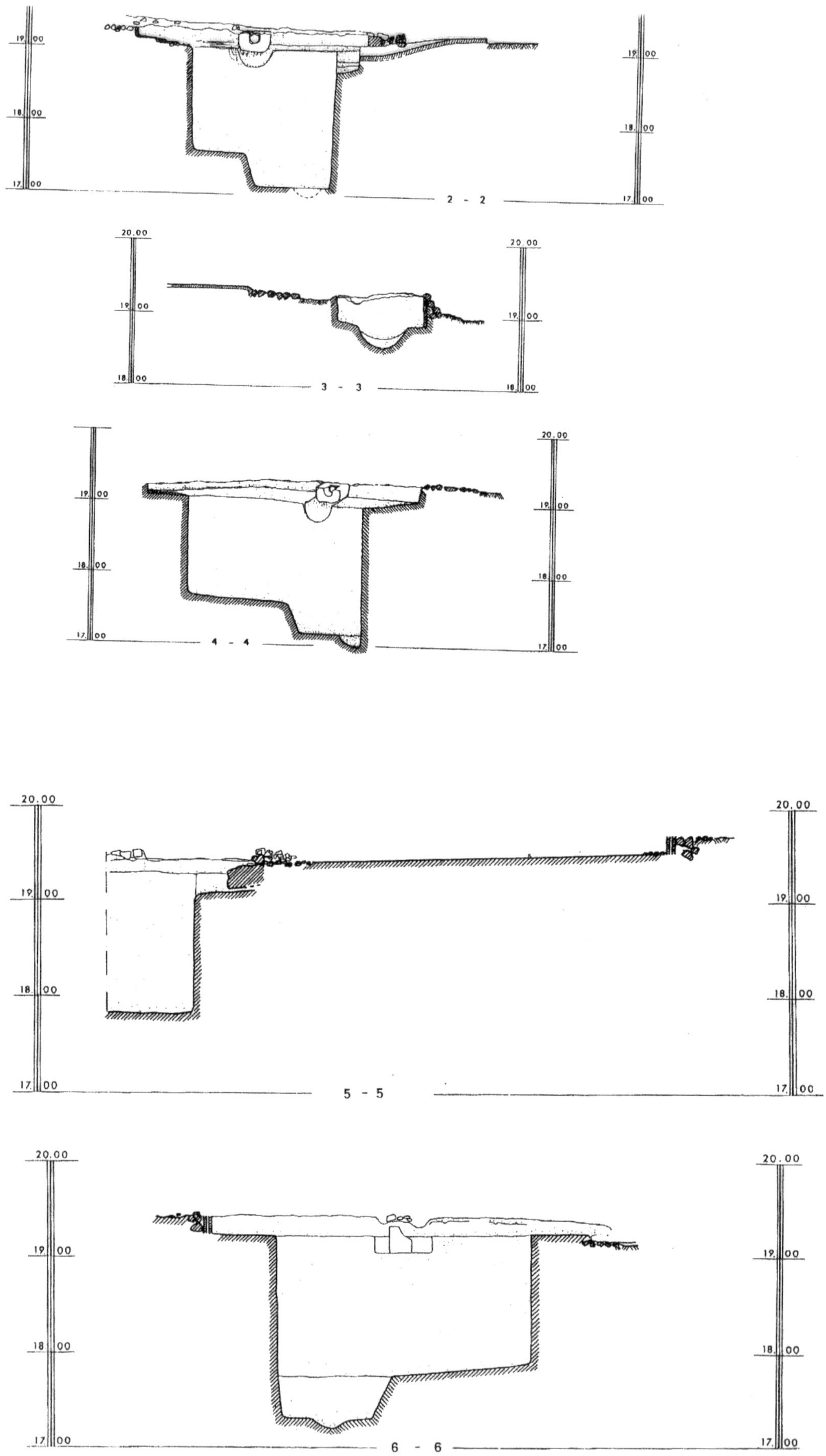

Fig. 7.4. Samariya, sections of Complex 409 (cont.).

Fig. 7.5. Samariya, Complex 409, general view to the north.

Fig. 7.6. Samariya, vat 409 and mosaic treading floor 410, view to the east. In background – floor 408 and vat 411.

8. The Oil Presses from Kh. el-Quseir

Rafael Frankel

The Site

The two small building complexes are situated on a low western spur less than 1 km to the north of the steep scarp of Mitlol Surim that divides Upper from Lower Galilee as defined by modern geographers (MR 1734/2603). The nucleus of the eastern complex measures 52×40 m. The other complex, 170 m to the west measures 45×40 m. Both are from the Byzantine period and a fragment of a stone screen found at the western complex suggests the presence of a church and the possibility that it was a monastery. The site was discovered by M. Aviam and Z. Lederman in a survey carried out in 1978.

The Presses

In 1986 two presses (A & B) were excavated in the eastern complex and one (C) in the western (Frankel 1992: 49–59). Although all three presses were built in the Byzantine period and were apparently all still in use at the time of the destruction of the site at the end of that period there are minor technical differences between them that suggest that they were built in the order A-B-C.

Press A

This press consists of two parts (Fig. 1). The eastern section (A1) is a lever and screw press made up of slotted piers at the fulcrum, a central collecting vat and further north two of the simplest type of screw weight, the Mi'ilia weight. To the north of these on either side of where the free end of the beam would have been there were two small holes. To the east of the press was a crushing basin with flat crushing surface and round sunken socket. The crushing stone (mill stone) was found standing between the two weights. In the one slotted pier that was *in situ* and complete the slot was open-ended and there was a small square upper hole to hold the beam during adjustment (compare to H. Karkara and H. Zabadi [Sites 2, 4]). The central vat combines the functions of the two components, press-bed and collecting vat used in lateral collection. The vat had radial grooves and the frails were pressed immediately above the vat, being placed on criss-cross slats or on a perforated board. The two Mi'ilia weights had a channel, dovetail in section cut half-way across their upper surface (Fig. 2). This type is one of several that lack a socket and work on the principal that the screw was fixed in the weight and the nut that was placed above the beam was turned. The two small holes at the northern end of the press almost certainly secured two uprights to which tackle was attached that served to raise the beam end after pressing.

The second part of press A (A2) was in a separate room to the west and connected to A1. The entrance connecting the two sections led directly to the central vat of A1. In A2 there were only two components, a central vat and a crushing basin. The central vat (Fig. 3) consisted of a free standing round vat and a separate stone cover with a central opening and circular and radial grooves. The central vat of A2 evinced finer workmanship than did that of A1. The crushing basin had flat crushing surface and protruding socket. The unusual feature of A2 was that there was no evidence for pressing equipment nor was there room for such equipment in the small space around the vat. A2 was apparently an installation for producing high quality oil that dripped from the olive pulp after crushing but without pressing. The olives were probably crushed in the crushing basin of A2. The frails of olive pulp were placed on the central vat of A2 till no more juice dripped out and were then transferred to A1 to press out the remaining liquid. While the frails were dripping on the vat of A2 press, A1 could have been used to press other olives that had been crushed in the crushing basin of A1. In Talmudic literature the "first oil", the finest oil used in the Temple rite is the oil that dripped from the crushed olives without pressure (Mishnah *Men.* 8: 4; Bab. Tal. *Men.* 86a). In Latin this type of oil was known as *lixivum* (Columella 12, 52, 11.)

Press B

Press B was, also in the eastern complex, less than 20 m from Press A. It consisted of two parallel lever and screw presses in one room with the crushing basin between them. The eastern press was excavated (Figs. 4–7). Two stages could be discerned in its construction. The fulcrum in the first stage was similar to that of Press A1, consisting of two slotted piers. At the second stage the space between the piers was blocked with ashlar blocks putting the slots out of commission. At the upper end of the piers, on their lateral faces, there were two round holes, open ended on one pier and closed on the other (Fig. 8). These were almost certainly cut at the second stage and served to hold a cross-piece to which the beam was attached replacing the slots after they were blocked. The central vat was similar to but slightly more sophisticated than those of Press A. As well as circular and radial grooves there were four depressions in the shape of right-angle triangles on the rim of the vat that together formed the outline of a square and also two deeper depressions one opposite the other. The square was clearly to hold a framework for criss-cross slats or a perforated board on which to put the frails and the two deeper depressions were for a cross-piece to give additional support to the framework or board. There was one much larger screw weight instead of the two of Press A1. It was of the Bet Ha-'Emeq type with closed dovetail mortice. This type of screw weight almost certainly functioned in a similar manner to the Mi'ilya weight of Press A1, the nut having been turned and not the screw. The Bet Ha-'Emeq weight probably developed from the Mi'ilya weight. As in Press A1 there were two holes below the beam end that probably held supports for tackle to raise the beam. One feature not present in Press A1 was a work pit between the central vat, the crushing basin and the weights pit (Fig. 9; compare to H. Karkara [Site 2]). In the wall of the work pit there were four projecting stones. These allowed the workers to climb down into the pit and leave it in two different ways. Two stones were above each other like in a ladder and three, including the upper of the pair, were arranged at an angle like stairs.

Press C

This press was in the western complex and was almost identical to Press B at its final stage (Figs. 10–11). In this case the western of two parallel presses was excavated as was the crushing basin, again in the centre of the room. There were some minor differences between Presses C and B. The fulcrum of C did not consist of two monolithic piers but was a perforated niche and the cross-piece instead of being inserted through an open-ended hole was put in place using an L-shaped groove. The central vat was simpler, as was the work pit that was not as deep as in Press B, a stone in the corner serving as a step. The screw weight was identical to that of Press B. The crushing basin similarly to the other three had a flat crushing surface but had a square sunken socket. The crushing stone was found in the weight pit as in Press A1 (playful destruction?). One improvement was that the two holes for the tackle posts were cut into one stone block instead of into several stones as in Presses A1 and B.

Conclusions

The three presses of Quseir were very similar. The four crushing basins all had flat crushing surfaces varying only in the form of the socket. Presses A1, B and C were all three lever and screw presses of the type in which the nut and not the screw is turned. In the more common type in which it was the screw that was turned, turning the screw in the opposite direction would raise the beam. In the Quseir presses this was not the case, hence the need for tackle to raise the beam and for holes in the floor to hold supports for this equipment. There are, however, certain differences between the presses that show a clear technical development. The Bet Ha-'Emeq weights of Presses B and C are a clear improvement on the Mi'ilia weights of Press A1. The other change is in the fulcrum: slotted piers in A1 and B stage 1, perforated piers in B stage 2 and a perforated niche in C. The technical difference is that the slotted piers allowed for adjustment of the height of the fixed end of the beam while in the case of the perforated piers or niche the height of the beam was fixed. In Israel in the Byzantine period perforations at the fulcrum are associated with lever and screw presses. Apparently the additional pressure brought to bear by the use of heavy screw weights allowed people to dispense with the advantages achieved by lowering the fixed end of the beam that had been done previously in order to keep the beam in a horizontal position. Not re-adjusting the beam end saved time. This was not the case in all countries, however, and slotted piers were in use to raise and lower the beam end in pre-industrial presses in Europe till recently.

It is of interest that the Quseir presses combine characteristics typical of the northern regions of the country with others that are found mainly in the south. The slotted piers are a Phoenician press component as are both the types of screw weights used at Kh. el-Quseir. Central oil collection and the central vat are a southern characteristic as are also perhaps the placing of the crushing basin in the centre of the room between two presses and the use of a niche and not monolithic piers at the fulcrum.

Bibliography

Frankel, R. 1992. "Some Oil Presses from Western Galilee". *BASOR* 286: 39–71.

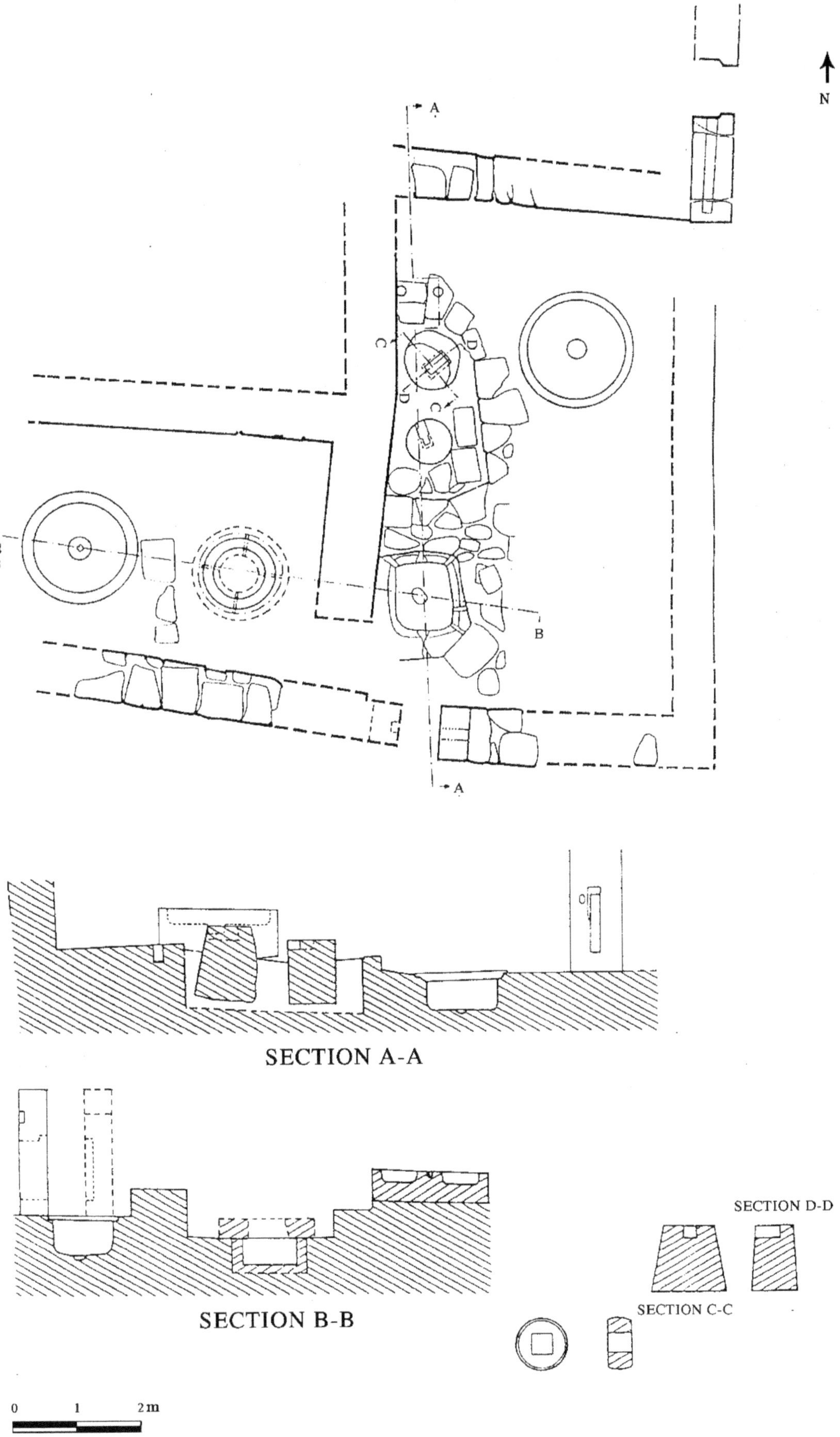

Fig. 8.1. Quseir, oil press A, plan and sections.

Fig. 8.2. Quseir, oil press A, screw weights.

Fig. 8.3. Quseir, collecting vat A-2.

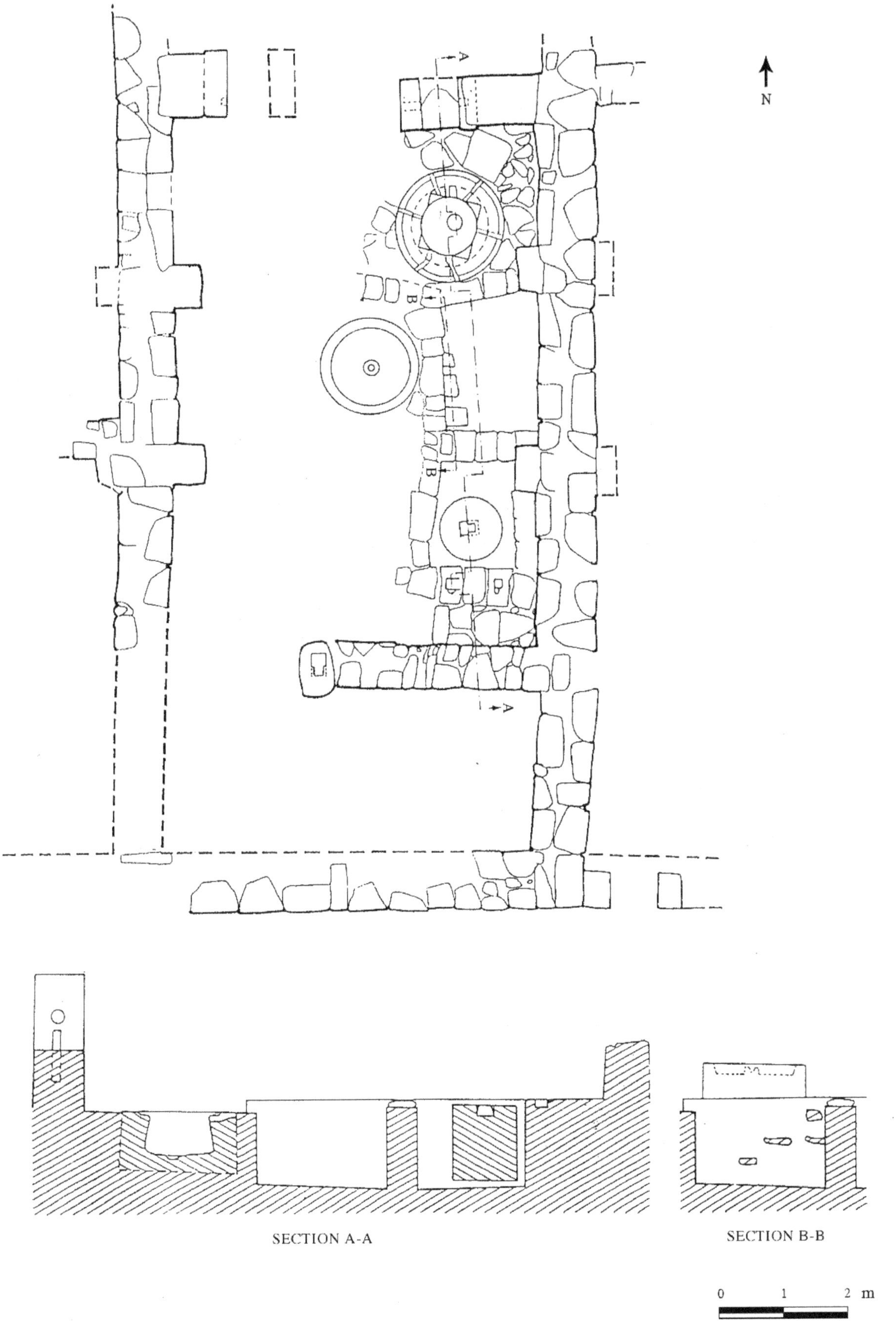

Fig. 8.4. Quseir, oil press B, plan and sections.

Fig. 8.5. Quseir, oil press B from the north.

Fig. 8.6. Quseir, oil press B from the south.

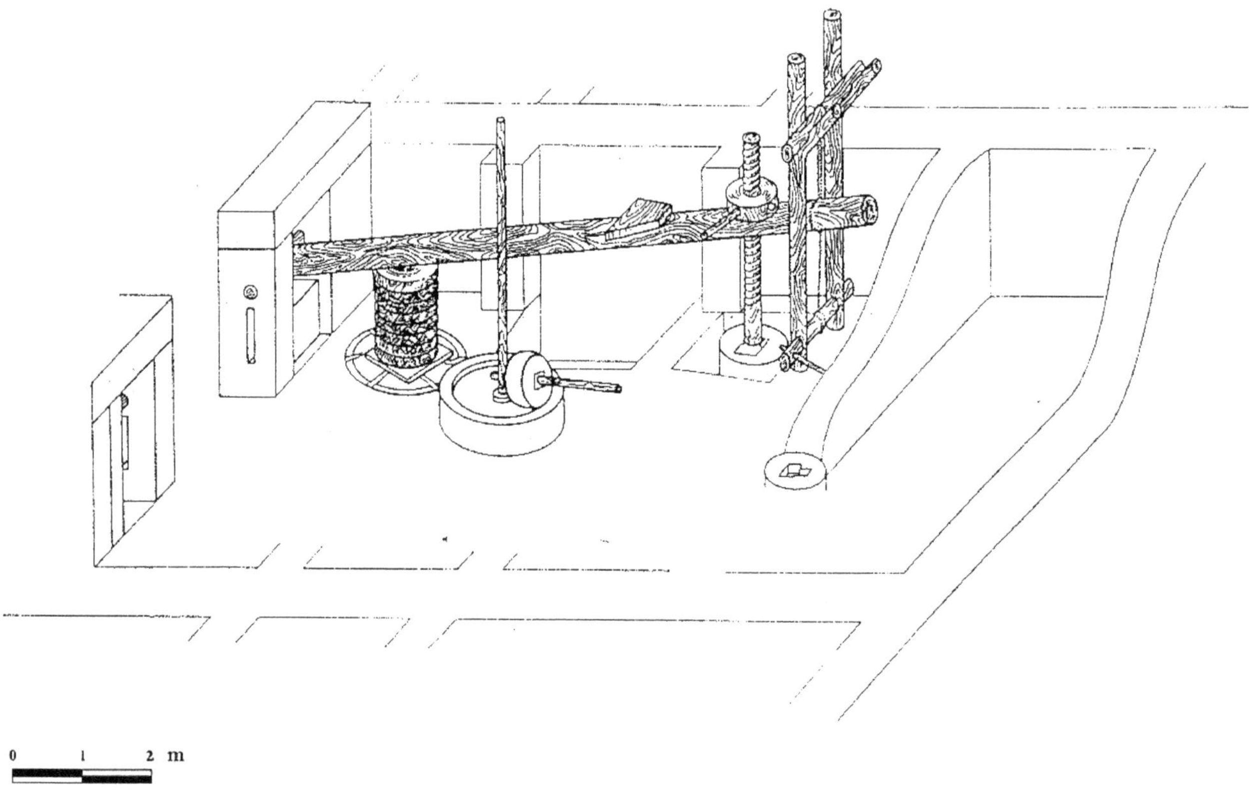

Fig. 8.7. Quseir, oil press B, recostruction.

Fig. 8.8. Quseir, oil press B, press piers.

Fig. 8.9. Quseir, oil press B, work pit.

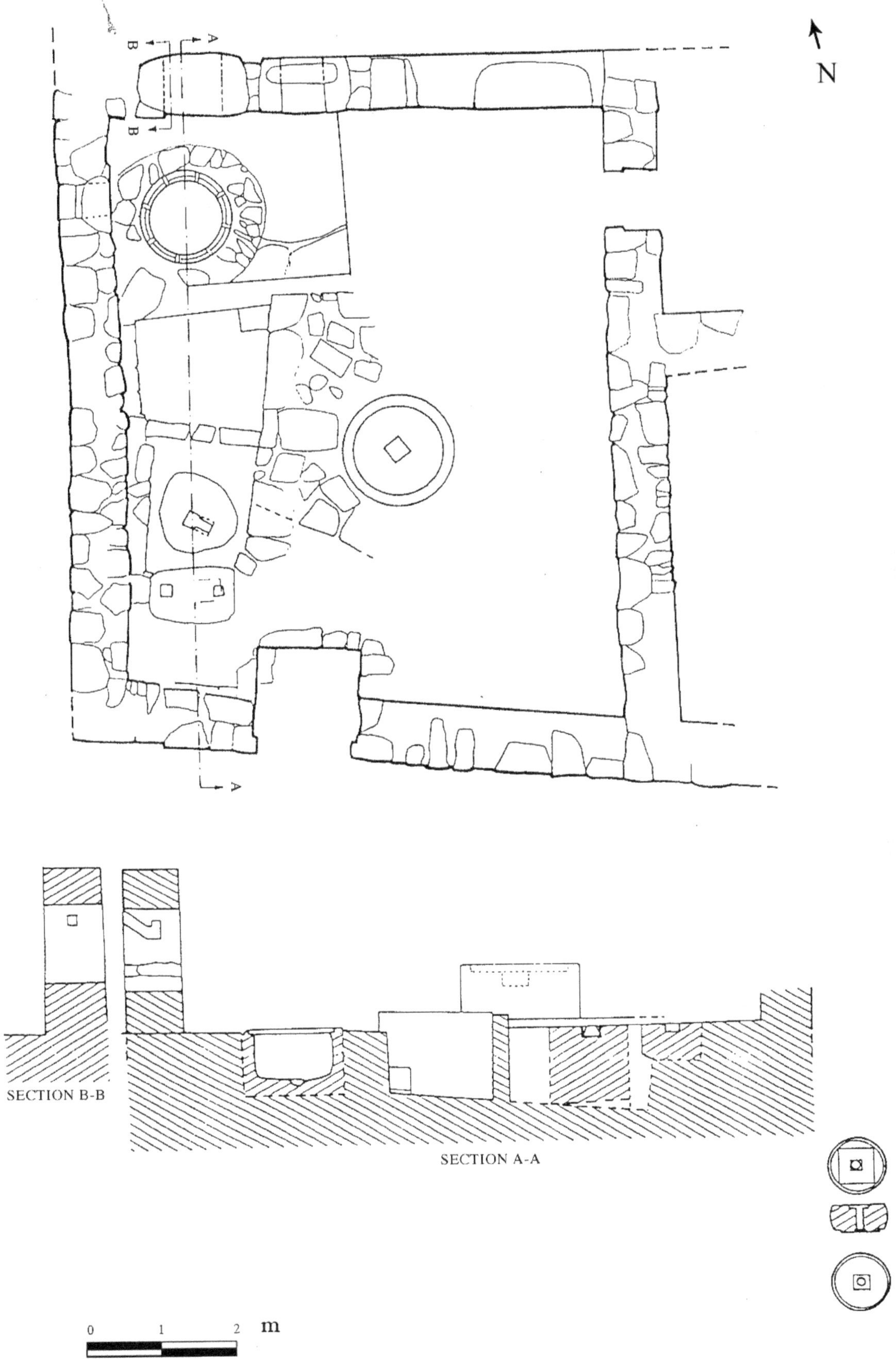

Fig. 8.10. Quseir, oil press C, plan and sections.

Fig. 8.11. Quseir, oil press C from the south.

9. An Oil Press at er-Rama

Etan Ayalon, Rafael Frankel and Amos Kloner

In 1976 excavations were carried out by Talia Shai and Vassilios Tzaferis in the village of er-Rama (MR 184/260). Roman baths and a Byzantine church were uncovered and an oil press was discovered about 100 m east of these buildings (Fig. 1). The press was in a square building (6.5×7.0 m). Close to the southern wall there were two screw bases approximately 1×2 m. In both presses there were round press-beds on either side of which were closed mortices to accommodate the press frame of a single screw press or the screws of a double screw press. The presses were connected to round collecting vats, the western to a small vat (diameter c. 0.75 m) and the eastern to a large one (diameter c. 1.7 m). Two additional screw bases were found in the vicinity. One of the four screw bases had square mortices but in the other three the mortices were central dovetail mortices, a device found in many different types of installations in Upper Galilee. It has been suggested to call this type of screw press "the Rama Press" (Frankel 1999: 135–136).

Bibliograsphy

Frankel, R. 1999. *Wine and Oil Production in Antiquity in Israel and Other Mediterranean Countries*. Sheffield.

Tzaferis, V. and Shai, T. 1976. "Excavations at Kafr er-Rameh". *Qadmoniot* 9: 83–85 (Hebrew).

Israel Antiquities Authority Archives.

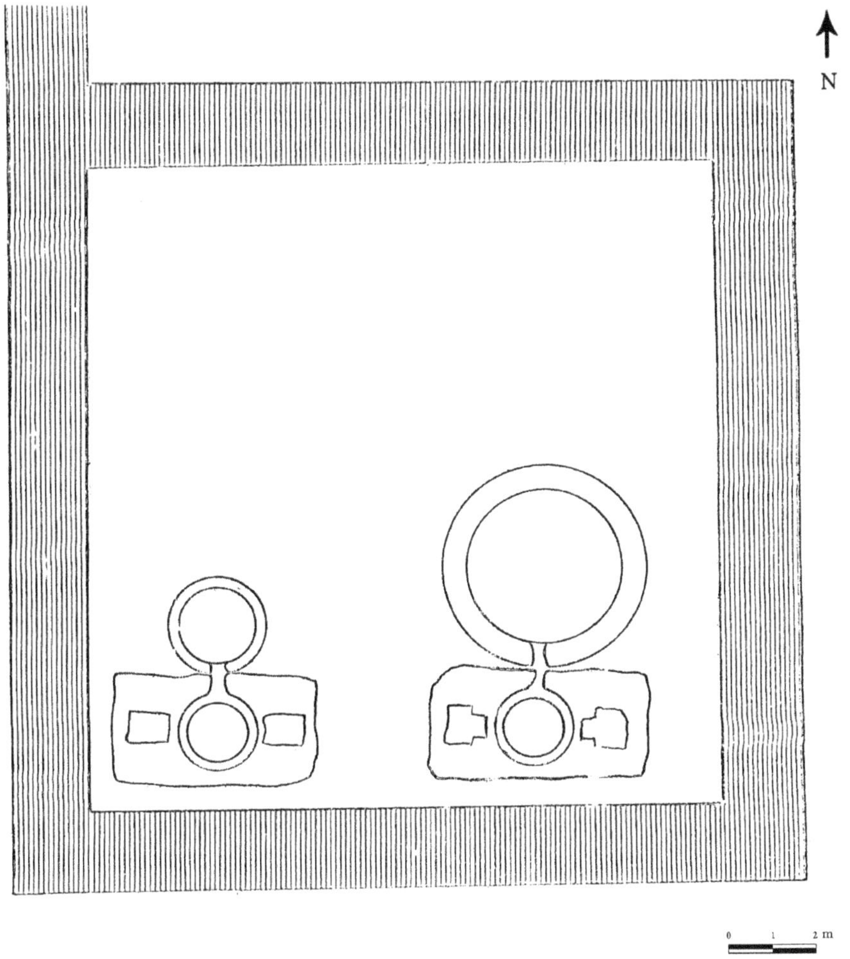

Fig. 9.1. Rama, plan of the oil press.

10. Roman-Byzantine Wine Presses on Mt. Hermon

Shimon Dar

On Mt. Hermon, large ancient wine presses were discovered of a type that is characteristic of the region, but differs in a few particulars from the wine presses found in other areas of the Land of Israel. In the Druze villages on Mt. Hermon, the tradition of viticulture and the extraction of the must in similar installations have been preserved almost until the present day. This situation makes it possible to explain some of the processes involved in the ancient wine industry.

The Wine Presses

Three ancient wine presses were surveyed by the author[1] on Jabal (Mt.) Nasbaya (1,289 m), one of the ranges of Mt. Hermon, north of the town of Rashaya al-Wadi (MR 7653/7123). Today, this area of the mountain is not cultivated, nor does any sign exist there of modern vineyards. However, ancient cultivation terraces, long abandoned, provide evidence for agriculture in the region in ancient times.

Two of the wine presses (Nos. 1 and 2) are adjacent to each other, while the third (No. 3) is located some 600 m to the north. The three installations have many common characteristics, but they differ in their measurements and in the location of their various components. The wine presses, hewn from the bed-rock, are surrounded by stone walls which remain partly preserved.

While there are even larger complexes in other parts of the region, each of these three wine presses covers an area of 90–100 m^2. Their basic components include (Figs. 1–2):

- Large treading floors, alongside of which were temporary storage surfaces for grapes brought in from the vineyards.
- Primary collecting vats.
- Secondary collecting vats.
- Channels for directing the flow of the must and basins for filtering it.
- A rear wall, hewn or constructed, in which was carved out a niche in which the beam of a lever and weights press was anchored.
- Stoves for boiling the must (described below).

Components of Mt. Hermon Wine Press No. 3 (See Fig. 1)[2].

No.	Component	Dimensions (Metres)
1	Main treading floor	3.2×9.4
2	Grape storage surface	1.0×3.2
3	Grape storage surface	1.8×3.2
4	Channels for must flow	
5	Filtration basins	Ø 0.25–0.40
6	Primary collecting vat	
7	Primary collecting vat	
8	Primary collecting vat	1.8×1.5×?
9	Primary collecting vat	
10	Primary collecting vat	1.3×1.1×?
11	Secondary collecting vat	1.3×1.2×?
12	Secondary collecting vat	0.8×0.7×?
13	Niche for end of press beam	

Similar wine presses have been surveyed at the village of 'Aiha (MR 766/710) that is near Rashaya al-Wadi, in the Yavsi valleys (MR 765/708), at the village of Beit Jan, that is within the borders of Syria, on the south-eastern slopes of Mt. Hermon (MR 769/687), and at the Druze village of Harfa (MR 771/684). The wine press in the Beit Jan area, long abandoned, was partly constructed of slabs of limestone integrated into the basalt surfaces. The wine press in Harfa was hewn from bed-rock. The evidence for the presence of a press was a mortice cut into the centre of the treading floor showing that instead of a lever and weights press the installation consisted of one fixed screw. At the time of the author's visit to Harfa, the villagers were still treading grapes in the ancient wine press! The must was intended for the production of grape honey, in Arabic *dibbes* (see below).

Judging by the area of the treading floors and the volume of the collecting vats, the wine presses of Mt. Hermon served large vineyards, and a considerable amount of must was produced here. The average content of a collecting vat is 1,000–1,500 litres, and every wine press had at least four or five primary vats and several secondary vats as well. In one of the wine presses of Jabal Nasbaya, three niches were found carved into the wine press wall, each of which accommodated the beam of a lever and weights press.

The Operation of the Wine Presses

Investigation of the components of the Mt. Hermon wine presses and comparison of the production processes of grape honey that were practiced in the Druze villages of the region, have made it possible to explain some of the working processes practiced in ancient times (Dar 1986: 147–164).

The grapes were trodden on the treading floor, and the must flowed via the channels and filtration basins into the collecting vats. From the must it was possible to produce various types of wine. Initial fermentation took 3–5 days (Tosefta *Ter.* 7, 15), at the end of which the must was transferred to jars. Subsequent activities – main fermentation, sedimentation, purification and the addition of spices – went on for about a year, until the wine was ready.

After the grapes had been trodden (barefoot, so as not to crush the seeds, that may spoil otherwise good wines), the remaining must was expressed from the pulp by lever and weights presses. The remains of the trodden fruit were loaded into sacks, moistened with water, and pressed once or twice. The must extracted in this way was generally used to produce inferior wines or grape honey.

The presence of cooking stoves, that represented an integral part of each of the Mt. Hermon wine presses, attests to the boiling of the must to produce grape honey or cooked wines (below).

Dating of the Wine Presses

An archaeological survey conducted on those parts of Mt. Hermon located in Israel and Lebanon indicated that the villages in which the wine presses were found already existed during the Hellenistic period, and continued to exist during the Roman and Byzantine periods as well. An inscription discovered at ʻAiha hints that the wine presses were already in operation during the 1st cent. CE. The inscription, dating to 9 CE and read by Professor S. Applebaum (1988: 38, 49), mentions Agripinus, who managed an estate in the Hermon area on which grapes may possibly have been raised and wine made.

Archaeological evidence points to continuing human habitation through the Early Arab period and into the Middle Ages, but the number and size of the settlements were diminished. The rise of Islam and the concomitant interdiction against drinking wine resulted in the cease of wine production in the Mt. Hermon villages, but the manufacture of grape honey remained. Viticulture on Mt. Hermon was apparently mentioned in Jewish liturgical poetry of the 8th-10th cent. CE (Rozenson 1987: 191–193), and has not completely ceased until the present.

Production of Warmed Wines and Grape Honey on Mt. Hermon

Mentioned above are stoves that existed in the Mt. Hermon wine press complexes for the purposes of boiling must and heating wine. The stoves were constructed of stones and clay, and were partially buried – or else surrounded by a low stone wall – to protect them from wind and other disturbances.

Archaeological excavations and ancient literary sources have taught us something about the various stages of heating and/or cooking of certain wines. The Mishnah (*Men.* 8, 6) mentions, among the wines forbidden to bring to the altar, cooked wine and smoked wine. The Roman naturalist Pliny the Elder (XVIII, 318) mentions varieties of grapes suitable for secondary extraction of the must, that can then be cooked. Apparently special wines were made on Mt. Hermon, too, that required heating or boiling at one or more stages of their production. The Babylonian Talmud mentions the words of Rabbi Ami: "...solidified wine brought from Snir (= Hermon), resembling dried figs" (*Sot.* 48b). Mention of the special wine from Mt. Hermon apparently indicates that it was an expensive commodity, although familiar. It should be noted that in accordance with Mishnaic language, solidified wine is wine that has thickened and become solid, has a high content of sugars, and maintains its preservation almost indefinitely. Pliny (XVIII, 80–86) also mentions concentrated wines used in medicine, to preserve fruits, and to complement other wines less sweet. The concentrated must frequently mentioned by Roman wine experts was obtained by boiling grape juice after treading of the grapes in the wine press.

Production of Grape Honey in the Hermon Region Until the Present

In villages on the Israeli section of Mt. Hermon and at its foothills, more than twenty installations have been found for the production of grape honey (Arabic *dibbes*), most of which are derelict today (Dalman 1928–1942, IV: 354–355). According to one of the local traditions, the production of *dibbes* in this area was revived only during the Ottoman period. These wine presses include the same basic components found in the ancient Mt. Hermon wine presses: treading floors, storage surfaces, primary and secondary collecting vats, channels through which the must flowed, filtration basins, niches for the press beam, and stoves. In the wine press surveyed by D. Urman (1974) were found both a treading floor and a small pit for the remnants of twice-trodden grapes.

Grape honey is made from select white grapes, trodden in wine presses that the Druze call *Ma'asara* (Arabic for "press"). After they were placed on the treading floors, marl was sprinkled on the grapes to adsorb dirt and clarify the must. According to Urman, 70–100 kg of grapes are trodden per person per hour. The must flows into the primary collecting vats, where more marl is added, and is allowed to remain there for 3–5 hours until the dirt and marl settle out. The clarified must is then drawn out of the vat into huge hemispherical tinned-copper containers (one such container was examined and measured, and its volume found to be 400 litres). The cauldron was placed on top of a stove partially buried in the ground and walled with stones. The fuel is Thorny Burnet (*Sarcopoterium spinosum*), and the must is boiled for 3–6 hours, during which time the foam that forms on top is continually removed. After boiling, the must is transferred to the vats of the wine press to cool for about six hours and then boiled again for about four more hours.

The *dibbes* produced represents 20–25% of the weight of the grapes, and laboratory analysis shows a sugar content of approximately 48%. It is a dark brown viscous liquid, and can be stored for years. In both the Mt. Hebron and the Mt. Hermon areas, the *dibbes* is used as a base for various foods, a beverage, and even medications. During the Ottoman period, the Druze would market *dibbes*, receiving payment in the form of various cereal grains and olive oil.

Notes

1. The research was conducted during the summer of 1984 under the leadership of the author.
2. Security conditions precluded excavation or accurate measurement.

Bibliography

Applebaum, S. 1988. "Selected Inscriptions from the Temples and Villages of Mt. Hermon". In: S. Dar (ed.). *The Hermon Settlements in Antiquity*. Tel Aviv: 33–53 (Hebrew).

Dalman, G. 1928–1942. *Arbeit und Sitte in Palästina*. 7 Vols. Gütersloh (Repr. Hildesheim 1964).

Dar, S. 1986. *Landscape and Pattern. An Archaeological Survey of Samaria 800 B.C.E. - 636 C.E.* (BAR IS 308). Oxford.

Dar, S. 1993. *Settlements and Cult Sites on Mt. Hermon, Israel* (BAR IS 589). Oxford.

Pliny the Elder, 1958–1963. *Naturalis Historiae*. 10 Vols. (Trans. H. Rackham). London and Cambridge.

Rozenson, I. 1987. "Vines on the Hermon – Literary Allusions to Jewish Agricultural Activity". *Cathedra* 46: 191–193 (Hebrew).

Urman, D. 1974. "Presses for the Production of Grape Syrup in the Golan". *Teva Va'Aretz* 16/4: 173–176 (Hebrew).

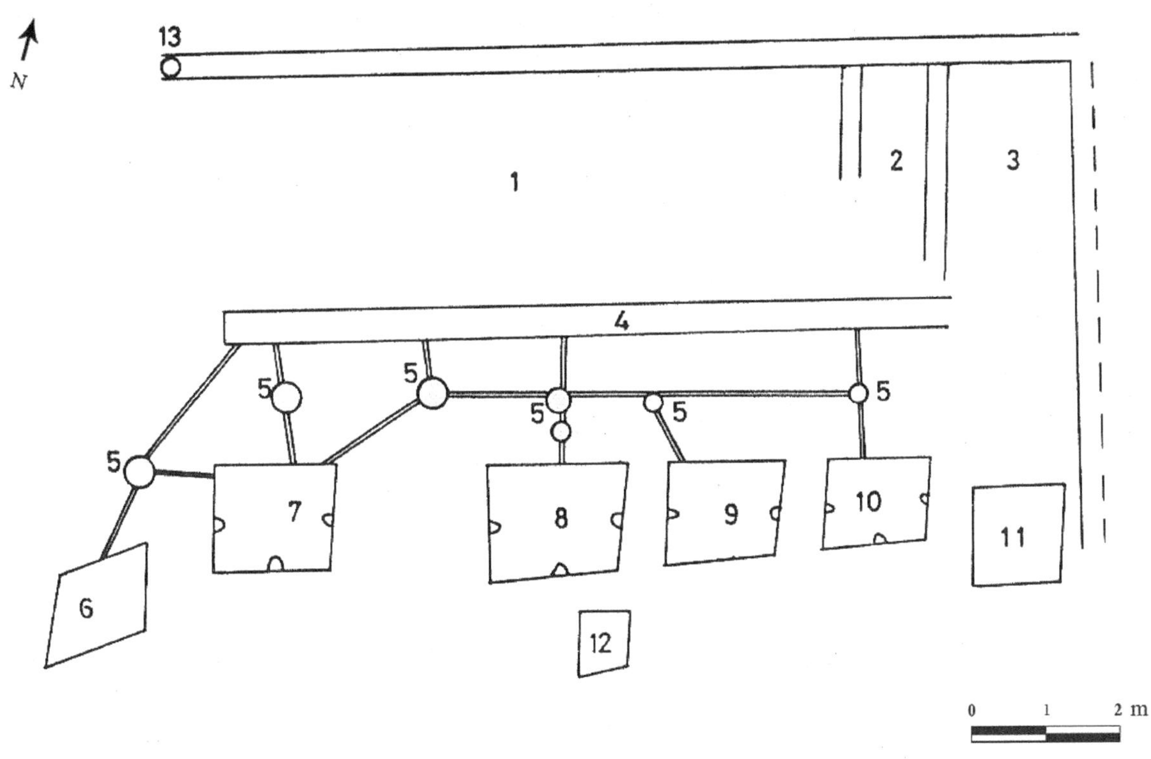

Fig. 10.1. Hermon, plan of wine press No. 3.

Fig. 10.2. Hermon, a wine press.

11. An Oil Press at H. ‘Ein Nashut[1]

Chaim Ben-David

The Site

H. ‘Ein Nashut (MR 2151/2686) is situated on the banks of the Meshoshim Valley in the central Golan Heights, c. 2.5 km north of Katzrin. In the opinion of Ma‘oz (1988: 128, No. 10), at the end of the 19th cent. the site was called Deir Rahib (Schumacher 1888: 119). In contrast to most Byzantine sites in the Golan, it was of no importance during the Middle Ages. In more recent times the site was robbed many times, a fact that was brought to light during the excavation. The synagogue at the site was excavated by Ma‘oz (1988; 1993), who suggests that it was built during the third quarter of the 5th cent.CE, and served the Jewish community until the end of the 6th or the beginning of the 7th cent., when it was abandoned.

According to the ceramic and numismatic evidence (Ariel 1987; Ben David 1998; 1999: 181) settlement at the site began in the Early Roman period and continued throughout the Roman and Byzantine periods reaching a peak during the years 250–450 CE.

The Oil Press Building

The oil press building was located among the outermost houses on the southern fringe of the village, 25 m south of the synagogue. Before the excavation, an irregularly rectangular room could be seen, containing a circular crushing mill (basin and crusher) in its southern section and, in the northern part, a press-bed with two open mortices for securing the press frame.

As already noted, the oil press installations were found in an asymmetrically rectangular room (Figs. 1–3). Each wall has a different length: southern 3.5 m, eastern 9.4 m, northern 4.1 m, western 9.2 m, i.e. an area of about 36 m^2.

North of the press room is an east-west corridor 1.25 m wide. The area west of the southern section of the room is apparently a courtyard and that west of the northern section another building. The southern wall of the room was the southernmost wall of all the buildings in the village and east of the oil press, but not adjacent to it, another building was found.

The western wall of the oil press underwent a number of changes. During the first stage there was apparently an opening 4.2 m wide in its southern part. Approximately in the centre of the opening a section of a column survived 0.4 m in diameter that originally supported a two-section lintel above the broad entryway. An additional column section, identical in diameter, was found in the rubble of the room. At a later stage, the width of the portal was reduced by continuing the construction of the western wall from the northern doorjamb southward, leaving only a narrow opening 0.8 m in width adjacent to the original southern jamb. The original broad doorway was located exactly opposite the crushing basin.

The floor space of the oil press can be divided into three areas: the northern, central and southern. In the centre of the northern area the press-bed was found, resting on the natural bed-rock. Adjacent to its western side, 0.35 m below the bottom of the press-bed, is the collecting vat. In the central area, at the level of the flooring discovered in the southern area, was found the crushing basin with its circular crushing stone still inside it. In the south-west corner of the room (L. 145), at a height of 0.1 m above the level of the bed-rock (in the northern space), a paved area was discovered measuring 1.8×2.0 m. The paving was opposite the building’s remaining narrow doorway, bounded by the structure’s southern and eastern walls, and a wall separated the southern area, where the flooring was found, and the area of the crushing mill. The wall was built of medium-sized stones, and its thickness was 0.65 m; it was preserved to a height of 0.3 m.

The Installations

All the oil press installations are made of basalt. They include:

- Circular crushing basin. Its inside diameter is 1.5 m, the thickness of the side-wall is 0.15 m, and its height above the crushing surface is 0.21 m.

- Discoid crushing stone 0.88 m in diameter and 0.37 m thick. In its centre is a square depression measuring 0.25×0.25 m continuing through the stone as a round hole 0.19 m in diameter and reaching a similar square depression on the other side of the stone. The two depressions apparently held square blocks probably of hard wood in the centre of which were round holes slightly narrower than that in the stone. These blocks would have acted as housings and the axle on which the crushing stone revolved that also acted as a handle would not come in direct contact with the stone but only with the wooden blocks (Callot 1984: Pl. 18).

• Press-bed. Dimensions: 1.48×1.08×0.30 m; pressing area diameter 0.82 m, circumscribed by a groove 4 cm wide and deep. The two mortices on the sides were 0.24 m across. The mortices probably served to secure the frame of a single rotating screw press but there is also a possibility that they served to secure the screws of a double fixed screw press (Frankel 1994: Fig 63, 5 & 6). The press was of the direct pressure frame type, with lateral collecting vat; presses of this type were found at 15 sites in the Golan Heights.

• Collecting vat. A rectangular trough measuring 0.5×0.8 m and 0.34 m deep, its volume is 136 litres. At the bottom of the vat is a sedimentation basin 11 cm in diameter and 6 cm deep. The discovery of a rectangular collecting vat alongside a press-bed *in situ*, first of its kind in the Golan Heights, can teach us something about the collecting system of this type of press-bed in the Golan. In this part of the country, hewing a collecting vat out of the basalt bed-rock was usually not even considered because of the difficulty of cutting this hard stone. A similar combination of a rectangular collecting vat adjacent to a press-bed was also found in one of the oil presses at Korazim (Yeivin 1984).

On the basis of the pottery assemblage and the numismatic finds it appears that this oil press operated from the 4th to the 6th cent. CE. This time frame is also compatible with the opinions of Frankel and Kloner, both of whom feel that the screw press came into use in the Land of Israel mainly in the Byzantine period (Frankel 1994: 50, 63; Kloner, "Judean Press", in this book).

Unique to the oil press at this site is the wide opening in the western wall, opposite the crushing basin, and the paved area (in the original plan). This is the place to which the olives were brought, so there is an advantage to easy access via a broad entrance. This phenomenon – a wide opening in the long wall – is not known from excavations of oil presses in Israel. However, in a number of oil presses in northern Syria, similar wide entrances were found in the long wall (Callot 1984: Pls. 112, 114, 116). At a later stage, apparently after the original broad opening had been damaged, its northern part was closed off and the oil press workers had to make do with a narrower entrance.

The internal partition of the oil press into separate crushing and pressing areas is well known, since it occurs in almost all such presses excavated in Israel. In contrast, the role of the paved floor surface in the southern part of the building is not clear. In the excavation of an oil press at Kh. edh-Dharia in Trans-Jordan (Villeneuve 1988: 464) a similar surface was found, but the excavators do not discuss its function. It should be mentioned that in the vicinity of the paved area at 'Ein Nashut were found many fragments of the so-called "Bet She'an Jars". In a study of Caesarea it was suggested that such jars should be considered oil storage vessels (Adan-Bayewitz 1986), and it is possible that the paved area was where the jars were filled with the oil produced by the press, and perhaps stored.

Notes

1. The oil press at H. 'Ein Nashut was first mentioned by Urman (1979: 64; see also Urman 1985). During the excavations conducted by Z. Ma'oz the oil press building was partially exposed. Continuation of the excavation and the right to publish it were passed on to me by Zvi Ma'oz, for which I thank him. Within the framework of the survey of oil presses in the Golan that I carried out in 1984, the installations in the excavated oil press were measured and recorded, as was the crushing basin belonging to another oil press at the site. The excavation of the oil press was conducted in August 1986 and supervised by the author, with the assistance of Gil'ad Peli and the youngsters of Moshav Keshet. Also assisting were Zvi Ma'oz, Moshe Hartal and Rogé Shefa. Measurements were taken by Israel Watkin, while the cross-sections and finds were drawn by Hagit Rosen. The coins were cleaned by Shula Lavi and identified by Donald T. Ariel.

Bibliography

Adan-Bayewitz, D. 1986. "The Pottery from the Late Byzantine Building and its Implications". In: L.I. Levine and E. Netzer. *Excavations at Caesarea Maritima 1975, 1976, 1979 – Final Report* (Qedem 21). Jerusalem: 90–129.

Ariel, D.T. 1987. "Coins from the Synagogue at 'Ein Nashut". *IEJ* 37: 147–157.

Ben David, H. 1998. "Oil Presses and Oil Production in the Golan in the Mishnaic and Talmudic Periods". *'Atiqot* 34: 1–61 (Hebrew).

Ben David, H. 1999. *Settlement in the 'Lower Golan' during the Hellenistic, Roman and Byzantine Periods*. Unpublished Ph.D. Dissertation, Bar Ilan University, Ramat-Gan (Hebrew).

Callot, O. 1984. *Huileries antiques de Syrie du Nord*. Paris.

Frankel, R. 1994. "Ancient Oil Mills and Presses". In: R. Frankel, S. Avitsur and E. Ayalon. *History and Technology of Olive Oil in the Holy Land*. Arlington and Tel Aviv: 19–89.

Ma'oz, Z.U. 1988. "Ancient Synagogues of the Golan". *BA* 52: 116–128.

Ma'oz, Z.U. 1993. "'Ein Nashut". *NEAEHL* 2: 412–414.

Schumacher, G. 1888. *The Jaulan*. London.

Urman, D. 1979. *The Golan During the Roman and Byzantine Periods*. Ph.D. dissertation. New York University.

Urman, D. 1985. *The Golan: A Profile of a Region during the Roman and Byzantine Periods* (BAR IS 269). Oxford.

Villeneuve, F. 1988. Fouilles à Khirbet Edh-Dharia (Jordanie), 1984–1987. *CRAIBL*. Paris: 458–479.

Yeivin, Z. 1984. "Korazim 1983–1984". *ESI* 3: 66–71.

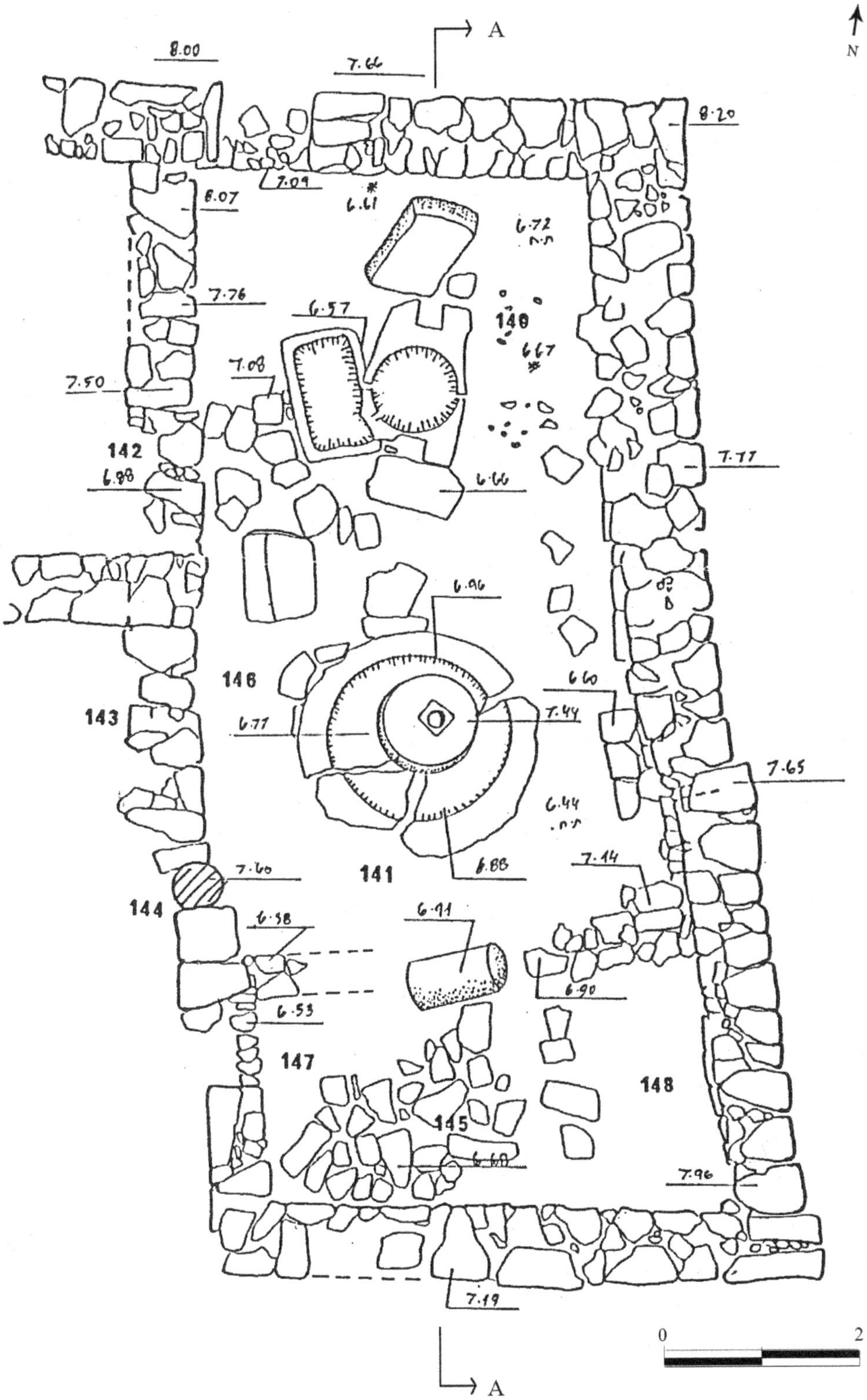

Fig. 11.1. 'Ein Nashut, plan of the oil press.

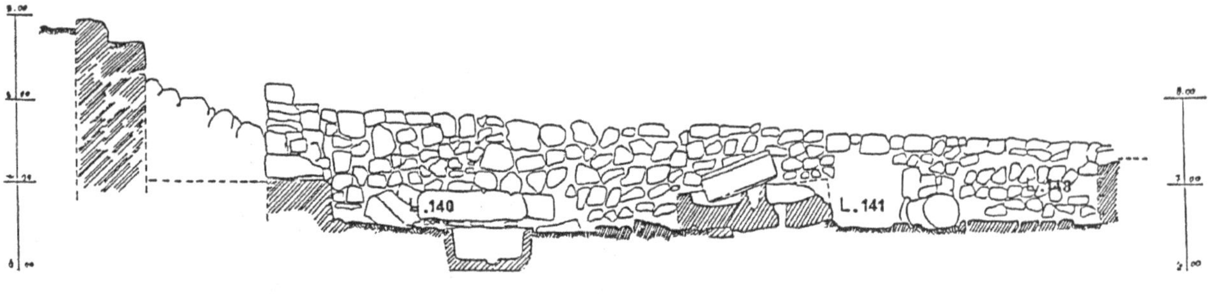

SECTION A-A

Fig. 11.2. 'Ein Nashut, section A-A through the oil press.

Fig. 11.3. 'Ein Nashut, the oil press after excavation: view to the south.

12. Oil Presses at Gamla

David Goren

The Site

Gamla (MR 2196/2564) is located in the southern Golan, on a rocky ridge that is very narrow at the top. The hill, shaped like the hump of a camel (*gamal* in Hebrew), separates Daliyot Stream from Gamla Stream and is connected by a narrow saddle to a higher plateau. The town was built on the southern slope of the hill that is wider and less steep than the northern slope. The identification of Gamla with this hill was first proposed by Y. Gal in 1968. It was confirmed by S. Gutman who after a preliminary survey conducted fourteen seasons of excavations at the site on behalf of the Israel Department of Antiquities (Gutman 1994). The excavations revealed a large thriving Jewish city from the Second Temple period with remains of public, private and industrial structures. The city participated in the great revolt against the Romans (66–73 CE). It was destroyed by Vespasian's legions in 67 CE and never settled again.

There are three main occupational phases on the site: Early Bronze II, Hellenistic (2nd-1st cent. BCE) and Early Roman (1st cent. CE).

The Presses

Two complete large twin oil presses were excavated at Gamla, one in area B and the other in area R. In both areas the oil presses were found in large well planned buildings with many artifacts and numismatic finds. The presses are well dated. The press in area B is the earlier one dated to the first half of the 1st cent. BCE. The press in area R functioned during the 1st cent. CE, when it was destroyed along with the rest of the city. The excavations of the two twin presses revealed a valuable regional and chronological picture.

Area B (1267)

The type of oil press found in area B is a twin complex lever and weights press (Frankel 1999: 76–77). The oil press is located in the centre of a residential area directly above a road that bisects the area running east to west (Gutman 1994: 112–117). The structure that housed the press is a large building, 10.8×7.5 m. Several architectural phases can be observed in the building. The oil press existed only during the first phase, after which it was abandoned. All of the press components were left in place and secondary walls were built on top of them (Fig. 1) except for the crushing stone that was found in secondary use in an adjacent room to the east. The roof of the original building was supported by a row of arches only the bases of which remain, however (a similar building technique was used in the oil press in area R and will be described below).

The components of the northern press A:

B A niche built in the western wall that served as the fulcrum for the beam (Fig. 2).

C An oval shaped central collecting vat for the expressed liquid (oil and watery lees) that was extracted from the olive pulp packed in frails stacked above the vat. On top of the vat two basalt slabs were inserted, intended to support the press-bed that was not found.

D Bases for the piers that stabilized the beam and the stack of frails.

E Three basalt weights found *in situ* in a pit dug into the oil press floor, two with a reversed T-shaped hole and the third with a horizontal hole.

F Plastered floor between central vat, fulcrum, collecting vat and northern wall.

G Secondary collecting vat that probably served as collecting vat for the "first oil" that was of the highest quality. It could have also be used as storage area for the expressed liquid extracted from the central collecting vat B before being collected into storage containers.

The components of the southern press B:

C1 A cylindrical lateral collecting vat with a wide rim around the opening. The southern part of the installation was found damaged. The vat was dug in the ground and plastered with high quality plaster.

D1 A pier broken into three parts was found in the weights pit. The pier fragments were placed in the pit in order to fill it up after the oil press stopped functioning.

E1 Three large basalt weights found in situ in a pit next to the southern wall of the oil press building. All three weights have reversed T-shaped holes.

G1 Trapezoid secondary collecting vat located next to collecting vat C1. The vat had a similar function as the secondary vat G.

Crushing Facilities:

H1 Large flat round basalt stone with a finely worked surface. The stone was found upside down in the eastern part of the oil press building. It seems probable that the stone served as a press-bed (Gutman 1994: 116) although due to later changes this explanation is not certain (it seems too small to be a crushing basin and too large for a crusher and no central protrusion or depression are distinguishable).

I Basalt crusher with a slightly concave profile although the crushing surface is flat. An interesting and possibly unique find are the remains of lead found along the inner flat walls of the socket. These are most probably the remains of a wedge guard that served as protecting sleeve between the wooden shaft and the crusher. Cato (234–149 BCE) recommends in his book *De Agri Cultura* (XX-XXI) to attach a similar device to the crushing stone and wooden shaft called *cunicae* (wedge guards) and *armillae* (ring washers) (White 1975: 228).

The Finds

In the excavation of the oil press many artifacts were found. Two complete cooking pots were found in the collecting vat G of the northern press and on the plastered floor of the oil press F. They may have been used in the oil processing (Frankel 1992: 48). The finds include iron nails, a complete oil lamp and fragments, and bronze artifacts. Over 160 coins were found in the press, all of which are from the 1st cent. BCE. and earlier (Syon 1994: 116). The numismatic evidence along with the pottery date the oil presses to the mid 1st cent. BCE.

Area R (5011)

The twin oil press is located in a large building (Fig. 3). The oil press building is part of a larger complex that included a second storey, a large public stairway and houses (Gutman 1994: 130–133; Wagner 1996: 301–306). The building is large, well designed and well built. It was planned to serve all the different aspects of olive oil production. The crushing facilities are located in the eastern half of the building and the press units in the western half. An impressive double arch supported the second storey that served as a storeroom and possibly a shop to sell the oil (Fig. 4). Columella (65 CE) writes: "...there ought to be a storeroom supported on arches, to which the fruit may be conveyed" (Columella XII, 52 3).

The press in area R is a twin complex lever and weights press as found in area B but not identical (the differences are discussed below in the conclusions). The oil press was destroyed along with the rest of the city in 67 CE. The destruction left it with all its components intact.

The components of the northern press A:

B A niche built in the western wall that served as the fulcrum for the beam.

C Two monolithic stone piers flanking the press-bed. The piers were found *in situ*.

D Cylindrical central collecting vat.

E Pit dug into the oil press floor for the weights and to enable the workers to stand next to them.

F Three large basalt weights found in the pit. All three weights have reversed T-shaped holes.

The components of the southern press B:

B1 A built niche in the western wall that served as the fulcrum for the beam.

C1 Two monolithic stone piers flanking the press-bed (Fig. 5). The piers were found *in situ*. The southern one still stands in place, with small stones attaching it to the southern wall of the building.

D1 Press-bed and central collecting vat. The press-bed is an elaborately carved basalt slab. It has six radial grooves connecting a circular groove with the central hole. Similar press-beds have been found in Kh. 'Azzun in the Sharon, Kh. el-Quseir (Site 8) in Upper Galilee and Meggadim at the foot of Mt. Carmel (Frankel 1994: 45–46). The press-bed had to withstand the extremely high pressure of the frails filled with olive pulp, the beam and the three weights attached to the beam. Therefore the inner side of the press-bed is quite thick (0.3 m) especially around the central hole although thinning out along the sides. The central collecting vat is also carved out of basalt and the press-bed sits flush on top of the vat. The area around the press-bed was plastered.

E1 Pit dug into the bed-rock of the oil press building. On both sides of the pit there are large basalt slabs for the workers to stand on and for holding the beam in place.

F1 Three large basalt weights with reversed T-shaped holes. The weights were found in the pit.

Crushing Facilities

H A large basalt crusher. The socket is square and has a round hole for securing the wooden shaft.

I Crushing basin. Large basalt slab with a raised square socket. The crushing basin is located just under the eastern arch, enabling a maximal radius of rotation for the wooden shaft.

Additional components

G *Mikveh.* A plastered pool (2.5×1.6 m) with four steps leading down from the floor of the main hall. Two pools located above the bath supplied the water. The presence of a *mikveh* in an oil press is to allow the production of ritually pure oil – "*Shemen Be-tahara*" (Gutman 1994: 132; Wagner 1996: 303–304). Additional ritual baths in oil presses have been found, for instance, at Qedumim (Magen 1982 [Site 34]) and Zur Natan (Site 32) in Samaria and at 'Ahuzat Hazzan in the Judean Shephelah (Kloner and Tepper 1987: 116–117 [Site 68]).

J A fireplace with ash and burnt material was found in the north-eastern corner of the room. It was built with medium size basalt slabs that are attached to the corner of the building. The fireplace was used to warm the oil press hall during the cold winter months when the press operated. Cato (65.2) and Pliny the Elder (15.22) both recommend heating the oil press building. Pliny adds that the best fuel is the press residue left over from the pressing, claiming that it was more economical and reduced pollution caused by soot. The heating of the area facilitates the separation of the oil and produces a higher yield of oil (Rossiter 1981: 354). Using the press residue for heating also diminishes the problem of removing and storing it.

The Finds

Many artifacts were found in the building on the floor and mixed in the debris. The oil press was burnt and destroyed along with the city in 67 CE. Numerous charred remains of wood and olives were found and were identified by N. Liphschitz (1986). The olives were identified as *Olea europaea.* The wood charcoal came from olive trees, *Atlantic pistachio* and *Acacia raddiana.* Olive trees and *Atlantic pistachio* trees grow today in the region. The presence of *Acacia raddiana* came as a surprise. This tree grows in arid areas and its most northern presence is recorded from Jericho in the Jordan valley. The Acacia can reach a height of 12 m and is known for its durability. We suggest, therefore, that the trees were brought to the Gamla oil press to be used as the press beam (*prelum*) and as the wooden pivot and shaft for the crushing mills.

Among the finds were two bone shovels (*tarvadot* – Mishnah *Kel.* 12: 17; Baruch 1999: 73–76). The shovels were shaped from scapula bones of cows and were used to collect the olive pulp from the crushing basin into the frails (Avitsur 1994: 99; Gutman 1994: 132). One of the shovels was completely worn down while the other seems to be brand new.

The numismatic finds from the olive oil complex are most impressive. Over 330 coins have been found in the area. They include a large percentage (70%) of Hasmonean coins (mostly Alexander Jannaeus), 6% Herodian coins (Herod the Great, Antipas and Archelaus), 16.5 % city coins, mostly Phoenician (Tyre and Sidon), coins of Roman procuraters, Nabatean coins, Roman imperial, and two coins minted in Gamla (Syon 1994: 151–171). In the corridor next to the oil press a small hoard of 27 silver coins was found. 20 are silver tetradrachms from Tyre and seven are Roman imperial tetradrachms of Nero minted in Antioch. The latest coin is dated to 64 CE, just two years before the revolt started. (Our assumption is that the hoard may be the proceeds from the olive oil and was dropped by the fleeing owner when the city was stormed by the besieging Roman army).

Many vessels were found in the oil press. Dozens of storage jars were found, over ten of which have been restored. These jars have the capacity of 22 litres and were used for storing and marketing the oil. These vessels were located on the second storey that served as a storeroom and collapsed into the oil press when the building was destroyed.

Fifty spouts of oil lamps have been found in the oil press building, all of which are of the Herodian type that is very common in Gamla (Gutman 1994: 142–144). The large number of lamps seems to indicate that the oil press was in use during the night as well (Frankel 1992: 46).

The Nature Reserve Authority and Y. Mintzker of the Israel Antiquities Authority have partially restored the oil press building (Fig. 4).

Conclusions

The two twin oil presses in Gamla are both complex lever and weights presses. The oil press in Area B was in use during the first half of the 1st cent. BCE. Later it was abandoned and the building modified. The oil press in Area R was built in the beginning of the 1st cent. CE and was in use until 67 CE when destroyed along with the rest of the city.

The main difference between the oil press in Area B and that in Area R is in the collecting vats. The northern press in area B has a central collecting vat while the southern press has a lateral one. Although the press-bed itself was not found in place we can reconstruct its location by the alignment of the weights and by the fact that the collecting vat is built from plaster and therefore could not support the press-bed that had to stand the pressure of the beam with its three massive weights. The twin oil presses in Area R are both with a central collecting vat with similar dimensions and both with large basalt weights of similar size and shape.

The oil presses in both areas were found in well built and well designed buildings. In Area B the oil press is located in the centre of the residential area but it is situated just above a road that bisected it. The oil press in area R is located on the southern perimeter of the site that enabled easy access to it and facilitated the removal of waste material (Goren 1994: 46).

The building in Area R (5011) is part of a large multi-functional building complex that combined the twin oil presses with storerooms, service rooms, water systems, dwellings and a large stairway with a paved area that may have served the merchants.

The two large oil presses indicate the importance of olive oil production to Gamla's economy.

Bibliography

Avitsur, S. 1994. "Olive Oil Production in the Land of Israel: Traditional to Industrial". In: R. Frankel, S. Avitsur and E. Ayalon. *History and Technology of Olive Oil in the Holy Land.* Arlington and Tel Aviv: 91–157.

Baruch, E. 1999. "Bone Tools in Talmudic Sources". In: E. Ayalon and C. Sorek. *Bare Bones, Ancient Artifacts from Animal Bones.* Tel Aviv: 73–76 (Hebrew).

Cato 1979. "Marcus Porcius Cato: De Agri Cultura". In: *Cato and Varro* (trans. W.D. Hooper). London: 2–157.

Columella 1969–79. *Columella, Lucius Junius Moderatus: De Re Rustica* (trans. H.B. Ash, E.S. Forster and E. Heffner), 3 vols. London.

Frankel, R. 1992. "Some Oil Presses from Western Galilee". *BASOR* 286: 39–71.

Frankel, R. 1994. "Ancient Oil Mills and Presses in the Land of Israel". In: R. Frankel, S. Avitsur and E. Ayalon. *History and Technology of Olive Oil in the Holy Land.* Arlington and Tel Aviv: 19–89.

Frankel, R. 1999. *Wine and Oil Production in Antiquity in Israel and Other Mediterranean Countries.* Sheffield.

Goren, D. 1994. *The Urban Character of Gamla during the First Century BCE and the First Century CE.* Unpublished MA Thesis, Bar-Ilan University, Ramat Gan.

Gutman, S. 1994. *Gamla – A City in Rebellion.* Tel Aviv (Hebrew).

Kloner, A. and Tepper, Y. 1987. *The Hiding Complexes in the Judean Shephelah.* Tel-Aviv (Hebrew).

Liphschitz, N. 1986. *Dendroarchaeological Research Report #143.* Tel Aviv University (Hebrew).

Magen, I. 1982. *The Archaeological Discoveries at Qedumim – Samaria.* Jerusalem.

Pliny the Elder, 1958–1963. *Naturalis Historiae* (Trans. H. Rackham). London and Cambridge (10 vols.).

Rossiter, J.J. 1981. "Wine and Oil Processing at Roman Farms in Italy". *Phoenix* 34: 53–70.

Syon, D. 1994. "The Coins of Gamla". In: S. Gutman. *Gamla – A City in Rebellion.* Tel Aviv: 151–172 (Hebrew).

Wagner, D. 1996. "Oil Production at Gamla". In: D. Eitam and M. Heltzer (eds.). *Olive Oil in Antiquity, Israel and Neighbouring Countries from the Neolithic to the Early Arab Period.* Padova: 301–306.

White, K.D. 1975. *Farm Equipment of the Roman World.* Cambridge.

Fig. 12.1. Gamla, area B oil press, view to the east. Centre: The central collecting vat with foundation for a pier on each side. Behind is a basalt weight in situ. *On the right – a secondary collecting vat partly covered by a later wall. Note the plastered floor.*

Fig. 12.2. Gamla, area B oil press after removal of later wall and some repair works. Behind – the niche that served as the fulcrum for the beam. View to the west.

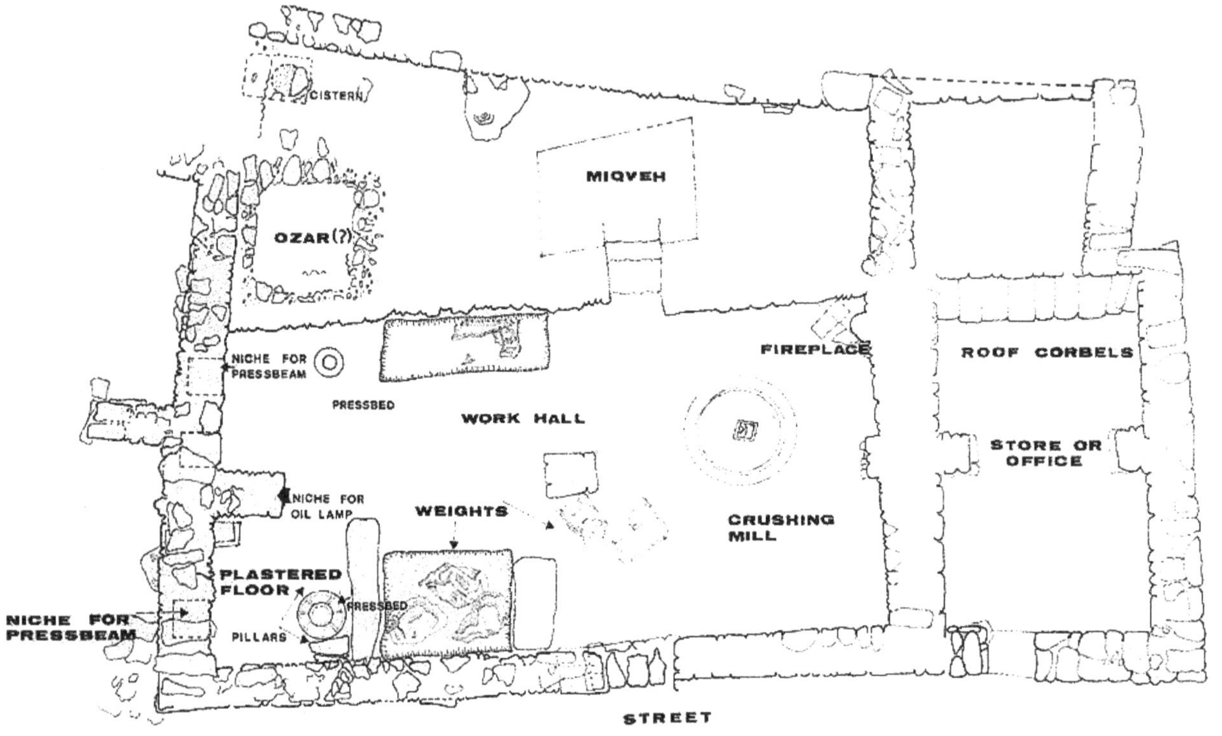

Fig. 12.3. Gamla, area R oil press, plan.

Fig. 12.4. Gamla, area R oil press after the arches that supported the roof had been reconstructed, view to north-west.

Fig. 12.5. Gamla, area R oil press, the south-western corner: In foreground the round central vat and one plain press pier in situ. *At the back the niche in the wall in which the press beam was anchored and to its left the other pier lying on the floor. The floor is plastered and the measuring rod stands in a plastered depression to which flowed surplus oil split on the floor.*

13. An Oil Press at Kursi

Etan Ayalon, Rafael Frankel and Amos Kloner

The monastery complex of Kursi, situated on the eastern shore of the Lake of Galilee (MR 2112/2478), was excavated by V. Tsaferis in the years 1970–1974. In the north-western wing of the church an oil press was uncovered. This room was apparently converted to an oil press in the 6th cent., at the second stage of the use of the church (Figs. 1–2). Only the screw base (1.8 m long) that had open mortices and the round collecting vat were found *in situ*. These were on a floor almost 2 m below the level of the floor of the church. The other components of the press were found amidst debris and ashes and apparently were originally in a second storey that was approximately at the same level as the church floor. These consisted of the crushing basin (1.7 m in diameter) that had a raised socket, the crushing stone (0.95 m in diameter) the bore of which was within a small square depression and other components the purpose of which is not clear.

Bibliography

Tzaferis, V. 1984. *The Excavations at Kursi-Gergasa* ('Atiqot 16). Jerusalem.

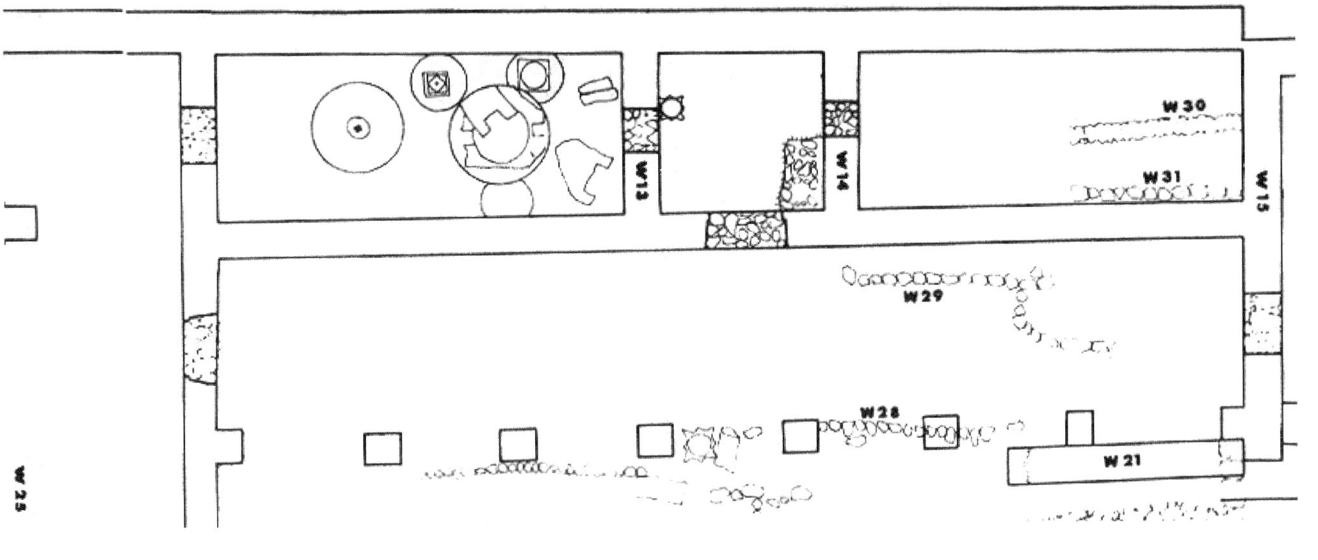

Fig. 13.1. Kursi, plan of the oil press.

Fig. 13.2. Kursi, The oil press, view to the south.

14. The Wine Press at Kh. el-Hutiyye

Claire Epstein

Kh. el-Hutiyye (MR 2151/2482) is a small tell between two seasonal streams in an area of calcareous rocks in the western reaches of Samakh Stream, south Golan. Surveys conducted at the site and in its immediate vicinity revealed sherds from the following periods: Chalcolithic, Early Bronze II, Late Bronze, Iron Age II, Hellenistic, Roman, Byzantine and Arab.

Soundings made in 1976 on the upper face of a limestone exposure in the north-east of the site revealed a plastered Byzantine wine press (Figs. 1–2). The treading floor (4.0×4.2 m, area 16.8 m^2) slopes gently towards the centre to a dovetail mortice (1.15×0.46 m; depth 0.55 m) for a single fixed screw press. A shallow channel enabled the must to flow directly into a collecting vat in the east through an aperture in the east wall; a similar aperture in the north wall led to a small circular vat only partially preserved. In the south wall is a second circular vat with a small hole in its back wall – perhaps pointing to a third, no longer extant. The east collecting vat (1.6×2.0 m) is plastered throughout. Close to the top of both the east and west walls is a corresponding group of three small cavities for the insertion of wooden poles to support some kind of covering.

The rock surfaces adjacent to the wine press are pitted with small-sized cup holes, some possibly connected to the wine-making process. Others could date to the Chalcolithic period, remains of which were not only present on the tell but in the collecting basin in the east, into which typical sherds and a broken basalt pillar figure had fallen long after the former had gone out of use.

Fig. 14.1. Hutiyye, plan of the wine press.

Fig. 14.2. Hutiyye, part of the wine press, view to the south-east.

15. The Oil Press at Giv'at Ha-Yi'ur

Chaim Ben-David

The Site

The Giv'at Ha-Yi'ur site is located on a hill overlooking Kibbutz 'En Gev, north of the 'En Gev Stream (MR 2108/2434). The site consists of the ruins of a village, covering an area of 21 dunams. Its present name was given it by the settlers of 'En Gev. Archaeological surveys were conducted at the site by Mendel Nun and Zvi Ilan, in which decorated stones from the Roman-Byzantine period were discovered (Ilan 1991: 113).

The Building (Figs. 1–2)

The oil press building is situated at the edge of the site, in the lowest row of buildings on the eastern slope of the hill (Ben David 1998). The installation itself is located within a symmetrically rectangular room (Fig. 3) aligned north-south (internal measurements 8.75×4.85 m, i.e. an area of 42.4 m^2). Its floor, hewn from the limestone bed-rock, was shown upon excavation to be divided into two sections each at a different height. The lower section, 1.8 m wide, is in the eastern part of the oil press where the beam and weights were located. The upper section, 3.5 m wide and 0.35 m higher than the other, is in the western part. The crushing basin was found on the upper section, as was the press-bed of a direct pressure screw press.

The installations of the oil press are concentrated in the northern two-thirds of the building. It is reasonable to assume that the southern third, opposite the doorway, was used for delivery and temporary storage of the olives before they were crushed.

Oil Press Installations

• Round crushing basin. Its inside diameter is 1.62 m, height 0.7 m. The wall is 0.11 m thick and projects 0.18 m above the floor of the basin. In the centre of the basin there is a sunken socket 0.5 m in diameter. The socket itself is square, measuring 0.23 m on a side and 0.15 m deep.

• Beam and weights press (Fig. 4). Three beam weights were found, of the type with bores in the shape of a reversed T, with a wide upper opening. Two of them, 0.74 and 0.88 m tall, were found in a weight pit (L. 207) 1.2×3.0 m in area and 0.3 m deep. An additional weight, 0.7 m tall, was discovered alongside the press-bed.

• Collecting vat (L. 106; Fig. 5). This vat was the one and only installation in the complex made of limestone and not basalt. It is 0.7 m in diameter and 0.45 m deep, with a volume of ~175 litres. Its wall thickness is 0.15 m. In the bottom of the vat there is a sedimentation depression 0.2 m in diameter and 8 cm deep. There are four slots cut into the rim of the vat. The two on the northern side are 6 cm deep and 3 cm wide, cut through the vat wall from the inside to the outside. The two on the southern side do not go through to the outside of the collecting vat wall.

• Piers of the beam press. Adjacent to the northern wall, two basalt piers were found standing 0.6 m apart and measuring 0.9×0.2 m, height 0.5 m. These piers apparently served to guide the beam and to support the stack of frails holding the olive mash. Above them in the northern wall is a niche for anchoring the end of the beam, measuring 0.7 m high, 0.55 m wide and 0.3 m deep. The original press-bed of the beam press was not found *in situ*. On top of the two piers was lying a press-bed in secondary use, measuring 0.55×0.78 m.

• Direct pressure screw press installation. This is a press-bed (Fig. 5) of the type that is characteristic of the region around the Sea of Galilee, with a central collecting vat and two roughly square open mortices for affixing the wooden press frame. The press-bed measures 1.70×1.03×0.39 m. The collecting vat opening, located in the centre of the press-bed, is 0.7 m in diameter and 0.34 m deep. A circular channel around the pressing area is connected to the central opening by six radial grooves. Each of the two lateral mortices measures 0.18×0.18 m.

The Finds

The ceramic evidence from the later phase when the direct pressure screw press was in use shows that this stage extended from the 4th to the beginning of the 6th cent. CE.

In the weight pit of the beam press, and in the level under the floor of the screw press, vessels were found typical of the 1st-3rd cent. CE. In contrast, at the bottom of the weight pit and on the bed-rock foundation of the building, a few shards were found, all of which were dated only to the period from the 1st cent. BCE to the 1st cent. CE. These represent the period of the installation's construction and initial operation showing that the early lever and weights press phase dates from the 1st cent. BCE to the 3rd cent. CE.

The discovery of sherds from broken lamps in this oil press attests to the possibility that it also operated at night, as Frankel deduced from his excavation of the oil press at H. Karkara (Frankel 1992: 46 [Site 2]).

Oil presses with lever and weights presses similar to that at Giv'at Ha-Yi'ur have been found at Yodefat (Aviam 1990 [Site 17]) and Gamla (Wagner 1996 [Site 12]), where they were dated from the 1st cent. BCE to the 1st cent. CE. It should be noted that at Gamla the floor was discovered to be plastered only in the pressing and collecting areas, exactly as in the early phase of the press at Giv'at Ha-Yi'ur.

Bibliography

Aviam, M. 1990. "Tel Yodefat, Oil Press". *ESI* 9: 106.

Ben David, H. 1998. "Oil Presses and Oil Production in the Golan in the Mishnaic and Talmudic Periods". *'Atiqot* 34: 1–61 (Hebrew).

Frankel, R. 1992. "Some Oil Presses from Western Galilee". *BASOR* 286: 39–71.

Frankel, R. 1994. "Ancient Oil Mills and Presses". In: R. Frankel, S. Avitsur and E. Ayalon. *History and Technology of Olive Oil in the Holy Land.* Arlington and Tel Aviv: 19–89.

Ilan, Z. 1991. *Ancient Synagogues in Israel.* Tel Aviv (Hebrew).

Wagner, D. 1996. "Oil Production at Gamla". In: D. Eitam and M. Heltzer (eds.). *Olive Oil in Antiquity, Israel and Neighbouring Countries from the Neolithic to the Early Arab Period.* Padova: 301–306.

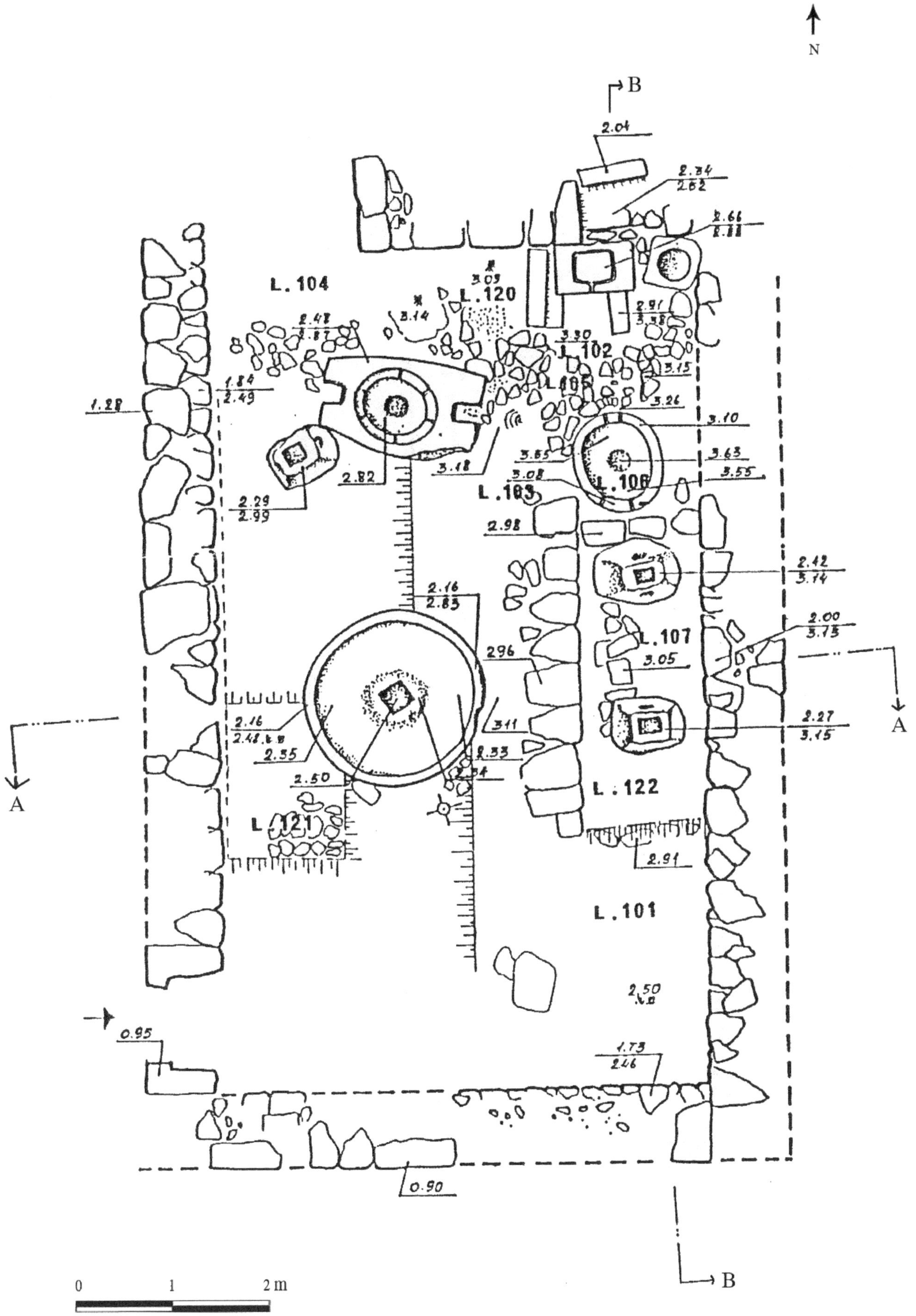

Fig. 15.1. Giv‘at Ha-Yi‘ur, plan of the oil press.

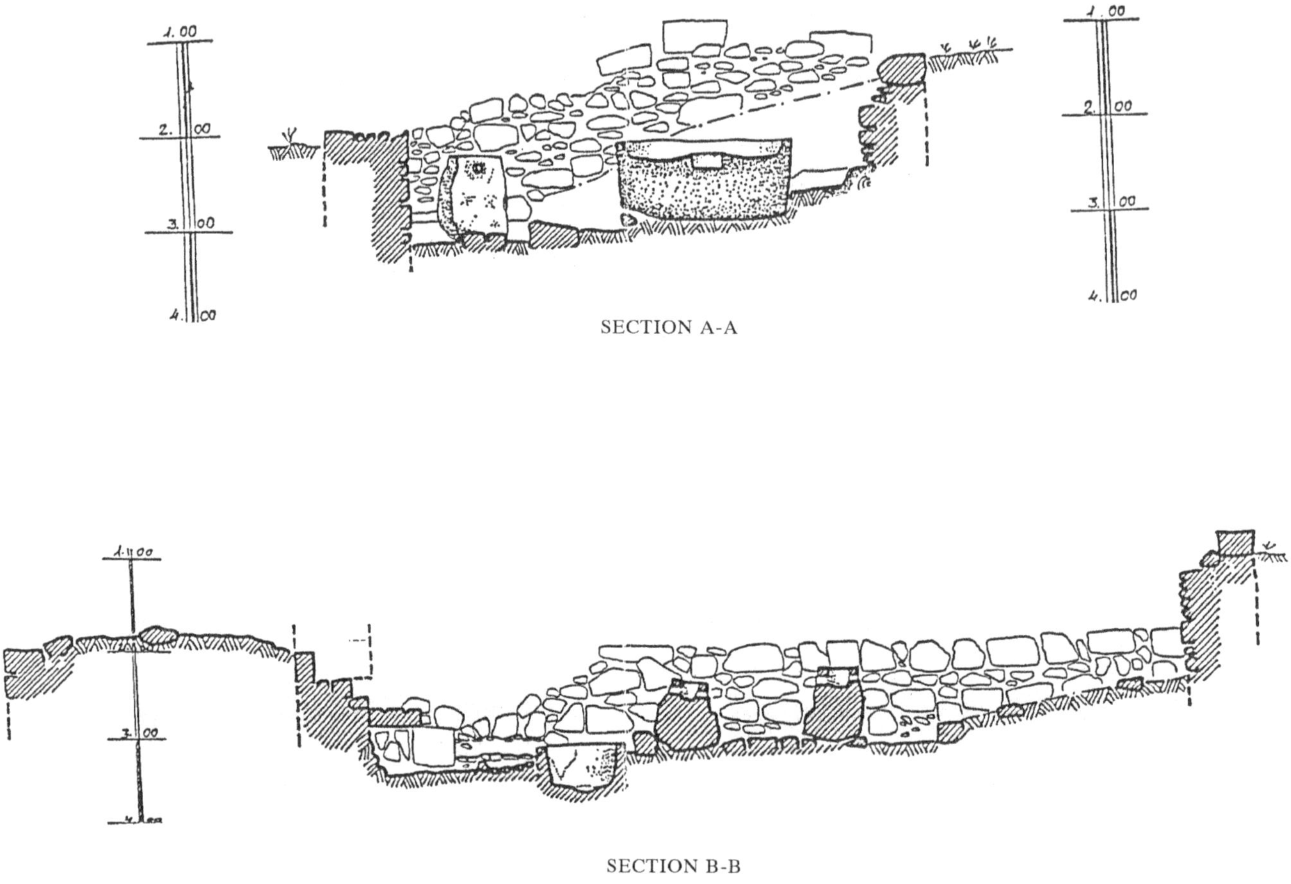

Fig. 15.2. Giv'at Ha-Yi'ur, cross-sections of the oil press.

Fig. 15.3. Giv'at Ha-Yi'ur, the oil press after excavation, view to the east.

Fig. 15.4. Giv'at Ha-Yi'ur, the weights, view to the north.

Fig. 15.5. Giv'at Ha-Yi'ur, press-bed of the direct pressure screw press (foreground). Behind it are the weights, collecting vat, and the stone piers of the lever and weights press.

16. A Survey of Oil Presses in the Golan Heights: Technological Aspects

Chaim Ben-David

In this investigation of oil presses in the Golan Heights (Ben-David 1998), the following items were listed: 88 crushing basins, 44 round crushing stones, 46 press-beds of various types, 61 beam weights, 13 collecting vats, 4 collecting installations for beam presses, and 4 screw weights.

Round Crushing Basin

In the survey, 62 crushing basins were measured (Ibid.: 38–39, Table 2). Their inside diameters range from 1.00–1.83 m. The diameters of 75% of the basins range from 1.3–1.6 m. In 27 of the basins, wall thickness is approximately 0.1 m, in 18 of them about 0.15 m and in 11 the width is some 0.2 m. There is an approximate ratio between the diameter of the basin and the thickness of its wall: the larger the diameter, the wider the wall. The height of the wall above the crushing surface ranges between 7–28 cm.

There were no clearly defined regional differences within the Golan area regarding the dimensions of the basins, although there were sometimes great differences within a single site. Thus, for example, at Dura basin diameters ranged from 1.0–1.7 m, and at Zamimra from 1.22–1.53 m.

In the centre of each basin there is a socket to accommodate the wooden pivot of the installation that propelled and guided the round crushing stone in its circular path on the working surface of the basin.

Frankel (1994: 32) differentiated between the sunken pivot socket, cut directly into the surface of the crushing basin, that in his opinion is typical of basins in the southern part of the country, and the raised pivot socket, cut into a sleeve or *jarun* (Arabic: "throat") that projected upward from the centre of the crushing surface (See introduction in this book). In the survey 45 basins were found to have raised sockets and only 13 to have sunken ones.

Among the 46 basins in which the internal shape of the socket was discernible, 20 were found to have round sockets and 26 to have square ones. Basins with round sockets were found mainly in the north-eastern section of the Lower Golan, while those with square sockets were found in the southern section of the Lower Golan and in the southern Golan (Ben-David 1998: 37, Map 9). In all of the eleven basins with visible sunken sockets found in the Golan, the socket is square. It should be noted that at Gamla (Site 12), two crushing basins were found: one with a square socket, a second with a round socket. The picture at Dabura and Dura was similar, but at none of the other sites were basins found with differently-shaped sockets.

The outside diameters of the sockets range between 0.2–0.5 m. In cases where the socket is square, the outside diameters are generally larger. The diameter of the round sockets is 9–13 cm, with one exceptional case measuring 0.21 m. In the square sockets, the length of the sides ranges from 9–25 cm; in 11 of these sockets, representing some 70% of those with data available, the length of the sides is 10–15 cm. The sockets are 5–19 cm deep, with no significant difference in the depth dimension between the two types.

At a number of sites there were sockets with special additions: in two round sockets found at Zamimra the bottom ends were ground down into conical depressions, suggesting that the wooden pivot may have been shod with a metal point. At Dir 'Aziz was found a square socket with its corners cut at an angle and a horizontal square hole hewn through its upper part. A similar straight-through hole was found in a square socket at the monastery in Kursi (Site 13). At el-Hasha a square socket was found with a round depression at the bottom. It appears that all of the foregoing are examples of locally developed improvements in, or repairs to, the form of the device that connected the installation's wooden parts (that turned the round crushing stone) to the socket of the crushing basin.

Round Crushing Stone

During the survey, 31 round crushing stones were measured (Ibid.: 40, Table 3). They range from 0.76 to 1.04 m in diameter, with 64% falling into the 0.9–1.0 m span. The stones are 0.24–0.39 m thick, with 60% in the 0.30–0.36 m range. There seems to be a definite correlation between the diameters of the stones and their thicknesses, i.e., the larger the diameter, the greater the thickness.

The crushing stone is pierced by a hole through its centre through which a wooden axle passes. The hole generally comprises a square depression through which the central opening – round or square – has been bored. Crushing stones have also been found with a plain square hole in the centre.

Out of 36 crushing stones (including photographs of items that had disappeared) in which the shape of the central hole could be identified, 19 were round and 17 square. In the southern Golan, seven out of seven were square, while both types were found in the lower Golan. Except for one place, crushing basins with round sockets were matched by round crushing stones with round central holes; similarly, crushing basins with square sockets were matched by round crushing stones with square central holes – in both cases, where the stone and the basin were found together. The square holes and depressions almost certainly served to hold square blocks of wood in the centre of which were round holes and that served as a housings in which the axle revolved thus avoiding direct contact between the wooden axle and the stone crusher (Callot 1984: 17–23, Pl. 14).

Pressing Installations

We will divide the pressing installations, as did Frankel (1999: 76–137), into three major types: lever and weights presses, lever and screw presses, and direct pressure presses.[1]

Out of 33 sites with identified pressing installations, nine had presses of the lever and weights type, ten had direct pressure screw presses, and 14 were found with both types (Ben-David 1998: 42, Figure 38). At only one site – Kh. Ra'abana in the northern Golan – the screw weight of a lever and screw press was discovered; at two other sites, clues were found pointing to the use of such a press. Conspicuous by their almost total absence were screw weights typical of western Galilee, that were also common in the foothills of Mt. Hermon – a fact already noted by Frankel (1999: 169).

In the Golan, the most widespread evidence attesting to the use of the lever and weights system is the 60 weights found at 23 sites. Weights with only a horizontal hole were found at Gamla (Site 12) and Dabura alone while those at all other sites have an opening shaped like a reversed T at the top, with a broad upper opening. The latter weights are almost completely restricted to the Golan and Lower Galilee (Ibid.: Map 16).

Installations for anchoring a beam press were found in excavations at Gamla and Giv'at Ha-Yi'ur (Site 15), as well as at Zamimra and Bathra. Collecting vats belonging to lever and weights presses were found only at Dabura. In the excavations at Giv'at Ha-Yi'ur were found a pit for the weights (with the weights still in it), a collecting vat, and a beam anchoring facility – all belonging to the same installation. In an oil press dating from the Roman period excavated at Gamla, a weight pit was discovered with the weights still in it, a press-bed with piers, a central collecting vat and a niche in which the beam was anchored.

Two main types of direct pressure screw press installations were found in the Golan, differentiated by the shapes of their collecting vats. Out of 23 sites with such installations, 18 had press-beds with two open mortices for piers and lateral collecting vats. Six sites had similar press-beds, but with central collecting vats.

During the survey, data were collected concerning 20 press-beds (Ben-David 1998: 41, Table 4). They range in length from 1.10–2.24 m; in 65% of them, the length is 1.7–1.9 m. They are 0.59–1.60 m wide, with 65% of them measuring 0.9–1.3 m. The diameter of the actual pressing area – encompassed by a circular groove – varied between 0.45–1.10 m. Five press-beds with small pressing areas were found in the southern Golan; the pressing area diameters range from 0.43–0.70 m. Out of ten installations in southern Golan 60% of them had diameters of 0.82–0.96 m. The drainage channels that encircled the pressing areas are 3–6 cm wide and 1–5 cm deep.

The square mortices in which the wooden posts for the press frame were fixed measure 0.16–0.32 m on a side, in 62% of the press-beds – 0.20–0.24 m. In some of the press-beds, step-like extensions were hewn into the bottoms of the mortices, so that the vertical section was L-shaped. The purpose of the extension was apparently to secure the press frame against pressure exerted by the screw on the press-bed. The height of the press-bed was measurable in only seven instances; these ranged from 0.30–0.42 m.

Collecting Installations

In four sites in the Golan Heights where oil presses were excavated, collecting vats were found, while in this survey, collecting vats were identified at only two sites (Ibid.: 42, Table 5). The main reason for this was the location of the collecting installations in the lowest (and most covered) levels of the complex; thus, in a survey in which nothing was excavated, their identification was virtually impossible. Other reasons might include their removal for secondary use as basins or containers, in contrast to other parts of the oil press.

In the oil press at Gamla (Site 12), excavated in Area R, collecting vats were found constructed of basalt, with volumes of 115 and 113 litres. At H. 'Ein Nashut (Site 11), a rectangular collecting vat with a capacity of 136 litres was found adjacent to, and lower than, the press-bed.

It is reasonable to assume that in most of the oil presses located on the basalt bed-rock of the Golan, vats were constructed rather than hewn, since it was almost impossible to carve a pit into this extremely hard igneous rock. At Giv'at Ha-Yi'ur, a 153 litre cylindrical collecting vat was found, constructed of limestone; a similar vat with a volume of 153 liters was discovered at Kursi (Site 13). As mentioned previously, some press-beds were found that had central collecting vats with capacities of 88–158 litres.

No significant differences were found in the volumes of collecting vats in the oil presses with lever and weights presses at Gamla and Giv'at Ha-Yi'ur (113–153 litres) and of those found alongside direct pressure screw presses at H. 'Ein Nashut and Kursi (136–147 litres). In addition, the volume of the central collecting vats in press-beds found in the southern Golan (88–158 litres) is also close to the foregoing values.

The volumes of the collecting vats found in the Golan are generally larger than those of the Hellenistic installations unearthed at Maresha (Kloner and Sagiv 1993; Site 65), the average capacity of which is 50 litres, and smaller than those discovered in western Galilee that measured 200–300 litres in volume (Frankel 1985, 1996). During various periods and with diverse pressing methods, there were differences in the quantities of olives placed in the pressing installation at any given time, and the collecting vat was hewn or constructed accordingly.

Oil Press Buildings and Their Locations

In this survey, data on oil press areas were recorded from eight sites (Ben-David 1998: 44, Table 7). They ranged from 16–60 m^2; the lowest figure belongs to the oil press at Kursi, located in a church. It may be assumed that it supplied oil to a small population, and did not operate as a large-scale "industrial" plant.

The area of the oil press of the Roman period at Gamla is 60 m^2 – largest of the oil presses measured in the Golan. It must be noted that two pressing installations operated there, compared with a single press in the other sites. This is apparently the reason for its relatively large size.

In five sites, the area of the oil press was 30–54 m^2. For the sake of comparison, oil press area is 40 m^2 at Korazim (Yeivin 1984), 45 m^2 at H. Sumaqa (Dar 1999 [Site 21]), and 50 m^2 each at H. Karkara and H. Zabadi in the Galilee (Frankel 1985 [Sites 2, 4]).

The survey recorded the location of 53 oil presses, almost all of which were found in the outermost buildings on the fringes of the sites, and not in their agricultural areas. Dar, Safrai and Tepper (1986: 95) and Aviam (1988: 205) have already shown that such installations were located on the edges of settlements. The oil presses were found in all directions from the main part of the village: 19% on the northern edge, 22% on the southern, 20% to the east and 39% to the west. No finds were excavated that clearly pointed to the existence of open-air oil presses, as were found in Samaria (Ayalon 1999; Site 30). All were found at settlement sites except for that at Kursi (Site 13), located in a church building. An exceptional oil press was found 500 m south of the village of Dabussiya. On the strength of other finds discovered in the area around it – e.g. wine presses, a cistern and other installations hewn from bed-rock – it appears that the press was installed in an agricultural building.

Oil Presses in the Golan Heights – Historical Geography

During the survey (Ben-David 1998) 109 oil presses were recorded. After the survey was completed and till 2005 30 more oil presses and four quarries for components of oil presses have been found, some by me and some by others. These 139 oil presses are, however, not distributed evenly throughout the Golan. Eight are found in its northern and eastern regions, 46 in the south and 85 in the western section of the central Golan. This large concentration in western Golan is in the stony area of the Dalwe Basalt in which the large exposures of bed-rock make the growing of field crops almost impossible but are suitable for the growth of olives. On the Cover Basalt in the south of the Golan few olives were grown for home consumption only and in the eastern area installations for the production of olive oil are almost completely lacking because of the drainage problems typical of Muweisse Basalt (Ibid.: 44–46).

As opposed to the difficulties that olive trees have to develop on the badly drained Muweisse Basalt soils these soils are particularly suitable for cereals and vegetables, and as the experience of modern farmers in the region has shown also for vineyards of grapes for wine. The vineyards of Ortal, 'Alone Ha-bashan and Kidmat Zvi are located on flows of Muweisse Basalt. It is true that typical wine presses have not been found in the basalt areas while many were found on the limestone slopes of southern Golan. Because of the basalt rock formations it is almost impossible to cut the typical rock-cut wine presses (also neither other rock-cut installations nor burial caves) typical of regions of limestone. Nevertheless there is archaeological evidence for the existence of improved wine presses from the Byzantine period in the form of screw bases of the type usually found in these installations. Today we know of 22 screw bases in the Golan, nine of which were found in its eastern part.

As showed in the past (Ibid.) the growing of olives and the production of olive oil was an important part of the economy of central Golan and was an ideal way to exploit the rocky soil of this region.

In our previous researches our quantitative estimates were based on the assumption that all the sites in western Golan with their oil presses were occupied at the same time. In our recent researches, however, we have carried out a detailed typological analysis of the pottery from the various sites in the region that has enabled us to distinguish between the settlement patterns of the various sub periods. As regards the oil presses themselves, only the lever and weights presses were in use in the Early Roman period while the direct pressure screw presses operated in the Roman-Byzantine period, although some of the lever and weights presses continued to be used also in that period. Our recent researches have shown that there were four and a half times as many oil presses in the Byzantine period than in the Early and Middle Roman periods.

In the Early and Middle Roman periods olive cultivation and oil production were on a small scale. However, in the Early Byzantine period and perhaps already in the Late

Roman period the western Golan became one of the largest oil producing in all the Land of Israel. According to our calculations (Ibid.: 50) the 64 oil presses that were in use at this period in the central Golan processed the olives from 20,000 dunams of olive groves and produced 640 tons of oil per year. The oil was exported mainly eastwards to the eastern Golan and further east to the Bashan and the Hauran in Provincia Arabia where almost no olive presses from this period have been found.

Notes

1. Frankel's third category was actually entitled "Direct-Pressure, Non-Lever Presses Using Screws or Wedges". None of the oil presses covered in this article utilized wedge presses.

Bibliography

Aviam, M. 1988. "Olive Growing and Viticulture in Upper Galilee in Ancient Times". *Israel – People and Land, Eretz Israel Museum Yearbook* 4 (22). Tel Aviv: 197–210 (Hebrew).

Ayalon, E. 1999. "The Open-Air Oil Plants of Western Samaria". In: Y. Eshel (ed.). *Judea and Samaria Research Studies, Proceedings of the Eighth Annual Meeting 1998.* Kedumim-Ariel: 115–126 (Hebrew).

Ben David, H. 1998. "Oil Presses and Oil Production in the Golan in the Mishnaic and Talmudic Periods". *'Atiqot* 34: 1–61 (Hebrew).

Callot O. 1984. *Huileries Antiques de Syrie du Nord.* Paris.

Dar, S. 1999. *Sumaqa, A Roman and Byzantine Jewish Village on Mount Carmel, Israel* (BAR IS 815). Oxford.

Dar, S., Safrai, Z. and Tepper, Y. 1986. *Um Rihan: A Village of the Mishnah.* Tel Aviv (Hebrew).

Frankel, R. 1985. "Western Galilee, Oil Presses". *ESI* 4: 110–114.

Frankel, R. 1994. "Ancient Oil Mills and Presses". In: R. Frankel, S. Avitsur and E. Ayalon. *History and Technology of Olive Oil in the Holy Land.* Arlington and Tel Aviv: 19–89.

Frankel, R. 1996. "Oil Presses in Western Galilee and Judaea – A Comparison". In: D. Eitam and M. Heltzer (eds.). *Olive Oil in Antiquity, Israel and Neighbouring Countries from the Neolithic to the Early Arab Period.* Padova: 197–218.

Frankel, R. 1999. *Wine and Oil Production in Antiquity in Israel and Other Mediterranean Countries.* Sheffield.

Kloner, A. and Sagiv, N. 1993. "The Olive Presses of Hellenistic Maresha". In: M.-C. Amouretti and J.-P. Brun (eds.). *Oil and Wine Production in the Mediterranean Area* (BCH Supplement 26). Paris: 119–136.

Yeivin, Z. 1984. "Korazim 1983–1984". *ESI* 3: 66–71.

17. An Early Roman Oil Press in a Cave at Yodefat

Mordechai Aviam

The Site

The site of Yodefat lies on an isolated hill surrounded by steep streams in western Lower Galilee (MR 1764/2485). The site is mentioned by Josephus as a town that was fortified by him before the Roman invasion. The town was besieged in the summer of 67 CE by the Roman army under the command of Vespasian while thousands of Jews were entrapped within its walls under the leadership of Josephus himself. The detailed heroic story is an important part of Josephus' account in his book (War III 141–340).

The site has been extensively excavated for seven seasons by M. Aviam, the first three with D. Adan (Adan Bayewitz and Aviam 1997). These excavations have proven that the site suffered from such a siege and was subsequently destroyed and abandoned in the mid-1st cent. It was never rebuilt.

The Press

In 1985 the survey team headed by M. Aviam entered a cave on the eastern slope below Yodefat and there discovered and registered the remains of an oil press. In 1988 the cave was disturbed by illegal excavations carried out around three stone weights associated with the oil press. In 1989 a salvage excavation was undertaken under the auspices of IAA and directed by the author with the help of Edna Stern. The cave was cleared and the weights pit was fully excavated. The earliest pottery from the pit was dated to the Hellenistic period but most of the material is dated to the Early Roman period and no sherds were found later than the first half of the 1st cent. CE.

The cave is about 13 m long and 8 m wide (Fig. 1). The main entrance was from the east where a well carved doorway is still visible but was not excavated. At the north-eastern corner, a narrow passageway connected the oil press cave to another cave in which no archaeological remains were found.

In the south-western section of the cave, 0.8 m from the western wall, a large crushing basin 1.6 m in diameter with a square sunken socket in the centre was found *in situ* on the rock floor. The pressing installation was found along the eastern wall and included eight elements. In the north wall, 0.8 m above the floor, a large rectangular niche was cut into the rock face. It measured about 0.4×0.9 m and probably served to anchor the beam. In the top of the niche a large heavy ashlar stone was inserted that apparently served to withstand the pressure exerted by the beam end. Below the niche is a plastered area measuring about 1.0×1.4 m that probably served as a working area. At the southern end of the plastered floor there was a shallow step, below which is a vat (0.4×0.6×0.6 m), apparently a central collecting vat, upon which the frails of olive pulp were placed during the pressing operation. The vat is surrounded by a narrow rim (0.1 m) in the south and north and a wide rim (0.4 m) to the east and west, all covered by a layer of high quality white plaster. About 2 m to the west, a large ashlar stone was found laying on the floor that was probably one of the two plain piers that stood on either side of the plastered vat, and that served to support the pile of frails. To the south, a long pit (3.7×1.0 m) was cut into the rock floor in which the three weights were found standing *in situ*. Each weight has a reversed T-shaped hole.

The plan of this oil press is typical of those found in sites of the Early Roman period such as those from Gamla (Gutman 1994: 130–133 [Site 12]), Mishmar Ha-'Emeq (Giveon 1988: 215–218) and others. The press is dated by the pottery from the pit as well as by the finds from the excavations of the town. This is the only oil press in a cave from Galilee so far published although some others are known from our surveys. In this region, the caves are wet from December to May and can only be used during the summer and autumn months, which perfectly coincides with the olive picking season.

The oil press cave at Yodefat lies directly under the city wall while its entrance is outside it. It is a fine illustration of the reference in the Mishnah (*Ma'as. Sh.* 3, 7) to an underground oil press with its cavity inside the town while its entrance is outside it.[1]

Notes

1. As a result of the last few seasons of excavations, I am completely convinced that the wall was erected shortly before the war, and by this the connection between the houses and the press was cut (Aviam 2002: 122–127).

Bibliography

Adan-Bayewitz, D. and Aviam, M. 1997. "Iotapata, Josephus and the Siege of 67; Preliminary report on the 1992–94 Seasons". *JRA* 10: 131–165.

Aviam, M. 2002. "Yodefat/Jotapata: The Archaeology of the First Battle". In: A.M. Berlin and J.A. Overman (eds.). *The First Jewish Revolt*. London and New York: 121–133.

Giveon, R. 1988. "Area 1 – An Oil Press and Domestic Buildings". In: B. Mazar (ed.). *Geva, Archaeological Discoveries at Tell Abu-Shusha, Mishmar Ha-'Emeq*. Tel Aviv: 215–218 (Hebrew).

Gutman, S. 1994. *Gamla – A City in Rebellion*. Tel Aviv (Hebrew).

Josephus, *Jewish War* (Trans. J. Thackeray). London 1927.

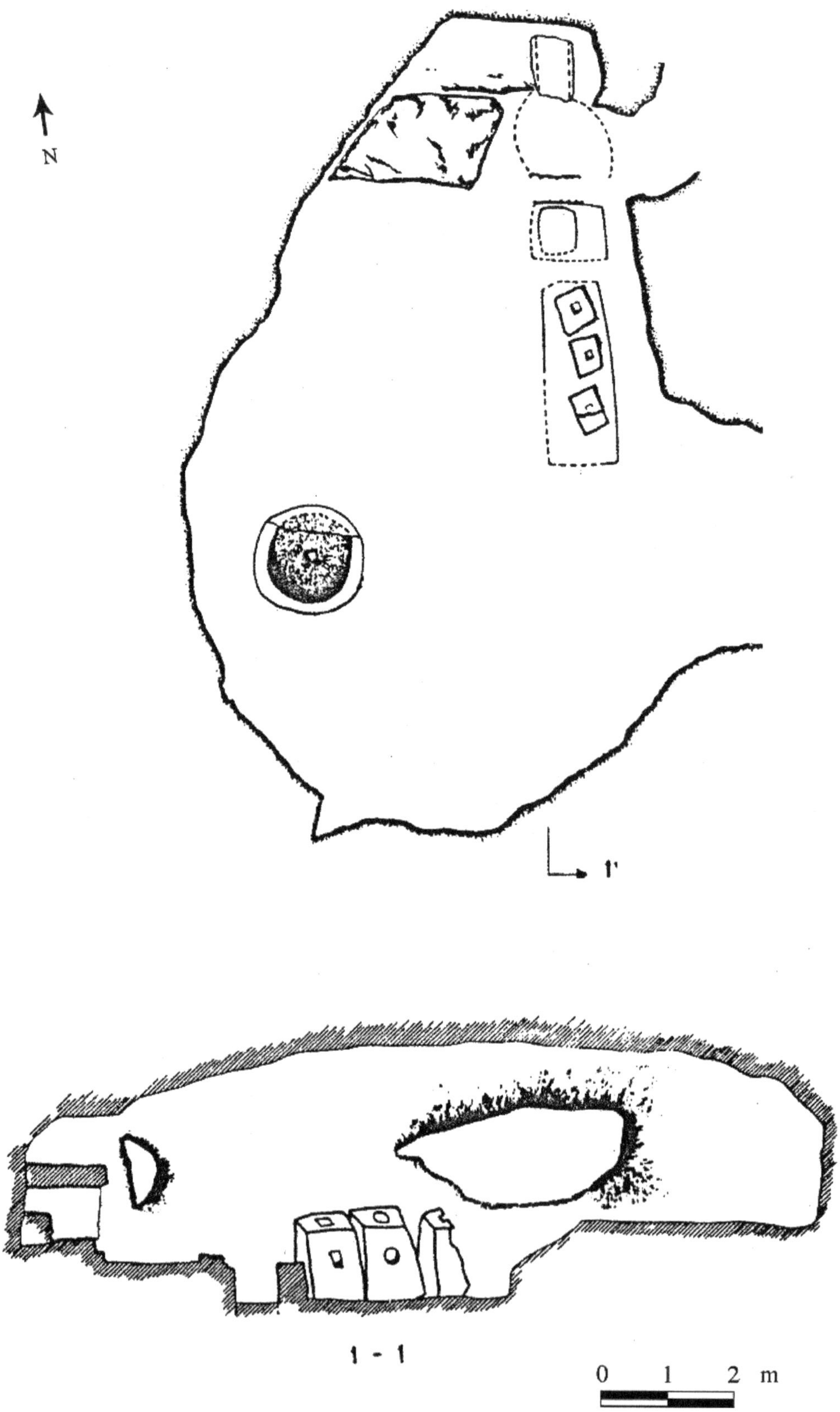

Fig. 17.1. Yodefat, plan and section of the oil press.

18. Olive Oil Press at Kh. Kafritha (Qiryat 'Ata)

Yosef Porath

The remains of several ancient olive oil presses were uncovered at the site of Kh. Kafritha (in the centre of the modern city Qiryat 'Ata; MR 1602/2455), on the foothills of western Galilee. The olive oil press discussed below (Figs. 1–2) was accidentally discovered during a construction project and was excavated by the writer in August 1988.[1]

The complex consisted of two units (L. 57 & L. 63), each comprising a crushing mill of which only the crushing basin has survived and probably two lever and weights presses.[2] The complex was gradually developed. The southern unit L. 57 was constructed first, and the northern one L. 63 was later added in place of an unroofed courtyard. The surface was lowered down to bed-rock during the preparation for the construction, creating a terrace on the hill's slope, thus clearing any earlier remains. It seems that the initial unit was built in the Late Roman period (late 4th cent. CE), according to the finds revealed west of W51, and was extended later. A layer of ash and charcoal entirely covered the floor and installations, evidence for the intensive conflagration of the violent destruction of the area. The pottery found in the debris dates to the end of the Byzantine period (late 6th-early 7th cent. CE), but it is not clear whether the destruction affected this complex only (an accident?) or covered the entire site (a catastrophe caused by the Persian or Arab conquest in 614 and 638 CE respectively).

The Crushing Basins

L. 82 of the northern unit L. 63 (1.9 m in diameter) is made of one block of limestone with a circular socket for the pivot (0.28 m in diameter, 8 cm deep). The crushing basin L. 93 of the southern unit L. 57 (1.72 m in diameter, Fig. 3) is made of radial segments of limestone and a raised constructed lip (preserved to a height of 0.12 m). A fragment of a marble column in secondary use (0.24 m in diameter, 0.285 m long, 0.12 m projecting from the basin's floor) served as the raised socket for the mill's pivot (Figs. 4–5). A square iron disc (9.5×9.5 cm, 4.8 cm thick) with a 3 cm conical hole was inserted into the centre of the top of the column and served as the socket in which the pivot revolved.

The Presses

In the southern unit L. 57 a lever and weights press was installed on each side of the crushing basin. In each case about 0.55 m south of wall W73 a central collecting vat was cut into bed-rock, each surrounded by a working area. The opening of the western collecting vat L. 72, in the centre of working area L. 66, is 0.6 m in diameter and its volume is 186 litres. The diameter of the opening to the eastern central collecting vat L. 76, in the centre of working area L. 71, is 0.5 m and its volume is 141 litres.[3] The openings of the vats are particularly wide as compared to those of other central collecting vats, and it seems that grills were placed above them. No remain of a stone grill was found under the destruction layer and it seems that it was made of wood, that had been burnt in the destruction.

Cylindrical stone weights with reversed T-shaped bores were found in rock-cut rectangular pits at the southern end of both presses (L. 75 & L. 81), located about 2.7 m from W73. Constructed basins paved in mosaic were located between the press-beds and the weights (L. 60 & L. 92). These apparently served to store the crushed olives (probably contained in frails) before the pressing process. The pressing was accomplished by a beam, one end of which was anchored in a niche in the wall (not preserved) and by a cylindrical stone weight that was hung from the other end. The relationship between the distance from the walls to the press-beds and the distance from the walls to the weights is 1:5.5 so that the mechanical advantage is 5.5.

The single excavated press in the northern unit L. 63 has a similar working area (here paved in mosaic) in the centre of which is a central collecting vat (L 58). To the north there are two cylindrical weights in a weights pit (L 64) and as in the other two presses between the pressing area and the weights there is a rectangular basin also paved in mosaic that doubtless served to store the olives (see above). The two rectangular holes (11×11×8 and 11×14×12 cm), cut in the northern edge of the weights pit, probably served to hold wooden uprights that guided the press beam and perhaps also held a device to lift it.

More than five kg of iron and bronze implements – 5 to 17 cm-long nails, chains and bands with rivets and hinges – were found in the destruction layer. Amongst them was a longitudinal iron point (0.21 m long, 4 cm thick, Fig. 6) soldered to a 0.34 m long band (3.2–3.4 cm wide, 1.1–1.4 cm thick) found in the debris of L. 60, west of crushing basin L. 82. The point possibly served as axis of the pivot of the crushing mill. The point was inserted into the pivot beam and the band encircled the beam and fastened them together, but it is uncertain whether it was the upper or lower axis. It probably rotated in a metal disk (like the one in crushing basin L. 93) or in a stone socket, some lubricating material (oil, fat or water) serving to minimize friction and to avoid the abrasion of the metal, wooden or stone components.

Dendroarchaeological investigation on the charred wood, carried out by Prof. N. Liphschitz of Tel Aviv University, identified the beams of the oil press as having been made of Cedar of Lebanon.

Each of the two oil press units at Kh. Kafritha is a typical example of a unit comprising one crushing mill and two lever and weights presses. One of the excavated crushing basins was constructed of radial segments while the other was monolithic.

Notes

1. The oil press was uncovered in salvage excavations that were carried out under IDAM license No. 1544 at 21 Hameyasdim St., Qiryat 'Ata. Additional oil presses were excavated at Kh. Kafritha by A. Siegelmann (1998).

2. Oil presses usually consist of two presses for each crushing mill and the unit of L. 57 was of this type. The oil press under discussion extended outside the limits of excavations, and it is assumed that the northern unit L. 63 had a second press in the unexcavated area. A circular depression L. 74 (1.6 m in diameter, 0.48 m deep), surrounded by construction, was excavated over the south-eastern portion of L. 63. It is apparently the poorly preserved remains of a crushing basin, related to a later oil press.

3. The total volume of the two collecting vats (327 litres) makes it possible to calculate the maximum amount of crushed olives that was pressed in each batch. Assuming that the olives were of the best quality (ca. 17% oil for the *Syrian* olive under modern conditions) and that the vats contained about 100 litres "pure oil" (about one third of the volume of the vats), each batch would have been about 588 litres olives (about 530 kg).

Editor's Note

These mosaic paved basins almost certainly served to separate the oil from the watery lees.

Bibliography

Siegelmann, A. 1998. "An Oil Press of the Byzantine Period in Qiryat Ata". *'Atiqot* 34: 101–106 (Hebrew).

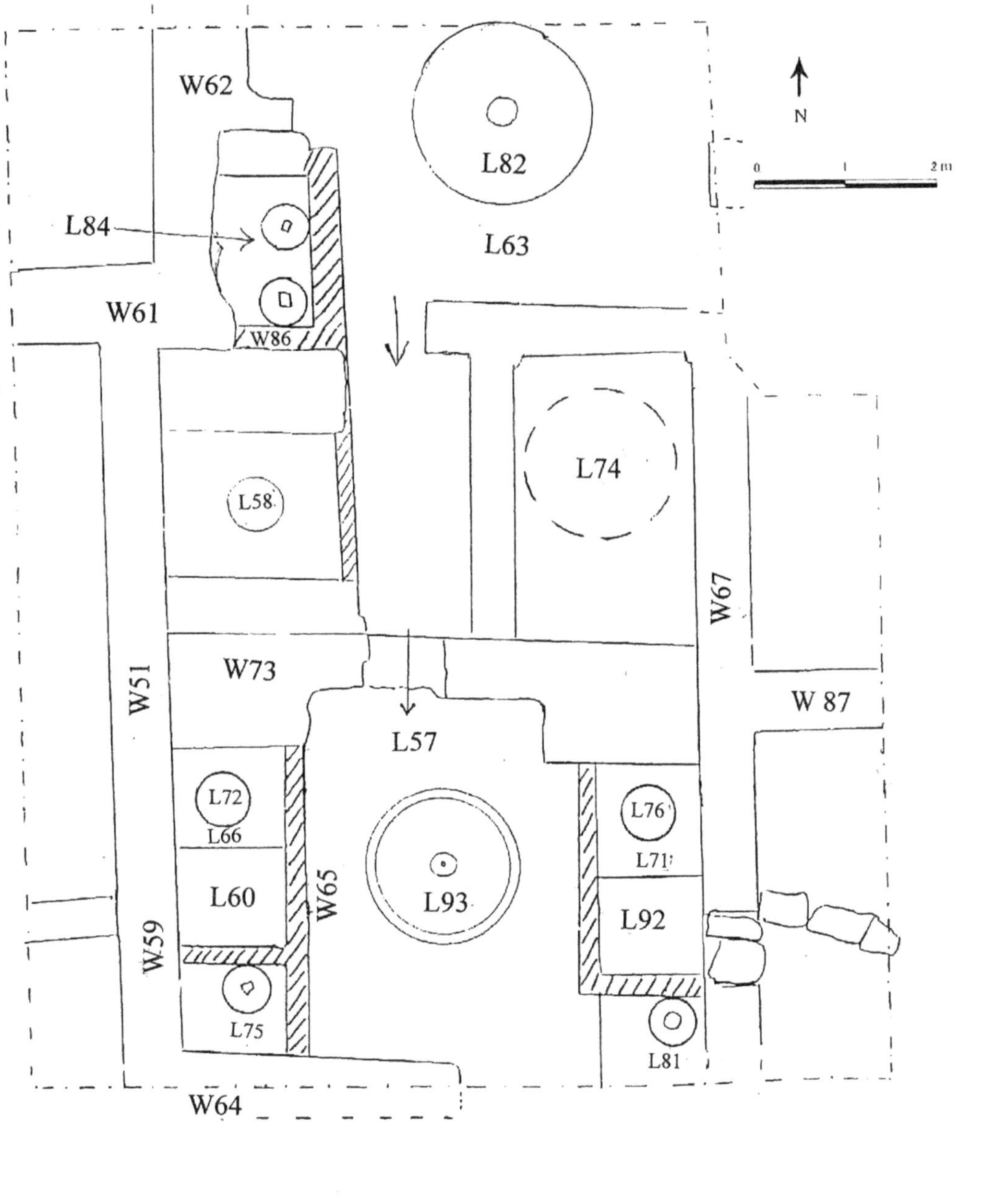

Fig. 18.1. Kafritha, plan of the oil press.

Fig. 18.2. Kafritha, general view of the oil press looking north.

Fig. 18.3. Kafritha, crushing basin L. 93, view to the west.

Fig. 18.4. Kafritha, centre of crushing basin L. 93 with iron disc in situ.

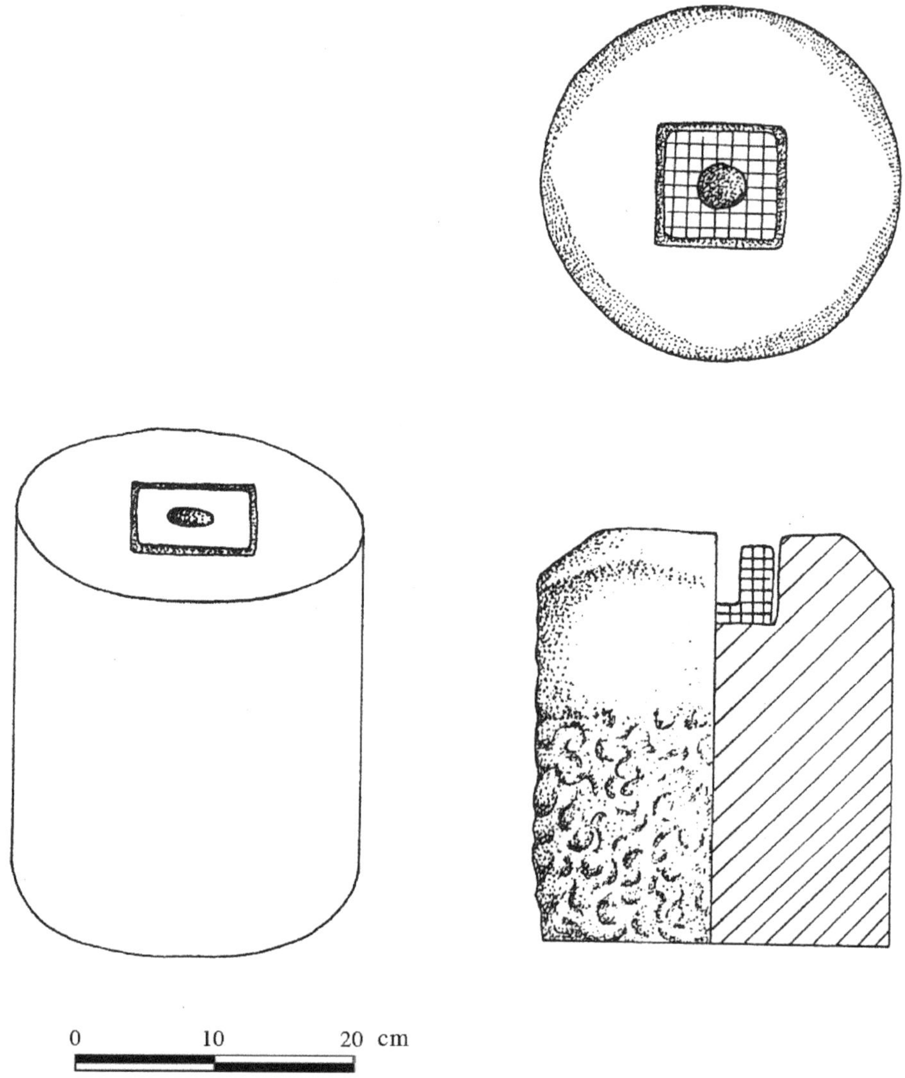

Fig. 18.5. Kafritha, marble column with iron disc installed in its centre.

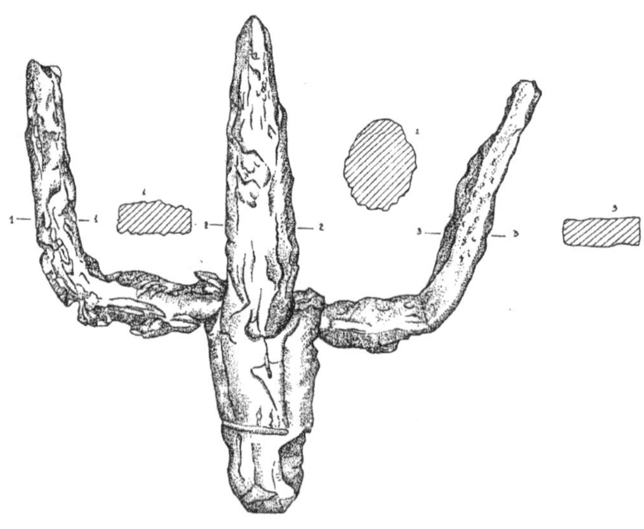

Fig. 18.6. Kafritha, iron pointed axis.

19. Wine and Oil Presses at H. Castra

Ze'ev Yeivin and Gerald Finkielsztejn

The Site

H. Castra lies at the base of the western flank of Mt. Carmel, south of Haifa and 1.5 km from the coast (MR 1475/2443). Excavations were carried out at the site by the Israel Antiquities Authority during the years 1993–1997 under the direction of the authors (Yeivin and Finkielsztejn 1999a; 1999b; 2001; 2008; Finkielsztejn 2005). A Christian town of the Byzantine period covering c. 250 dunams was revealed dating from the 4th-7th cent. CE. An additional excavation took place outside the south-east corner of the main excavation in 2000–2001, directed by Gerald Finkielsztejn. The ethno-religious nature of the site precludes its traditional identification with the Samaritan settlement of Castra, and is rather in favor of the largest Christian settlement in the area to appear in the written sources: Porphyreon of the South.

The Wine Presses

In and around the town and at its outskirts, 14 wine presses were uncovered, evidence for a large scale wine industry. Some of them survived almost completely but others were found partially destroyed, with only collecting vats, sections of treading floors or bases of screw presses remaining. In some cases several structural stages could be discerned in the wine presses including some in which the plan was changed. Two could be assigned to the Hellenistic and maybe the Roman periods but all the others were contemporary to the Byzantine city.

Some of the installations were large and complex, including, in addition to a central treading floor, several "secondary working surfaces" (most probably actual treading floors; see below) and more than one collecting vat, while others were smaller, consisting of only one treading floor and one collecting vat. All the wine presses without exception included a sieving vat between the treading floor and the collecting vat and in the floors of the collecting vats there were small shallow sumps to drain the liquid. The sieving vats were 0.4–0.5 m deep while the sumps were 0.15–0.20 m deep. Ceramic or lead pipes connected the treading floors to the sieving vats and the sieving vats to the collecting vats. In all the collecting vats there were steps attached to one of the walls. The steps led from the surface to the floor of the vat enabling descent into it in order to take out the must, to clean the vat or for any other purpose. In all the wine presses except one (No. 1 below) the treading floors, the "secondary working surfaces", the floors of the vats and the steps leading down into them were paved in white large mosaic tesserae.

Hereunder we present four wine presses that are representative of the types found at the site.[1]

1. The Plastered Wine Press D

This installation was found in the western part of the main excavations, on the outskirts of the town. It was covered with debris 2 m deep consisting of earth and pebbles and was apparently one of the earlier wine presses of the site. At the time of the discovery, it was dated to the first stage of the town's existence, the end of the 3rd and the beginning of the 4th cent. CE. However, a similar structure was uncovered in the east of the site during the 2000–2001 excavations, and objects abandoned *in situ* in the bottom of the collecting vat dated the end of its use to the late 2nd-early 1st cent. BCE, i.e. the Late Hellenistic period, but in spite of their similarity the two installations were not necessarily of the same date. As regards the character of the site prior to the Byzantine period, the present evidence suggests the existence of a reduced settlement in the Hellenistic period (a farm or an estate?), and a small Jewish village (with ritual baths or *mikva'ot*), in the Roman period.

The wine press presented here was completely plastered. The treading floor was trapezoidal in shape (5.4×5 m, area 27 m^2) and lacked any trace of a screw base. To the west of the treading floor was the rectangular collecting vat (2.5×2.0 m, depth 1.7 m, volume 8.5 m^3). Against the eastern wall of the vat was a semicircular sump (diameter 0.65 m). In the south-western corner of the vat there were two steps. Between the treading floor and the collecting vat there was a round sieving vat (diameter 0.5 m, depth 0.55 m). A pipe (diameter 9 cm) connected the treading floor to the sieving vat and a pipe of the same diameter connected the latter to the collecting vat, leaving the sieving vat at the level of the floor and entering the collecting vat 0.5 m below the edge. A shallow square surface quarried in the bed-rock is located to the east of the treading floor, but the relation of the structure that may have been standing there and the wine press cannot be understood.

2. The Eastern Wine Press G (Fig. 1)

This press was found in the north-eastern part of the excavations, in the north-eastern corner of a large dwelling building, delimited by two streets. The treading floor is trapezoidal in shape (5.6×5.0 m, area 28 m^2) and paved with two superimposed layers of white mosaic tesserae. The upper floor was laid after the earlier one had sunk 0.22–0.30 m and could no longer be used. The number of tesserae by dm^2 was 25 for the lower floor and 9 for the upper floor. In the centre of the treading floor there was a large round stone (diameter 1.5 m, height 1.6 m), with a square mortice in the centre that served to secure a fixed wooden screw. The collecting vat was square (2.5×2.5 m, depth 1 m, volume 6.25 m^3) and situated to the west of the treading floor. A round sump (diameter 0.8 m, depth 0.4 m) was located in the north-eastern corner of the floor of the vat. The floor of the vat was also paved twice, the second floor being made of white large mosaic tesserae (9 by dm^2) and having been laid together with the upper layer of the treading floor. The laying of the second floor reduced the size of the vat, but made the sump deeper. Four steps were attached to the southern wall of the vat, leading westwards to the latter's floor. The steps were paved in mosaic, the tesserae of which were similar in size to those of the first layer of the treading floor (25 by dm^2). Between the treading floor and the collecting vat there was apparently a square sieving vat but all that remained of it was a small section of its mosaic floor. No trace was found of the pipes connecting the treading floor to the sieving vat and the latter to the collecting vat.

3. The Large Wine Press A (Fig. 2)

This is the largest and most complex of the wine presses of Castra. It was found in the western area of the excavations on the outskirts of the town, between the buildings and the burial area (Yeivin and Finkielsztejn 1999a: picture p. 64; 1999b: 35, Fig. 45).

In the centre of the whole wine press is the square main treading floor (5.5×5.5 m, area 30.25 m^2), paved with medium sized white mosaic tesserae (25 by dm^2), with the design of two concentric rectangular frames, each made up of two rows of blue-gray tesserae. In the outer part of the mosaic floor the tesserae were laid in lines parallel to the edges, whereas those in the centre were laid diagonally. In the centre of that main treading floor (or rather, maybe, the central working area; see below) was a square stone (1.6×1.6 m) with a rectangular mortice, widening at the bottom in three directions, that served as a base for a fixed wooden screw. The floor sloped slightly towards two square collecting vats that were placed side by side, each measuring 2.5×2.5 m, depth 1.65 m, volume 10.3 m^3. There were relatively wide platforms around the vats that doubtless served as work areas. The platforms, the floors of the vats and the steps leading down to them were all paved with large white mosaic tesserae of various sizes. Altogether, three vats were uncovered, evidence for two different stages of building. Only two were in use in the second and last stage, as a previous one was later turned into a tomb (Fig. 3; see below). In the north-western and south-western corners of the flat frame that surrounds the southern one of these two vats, there were crosses designed in the mosaic floor. The northern of the two was combined with a chrism, with an *alpha* and an *omega* on its sides.

One round sieving vat served both collecting vats of the last stage (diameter 0.5 m, depth 0.35 m). A pipe led from the central treading floor to the sieving vat and two ceramic pipes led from the bottom of the sieving vat to each collecting vat. The remains of a square sieving vat were found to the east of the northern collecting vat that was apparently originally connected to an early treading floor, sections of which could be discerned under the northern "secondary working area" (see below). To the south of the two later collecting vats, the third one appeared to have had been part of an earlier wine press too. It went out of use around the middle of the 4th cent. CE and turned into a tomb, that was cut into its eastern wall. The sump of this vat was semicircular in shape and adjacent to the same eastern wall, while two steps attached to the opposite wall descended from south to north. Remains of the treading floor that was once connected to this vat were found to the south, below the southern "secondary working areas" of the late phase (see below).

Adjacent to the northern, southern and eastern sides of the central treading floor seven compartments that served as working areas, storing vats or, rather it seems now, treading floors, were found. Each of them is connected to the main central treading floor by a pipe, evidence for must having to run through it. These compartments are paved with mosaic tesserae arranged in a similar manner to that of the central treading floor: the outer lines parallel to the edges of the floor and the central part in diagonal lines.

Adjacent to the northern side of the central treading floor is a rectangular compartment oriented east-west (4×2 m, area 8 m^2). In its south-western corner there is a small shallow sump (diameter 0.4 m). Adjacent to the southern side of the central treading floor there are three more compartments. They were laid over the remains of the treading floor of the earlier wine press mentioned above. The sump of the western of the three compartments is close to the centre of its northern wall, while those of the other two are situated in their north-western corners.

Adjacent to the eastern side of the central treading floor there are also three compartments. The two northern ones are roughly rectangular (3×2 m, area 6 m^2). The south-western corner of the southern compartment is covered by the north-eastern corner of the eastern compartment on the southern side. Besides its modified form, the southernmost compartment on the east side is exceptional in that there is a square carpet decorated with a geometric colorful pattern in its otherwise white mosaic floor and the sump is also framed by a row of red tesserae.

33 coins dating in the end of the 7th-beginning of the 8th cent. CE were found scattered in the fill of the northernmost vat. The filling appears thus intentional, suggesting that the

wine press went out of use as a consequence of the Arab occupation. This may be interpreted as a consequence of some expulsion of the inhabitants of a city-village (a *komopolis*; see Finkielsztejn 2005) that was probably a centre of the economic and religious life of the area, and not just a mere remote monastery, where production of wine could have continued into the 8th cent. CE.

4. Wine Press E with the Inscription

This wine press is one of the smallest at the site. The treading floor is square (4×4 m, area 16 m^2) and in its centre is a round stone base (diameter 0.8 m), in the centre of which is a square mortice to secure a screw. The floor is paved in white mosaic tesserae. Those around the screw base were arranged in a circle within a square frame. The collecting vat (2×2 m, depth 1.8 m, volume 7.2 m^3) is to the west of the treading floor. On the top of the partition wall separating these two units, there is a Christian inscription in Greek introduced by a cross, which is clearly a dedication of the mosaic (and not of the whole press) by a donator, and where the date is unfortunately only partly preserved. Because the inscription occupied the central part of the partition between the treading floor and the vat, the sieving vat is at its northern end. A sump is situated in the north-eastern corner of the collecting vat and five steps were built against its western wall leading from the north southwards.

To conclude on the wine presses, we would suggest that the structural differences between the wine presses are of chronological significance, as evidenced by the study of all the installations in H. Castra. There are certain elements in the collecting vats that are characteristic of the early wine presses: a) the sump is next to one of the walls of the vat and is semicircular in shape; b) the sieving vat in the early wine presses was square or rectangular; c) the number of steps leading down to the collecting vat was small starting 0.6–0.7 m below the edge as, for example, in the plastered wine press D and in the southern collecting vat in wine press A.

The later wine presses are characterized by large treading floors and large square or rectangular screw bases, round sumps, usually in the corner of the vat and steps starting from the surface and leading down to the floor (wine press G).

The latest of the wine presses are smaller and the screw bases are round and smaller. The collecting vats are smaller but deeper and have five steps (wine press E).

On the basis of the results of an experimental fabrication of wine, which was undertaken by Yehoshua Dray in the largest wine press A (Fig. 4; Dray 2003), it appears that the compartments surrounding the central area (the aforementioned "secondary working surfaces") should be actually identified as treading floors. After the treading and initiation of the fermentation, the must would be freed and run through the central area down into the collecting vats. The central area (the aforementioned "main treading floor"), where the screw press is located, should be identified as mainly a working area. There, treading was certainly done in the case of small wine presses, but did not necessarily take place (if at all) when compartments surrounded it. In all cases, the second pressing of the grapes, using the screw and a built barrel, would have been done in the central area.[2]

The Oil Presses

Oil presses were found throughout the town of the Byzantine period. Some were part of dwellings while others were set in buildings reserved only for oil production. No industrial zone was identified in the town. However, a group of five oil presses were concentrated in a small area in the south-eastern part of the excavations.

The vast majority of the presses were beam presses: most were lever and weights presses, and some were lever and screw presses. Only two were operated by direct pressure. It should be stressed that both the latter are connected to a context that is earlier than that of the beam presses.

In two cases only, one olive mill (or crusher: crushing basin and crushing stone or millstone) served two presses, while in all the others one mill served three presses. Usually the crushing mill was close to one wall of the building and the three presses were placed one along each of the other three walls, in a "U" (or Hebrew *het*) shape. The presses were found generally complete, all the stone elements *in situ*, just missing the wooden parts. In most cases only parts of the olive mills were found, sometime scattered or reused in poor walls, one of them even covering a filled collecting vat.

We have chosen four oil presses to illustrate those found at the site.

1. Direct Pressure Screw Oil Press F (Fig. 5)

The press-bed of a direct pressure screw press was found *in situ* to the north of oil presses E (see below) and D, and west of oil press G. To the north of the press-bed was a round rock-cut depression presumably where the crushing basin originally stood. In the centre of the press-bed was a circular groove with square mortices on either side that held the wooden press frame. To the north-east of the press-bed was a square collecting vat quarried in the bed-rock. A short channel connected the circular groove to the vat.

This oil press was earlier than the other oil presses at H. Castra and ceased to function after these were built (see below, oil press C, for a reused press-bed).

2. The Burnt Oil Press B (Figs. 6–7)

This oil press was found in a rectangular room (6×4 m)

in the north-eastern corner of a large building (Yeivin and Finkielsztejn 1999a: cover; 1999b: 36, Fig. 46). The crushing basin (diameter 1.8 m, height 0.65 m) that stood in the northern part of the room was found cracked as the result of a conflagration that destroyed the installation and after which it no longer functioned. In this press there were one crusher and two presses, one along the eastern wall and the other along the western one.

The presses were entirely quarried in the bed-rock, and the weights were also extracted from the bed-rock. In the eastern press the rectangular weights pit (1.5×0.7 m) was at the northern end and two weights were found *in situ* inside it (each 0.5×0.5 m, height 0.7 m). In the upper part of each weight there was a reversed T-shaped bore. A rope that was inserted through the vertical hole would be tied to a rod that was inserted into the horizontal bore. To the south of the weights pit was a central collecting vat (diameter 0.8 m, depth 1.1 m, volume 2.2 m^3), quarried in the bed-rock. Grooves on the circular upper edge of the vat apparently accommodated iron or wooden rods placed criss-crossed, on which the pile of frails full of olive mash were placed, directly above the vat. Therefore, the edge of the vat served as a press-bed. On either side of the vat there were massive stone piers that apparently served to keep the frails and the beam in place. To the south of the collecting vat was a trapezoidal work pit (in the western press it was rectangular) that made it possible to work below the beam.

The western press was almost identical to the eastern one.

The beams were anchored in the southern wall of the building (preserved to one course only, as was the case for most of the buildings of the site). The weights were hung from the free northern end of the beams, exerting pressure on the frails that were laid over the collecting vat. It should be noted that in this oil press, as in almost all the other ones at the site, the central collecting vat, that served also as the press-bed, was half way between the anchoring point of the beam (the fulcrum) and the free end of the latter, where the weights exerted pressure. As a result the effective force applied was twice that of the actual weight of the weights.

3. The "Standard" Oil Press E (Fig. 8)

This installation was found in the south-eastern part of the excavations, one of five oil presses found in this area. It was located in a square building (8.5×7.5 m internal measurements). Its north-eastern corner touches on the south-western corner of another oil press (D). The olive crushing mill was in the northern part of the room and around it there was an area 1.0–1.5 m wide that made it possible to turn the crushing stone in the basin. The crushing basin was built of several parts. The olives were crushed on a round horizontal slab in the centre of which was the socket of the pivot around which the crushing stone revolved. Around this slab stood several vertical slightly rounded stones, the tops of which were stepped, the whole recalling a sliced tart or quiche. These formed the outer face and the rim of the basin. Three presses quarried in the bed-rock were located along the remaining three walls – on the eastern, western and southern sides of the building. In each press there was a niche to anchor the beam, a work pit, a central collecting vat that was also the press-bed, and a pit for the weights, in each of which two weights were still standing *in situ*. The diameter of the vats was larger at the bottom than at the top. Their walls were coated with well jointed stones and the bottom paved with white mosaic tesserae laid in concentric circles, as was the case in the other oil presses of this type (see Yeivin and Finkielsztejn 1999a: picture p. 65). In the centre of the floors of the vats there were small sumps to facilitate the collection and removal of the solid lees. Along the eastern wall of the western press there is a narrow channel that descends from north to south to a fourth vat, similar to those in the presses. We have no explanation for that addition.[3]

4. The Improved Oil Press C (Figs. 9–10)

This installation is similar to oil press E. It also has a crushing mill in the north, three presses with central collecting vats along the other three walls and a fourth vat close to the western press. At a second stage in the use of this press the fourth vat was widened, deepened and converted into a water cistern. This press differs from the others at the site in that the pressing method used in the presses is not uniform. It actually displays evidence for changes in the methods, providing a dynamic picture of the evolution of pressing techniques.

The western press is similar to other presses at H. Castra, having two beam weights with a bore in the form of a reversed "T". In the southern press there was one large cylindrical screw weight with a socket in the top for inserting a screw and two lateral mortices, and a regular beam weight, the bore of which was filled with a stone. It seems that the latter went out of use, following a change in the pressing method, and it was left *in situ* to be used as a working surface. The eastern press was also a lever and screw press but the screw weight was exceptional. It consisted of two stones standing side by side that served as one weight. These stones were the two (most probably intentionally) broken parts of the base of a direct pressure press in secondary use (Yeivin and Finkielsztejn 1999b: 36, Fig. 47). Originally, it consisted of a rectangular stone slab with a carved circular groove in the centre and two square holes on each side that secured the press wooden frame. The two parts were most probably tied together to be used as one weight. The socket for the screw end was cut in both stones at the point where they were in contact, and the two lateral mortices were cut one in each stone. These finds are evidence for the (partial) transition from the beam and two weights pressing system to the beam and screw one, reusing structural elements of the definitely abandoned old fashion (or, rather, less effective) direct pressure press system.

Notes

1. The plans of the excavations were drawn by O. Cubani.
2. Editors' note: As regards work procedures in wine presses such as press A at H. Castra see section on developed wineries and note 2 in the Introduction, where a different explanation is presented.
3. Editors' note: This vat was probably to produce the high quality oil that seeped out of the olive mash before pressing. High quality first oil was mentioned in Talmudic literature and was known in Latin as *lixivium* (see Introduction).

Bibliography

Dray, Y. 2003. "Wine Production in the Improved Byzantine Wine Press". In: Y. Eshel (ed.). *Judea and Samaria Research Studies* 12: 219–228 (Hebrew).

Finkielsztejn, G. 2005. "Les mosaïques de la komopolis de Porphyreon du Sud (Kfar Samir ; Haïfa, Israël): un évêché (?) entre village et cité". In: H. Morlier (ed.). *La mosaïque gréco-romaine* IX. *Collection de l'Ecole Française de Rome* 352 (2 Vols.). Paris: 435–452.

Yevin, Z. and Finkielsztejn, G. 1999a. Catalogue of the exhibition *Castra at the Foot of Mount Carmel, the City and its Secrets*. The National Maritime Museum. Haifa.

Yevin, Z. and Finkielsztejn, G. 1999b. "Horbat Castra – 1993–1997". *HA-ESI* 109: 23*-27*.

Yevin, Z. and Finkielsztejn, G. 2001. "Kfar Samir", "Castra Samaritanorum" and "Porphyreon of the South". In: A. Negev and S. Gibson (eds). *Archaeological Encyclopedia of the Holy Land*. New York-London: 116, 279–280, 395.

Yevin, Z. and Finkielsztejn, G. "Kfar Samir". *NEAEHL* 5:1895-1899.

Fig. 19.1. Castra, Wine Press G to south.

Fig. 19.2. Castra, areal view of Wine Press A.

Fig. 19.3. Castra, Wine Press A, the south vat as a tomb, looking east.

Fig. 19.4. Castra, Dray and Salmon pressing grapes in Wine Press A.

Fig. 19.5. Castra, Oil Press F, looking east.

Fig. 19.6. Castra, Oil Press B, plan.

Fig. 19.7. Castra, Oil Press B, looking south.

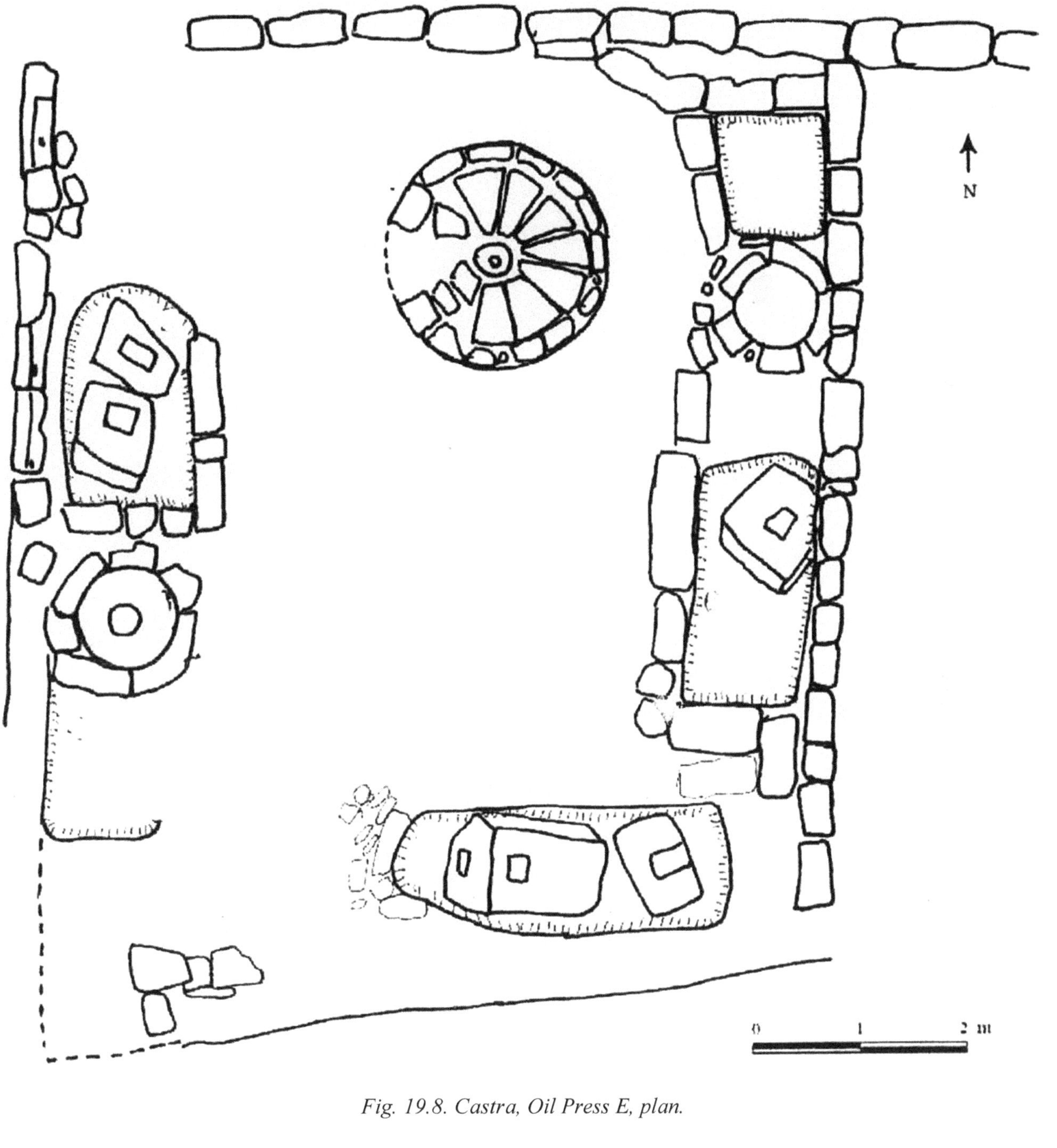

Fig. 19.8. Castra, Oil Press E, plan.

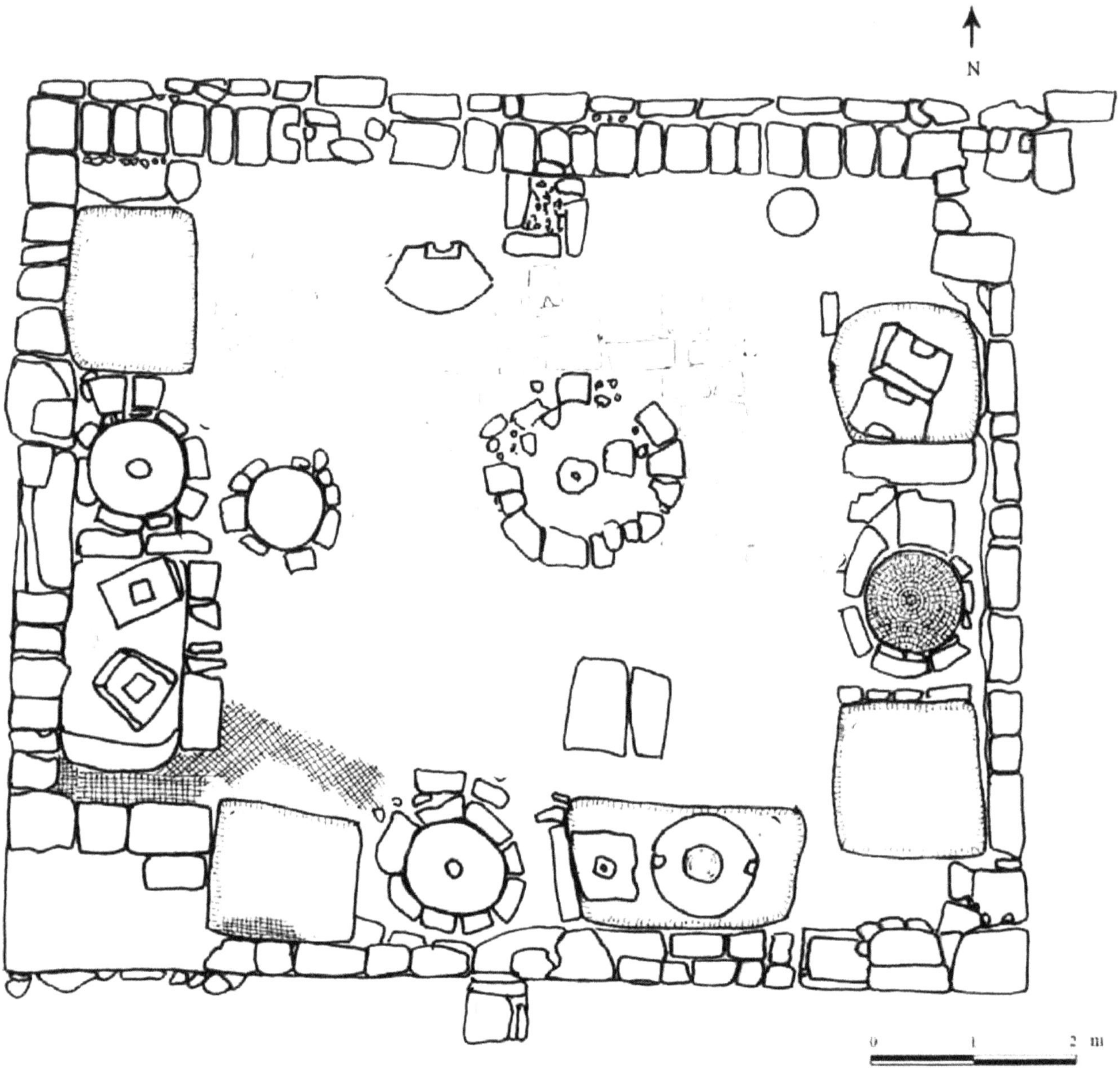

Fig. 19.9. Castra, Oil Press C, plan.

Fig. 19.10. Castra, Oil Press C: reused stone as a weight, looking south-east.

20. Wine and Oil Presses at Kh. Jalame

Etan Ayalon, Rafael Frankel and Amos Kloner

The Site

Kh. Jalame is situated at the foot of the northern slopes of Mt. Carmel. The site (MR 1589/2368) was excavated by the Corning Museum of Glass and Missouri Museum (Davidson Weinberg 1988).

The excavators distinguished four periods:

- Period I (75–125 CE). No actual buildings were attributed to this period but ashlar blocks incorporated into later structures probably derived from this stage.
- Period II (275–350). At this stage the site was a villa or country house.
- Period III (351–383). The main activity at the site during this period was glass manufacture.
- Period IV (383-early 5th cent.). Glass production ceased but the site was reconstructed.

A wine press and an oil press excavated at the site were both attributed to the last stage of Period II i.e. 335–350 CE.

The Wine Press

The wine press was at the south-eastern end of the villa (Fig. 1). Building the installation entailed destroying part of one wall of the building. The excavators suggest that the wine press was constructed in two stages and that it was only the building of the second stage that entailed damaging the structure of the villa. When the installation was first uncovered it was thought to be a nymphaeon. At a later stage it was suggested that it was a nymphaeon that was turned into a wine press. In the final report, however, these suggestions were abandoned and it is described as a wine press.

The installation consists of a treading floor paved in mosaic measuring 4.75/5.0×4.2 m (area 20.5 m^2). To the west was the small intermediary vat with an outlet at the bottom leading to the collecting vat, both plastered. The intermediary vat measured 0.45×0.50 m, depth 0.4 m (volume 0.09 m^3). Four steps lead down into the collecting vat and in the south-west corner is a small sump 0.80×0.55 m and 0.4 m deep. The vat measures 2.3/1.9×2.0 m and is 1.7 m deep (volume c. 7.1 m^3). To the east of the treading floor there are two auxiliary treading floors paved in cobblestones. The northern was found complete, was trapezoidal in shape and measured 3.0×3.4/2.5 m (area 8.85 m^2). Between the auxiliary treading floors and the main one there were small compartments ("niches") with semi domical roofs and mosaic floors "apsidal" in shape, with three straight walls and one semicircular one. The northern niche was 0.52 m wide and 0.58 m long, the southern 0.48 m wide and 0.52 m long. The first must almost certainly flowed from the auxiliary floors to the small compartments although exactly how this took place is not clear. From the compartments this first must could be obtained separately or it could flow from there to the main floor. It was this eastern section, the auxiliary treading floors and niches that in the excavator's opinion was added in the second stage. In the centre of the main treading floor was a square screw mortice that widened on one side and was cut right through the block. The block measured 0.85×0.80 m and the mortice was 0.3 m each side. A plastered channel led from the mortice to the intermediary vat. During the last stages of the use of the wine press the screw base was apparently still used but the channel was not.

The Oil Press

The oil press is at the north-eastern corner of the villa and consists of one crushing basin and two screw press bases, all three cut in bed-rock (Fig. 2). The excavators suggest that it was constructed at about 350 CE and was probably in use till the end of the century. The crushing basin has a flat crushing surface and sunken socket. Around the circumference of the basin a channel was cut in which stones were placed. This could have been a repair or possibly harder stone was emplaced in the basin from the start. The basin is 2 m in diameter and 0.52 m high. The presses have round press-beds and round collecting vats. The diameter of the southern press-bed is 0.82 m, of the northern one 0.85 m, that of the southern vat 0.82 m and of the northern one 0.95 m. The depth of the southern vat is 0.80–0.84 m and that of the northern one only 0.65 m. The mortices on either side of the press-beds were re-cut at least once. At the first stage they were apparently of the central dovetail type but at the second they were apparently small square mortices that widened in one direction similarly to the mortice in the wine press but with the addition of an approach channel.

Bibliography

Davidson Weinberg, G. (ed.). 1988. *Excavations at Jalame, Site of a Glass Factory in Late Roman Palestine.*

Columbia.

HA 11 (1964): 15–17. "American Expedition to Study Ancient Glass in Israel" (Hebrew).

HA 16 (1965): 10–12. "The Expedition to Study Ancient Glass: Kh. Jalame ('Asafne)" (Hebrew).

HA 21 (1967): 13–14. "The Study of Ancient Glass" (Hebrew).

HA 25 (1968): 8–9. "The Expedition to Study Ancient Glass" (Hebrew).

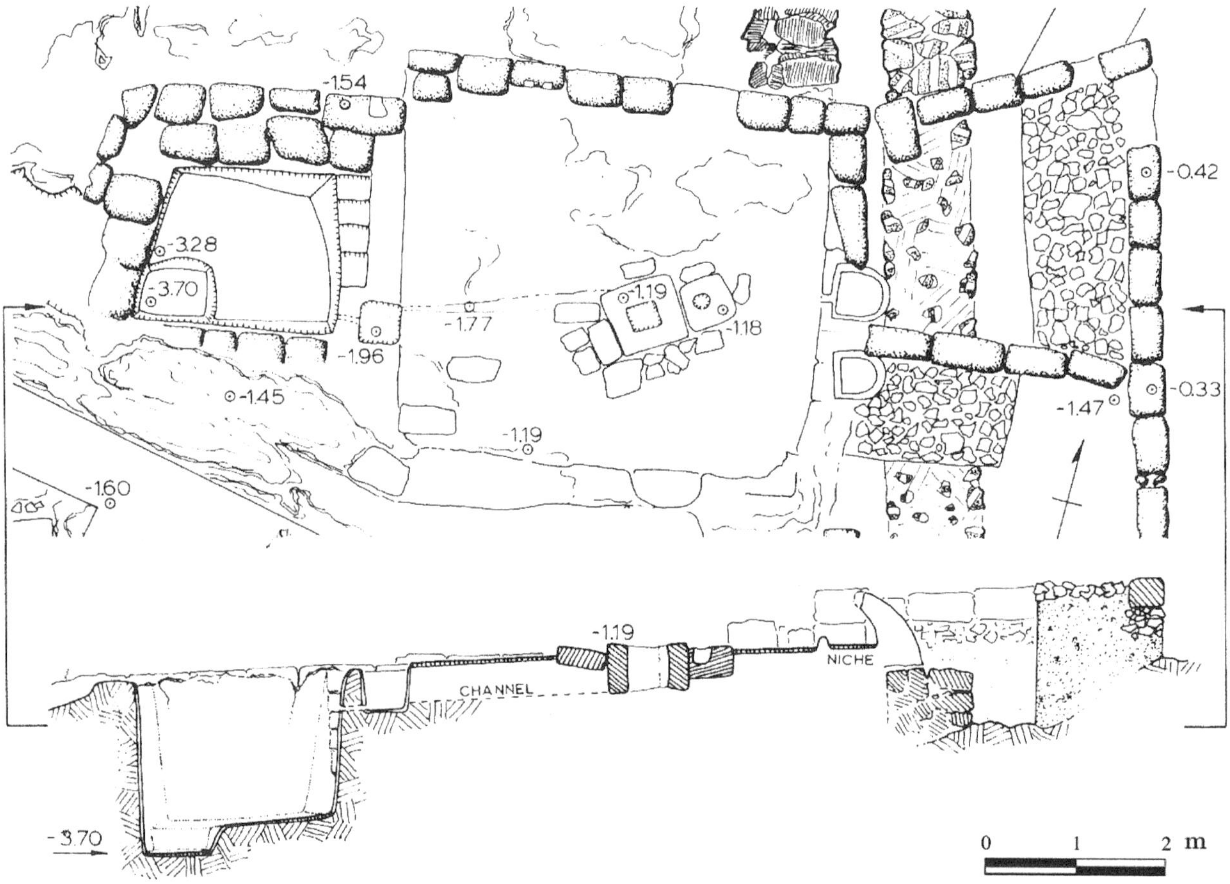

Fig. 20.1. Jalame, plan and section of the wine press.

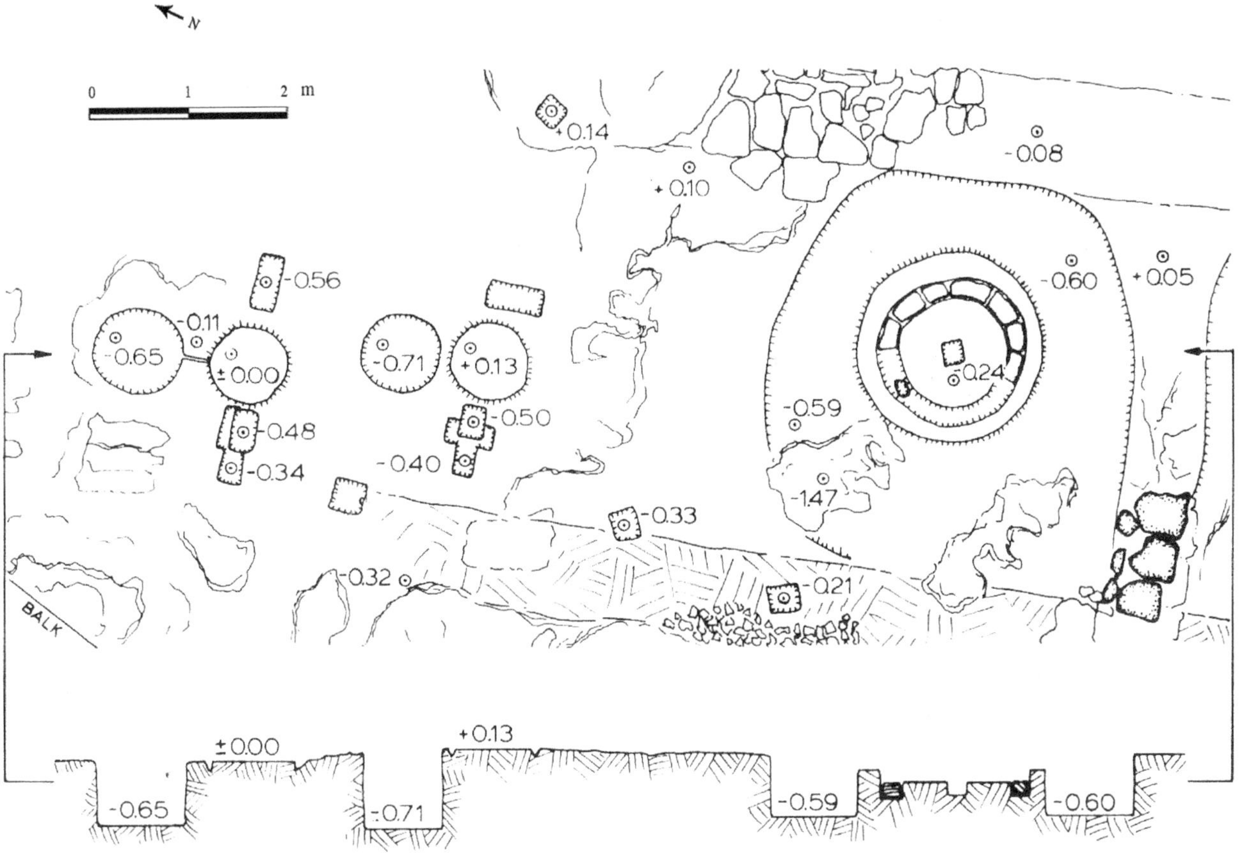

Fig. 20.2. Jalame, plan and section of the oil press.

21. An Early Byzantine Oil Press at H. Sumaqa

Shimon Dar

The Site

H. Sumaqa was a Jewish village of the Roman and Byzantine periods, situated on Mt. Carmel (MR 1539/2307). Its area was about 30 dunams, and excavations have been carried out there by the author.[1] The finds revealed by the excavations include a synagogue, dwellings, industrial and commercial structures, many installations (including a number of wine presses), reservoirs and burial caves (Dar 1999). Most of the dateable artifacts from the site are from the 3rd to the beginning of the 5th cent. CE.

The Oil Press

The oil press (Building 4), discovered on the southern edge of the residential quarter approximately 50 m from the synagogue, was excavated in 1984–1985.[2] A subterranean reservoir was found south-west of it, as was a cave that may have been used as a storehouse. North of the oil press and at a higher level was found a room measuring 5×6 m, and east of it, remains of additional walls. All of these were probably part of the oil press complex.

The oil press was not excavated in its entirety (Figs. 1–3). The building apparently comprised two storeys: the installations of the oil press were located on the ground floor, parts of which were hewn from the natural bed-rock, making a kind of cellar. The upper floor was not preserved, but the remains of a flight of stairs discovered outside, on the eastern side of the building (L. 64) hint at its existence. The external measurements of the complex are 4.5×10.0 m and the internal measurements of the press room (L. 62) are 4×8 m. The height of this chamber was about 3 m; the ceiling was supported by stone arches, the bases of which have survived. The floor was made of a layer of soft limestone, ground and well compacted; in the vicinity of the installations, it was coated with plaster.

The chamber that was excavated included only the pressing installation, of the lever and weights type. The crushing mill was probably – as was customary in complexes of this type – installed in a side room that has not yet been excavated. Likewise, the entrance to the building has also not been excavated.

The press includes the following components, from west to east:

- The niche for anchoring the back end of the beam was built into the western wall (W4) that was partially hewn from bed-rock – a factor that doubtless contributed to its stability. The niche, the square lower part of which measures 1.1×1.1 m, was built of large stones. Since the capstone was not preserved, however, the height of the niche is not known.
- A central collecting vat, typical of pressing installations in the centre and south of the country, was found cut into the bed-rock floor of the hall at a distance of about 1.6 m from wall W4. It measured 0.8×0.6 m and was 1.2 m deep. The vat was well plastered and had a sedimentation depression in the centre of its floor. Two small cuts in the lip of its aperture facilitated the flow of the expressed liquid into the vat.
- The cover that originally rested on top of the collecting vat was not found *in situ*. However, south of the collecting vat and the piers (see below) a broken piece of a press-bed was found; it measured 0.5×0.9 m, and was 0.3 m thick. It was set on a raised, plastered stone platform, apparently having been re-used as a "table" for convenience in performing various tasks such as filling the frails with olive mash.
- Two stone piers coated with light-colored plaster were found, one on either side of the collecting vat, 0.95 m apart. The northern pier was set adjacent to the wall of the room; it measured 0.9 m wide, 0.7 m deep and 1.85 m high. A slot measuring 0.10×0.25 m had been carved across the top of the pier, apparently for holding a wooden board. The southern pier measured 1.2 m wide, 0.4 m deep and 2 m high. On its southern (outer) side, a groove had been carved for almost its full height – again, apparently for holding a wooden rod. In light of the differences in dimensions between the two piers and the fact that their grooves are totally unrelated in size, shape and location, the piers found here were probably in secondary use. The difference in size, however, had no effect on their use here that was simply to keep the beam and frails from moving sideways. The plaster coating of the piers apparently is not original, but belongs to the later stage of their use, described here.
- The weights: two weights were found *in situ* along the beam line, 4.5 m and 5.5 m respectively from the western wall. They were located in a shallow (0.3 m deep) rectangular pit, hewn out of the bed-rock floor

expressly for this purpose. The pit was designed from the start to accommodate two weights, despite the series of three weights more common in presses of this type. The weights were in the shape of truncated cones, with a hole shaped like a reversed T cut into the top of each for tying the suspension rope. The western weight is 0.60×0.65×0.85 m and 1.1 m high. The eastern weight is 0.55×0.60×0.80 m and 1.07 m high.

- At the eastern end of the press room a compartment measuring 1.80×1.35 m and 1.2 m deep – i.e. a volume of 2.9 m^3 had been hewn from the bed-rock. Its walls and floor (in which there was a small sedimentation depression) were plastered. The compartment could have been used for storing olives before crushing, storing olive mash before a second pressing, separating the oil from the watery lees, or perhaps even for storing olive oil before it was packaged in ceramic jars.

Conclusions

The installation is a lever and weights press. The large conical weights with a hole in the top shaped like a reversed T have been used in oil presses continuously from the Hellenistic period (such as those in Maresha [Site 65]; see Kloner and Sagiv 1991: 61–68) up until recent times. Because they were too heavy to raise and lower by hand, and in light of ethnographic parallels from Arab oil presses (Avitsur 1994: 115–116), they were apparently operated with the aid of a winch. The winch was hung beneath the beam, and the rope was wrapped around it and tied to the weight. Turning the winch tightened the rope and pulled the beam downward, affecting the pressing. At a certain stage, as the rope continued to be tightened, the weights were lifted and hung from the beam. The distance from the back of the niche to the end of the weights pit – i.e. the length of the beam – was almost 7 m. The weights weighed approximately 1,200 kg, and in accordance with the principle of the lever, theoretical calculation shows that a pressure in excess of 7 tons could be exerted upon the olive frails.[3]

The stone piers were used both to support the stack of olive frails during pressing and to help hold the beam when it was lifted for purposes of changing the frails or drawing the expressed liquid from the collecting vat. A comparable installation, with similar piers and weights, and with vats or tanks for storing oil, was discovered in a Byzantine (?) oil press in a cave at Tell Abu-Shusha (Geva) near Kibbutz Mishmar Ha-'Emeq, not far from Sumaqa (Giveon 1988: 215–218).

The finds unearthed within the limits of the oil press[4] included ceramic vessels, many glass vessels, and coins of the 4th and early 5th cent. CE. Also discovered were broken pieces of stone vessels typical of Jewish settlements, and a lead weight decorated with a design consisting of a double trident. The mass of the lead weight was 308 gr., close to that of the Roman *libra* (pound) that weighed 328 gr. (Manns 1984: 13, No. 1; 40). All of the foregoing attest to the time span during which the oil press at Sumaqa was in operation. Some charred olive pits were also found in the press; many others were found in Complex 2, north of it, that was probably used for storage.

Notes

1. Azriel Siegelmann, Yigael Ben Ephraim and Johanan Mintzker assisted in conducting the dig that was carried out from 1983–1995 on behalf of the Department of The Land of Israel Studies at Bar-Ilan University. Anat Peled and Nachum Sagiv supervised the excavation of the oil press.
2. Additional oil press facilities were found in Building 31, north of this installation, and broken crushing stones were discovered in Complex 6.
3. The calculations were performed by Engineer Moshe Levy of Kibbutz Ma'abarot, for which many thanks.
4. The pottery from Sumaqa was prepared for publication by Dr. Sean Kingsley, the glass vessels by Gusta Lehrer Jacobson, and the coins by Dr. Arie Kindler (the latter two from the Eretz Israel Museum, Tel Aviv).

Bibliography

Avitsur, S. 1994. "Olive Oil Production in the Land of Israel, Traditional to Industrial". In: R. Frankel, S. Avitsur and E. Ayalon. *History and Technology of Olive Oil in the Holy Land.* Arlington and Tel Aviv: 91–157.

Dar, S. 1999. *Sumaqa, A Roman and Byzantine Jewish Village on Mount Carmel, Israel* (BAR IS 815). Oxford.

Giveon, R. 1988. "Area I – An Oil Press and Domestic Buildings". In: B. Mazar (ed.). *Geva, Archaeological Discoveries at Tell Abu-Shusha, Mishmar Ha-'Emeq.* Tel Aviv: 215–218 (Hebrew).

Kloner, A. and Sagiv, N. 1991. "The Technology of Oil Production in the Hellenistic Period at Maresha, Israel". In: M.-C. Amouretti, J.-P. Brun and D. Eitam (eds.). *Oil and Wine Production in the Mediterranean Area from the Bronze Age to the End of the XVIth Century* (*Pré-Actes.* Aix-en-Provence-Toulon: 61–68.

Manns, F. 1984. *Some Weights of the Hellenistic, Roman and Byzantine Periods.* Jerusalem.

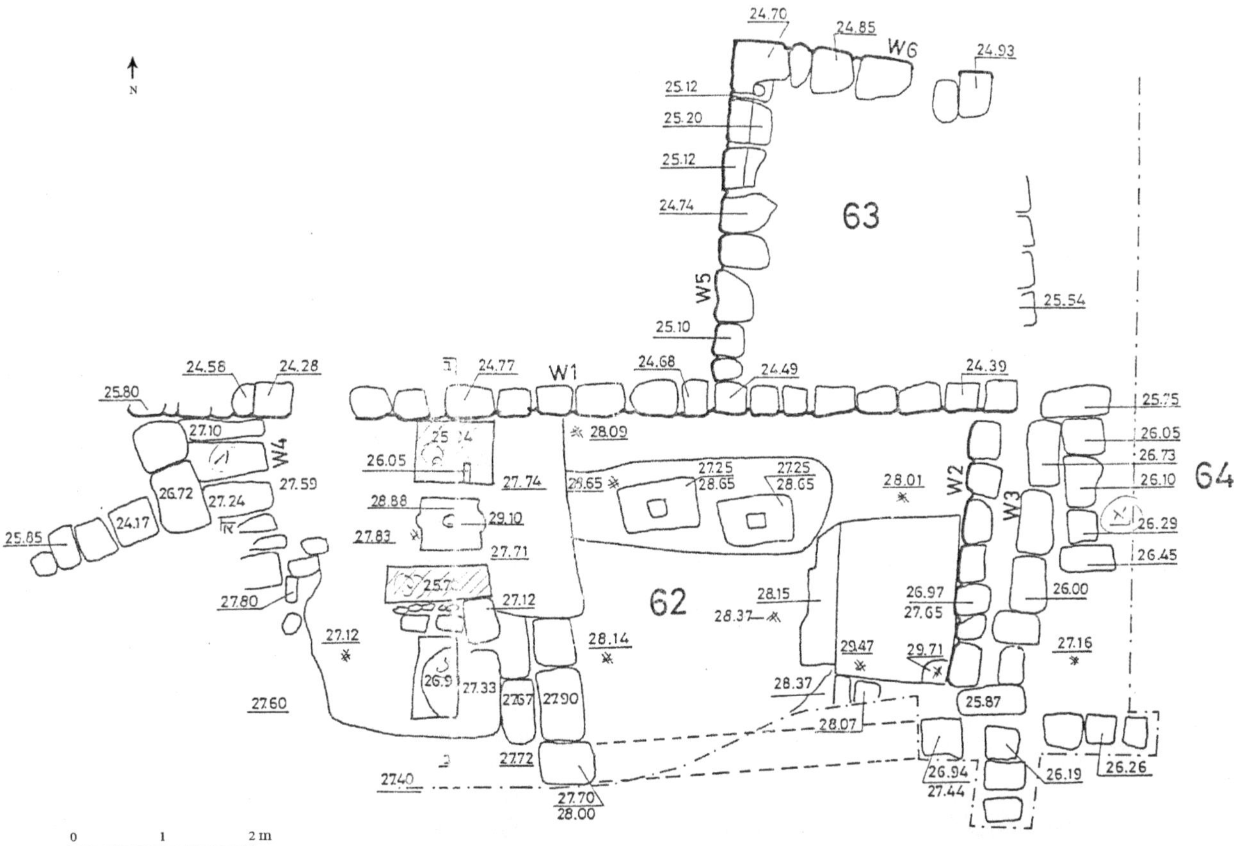

Fig. 21.1. Sumaqa, plan of the oil press.

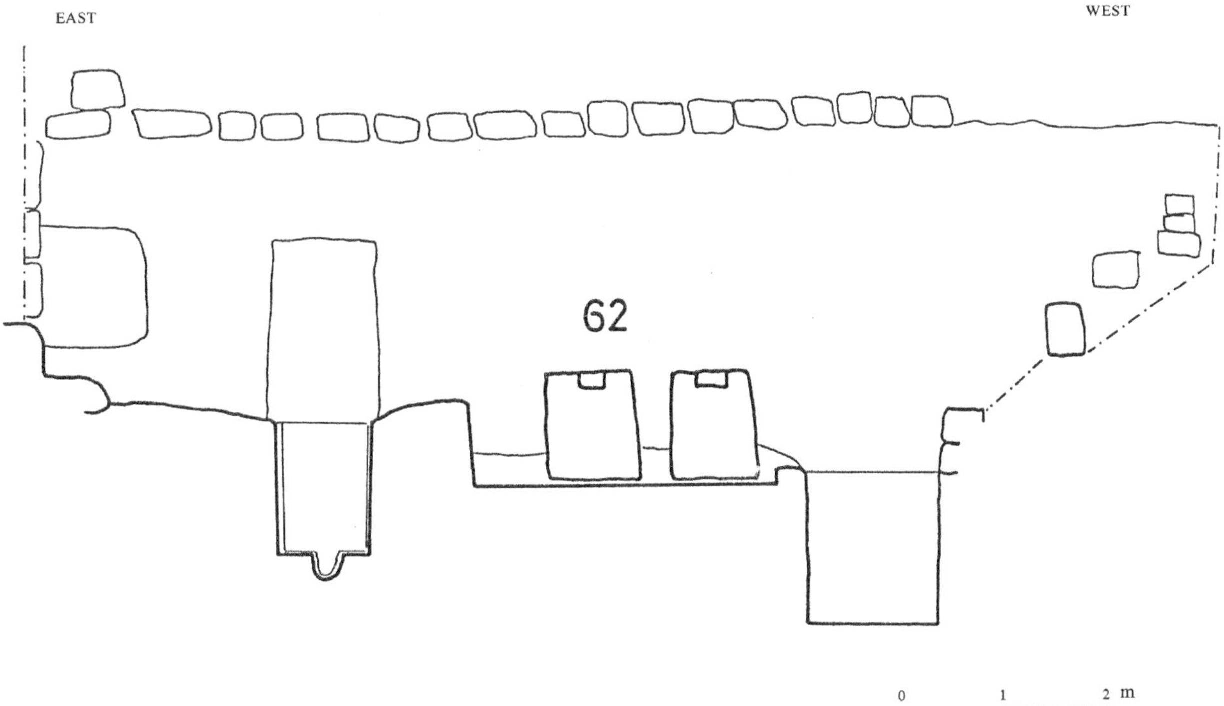

Fig. 21.2. Sumaqa, section of the oil press.

Fig. 21.3. Sumaqa, the oil press, view to the west.

22. The Oil Press from Tel Safsafot

Rafael Frankel

The Site

Tel Safsafot is a tell in Lower Galilee between Mt. Tabor and Giv'at Ha-Moreh (MR 1869/2276). Sherds from all periods from the Chalcolithic to the Byzantine have been found at the site and it has been identified with the Biblical En-Dor (ISam. 28: 7; Ps. 83: 11; Josh. 17: 11). The only excavation to have been carried out at the site is that of August 1987 in which the oil press was uncovered by the author (Frankel 1988–1989).

The Oil Press (Figs. 1–2)

The press is to the north side of the tell and the entrance is to the north. The crushing mill was in the front area of the building that was paved with stone slabs. A line of slightly protruding stones divided the front area from that in the back in which stood three direct pressure screw presses, one parallel to each of the walls – west, south and east. The back area was paved in white mosaic with a black and white design in the centre consisting of a circle within a square. There were remains of two arches that served to support the roof, one placed between the crushing and pressing areas and one between the southern press and the other two.

The crushing basin stood to the left of the entrance to the building and consisted of two parts, each of a different type of stone: the round flat crushing surface of finely ground basalt with sunken square socket in the centre, and the rim and wall of the basin of eight neatly fitting suitably shaped limestone segments.

The three presses were almost identical each consisting of two grooved piers, one on either side of a central collecting vat (Fig. 3).

The western pier and the vat cover of the southern press were robbed in late antiquity. The robber's pit full of large boulders was clearly discernable in the baulk and floor. It also destroyed part of the southern wall of the building and a section of the floor.

The collecting vats, sunken into the floor, were each cut in one piece. The vat in the southern press was rectangular, while those in the other two presses were round. These were closed by a stone cover with a round hole in the centre about 0.4 m in diameter. In the eastern press this cover was made up of two stones, in the western press of one, while the cover of the southern vat had been robbed in ancient times. In both the eastern and western presses there are small holes to the back and front of the cover that connect to the vat. The small grooves on the rim of the southern vat are evidence for the existence of similar holes beside the missing cover of this vat also.

The northern pier of the western press was made in two sections, probably held together by a wooden peg inserted into holes in their contiguous surfaces. The upper section was found but not *in situ* and was returned to its original position.

The piers of all three presses, to the extent that they survive, are similar in character but in each case the left-hand pier was slightly different from the two surviving right-hand piers that, in turn were also identical. The piers were cut from a third type of stone, a limestone-breccia incorporating basalt fragments.

The outer face of all the piers was smooth, whereas there was a vertical groove, semicircular in section, on each inner face (Fig. 4). Below this groove there was a small hole in the floor. On the anterior and posterior faces were dovetail mortices that were connected at the lower ends by a horizontal bore. The dovetail mortice on the left-hand pier was high and that on the right-hand low. The piers were not sunken in the floor and in some places stones had been placed underneath to support and align them.

The Safsafot press functioned in a manner similar to that of the Judean Grooved Pier Press (see chapters 66 and 72 in present volume). The mortices and bores secured a threaded board that was fixed above the piers and in which a female thread was cut (Fig. 5). A screw that was turned in this thread pressed on a pressing board that in turn pressed frails full of olive pulp that had been previously crushed in the crushing basin and then placed on slats or on a perforated board above the openings in the vat cover. The vertical grooves on the inner faces of the piers held rods, the top ends of which were secured in holes in the thread board and the bottom ends in the small holes in the floor. Their function was to stabilize the press and primarily to resist lateral pressure.

The Safsafot press is very similar to the standard Judean Grooved Pier Press:

1. There are two piers on either side of a central vat.
2. On each pier on two opposite faces there are mortices

connected by a horizontal bore and on other faces vertical grooves.

The Safsafot press is, however, different from the Judean press in several ways:

1. In the Judean press there are vertical grooves on both the anterior and posterior faces, on the Safsafot press there is a vertical groove on the inner face only.
2. In the Judean press the mortices are on the lateral faces, on the Safsafot press theyare on the anterior and posterior faces.
3. The Judean mortices are square, those of the Safsafot press dovetailed.
4. The Judean mortices are all the same length while in the Safsafot press the right- hand one was in each case longer than the left-hand one.

There are many examples of the Judean press and it is the dominant screw press of Judea. The Safsafot press, however, is at present unique. Therefore it is to be regarded as a variant of the Judean press and the differences are to be explained in those terms.

We suggest that the use of a dovetail mortice at Safsafot is to be explained in that this was the mortice typical of Galilee just as the square mortice was typical of Judea.

The other differences are technical, however, and we suggest that the mortices were moved from the lateral faces to the anterior and posterior faces in order that these mortices and the bores that connected them could serve a secondary function as well as the basic one of securing the threaded board. This secondary function was almost certainly to hold drums or capstans to turn the screw and thus provide additional mechanical advantage. The higher drum would be used when the frails of olive pulp were still full and the lower one when they had been partly pressed and the height of the pile of frails had decreased. Capstans of this type are common in pre-industrial screw presses (Humbel 1976: Pls. XXIII, XXIV, XXIX). They are not mentioned in ancient written sources nor have archaeological remains been explained in this way. Capstans were, however, known at the time these presses were in use and the mechanical advantage they provide understood. It is also probable that a stone with a mortice found next to the bases of two screw presses at Dabussiya in the Golan (Ben-David 1998: Fig. 35) served the same purpose.

Chronology

No complete vessels were found on the floor but the complete skeleton of a dog was found in one of the vats suggesting it was deserted and became a rubbish dump. The latest sherds actually on the floor are from the Early Arab period. These show when the press was deserted but it is probable that it was built earlier, probably in the Byzantine period.

Conclusions

We have shown the Safsafot press to be a very sophisticated variant of the Judean Grooved Pier Press. The press was carefully planned using two types of flooring and three types of stone. The arches that supported the roof were placed in such a way as not to interfere with the work. The technical changes from the original prototype made in the press itself were even more unusual, involving basic changes in the structure of the press in order to attach an additional device that improved its mechanical capacity.

Bibliography

Ben-David, H. 1998. "Oil Presses and Oil Production in the Golan in the Mishnaic and Talmudic Periods". *'Atiqot* 34: 1–61 (Hebrew).

Frankel, R. 1988–1989. "An Oil Press at Tel Safsafot". *Tel Aviv* 15–16: 77–91.

Humbel, X. 1976. *Vieux pressoirs sans frontières*. Paris.

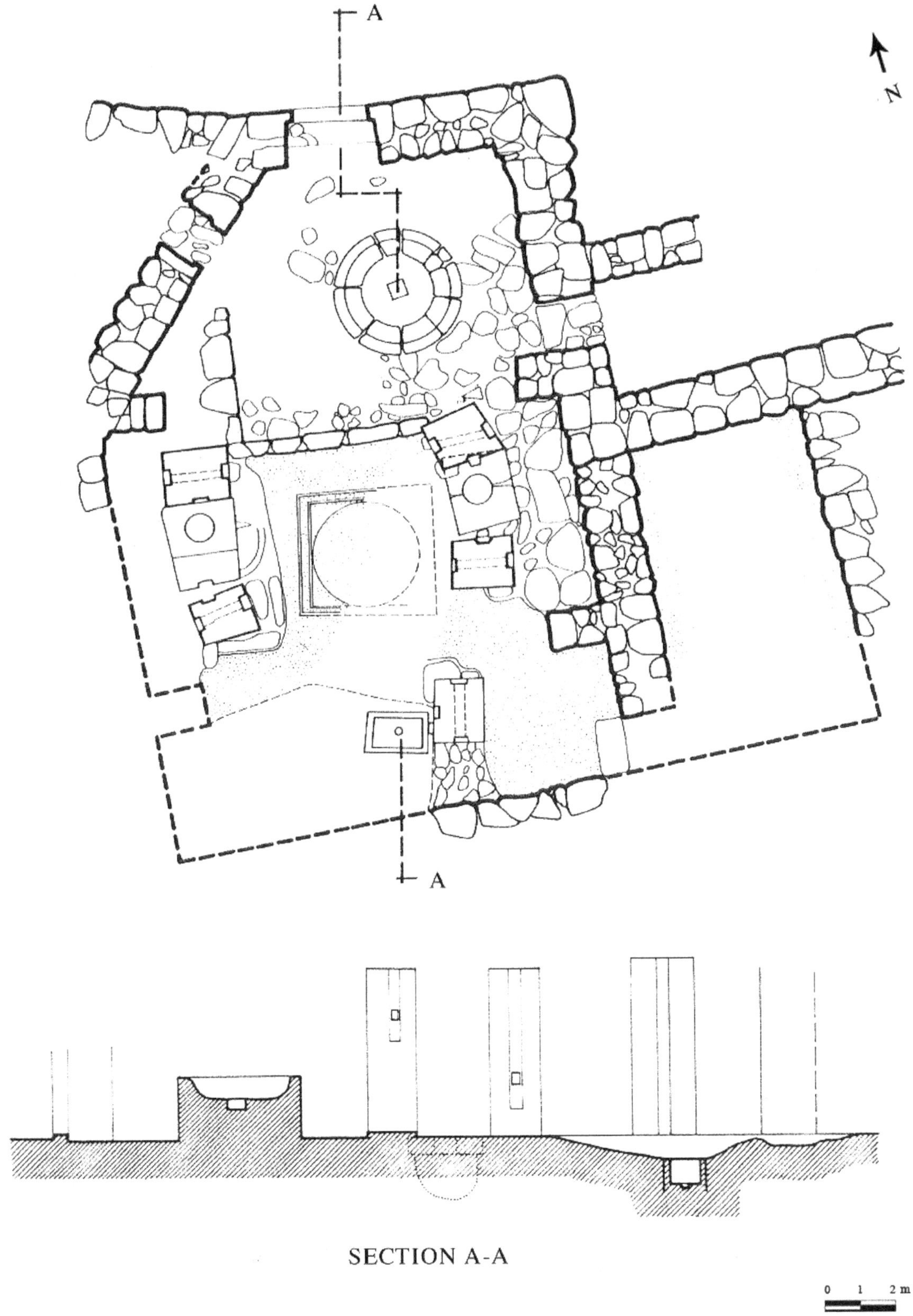

Fig. 22.1. Safsafot, plan and section of the oil press.

Fig. 22.2. Safsafot, the oil press from north-east.

Fig. 22.3. Safsafot, the eastern press and remaining pier of the southern press, looking east.

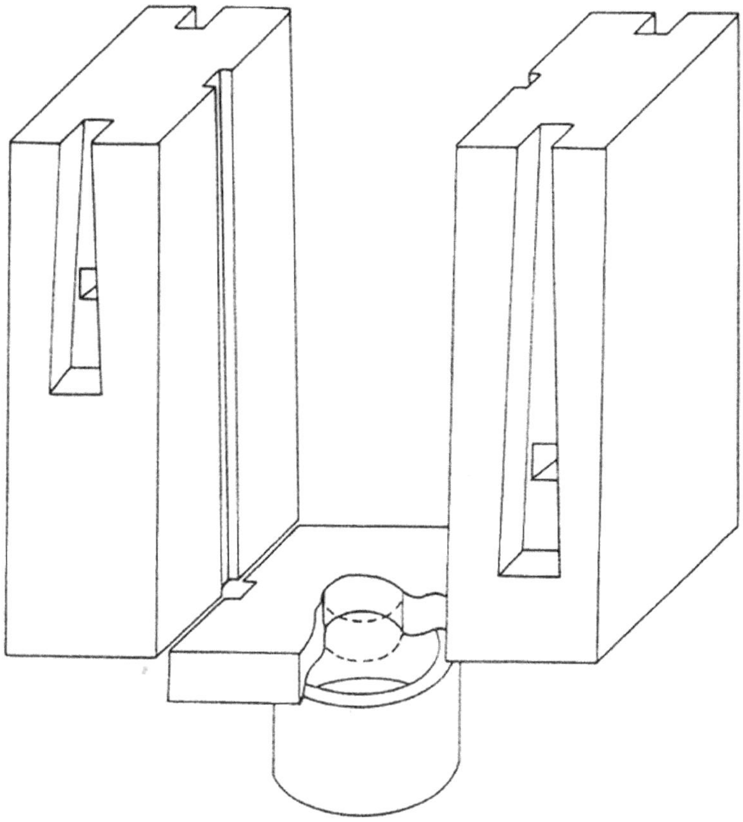

Fig. 22.4. Safsafot, grooved pier press.

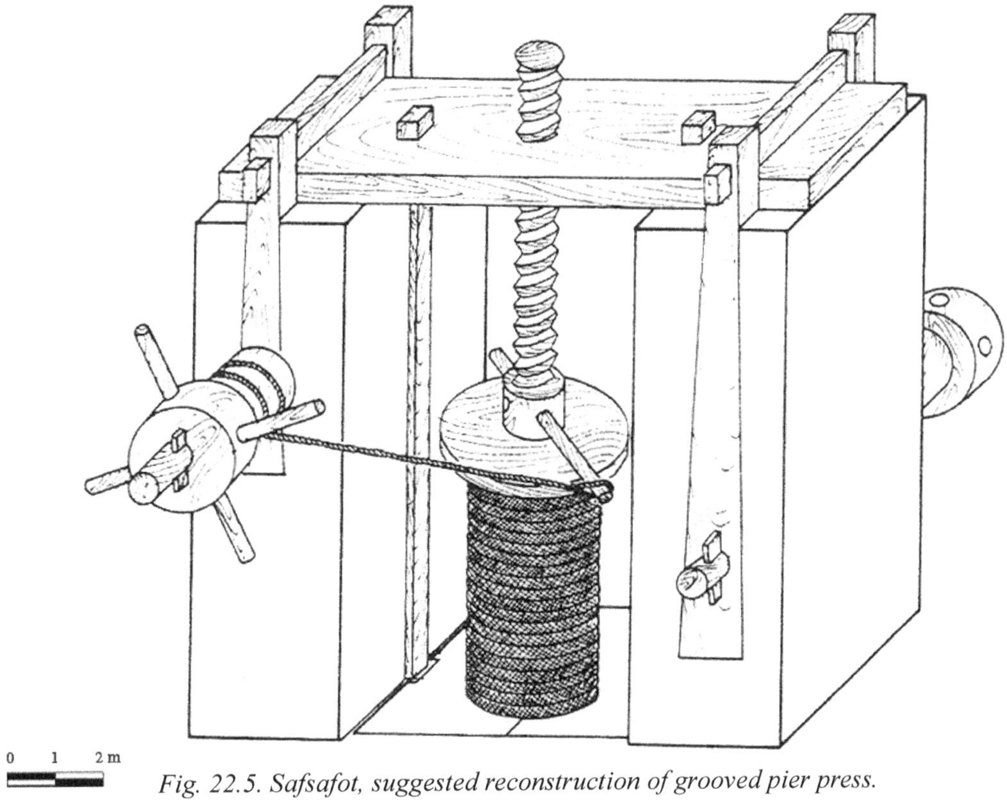

Fig. 22.5. Safsafot, suggested reconstruction of grooved pier press.

23. An Oil Press at Nahal Haggit

Jon Seligman

The Site

The site (MR 1539/2245) is set on a hill between Nahal Haggit (Haggit Stream) and Nahal Tut (Tut Stream), 4 km north-east of Bat Shlomo and commanding a good view over the Zichron Ya'acov–Yokne'am road. This was probably also a major route in ancient times. Excavations were conducted here during 1993 due to the construction of power station[1] (Seligman 1997; forthcoming).

Three levels of settlement were uncovered, the oil press belonging by the coins and pottery to the second level that is dated to the 2nd-early 4th cent. CE. The remains of the site during this period consist of a square enclosure wall (90×90 m) enclosing a number of separate residential units, rooms and an oil press complex at the centre of the site. While this site seems to be an agricultural complex it clearly shows central organization and may well be a settlement of Roman army veterans.

The Oil Press

The oil press consisted of large crushing basin, two separate pressing installations of different types and a number of separation vats (Figs. 1–2).

The crushing basin was cut from a single block of limestone with a diameter of 2.2 m. It had a concave profile and raised socket.[2] The basin was set on a base of stones raising its surface 1 m above the bed-rock floor. No crushing stones were found though we suggest that these probably had rounded edges to match the opposing profile of the basin surface.

The use of a friable limestone probably meant that the olive pulp resulting from the crushing contained grains of the rock that certainly reduced its quality. The working surface of the basin was heavily worn, showing a long period of use.

10 m east of the crushing basin was the first of two pressing units (Fig. 3). All its components were also carved in the bed-rock. The press-bed was an oval shaped stone slab 0.8–1.0 m in diameter with radial grooves leading to a hole in its centre.[3] The slab was set above a pear-shaped collecting vat some 0.6 m in diameter and 0.6 m deep. No evidence of press piers was found in the vicinity of the press-bed. To the west was an elongated pit with tapered sides, 3.1 m long and 0.7 m deep. This pit contained the weights that were hung on the beam. Two such weights were found (1.3×0.95 m and 0.3–0.4 m thick), one alongside the oil press and one at some distance from it (Fig. 4). The weights were slabs of rock with a small perforation on the top by which a rope attached them to the beam. At the western end of the weight pit were two small sockets in the bed-rock in which were fixed uprights that prevented the beam from moving from side to side during operation. This was a lever and weights press, the beam probably anchored in the wall east of the press-bed and weighted down by stone weights that were hung on it.

The second pressing installation was of a different type (Fig. 5). It was located 3.5 m north of the crushing basin. All that remained of the press were the parts of the installation that were hewn in the bed-rock. The press-bed consisted of a flat area of bed-rock with a diameter of 0.95 m, surrounded by a shallow channel. The channel branched to the south towards a small lateral collecting vat cut into the bed-rock (diameter 0.4 m, depth 0.5 m). The collecting vat was covered by a stone with a hole in its centre that was mortared to the bed-rock. On either side of the press-bed were key-hole shaped central dovetail mortices hewn into bed-rock.[4] Wooden posts were fixed by placing them in the round holes and were then slotted into the narrower rectangular part. It is clear that this was a direct frame screw press. The posts supported a wooden screw that provided the pressure on the frails of olive pulp placed on the press-bed.

In the immediate area of the presses were three round pits. Each was plastered to prevent the seepage of any liquid stored within. Large broken amphorae of Spanish origin were found in the pits. We suggest that these were vats in which the oil was separated from the watery lees.

Discussion

The crushing basin with a concave profile is extremely uncommon in Israel. From the many examples of crushing basins listed by Frankel (1984: 31–35; 1999: 68–72) only a very limited number have this profile. Two examples of basins with a concave profile and a raised central socket are listed: Kh. Clil (Frankel 1984: 332) and Kh. Hamoud (Ibid.: 334). To this should be added the basin from Khorazim (Yeivin 1966: 58). Most crushing basins in Israel have a flat crushing surface, though it must be noted that in the Aegean and Italy the concave profile was predominant and is found from the Hellenistic period onwards (Frankel 1984: 124; 1999: 73). The crushing stone at Nahal Haggit was probably not convex, as in many Aegean examples, but rather had the usual wheel shape with rounded edges.

The lever and weights press was common from the Hellenistic period until the modern day. The press-bed with radial grooves and central collection has no clear regional distribution and this type is known from a number of sites.[5] The nearest known examples are from Meggadim, Shalallah and Bet She'arim (Frankel 1984: 38–39; 1999: 85).

Pliny (317, 74, 18) notes direct frame presses in the 1st cent. CE and it is thus clear that this type existed outside Israel during his time. The presses in Israel are dated mainly to the Byzantine period (Frankel 1984: 178). Indeed Frankel (1999: 117) is most forthright on this chronological foundation and states that "...no screw weight has been recorded that dates to the Roman period...". Gichon (1979–80: 232–241) suggests that the related upright pillar press is associated with military settlement from the 2nd cent. CE. The very organized nature of the Nahal Haggit enclosure suggests a military connection and this may strengthen the argument and explain the early date for our press. It should be noted, though, that Gichon's evidence for this connection and dating is circumstantial and not stratigraphic.[6] The earliest press in Israel of this general type from a stratigraphic context is found at Khorazim and is dated to the 2nd-4th (Yeivin 1966: 52) or to the 3rd cent. CE (Ibid.: 6*). Of our specific type the earliest example known previously is from Kh. Jalame (Site 20) and is dated from 350 CE (Davidson-Weinberg 1988: 15). Thus the direct frame press found at Nahal Haggit is the earliest of its specific type found till now in Israel.

This type of press has a wide distribution (Frankel 1984: 65–66, 175; 1999: 130–137) though the specific type with a central dovetail frame socket is known from a limited number of sites: two oil presses at Jalame (Davidson-Weinberg 1988: 15–16, Fig. 2–5), Mishkena (Frankel 1984: 65, 371; 1999: 135–136) and er-Rama (Frankel 1984: 65, 371; 1999: 135 [Site 9]). All these sites are in the Galilee, Jalame located not far from Nahal Haggit. It should be noted that the central dovetail socket is often used to fix single screws of wine presses, especially in the north of the country (Frankel 1984: 176; 1999: 144–145).

A number of questions arise from the existence of two pressing units of different types in the same pressing complex. Are the two units contemporary? Certainly they belong to a single layer that ceased its existence in the early 4th cent. CE. If they were built at the same time then why use two different pressing technologies? As Frankel (1984: 178) notes there are other examples of different pressing technologies found together in a number of sites and he suggests that the beam and weights press was used for initial pressing while the screw press was utilized for secondary processing.

It has been suggested that central and lateral collection of the expressed liquid was a regional feature, lateral collection being predominant in the north and central collection being a feature of Judaea (Frankel 1996: 205). The lever and weights press at Nahal Haggit displays a southern characteristic as the collection vat is central while the collection in the screw press is lateral. Thus interestingly, both collection types appear together, a fact that needs further explanation.

The direct frame press found here is out of its usual chronological framework. It is clear, as stated above, that the press went out of use in the 4th cent., thus making it the earliest known press of this type in the country.

Notes

1. The excavation at Nahal Haggit was conducted throughout 1993 under the auspices of the Israel Antiquities Authority (permits 1935 and 1984). The excavation was directed by the author with the assistance at various times of: Bella Davidson, Iskander Jabbour, Maha Derouche, Camil Sari, Nurit Shaharon, Gideon Surger, Samuel Wolff and Vladimir Zbenovich.
2. The crushing basin with a concave profile and a raised central socket is defined by Frankel (1984: 34, 1999: 69–70) as type 3.3.2.
3. The press-bed with radial grooves and central collection is defined by Frankel (1984: 39, 130) as type 4.1.8.2.
4. The press-bed with two closed frame mortices is defined by Frankel (1984: 17, 65, 1999: 135–136) as type 7.4, of which ours is of sub-type 7.4.3 that has key-hole shaped mortices, a sub-type named the Rama and Mishkena Presses.
5. Frankel (1984: 39) lists eight examples of this type of press-bed from all over the country.
6. Indeed Kloner (1989: 74) gives a 4th cent. CE date for the same pillar presses.

Bibliography

Davidson-Weinberg, G. (ed.). 1988. *Excavations at Jalame, Site of a Glass Factory in Late Roman Palestine.* Columbia.

Frankel, R. 1984. *The History of the Processing of Wine and Oil in Galilee in the Period of the Bible, the Mishna and the Talmud.* Unpublished Ph.D. Thesis. Tel Aviv University (Hebrew).

Frankel, R. 1996. "Oil Presses in Western Galilee and Judaea – A Comparison". In: D. Eitam and M. Heltzer (eds.). *Olive Oil in Antiquity, Israel and Neighbouring Countries from the Neolithic to the Early Arab Period.* Padova: 197–218.

Frankel, R. 1999. *Wine and Oil Production in Antiquity in Israel and Other Mediterranean Countries.* Sheffield.

Gichon, M. 1979–80. "The Upright Screw-Operated Pillar Press in Israel". *SCI* 5: 206–243.

Kloner, A. 1989. "The Upright Pillar Press in Caves in Judaea". *NZ* 15: 66–74 (Hebrew).

Pliny the Elder, 1958–1963. *Naturalis Historiae* (10 vols.) (Trans. H. Rackham). London and Cambridge.

Seligman, J. 1997. "Nahal Haggit". *ESI* 16: 61–63.

Seligman, J. Forthcoming. "Nahal Haggit – A Roman and Medieval Farming Community in the Southern Carmel". '*IAA Reports*. Jerusalem.

Yeivin Z. 1966. "Two Ancient Oil Presses." '*Atiqot* (HS) 3: 52–3 (Hebrew).

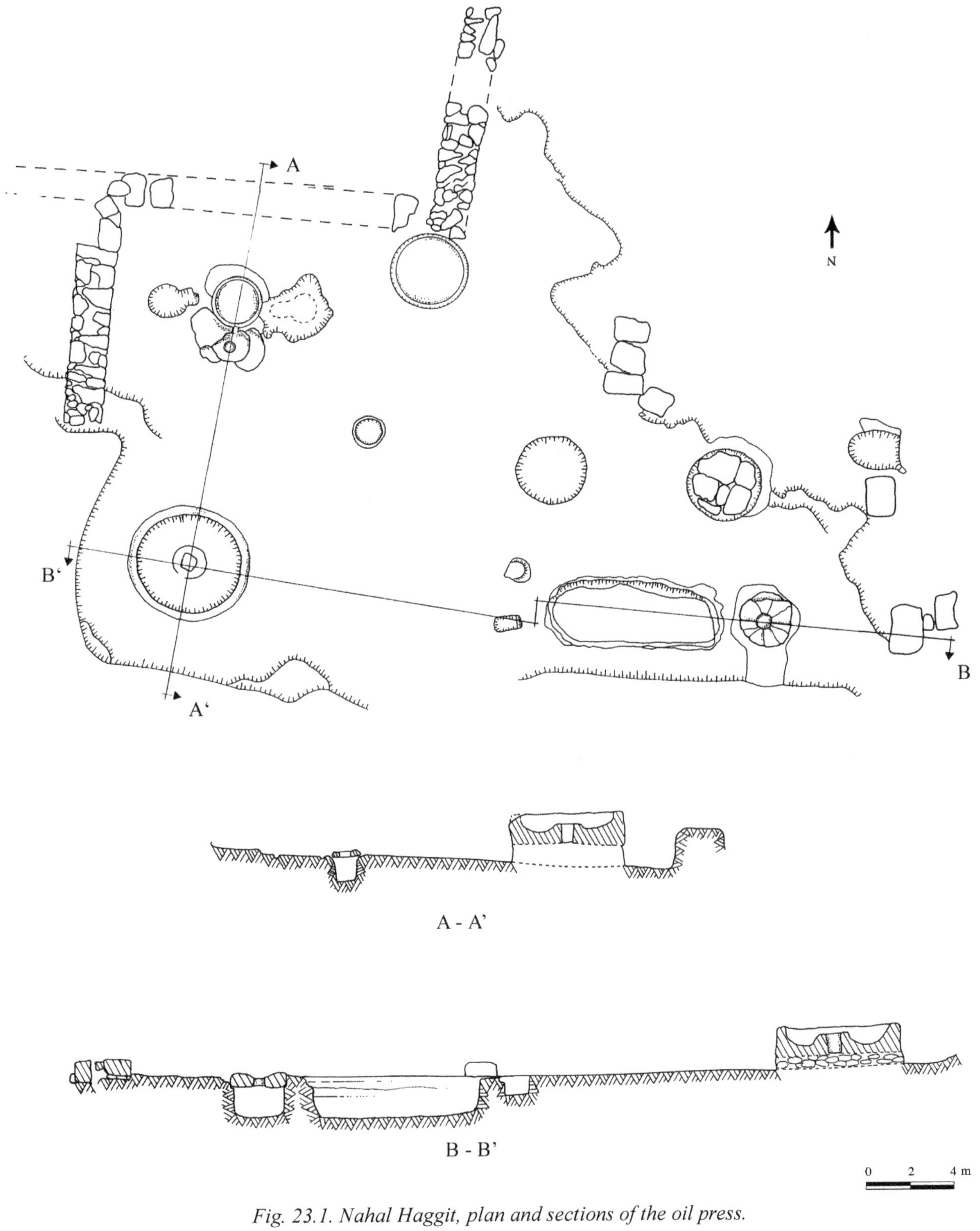

Fig. 23.1. Nahal Haggit, plan and sections of the oil press.

Fig. 23.2. Nahal Haggit, general view of the oil press looking south.

Fig. 23.3. Nahal Haggit, the oil press, view to the west.

Fig. 23.4. Nahal Haggit, oil press weight.

Fig. 23.5. Nahal Haggit, the northern oil press, view to the south.

24. A Wine Press at Bet Ha-Shitta

Zvi Gal

The wine press is located on a basalt hill-side near Kibbutz Bet Ha-Shitta on the northern fringes of the Harod Valley, 10 km west of Bet She'an (MR 1918/2175) (Fig. 1).[1] The wine press is far from any known site, the closest being an isolated 6th cent. CE monastery about 500 m to the south (Aharoni 1954). The wine press had a long life span during which some changes took place.

The Early Phase (Fig. 2)

At this initial stage the press was constructed having a 42 m^2 treading floor paved with a white simple mosaic (Fig. 2: 1), of which half has been destroyed by a bulldozer. The treading floor was built on sloping ground so that its northern and eastern walls were cut into the basalt bed-rock while the other two were free-standing walls, built of field stones. The walls were well plastered. In the eastern wall there were two niches (Fig. 2: 2). Near the south-western corner of the treading floor there was shallow rock-cut depression (Fig. 2: 3) in which fragments of white material was found. This was probably marl that was spread on the must in order to settle the remains of grape skins and other solids.

In the centre of the western wall of the treading floor was an outlet through which the must flowed into a settling vat (Fig. 2: 4). This was a rock-cut underground irregularly shaped and well plastered tank. From there the must flowed to a rock-cut underground and plastered bell-shaped collecting vat 5 m^3 in capacity (Fig. 2: 5). A small sump was cut in the floor of the vat to facilitate in removing the last remaining must (Fig. 2: 6). The vat was approached from the north by series of stairs cut in the rock and plastered (Fig. 2: 7).

The must was collected into jars that were transferred via another series of rock-cut plastered stairs (Fig. 2: 8) to a storage cave also cut into the basalt rock and plastered (Fig. 2: 9).

The first stage of the wine press terminated when the ceiling of the underground collecting vat collapsed. Some Pottery vessels and coins were found in the debris and on the floor of the vat, dating the end of this phase to the first half of the 4th cent. CE.

The Later Phase (Fig. 3)

The wine press of this phase continued to use the original treading floor. Near the western side of the floor a rock-cut rounded tank was found (Fig. 3: 10), that probably held water for washing the feet of the workers who trod the grapes. A new plastered pipe led the must from the treading floor to a small square settling vat (Fig. 3: 11) from which the must flowed into a new collecting vat (Fig. 3: 12) that replaced the former vat that had collapsed. The new vat consisted of the cavity of the stair-case that had led to the underground collecting vat of the early phase. The new square open vat had a capacity of 5.5 m^3. Its southern wall was built and narrow plastered steps led down into it (Fig. 3: 13). In the floor of this vat there was also a small sump. The vats of this stage were paved with white mosaics. The stairs that led to the storage cave continued in use during the later phase. A beautiful relief of a palm-tree was made on the new layer of plaster to the right of the entrance to the cave (Zori 1977: 85). A coin found in the new vat dates this phase of the wine press to the end of the 5th or the beginning of the 6th cent.

After the wine press had been abandoned the storage cave was converted into a water cistern by blocking the entrance and opening a hole in the ceiling. The pottery associated with this phase shows that this change took place during the 15th-16th cent.

The various parts of the wine press make it possible to suggest the different stages of production. Before treading the workers washed their feet using water from the tank (Fig. 3: 10). Then the grapes were trodden on the floor. The two niches in the eastern wall of the treading floor (Figs. 2: 2, 3: 2) show that after treading the grape skins and stalks were pressed using beam presses anchored in the niches. The must flowed through the settling tank where marl that had been stored in a tank (Figs. 2: 3, 3: 3) was spread on it so that the solid matter settled, and from there to the collecting vat from which the jars were filled and then stored in the cave.

The large dimensions of this wine press show that it served either many vineyards or a large central (royal?) vineyard. The results of surveys conducted in the region (Gal 1986) as well as the vintage scenes showing the various stages of wine production that appear on the mosaics of the Lady Mary monastery at Bet She'an (Fitzgerald 1939) show that wine production played an important part in the agriculture of the region during the Roman-Byzantine period.

Notes

1. The wine press was excavated by the author in 1975 on behalf of the Israel Department of Antiquities.

Bibliography

Aharoni, Y. 1954. "Byzantine Monastery near Beit Hashita". *BIES* 18: 209–215 (Hebrew).

Fitzgerald, G.M. 1939. *A Sixth Century Monastery at Beth Shan*. Philadelphia.

Gal, Z. 1986. "Vineyard Cultivation at 'Emek Harod and its Vicinity During the Roman-Byzantine Period". *Israel – People and Land, Haaretz Museum Yearbook* 2–3 (20–21): 129–138 (Hebrew).

Zori, N. 1977. *Nahalat Issachar*. Jerusalem (Hebrew).

Fig. 24.1. Bet Ha-Shitta, the wine press, view to the north.

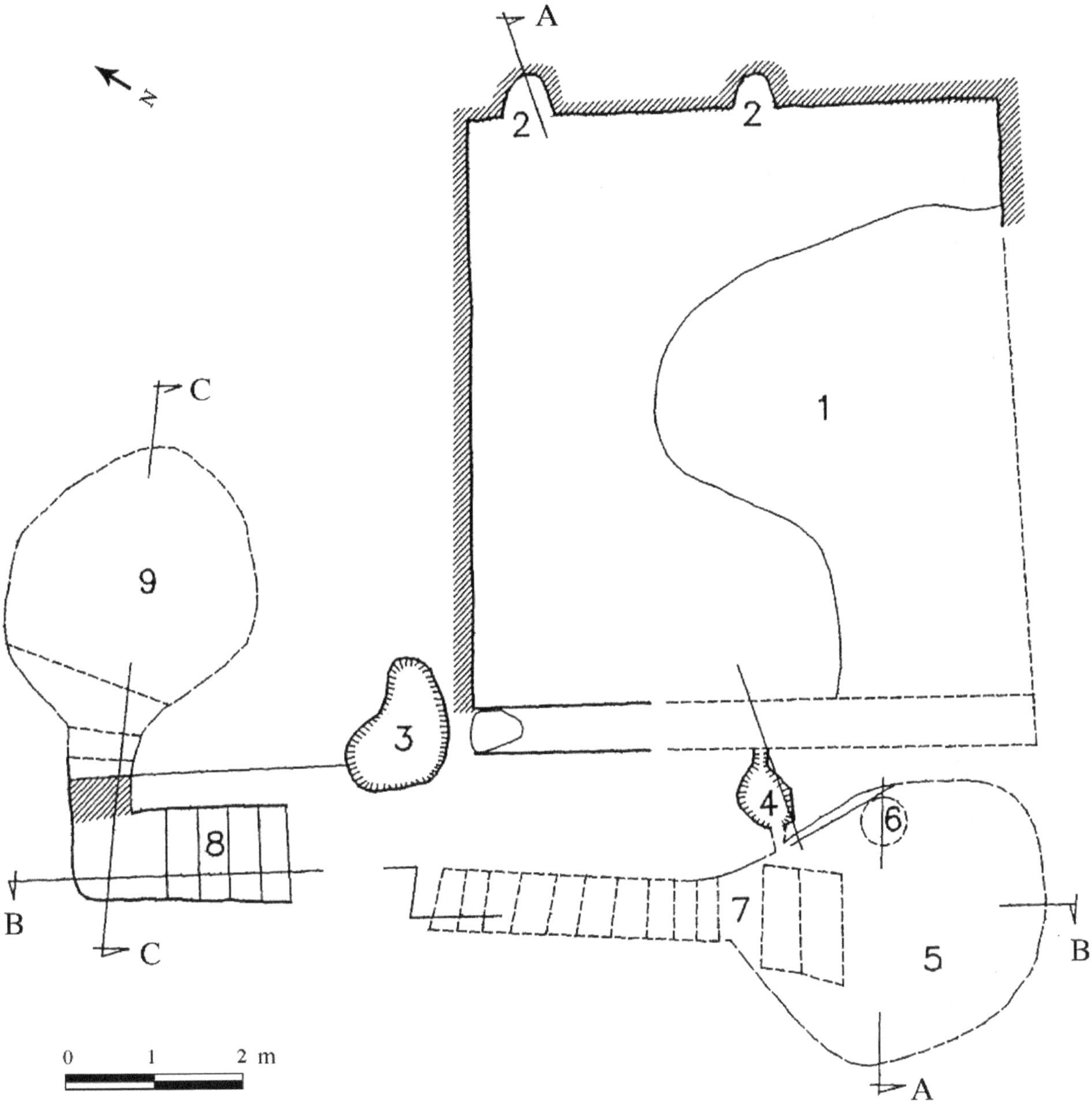

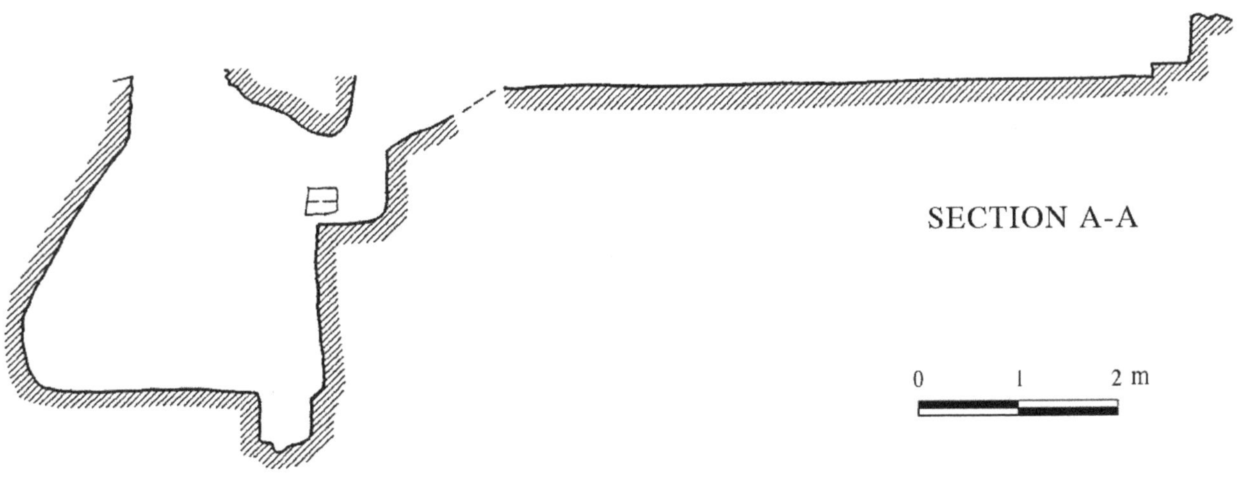

Fig. 24.2. Bet Ha-Shitta, plan and sections of the early phase of the wine press.

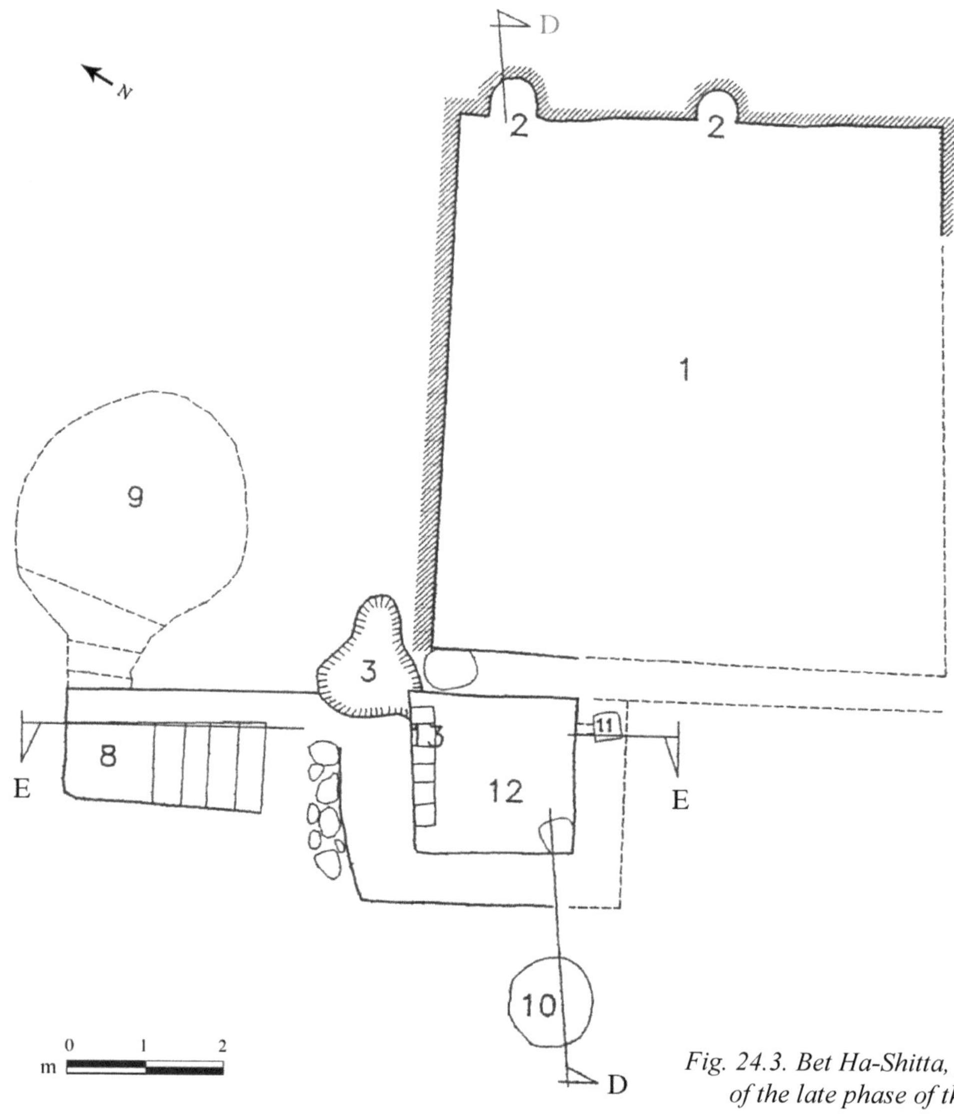

Fig. 24.3. Bet Ha-Shitta, plan and sections of the late phase of the wine press.

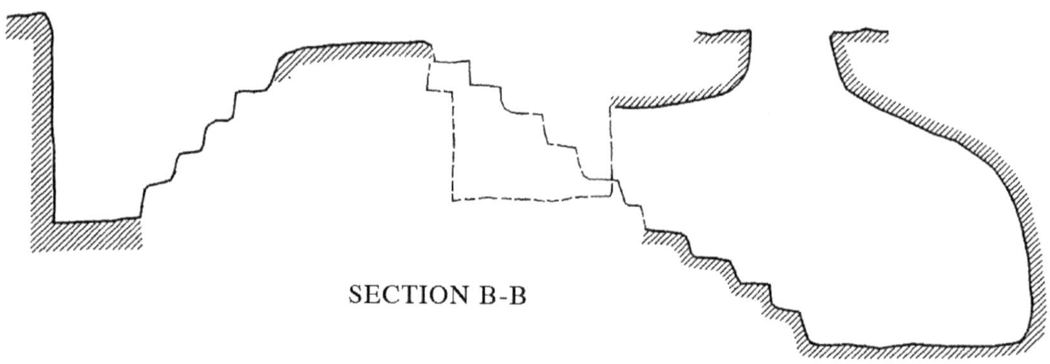

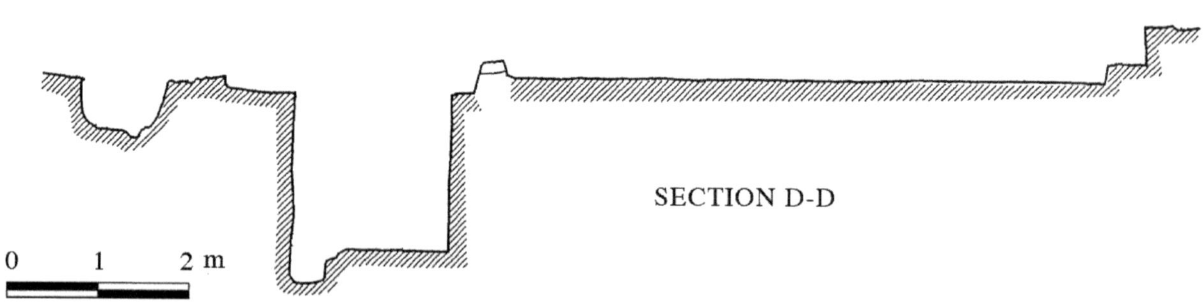

25. A Byzantine Oil Press at Shuni

Eli Shenhav

The Site

Shuni (Miamas) is situated at the foot of the Carmel Ridge near its southern end, adjacent to the Shuni springs that supplied water to Caesarea (MR 1453/2157). The excavation of the site[1] revealed a Roman theater that had undergone changes and repairs during the Early Byzantine period; a local aqueduct leading to a group of reservoirs; installations and remains of buildings adjacent to the theater; a paved street east of the theater lined with shops and other buildings; and a section of a large aqueduct probably leading to Caesarea. All these remains are from the Late Roman and Byzantine periods (Shenhav 1993: 1382–1384).

The Oil Press

The oil press was built during the Late Byzantine period on the southern edge of the theater that was no longer in use at that phase. The installation is almost complete (Fig. 1), although parts of it – mainly on the southern fringe – were damaged as a result of construction during the Crusader and Ottoman periods. Its interior measurements (without the oven D, see below) are approximately 11×12 m. The installations were in a building constructed of ashlar masonry, the main entrance of which has not been identified with certainty but that was probably in the western wall, where a door from a later period was found. The floor of the building is paved with small stones, except for the sections around the pressing areas that are floored with well-cut stone slabs, nicely laid and closely fitted.

The complex included the following installations:

A. A crushing mill, located in the centre of the building between the two pressing installations – an arrangement common in assemblages of this type. The crushing basin was a single whole unit made of limestone, emplaced on a base built of cut stones (Fig. 2). Its overall diameter is 1.83 m and its overall height 0.37 m. These overall measurements include raised edges 0.15 m wide and 0.13 m high. There was a pivot socket about 0.25 m square in the centre of the basin. The crushing stone, made of hard limestone, was discovered about 1.5 m from the basin. Its diameter is 1.24 m and its width 0.27 m. A hole 0.22×0.22 m was cut through its centre for the shaft.

B. The western pressing installation, of the lever and screw type, was preserved in its entirety except for the wooden parts (Fig. 2). In the rear wall (W10) that was covered by later construction the niche in which the rear end of the press beam was inserted was well preserved; it measured 0.55 m high, 0.35 m wide and 0.85 m deep. The pressing installation had a central collecting vat, built below floor level of cut stones and measuring 1.03×0.92 m and 0.95 m deep. On top of it was placed a large press-bed comprising a single round slab of hard limestone 0.25 m thick, with an overall diameter of 2.1 m. On the upper surface of the slab, near the edge, is a circular groove 10 cm wide and 4 cm deep. Four radial channels, each 6 cm wide and 5 cm deep, connect the circular groove to the central opening of the slab (diameter 0.51 m) through which the expressed liquid flowed into the collecting vat. On the top of the press-bed, around the central opening, is cut a rectangular depression measuring 0.55×0.60 m and 3 cm deep. A perforated wooden plank or perhaps slats layed criss-cross were set into the depression, and the frails of olive mash placed on them to be pressed (compare Frankel 1994: Fig. 41B). The sophisticated press-bed was well integrated into the surrounding floor. Between the press-bed and the weight at the distal end of the beam a low wall was found in which a large stone (not found *in situ* – see below and Fig. 6) with two sockets carved in its top, was probably incorporated. Apparently two wooden poles were set into these sockets with their ends inserted into two matching sockets in the ceiling above – one on each side of the beam, to prevent it from moving sideways.

The cylindrical limestone screw weight, found *in situ*, measures 1.23 m in diameter and 0.92 m high; it weighs approximately 3.14 tons. It is located in a pit 1.85 m in diameter, walled with stones. A socket 0.35 m in diameter and 0.3 m deep is cut in the top of the weight, and in it the bottom end of the wooden screw revolved. On the sides of the weight (today not perpendicular to the line of the beam) are carved two trapezoidal mortices. They measure 0.55/0.60 m high, 0.22/0.15 m deep, width at the top 0.15/0.13 m and at the bottom 0.22/0.22 m respectively. The distance from the back of the anchoring niche to the far side of the weight, i.e. the estimated length of the press beam is 8.6 m.

C. The eastern pressing installation, of which the rear section (wall and niche) is missing, is almost identical to the western installation (Fig. 3). The dimensions of the collecting vat are 1.05×0.92 m; the depth is unknown because the vat is presently blocked. The press-bed is 0.35 m thick and 1.66 m in diameter (Fig. 4). The diameter of the central opening is 0.43 m and the square depression for the perforated plank or slats measures 0.54×0.52 m and 3–4 cm deep. The circular groove and radial channels are 6–7 cm wide and 3–4 cm deep. A low wall (W13) was found between the press-bed and the screw weight; apparently it once contained a stone with two sockets as described in press B above. The screw weight was found *in situ* in a pit about 1.7 m in diameter, and walled with stones. The limestone weight was 1.1 m in diameter and 1 m high, thus weighing about 2.64 tons. A socket 0.34 m in diameter and about 0.3 m deep is cut into the top of the weight for the screw end. In this weight, there were four trapezoidal mortices. At first there were apparently the usual two but at a later stage, when these wore out, a second pair was added at right angles to the first. The dimensions of the two mortices found along the beam line are: height 0.70/0.73 m, depth 0.12/0.13 m, width at the top 0.12/0.12 m and at the bottom 0.20/0.24 m respectively. In the eastern installation, the distance from the centre of the press-bed to the centre of the screw weight is 3.7 m compared with 4.2 m for the western press. We can, therefore, deduce that the length of the beam in the eastern installation was approximately 7.5 m. Because of its proximity to the oven (D; see below) it is reasonable to assume that the eastern installation was also used for secondary pressing of the residue from the primary pressing after it had been heated. It is probable that the opening that exists today in the wall between the press and the oven also existed in the original structure.

D. The oven was discovered east of the oil press (not on the plan, Fig. 1); it was built in line with the entrance (*parodos*) to the Roman theatre, and slightly inside it (Fig. 5). The oven was constructed of a mixture of stones and fired bricks, and included two vaulted rectangular compartments, the entrances of which were 0.6 m wide and faced north, toward the orchestra of the theater. The approximate inside measurements of each compartment were 1.15×2.40 m and about 1 m high. The location of the fire compartment is not clear. Inside the oven was found a large quantity of carbonized olive pits, attesting to the fact that it was fueled with olive mash residue that, after the oil was extracted, had been dried for this purpose – a custom widespread in Arab villages until recently. It is reasonable to assume that the oven was used to heat olive mash that had already undergone cold pressing, in order to extract an additional quantity of oil (that was, of course, of inferior quality). This is one of the very few known discoveries of a large oven in an ancient oil press.

E. Under the stage of the theater, north-east of the oil press, five storage compartments were discovered built with double-faced stone walls and floored with coarse white mosaic. The floor of the compartment closest to the oil press contained a depression, probably a sedimentation basin, since the compartments were apparently used for storing the oil produced in the press. This hypothesis is borne out by their similarity to the storage facilities found in great numbers at nearby Caesarea (Levine and Netzer 1986: 61–63).

F. Two large stones were discovered not *in situ* but in light of the fact that each had two sockets carved into it (Fig. 6), they may have originally been used to stabilize the press beams with the aid of a pair of vertical poles between each stone and the ceiling above it, as described in press B above. The first stone was rectangular, 1.30×0.52×0.30 m, and its sockets measured approximately 0.15×0.33 m, 0.12 m deep. The second was a piece broken off of a column or a roller used for crushing olives (Frankel 1994 Fig. 19), 0.79 m long and 0.42 m in diameter. The sockets measured 0.13×0.12 m and 0.11 m deep, and the distance between them is 0.4 m.

Conclusions

The oil press was established during the Byzantine period (5th-6th cent. CE) as shown by a) the finds from the press, b) its parallels (see below), and c) the fact that it was constructed on top of the late phase of the theater, after it went out of use. During the Umayyad period (7th-8th cent. CE) it was expanded on its western side by the construction, on the other side of wall W15, of storerooms or platforms where the olives were kept while waiting to be processed. The complex went out of use at a later stage of the Early Arab period, apparently at the end of the 8th or beginning of the 9th cent. By the Crusader period, other buildings had already been constructed over it.

Oil presses that included similar pressing installations were unearthed at a few sites in Samaria and the Sharon area, all of which date from the Byzantine period: Kh. ‘Azzun, Ra‘anana (erroneously identified and published as a wine press – Roll and Ayalon 1981), Zur Natan (two assemblages: Ayalon, Neidinger and Matthews 1989/1990: 137; 1993: 46; Site 32), and Shiloh (Anderson 1985: 20, Fig. 3, Plan C). A similar installation, hewn into bed-rock, was discovered at Megaddim in the Carmel (Frankel 1994 Fig. 40). Because of the distribution of this model in the areas where the Samaritans settled during this period, it is possible that the installation under discussion represents a “Samaritan” subtype (compare Ayalon 1996: 149–156).

Notes

1. The site was excavated by the author on behalf of the Jewish National Fund and the Israel Antiquities Authority.

Bibliography

Anderson, F.G. 1985. *Shiloh II. The Remains from the Hellenistic to the Mamluk Periods*. Copenhagen.

Ayalon, E. 1996. “The “Samaritan” Olive Oil Press”. In: D. Eitam and M. Heltzer (eds.). *Olive Oil in Antiquity, Israel and Neighbouring Countries from the Neolithic to the Early Arab Period.* Padova: 149–156.

Ayalon, E., Neidinger, W. and Matthews, E. 1989/1990. “Horvat Migdal”. *ESI* 9: 137–138.

Ayalon, E., Neidinger, W. and Matthews, E. 1993. “Zur Natan (Horvat Migdal)”. *ESI* 13: 45–46.

Frankel, R. 1994. "Ancient Oil Mills and Presses". In: R. Frankel, S. Avitsur and E. Ayalon. *History and Technology of Olive Oil in the Holy Land.* Arlington and Tel Aviv: 19–89.

Levine, L.I. and Netzer, E. 1986. *Excavations at Caesarea Maritima 1975, 1976, 1979 – Final Report* (Qedem 21). Jerusalem.

Roll, I. and Ayalon, E. 1981. "Two Large Wine Presses in the Red Soil Regions of Israel". *PEQ* 130: 111–125.

Shenhav, E. 1993. "Shuni". *NEAEHL* 4: 1382–1384.

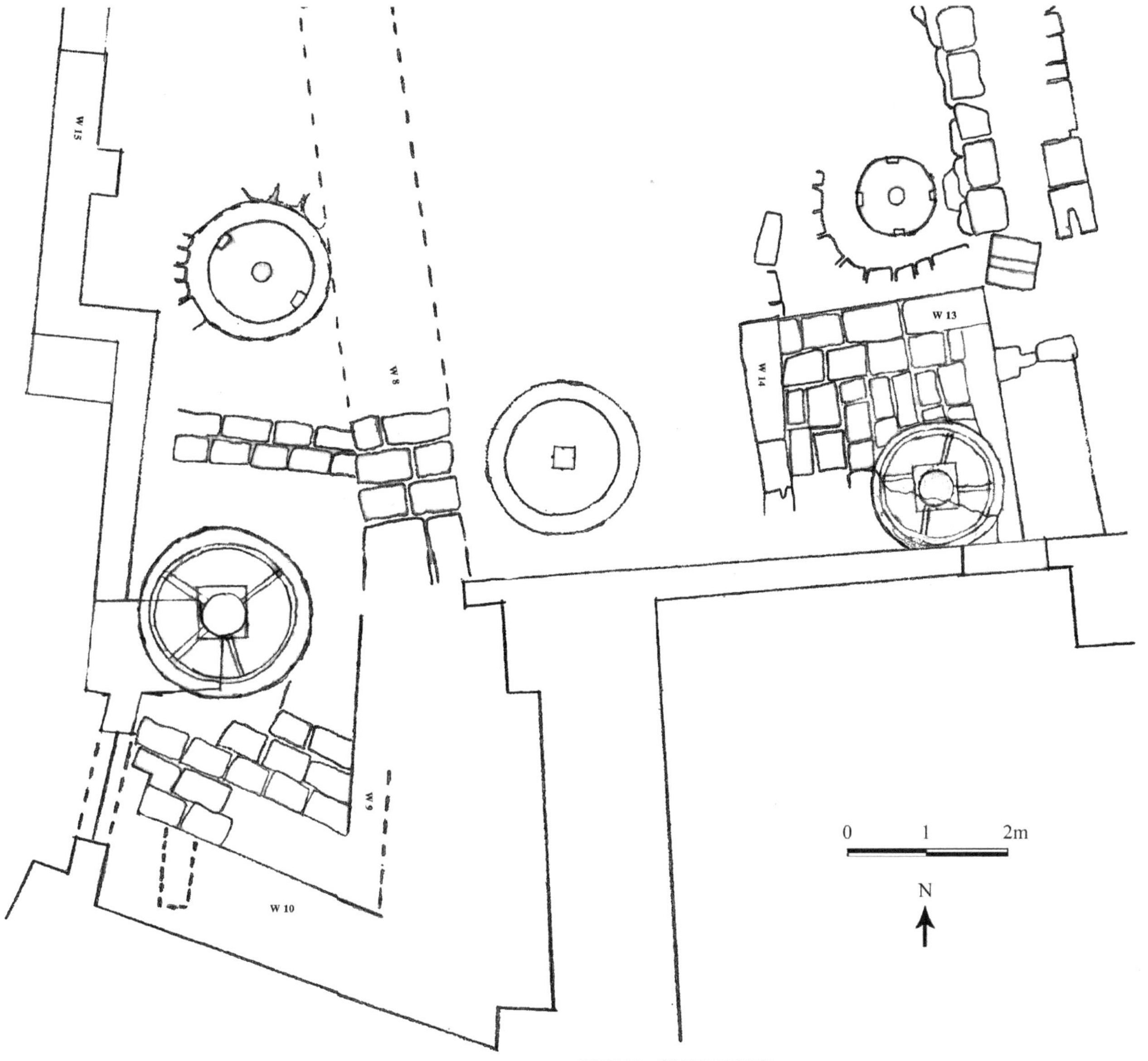

Fig. 25.1. Shuni, plan of the oil press (excluding the oven and storage compartments).

Fig. 25.2. Shuni, the crushing basin (left) and the western pressing installation (right), looking south. Note the beam anchoring niche in the back wall.

Fig. 25.3. Shuni, the eastern pressing installation, looking south.

Fig. 25.4. Shuni, the press-bed of the eastern pressing installation, looking south.

Fig. 25.5. Shuni, The oven, looking south.

Fig. 25.6. Shuni, a stone equipped with two sockets in which poles were installed that stabilized the press beam.

26. A Byzantine Wine Press at Tel Hefer

Eli Yanai

The Site

Tel Hefer (Tell el-'Ifshar) is located in the centre of the Sharon area, close to where the Alexander River cuts through the easternmost of the three sandstone ridges (MR 1415/1976). The tell, approximately 40 dunams in extent, was settled from the Bronze Age to the end of the Byzantine period, and during these periods the settlement expanded northward and eastward. The tell was excavated on behalf of the University of Buffalo and the Israel Antiquities Authority (Paley and Porath 1993). In previous rescue excavations, two large wine presses had been discovered east and north-east of the tell. A third wine press, described below, was unearthed by the author in 1994 in a rescue excavation on behalf of the Israel Antiquities Authority.

The Wine Press

At the foot of the tell, on the north-eastern side, an industrial complex from the Roman and Byzantine periods was discovered. In it were buildings, storehouses, ovens or kilns, and two wine presses. One of these, the subject of this paper, was built in the same orientation as the other buildings of this complex in its northern part, and separated from them by a narrow alley. The wine press measures approximately 11×11 m and consists of the following components (Figs. 1–2):

- A large treading floor (L. 583, 5.8×6.0 m, an area of 35 m^2), in the centre of which was a pit from which a stone base for a wooden screw was apparently robbed in a later period. The must, produced by both treading and pressing, drained from the floor through a large, plastered open channel into the collecting vat.

- A collecting vat (L. 584, Fig. 3), only the western part of which was unearthed. Its inside length, north to south, was 2.8 m. It was about 1.6 m deep. Presuming that the vat was square its volume can be estimated as 12.5 m^3. Around the vat was a working platform, approximately 0.6 m wide. In the south-east corner of the vat there were two steps in the shape of quarter-circles.

- To the south of the treading floor were a series of three rectangular compartments (L. 411, 412, 461, Fig. 4) built adjacent to each other and sharing common walls. Each measured approximately 3.3×2.3 m and 0.3 m deep, the area of each 7.6 m^2. The floor of each compartment sloped slightly southwards away from the main treading floor, and drained via an open channel into a small round vat, diameter 0.55 m, depth 0.5 m. The three vats were built into a common surface at the southern end of the wine press (L. 408), the length of which was equal to the combined width of the three compartments, and its width was about 1 m.

All the walls of the wine press were built of stones and mortar and covered with a heavy layer of plaster, and all its floors were tiled with coarse white mosaic.

This wine press is unique in its complex arrangement of three compartments and their three vats unconnected to the main treading floor (compare: Peilstöcker and Kapitaikin 2000). These probably served to produce "first must": The grapes were first placed in the compartments and the must that seeped out because of their own weight drained into the separate vats (in most wine presses of this type, pipes connected the storage compartments directly to the main treading floor). From this must the choicest wine was prepared. In the second stage, the grapes were trodden on the treading floor, and in the third stage the remaining must expressed from the rape was squeezed using the wooden screw press that stood in the centre of the treading floor. The must from the latter two operations accumulated in the large collecting vat.

Bibliography

Paley, S.M. and Porath, Y. 1993. "Tel Hefer". *NEAEHL* II: 609–614.

Peilstöcker, M. and Kapitaikin, A. 2000. "Bet Dagan". *ESI* 20: 59*-60*.

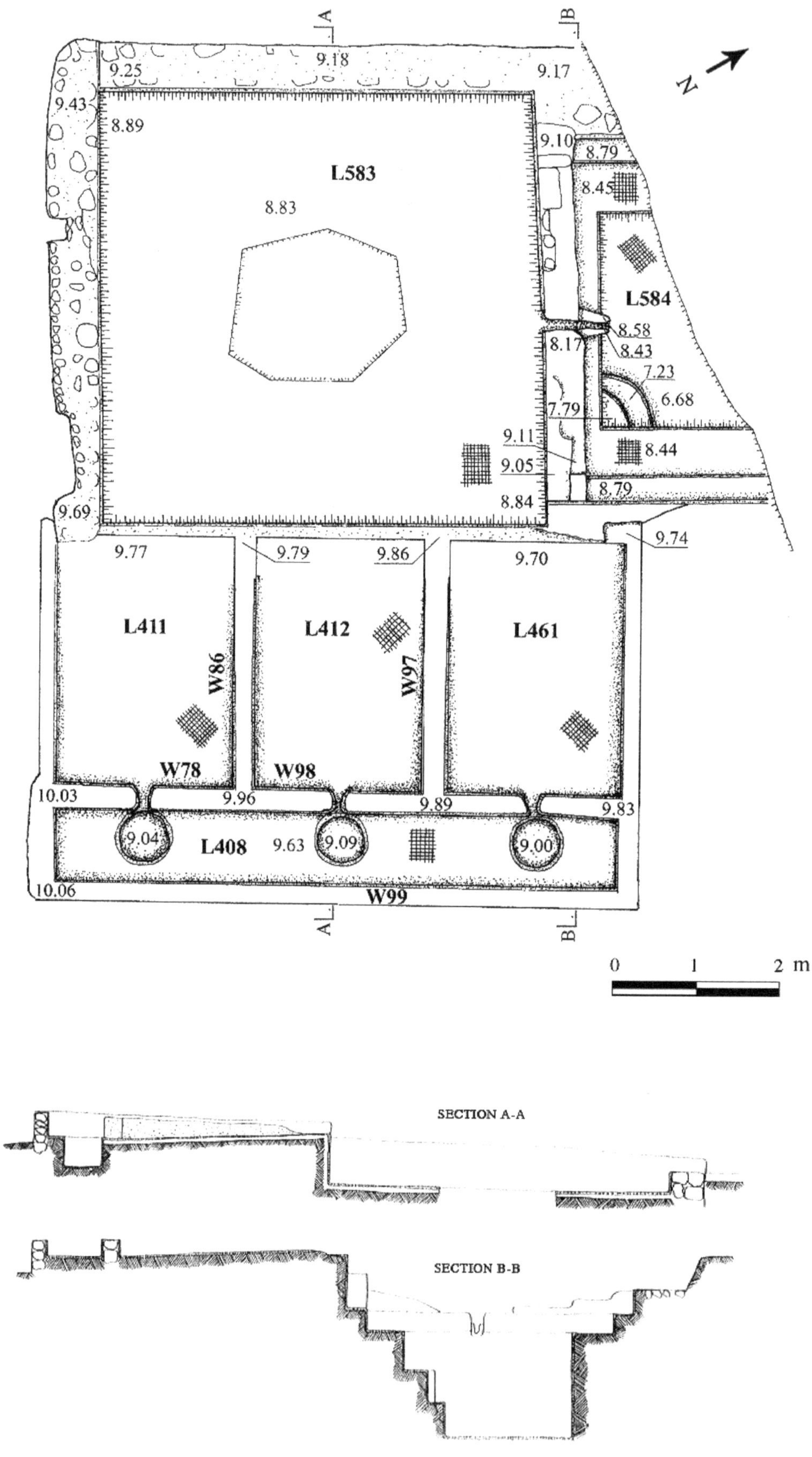

Fig. 26.1. Hefer, plan and sections of the wine press.

Fig. 26.2. Hefer, the wine press, view to the south-east.

Fig. 26.3. Hefer, the collecting vat, view to the south-east.

Fig. 26.4. Hefer, the side compartments, view to the south-east.

27. Wine Presses at Kh. Fardisya (et-Taiyiba)

Boaz Zissu, Haim Moial and Amir Ganor

The Site

Kh. Fardisya is located in western Samaria; the modern city of et-Taiyiba was built on its remains. The ruin is situated on a moderate and rocky westward-facing spur, adjacent to the fertile cultivated lands of the Sharon region. At the site, finds were unearthed covering a time span from the Early Bronze Age to the Ottoman period.

Rescue excavations conducted on two spurs north of Kh. Fardisya revealed many finds such as wine presses, quarries, lime kilns and tombs hewn into bed-rock.[1] In Area A (MR 1516/1872) two groups of wine presses were found, only 6 m apart (Figs. 1–2).

Installation No. 1

This installation is made up of three adjacent wine presses cut into a rock outcrop (Fig. 3).

The Eastern Wine Press

This installation consists of a round treading floor with external dimensions of 2.0×2.1 m – 4.2 m^2 and 0.4 m deep that drains into a horseshoe-shaped collecting vat measuring 0.65×0.65 m across and 0.4 m deep. Its approximate volume is 150 litres.

To the north of this wine press, a shaft measuring 0.8×0.9 m has been hewn into the hard *nari* bed-rock. It leads into an irregularly-shaped artificial cave cut into the soft limestone, measuring some 5×7 m. The ceiling of the cave had collapsed, and a probe about 2.5 m deep into the rubble layer revealed no artifacts.

The Central Wine Press

It is the largest of this group. The treading floor is almost square measuring 2.2×2.4 m – 5.3 m^2 in area and 0.25 m deep. In the centre of the floor a square mortice (0.6×0.6 m, depth 0.4 m) was cut to accommodate the wooden screw of a press. North of the treading floor was a round depression on which were remnants of pink plaster that contained crushed sherds. In its centre was a square collecting vat measuring 1.1×1.2 m with an estimated depth of 0.4 m, i.e. an approximate volume of 530 litres. The treading floor is connected to the vat by a deep fissure in the rock that continuously widens in the direction of the vat. North of this vat is hewn a horseshoe-shaped collecting vat very similar to that of the easternmost wine press, measuring 0.65×0.65 m across and 0.4 m deep, with an approximate volume of 150 litres.

In the collecting vat a shaft was cut at a later stage that led down to a man-made cave similar to that north of the eastern wine press.

The Westernmost Wine Press

This installation includes an almost square treading floor measuring 2.0×2.1 m and 0.3 m deep. In its centre a square mortice (0.27×0.26 m, depth 0.27 m) has been carved out to accommodate a press screw. The bottom of the mortice widens out on one side. The western wall of the treading floor was pierced by an opening that was apparently used for hitching an animal. To the north, a rectangular collecting vat 0.6×0.8 m in area and 0.5 m deep – volume 240 litres – was cut into the bed-rock, connected to the treading floor by a narrow open channel 0.2 m wide and 0.3 m deep.

Installation No. 2

This installation, located 3 m north of installation No. 1 was a single large wine press with a treading floor that has been almost totally destroyed. Its dimensions can be reconstructed as being about 4×4 m. In its centre was a square mortice and next to it is a fragment of a round stone screw base.

West of the treading floor was a rectangular collecting vat hewn into the bed-rock. It measures approximately 2.4×2.8 m and 1.5 m deep, i.e. volume of about 10 m^3. A flight of stairs next to the western wall, made of fieldstones and about 0.5 m wide, enabled workers to descend into the vat but only a few of the steps remain. The vat is floored with white large (3×3 cm) mosaic tesserae. These were laid upon a sub-stratum that leveled off the rock-face and consisted of two layers of light-colored mortar mixed with pebbles and a few shards from ribbed jars. An elliptical sedimentation basin measuring 0.6×0.9 m was found in the south-east corner of the floor. The upper parts of the vat are hewn from the bed-rock, while the lower parts are constructed of small fieldstones in a matrix of light-colored mortar and smoothly coated with light-colored plaster.

The pottery shards embedded in the plaster below the floor of this vat are typical of the Late Roman and Byzantine periods. After it went out of use, the vat was filled with brown earth containing many shards of glazed pottery characteristic of the 12th-14th cent.

A square filtration basin measuring 0.5×0.5 m and about 0.14 m deep was hewn into the bed-rock north of the collecting vat. Its bottom was floored with rough mosaic tesserae. A hewn channel connected the basin to the collecting vat.

Reconstruction of the History and Operation of the Wine Presses

On the presumption that the wine presses were in continuous use and because of the physical proximity of the two installations, the changes that took place in the production of the must may be reconstructed in two stages (Fig. 3).

Stage I

Installation No. 2 functioned as a typical wine press, including the passage of the must through a filtration basin and pressing the once-trodden pulp in a central screw press. At this stage apparently the square treading floor of the central wine press of installation No. 1 did not yet exist and the round depression of this press in which the square collecting vat was later cut acted as the treading floor of the wine press that was then almost identical to the eastern wine press. We suggest that at this stage the eastern and central wine presses of installation No. 1 served as grape storage areas[2] for installation No. 2. Bunches of grapes were placed in these areas, and from them the choicest must – juice that seeped from the grapes under their own weight – flowed into the adjacent small collecting vats.

Stage II

Installation No. 2 ceased to be used, and south of it installation No. 1 was subsequently enlarged and replaced it. A large treading floor with a screw press was cut in the bed-rock south of the central wine press and a collecting vat was cut in the centre of the former round treading floor that as a result no longer served as such. This then became the main wine press of the complex.

The eastern wine press continued to function as a grape storage area. The western wine press was probably added at this stage and apparently used mainly for pressing the trodden pulp as shown by the comparatively small size of its collecting vat, into which the small quantities of resulting "Grade B" must flowed. The extraction was apparently carried out using a wooden screw press installed in the mortice cut into the centre of the treading floor and secured in position in the widened section at the bottom of the mortice. Square screw mortices that widen in one direction only were especially common in the Samaria region (Frankel 1999: 142).

Abandonment of the Installations

After cessation of wine production here, apparently at the beginning of the Early Arab period (Ayalon 1997: 149–166), a shaft was hewn in the bottom of the collecting vat of the central wine press of installation No. 1, possibly for quarrying limestone. Shafts with subterranean cavities beneath them, probably used for similar purposes, were found to the east of this one and in other places on the excavation site.

Summary

The excavations of the installations did not bring to light any finds that can aid in dating them, but comparison with dated wine presses makes it possible to suggest a tentative date. They are typical examples of medium-to-large wine presses (Ayalon, Charvit and Qidron 1990: 93–120) used during the Late Roman and Byzantine periods.

It should be noted that the installations were situated at the outskirts of the village adjacent to the cultivated area as is typical of wine presses that were usually in the vineyards.

It is of interest that a filtration basin was built in installation No. 2, while the later use of installation No. 1 is marked by the absence of such a basin. Wine presses lacking a filtration basin are common throughout Samaria (i.e. Neidinger, Matthews and Ayalon 1990: Pl. 22).

The finds demonstrate how farmers who made wine over a period of several generations continuously, needed and found alternative solutions for production installations that went completely out of service. Also typical is the phenomenon that occurred after the cessation of winemaking in the area, whereby they continued to exploit the space of the collecting vat for other purposes, in this event, for quarrying limestone.

Notes

1. The excavations were carried out in August 1995 on behalf of the Israel Antiquities Authority and financed by the Ministry of Housing. The authors wish to express their gratitude to Etan Ayalon, David Amit and all the others who contributed to the success of the dig. Our special thanks to Nili Graicer for her cordial assistance in the preparation of the report.
2. A Mishnaic expression, *mishtach shel 'alim*, literally meaning "area of leaves", refers to such a temporary storage area (see Hirschfeld 1983: 207–218).

Bibliography

Ayalon, E. 1997. "The End of the Ancient Wine Production in the Central Coastal Plain". In: Y. Friedman, Z. Safrai and J. Schwartz (eds.). *Hikrei Eretz, Studies in the History of the Land of Israel Dedicated to Prof. Yehuda Feliks*. Ramat Gan: 149–166 (Hebrew).

Ayalon, E., Charvit, Y. and Qidron, A. 1990. "Grouped Installations from the Roman-Byzantine Period in the Fields of Zur Natan". *Israel – People and Land, Eretz Israel Museum Yearbook* 5–6 (23–24). Tel Aviv: 93–120 (Hebrew).

Frankel, R. 1999. *Wine and Oil Production in Antiquity in Israel and Other Mediterranean Countries*. Sheffield.

Hirschfeld, Y. 1983. "Ancient Wine Presses in the Park of Aiyalon". *IEJ* 33: 207–218.

Neidinger, W., Matthews, E. and Ayalon, E. 1990. *Preliminary Report on the Excavations at Zur Natan, 1989 and 1990 Seasons*. Houston.

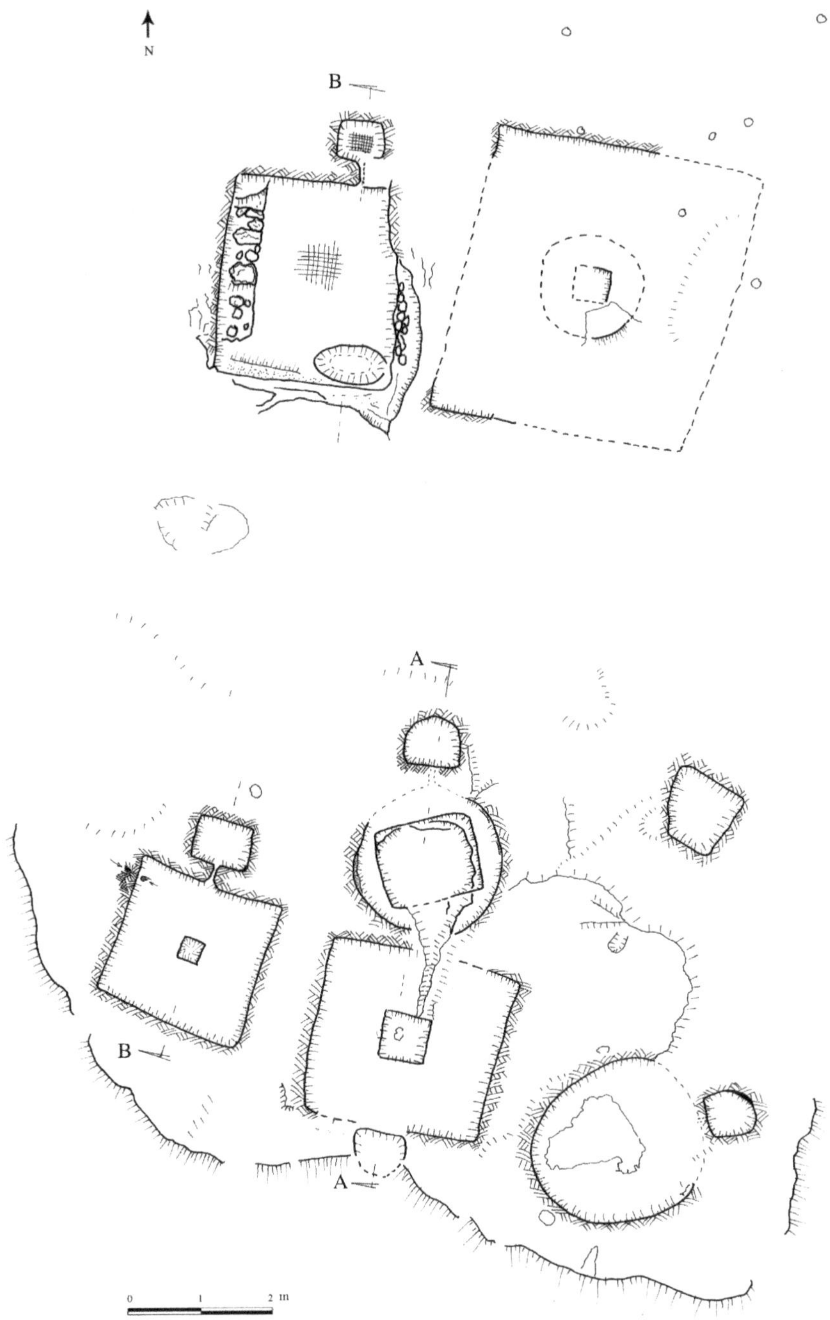

Fig. 27.1. Fardisya, plan of the wine presses.

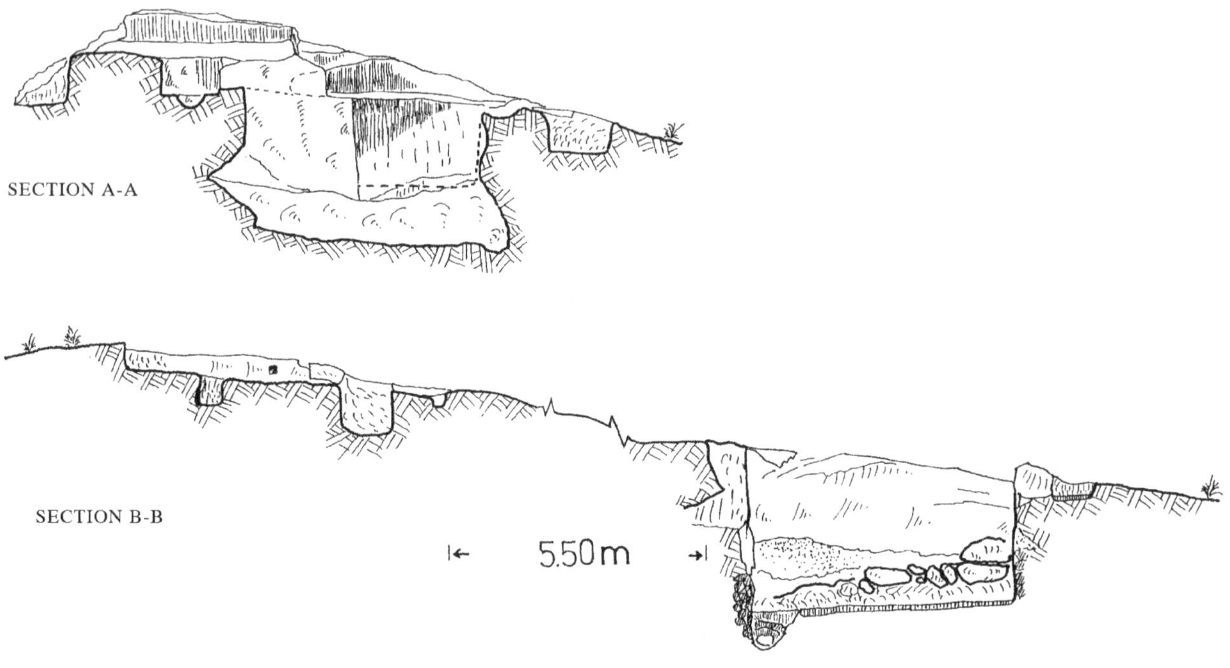

Fig. 27.2. Fardisya, sections of the wine presses.

Fig. 27.3. Fardisya, Installation No. 1, view to the south.

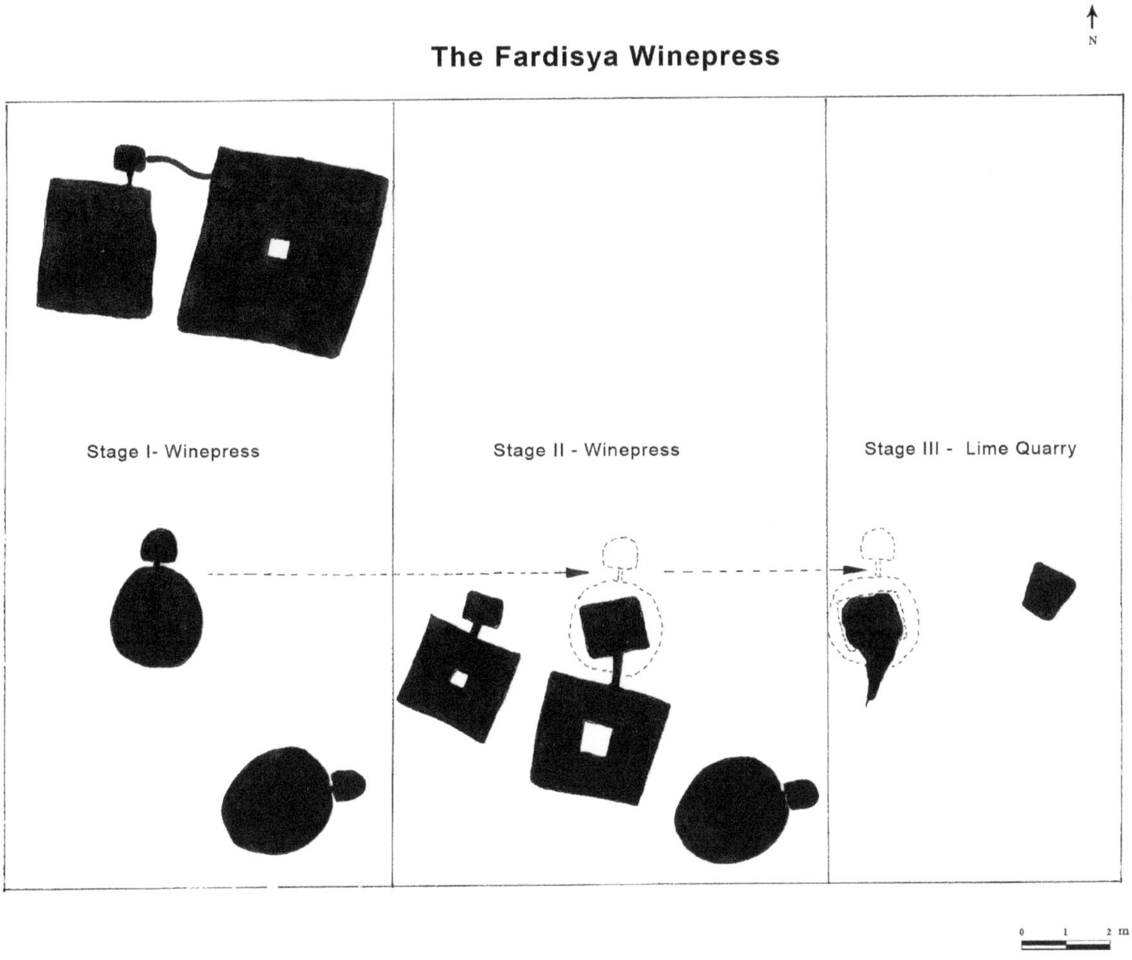

Fig. 27.4. Fardisya, reconstruction of the development stages of the wine presses.

28. A Byzantine (?) Oil Press in Samaria-Sebaste

Etan Ayalon, Rafael Frankel and Amos Kloner

The oil press was unearthed in the 1930s during the excavations of the Joint Expedition to Samaria (MR 168/187). It was described in the report in a short paragraph accompanied by two illustrations (Crowfoot, Kenyon and Sukenik 1942: 70, Figs. 32, 35). The installation (Fig. 1) was discovered in the south-eastern quarter of the upper city, between the Roman forum to the north and a Roman street running east-west, to the south. During the Byzantine period and the beginning of the Early Arab period the area was residential and industrial. In addition to the oil press and dwelling units, a factory for the production of glass vessels was discovered west of the oil press.

Only the northern part of the installation was exposed during the excavation and this consisted of a room containing one complete pressing installation of the lever and screw type. Adjacent to it to the west two additional rooms were found; all three were connected by openings. In one of the rooms was a cistern that may have served the oil press; the second was a domed room only partially unearthed. The unexcavated area between the press room and the street to the south was large enough to accommodate the usual two additional installations: a crushing mill and a second beam press.[1] The entrance to the oil press was probably from the street to the crushing mill, from which the olive mash was transferred to the pressing installations probably through an opening 0.8 m wide that was discovered in the southern wall of the press room.

The room measured 4.8×2.4 m and included two sections:

- The western section, that was the pressing area, was floored with stone slabs. In its western wall there was a niche (0.7×0.8 m) to accommodate the back end of the press beam (fulcrum). Opposite the niche, approximately in the centre of the room, a stone slab was integrated into the floor. Its centre was chiseled out to form a round press-bed 0.9 m in diameter, with a circumferential groove that directed the expressed liquid through a small hole to a stone collecting vat placed beside the press-bed and below the floor level. It is not clear whether the vat was round or rectangular – both types are well known. The width or diameter of the vat, according to the illustration in the report, was 0.75 m and it was 0.5 m deep. It was covered with a stone slab integrated into the floor, in which there is an opening 0.35 m in diameter, through which the expressed liquid was drawn from the vat. At the eastern end of the pressing area there was a low, narrow stone wall 0.35 m wide and 0.3 m high, in which two small cavities have been cut 0.7 m apart. Wooden poles were set into these holes (the top ends of these poles may have been inserted into similar holes in the ceiling) that guided the press beam from both sides, keeping it in line with – and directly above – the press-bed. Similar guide-pole sockets have been discovered at Zur Natan (Site 32), Shuni (Site 25), H. Beit Loya (Site 67) and other places.

- As in many other presses, the screw weight was installed in a pit 1 m deep that was to the east of the collecting vat. The weight was of a type common throughout the central part of the country, and therefore called a "Samaria Weight" by Frankel (1999: 111–113). It was 1.6 m high and 1 m in diameter. On the assumption that it was of limestone its weight can be estimated as being approximately 3.5 tons. A cylindrical socket 0.35 m in diameter and 0.3 m deep was cut into the top of the weight, in which the bottom end of the screw revolved. Mortices were carved on opposite sides of the upper part of the weight, into which were fitted the planks that connected the screw to the weight. The mortices are not in the usual dovetail form, but are cut in the shape of reversed T. Their overall length is 0.85 m.

The method of operating the installation is perfectly clear because of its many parallels, both ancient and pre-industrial (the excavators of Samaria note that only "lately" a similar installation went out of use in the adjacent village of Sebaste, a fact that testifies to the preservation of local technological traditions over a span of many generations). The olive mash, packed into frails, was placed upon the press-bed. The back end of the beam, the overall length of which was approximately 5 m, was set into a niche in the wall, and the forked front end was attached by a threaded board to the screw that was fastened to the weight. As the screw was turned, the beam was lowered and pressure brought to bear on the frails. The expressed liquid flows through the small hole into the stone vat under the floor, from which it was drawn through the larger hole using a jug or ladle. Many similar installations have been discovered in Samaria and the Sharon area in sites such as Qedumim (Magen 1993b; Site 34), Kfar Saba (Pifano 1983: 27), Apollonia (Roll 1999: 26, Figs. 1.12–1.14), Petah Tiqwa, etc. Therefore this press can possibly be seen as the type used by the Samaritan community who flourished in the area (Ayalon 1996).

The excavators dated the installation to the Early Arab period, no earlier than the Umayyad period, and indeed in the catalog of finds appears a lamp of the "Mafjar" type (8th-9th cent. CE) from the oil press (Crowfoot, Crowfoot and Kenyon 1957: 376, Fig. 89: 9). However, finds discovered in any structure are generally from the stage of its abandonment, and an oil press of this type could have been used for hundreds of years. For this reason, and because of similar parallel installations, the building of this oil press should probably be dated to the Byzantine period.

Notes

1. About 25 m south-west of the oil press, an "olive mill" was discovered (that according to Fig. 35 comprised only a crushing stone), but it was too far away to be considered part of the complex under discussion here; in addition, it may not have been *in situ*.

Bibliography

Ayalon, E. 1996. "The "Samaritan" Olive Oil Press". In: D. Eitam and M. Heltzer (eds.). *Olive Oil in Antiquity, Israel and Neighbouring Countries from the Neolithic to the Early Arab Period.* Padova: 149–156.

Crowfoot, J.W., Crowfoot, G.M. and Kenyon, K.M. 1957. *Samaria-Sebaste III. The Objects from Samaria.* London.

Crowfoot, J.W., Kenyon, K.M. and Sukenik, E.L. 1942. *Samaria-Sebaste I. The Buildings at Samaria.* London.

Frankel R. 1999. *Wine and Oil Production in Antiquity in Israel and Other Mediterranean Countries.* Sheffield.

Magen, Y. 1993b. "Qedumim – A Samaritan Site of the Roman-Byzantine Period". In: F. Manns and E. Alliata (eds.). *Early Christianity in Context, Monuments and Documents* (Studium Biblicum Franciscanum, Collectio Maior, 38). Jerusalem: 167–180.

Pifano, S. 1983. "Kefar Saba". *HA* 83: 27 (Hebrew).

Roll, I. 1999. "Introduction: History of the Site, Its Research and Excavations". In: I. Roll and O. Tal. *Apollonia-Arsuf, Final Report of the Excavations Vol. I. The Persian and Hellenistic Periods* (Tel Aviv University Institute of Archaeology Monograph Series No. 16). Tel Aviv: 1–62.

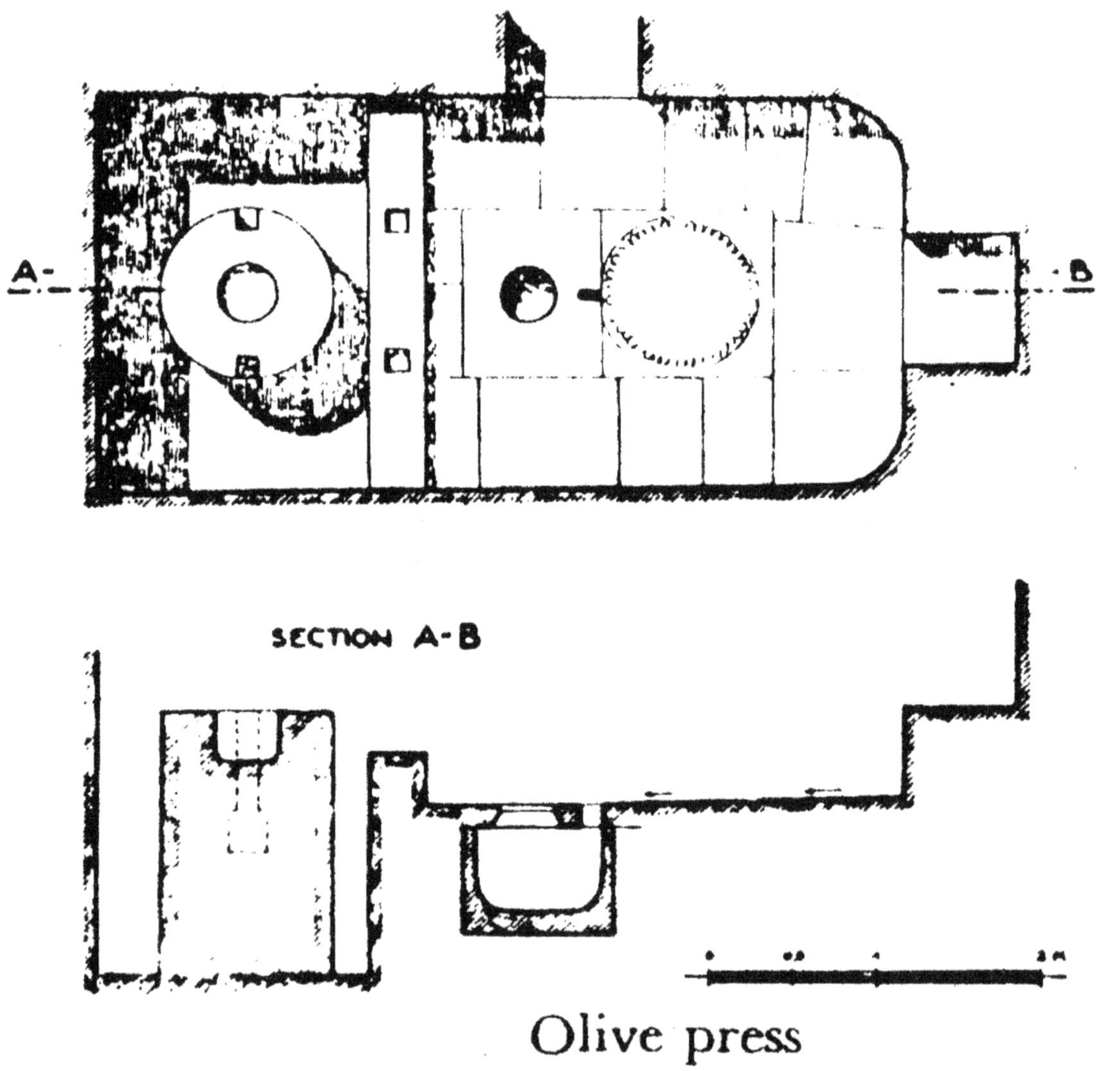

Fig. 28.1. Sebaste, plan and section of the oil press.

29. A Byzantine Wine Press at Kfar Ya'abez

Etan Ayalon

The Kfar Ya'abez (Kilodia) site is a rural settlement on the eastern fringes of the Sharon Region (MR 1468/1864) that existed during the Roman, Byzantine and Early Arab periods (approximately the 1st-10th cent. CE). The wine press was discovered in 1982 on the western edge of the site,[1] that had been damaged during agricultural cultivation.

The installation (Figs. 1–2), large even when compared with the usual "public" wine presses of the period, consisted of the following components:

- A treading floor measuring 7.00×7.25 m (50.75 m^2), surrounded by a partly damaged wall 0.6 m thick. In the centre of the floor a round stone screw base was found, 1.4 m in diameter in which a square mortice measuring 0.5×0.5 m at the surface and 0.45 m deep had been cut out to hold a wooden fixed screw. Two adjacent sides of the mortice widened out at the bottom for the purpose of securing the screw, as is usual in wine press screw mortices of the Coastal Plain (Frankel 1999: 140–142). No signs were found around the wine press of storage compartments, that are a common element in wine presses of this type, although they may have once been present and been destroyed over the years. The treading floor was connected to the filtration basin by a stone-cut channel, inner width and depth 0.15 m, outer width 0.35 m and length 0.6 m.
- A circular filtration basin, diameter 0.7 m, depth 0.5 m, from the bottom of which a ceramic pipe 10 cm in diameter leads to the collecting vat.
- A large circular collecting vat, diameter 3.4 m, depth 1.55 m, i.e. volume of slightly over 14 m^3. Three steps built at its eastern edge lead down to the floor of the vat in which there was a sedimentation basin, also round, diameter 0.8 m, depth 0.5 m.

The vat and basin are surrounded by a floor that was flush with the edge of both, on which the wine press workers could stand during the performance of their duties. This working floor is enclosed by low walls. The main walls of the wine press were not preserved to any significant height. The sides of the vat and basin were covered in hard gray plaster, and their floors, as well as the treading and working floors, were tiled in coarse white mosaic.

The wine press is amongst the largest of such installations from the Byzantine period known in Israel. It differs slightly from the type most common in the Sharon coastal plain in that its three components are arranged in a line (as is common, for example, in the Galilee – Frankel 1999: 152–153) and that its basin and vat are round (compare Sidi, Amit and 'Ad 2003: 262–263, Fig. 1). Generally speaking, the wine presses scattered throughout this area were built on the "four-square" plan, in which the vats and basins are also square (Frankel 1999: 149–150).

In light of the form of the wine press, its dimensions and components (plaster, mosaics, screw), there is no doubt that it was constructed and operated during the Byzantine period. There is a clearer testimony regarding the period during which it went out of use. Part of its area was used for cooking, as attested by accumulations of ashes, shards of ovens and pots, and animal bones; the collecting vat was used as a trash pit. The finds discovered consisted of a rich assemblage of vessels all from the Early Arab period – the 8th-9th cent. CE (Ayalon 1997: Fig. 4). It includes plain and glazed bowls, jugs and lamps of the "Mafjar" type, and the handle of a large jar impressed with a seal in Arabic. This evidence is important in determining the time of the breakdown of wine making that had been so important during the Roman-Byzantine period. According to the findings from this site and from other large wine presses, such as the one in Kh. Duran (Rehovot [Site 45]), it appears that this crisis took place at the beginning of the Early Arab period (Ibid.: 149–166). One reason for this change was without doubt the prohibition imposed by their religion upon Muslims regarding the production and drinking of wine. Other possible reasons were the loss of export markets and a possible change in climate during this period.

Notes

1. The wine press was unearthed by the author in a rescue excavation on behalf of the Israel Department of Antiquities.

Bibliography

Ayalon, E. 1997. "The End of the Early Wine Industry in the Central Coastal Plain". In: Y. Friedman, Z. Safrai and J. Schwartz (eds.). *Hikrei Eretz, Studies in the History of the Land of Israel Dedicated to Prof. Yehuda Feliks*. Ramat Gan: 149–166 (Hebrew).

ESI 2 (1983): 61: "Kefar Ya'bes (Kiludiya)".

Frankel R. 1999. *Wine and Oil Production in Antiquity in Israel and Other Mediterranean Countries*. Sheffield.

Sidi, N., Amit, D. and 'Ad, U. 2003. "Two Winepresses from Kefar Sirkin and Mazor". *'Atiqot* 44: 253–266.

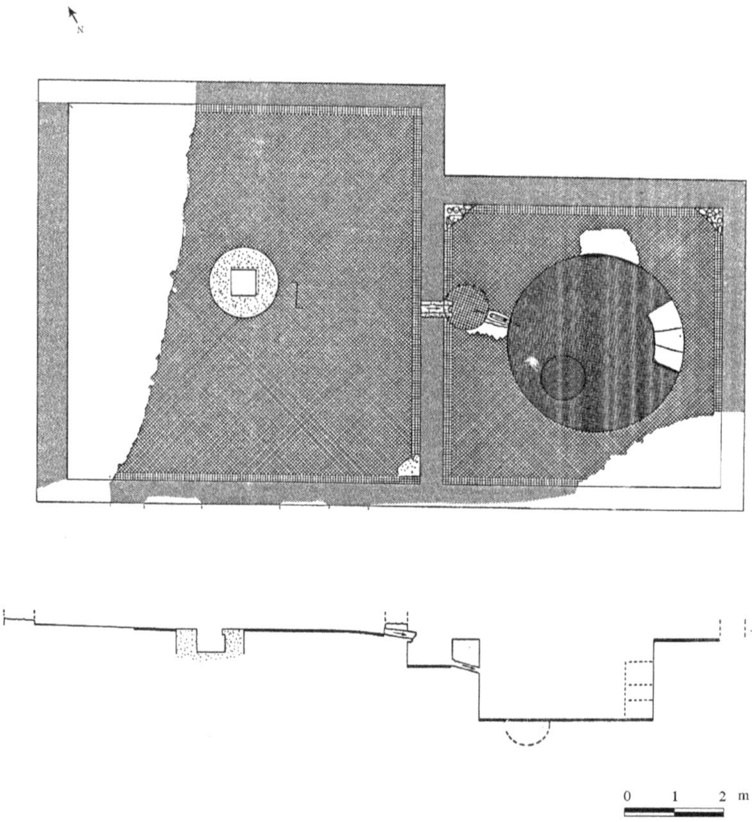

Fig. 29.1. Kfar Ya'abez, plan and section of the wine press.

Fig. 29.2. Kfar Ya'abez, the wine press, view to the south-east.

30. The Open-Air Oil Presses at Zur Natan

Etan Ayalon

Ancient and pre-industrial olive oil presses have almost always been installed inside buildings or caves for several reasons: the occasional rain during the olive harvesting season, the need to shelter and protect the equipment, and the necessity of lighting the complex for operation at night (Rosenson 1996: 115–118). However, the remnants of open-air oil presses that operated in the Hellenistic(?), Roman and Byzantine periods, abound in western Samaria (Dar 1986: 170–174) and in smaller numbers in other regions.

At least 25 such installations have been discovered on the spur of Zur Natan in western Samaria (central MR 1509/1831). The spur covers an area of about four km^2. The local bed-rock is suitable for carving installations and the rich soil is ideal for growing olive trees (the Arabs of the area still grow olives there). Oil presses were generally hewn out close to small level areas where the soil was less rocky and olive trees were probably cultivated. Most of them were discovered within "clusters" of installations scattered over the spur about 200 m apart. The clusters that surround the ancient settlement (H. Migdal, ancient name – *Antesion*) usually included a wine press, an oil press, a water cistern and a tomb, in addition to elements of secondary importance. It appears that each family hewed its own installations out of the bed-rock on its own agricultural estate. There it processed the produce of the field, mainly must from grapes and oil from olives (Ayalon, Charvit and Qidron 1990).

An open-air oil press (Fig. 1) normally contains two components (Ayalon 1999):

- The crushing mill, of which only the crushing basin, hewn in bed-rock, remains. It is round and flat, 2.0–2.2 m in diameter and 0.4–0.5 m high, with a raised sidewall 0.1–0.2 m in both width and height. Sometimes the wall had a narrow channel cut into it that played no part in oil production, since no oil seeps out during the crushing process. The channel was apparently for draining off rainwater so that work could continue in the open-air installation even on rainy days.

A socket was carved into the centre of the crushing basin to hold the pivot. It was generally square, 0.20–0.35 m in all three dimensions. Sometimes it was elliptical or round, ~0.3 m across and 0.4–0.6 m deep. Sometimes the socket would widen toward the bottom, and sometimes it was surrounded by an elevated cylinder – both designed for better support of the pivot. A wooden power shaft with the crushing stone at one end was affixed to the top of the low pivot (compare Avitsur 1994 Fig. 110). A beast of burden harnessed to the shaft operated the mill by walking around a track 1 m wide chiseled into the rock.

In none of the installations at Zur Natan were crushing stones discovered; it appears that they had been removed for secondary use in construction, or had been converted into lime or plaster.

- Cup-shaped depressions (Fig. 2) used for pressing the olive mash and separating the oil from the watery lees. One to three such depressions were found within a few meters of most of the crushing basins. The diameters and depths of the smaller among them were 0.2–0.3 m, the larger 0.4–0.5 m. Sometimes a small sump was chiseled into the bottom. Some had flat edges on which a cover or a perforated wooden pressing board could be placed. In a few isolated cases a shallow press-bed had been hewn out alongside the depression, upon which the frails of olive mash were placed to be pressed.

The absence in the area of "industrial-type" pressing fixtures using a lever, weights or a screw, and the proximity of the cup-shaped depressions to the crushing basins, suggest that pressing was carried out in these installations. The operating system has not changed in the Land of Israel for thousands of years (Frankel 1994: 28–31; Avitsur 1994: 98–102). Several sticks or a perforated plank were placed across the depression, and a sack or frail containing the olive mash was placed on top. Heavy stones were piled on, and the expressed liquid flowed into the depression. It was left for a short time, during which the oil separated from the watery lees and floated to the top, where it was skimmed off.

This combination of a large, advanced crushing installation with small, "primitive" pressing apparatus is rather exceptional in the annals of technology. Dar (1986: 172) proposed that pressing in these open-air installations was done by portable wooden screw presses that have long since disappeared. However, the use of such devices in the field far from the village, and in a place where carving out permanent installations would have been easy, is not convincing. Eitam (1997, 1999) suggests (in light of installations surveyed in eastern Samaria and ethnographic parallels from Turkey) that these installations were used for industrial production of wheat groats. However, no threshing floors have been discovered in the fields of Zur Natan, and the relatively rocky area is far more compatible with growing olives than grain (unlike eastern Samaria, where grain was the main crop). Further, it is illogical in light of the large numbers of installations to assume that so small a village would produce such immense quantities of groats, while wine and oil were natural export merchandises of the region.

It appears, then, that the depressions were indeed the pressing installations of these oil presses. Apparently the growers wanted to produce a choice, high-priced oil, even in relatively small quantities. Crushing the olives in a sophisticated mill made the process more efficient, but did not reduce the quality of the oil. Pressing in a simple installation, however, gave the farmer a relatively smaller quantity of oil, but of the finest quality, and in addition saved payment to the owner of an "industrial" oil press.

It is not easy to date an open-air installation in which few finds were discovered. Dar (1986: 173–174) claims that the open-air oil presses he examined in western Samaria were in operation during the Hellenistic and Roman periods (4th cent. BCE to 2nd cent. CE), and were then abandoned in favor of the advanced oil presses erected in the villages. He established the dating indirectly, according to finds unearthed in sites and field towers adjacent to the oil presses. In the installations at Zur Natan we found no evidence of their operation in so early a period. Most or all of them were integral parts of "clusters" of installations the main elements of which were characteristic of the Roman and Byzantine periods such as large wine presses with mosaics and screw presses and burial tombs with *kokhim* or *arcosolia.* The few potsherds found in the oil presses were from the Roman, Byzantine and Early Arab periods. Also, the small scale of the nearby settlement (H. Migdal) in the Hellenistic period is incompatible with the distribution of those oil presses throughout the spur. We suggest, therefore, that these installations were primarily from the Roman and Byzantine periods.

Bibliography

Avitsur, S. 1994. "Olive Oil Production in the Land of Israel, Traditional to Industrial". In: R. Frankel, S. Avitsur and E. Ayalon. *History and Technology of Olive Oil in the Holy Land.* Arlington and Tel Aviv: 91–157.

Ayalon, E. 1999. "The Open-Air Oil Plants of Western Samaria". In: Y. Eshel (ed.). *Judea and Samaria Research Studies, Proceedings of the Eighth Annual Meeting 1998.* Kedumim-Ariel: 115–126 (Hebrew).

Ayalon, E., Charvit, Y. and Qidron, A. 1990. "Grouped Installations from the Roman-Byzantine Period in the fields of Zur Natan". *Israel – People and Land, Eretz Israel Museum Yearbook* 5–6 (23–24): 93–120 (Hebrew).

Dar, S. 1986. *Landscape and Pattern. An Archaeological Survey of Samaria 800 B.C.E. - 636 C.E.* (BAR IS 308). Oxford.

Eitam, D. 1997. "Mass Production of Cereal Groats in Ancient Israel during the Roman and Byzantine Periods". In: Y. Eshel (ed.). *Judea and Samaria Research Studies, Proceedings of the Sixth Annual Meeting.* Kedumim-Ariel: 191–202 (Hebrew).

Eitam, D. 1999. "The Bread of the Bible – of Groats or of Dough". In: Y. Avishur and D. Deutsch (eds.). *Michael – Papers in Honour of Michael Heltzer.* Tel Aviv: 119*-125* (Hebrew).

Frankel, R. 1994. "Ancient Oil Mills and Presses". In: R. Frankel, S. Avitsur and E. Ayalon. *History and Technology of Olive Oil in the Holy Land.* Arlington and Tel Aviv: 19–89.

Rosenson, I. 1996. "The Essence and Siting of Oil Producing Facilities in the Mishnaic and Talmudic Periods based on Authentic Sources". In D. Eitam and M. Heltzer (eds.). *Olive Oil in Antiquity, Israel and Neighbouring Countries from the Neolithic to the Early Arab Period.* Padova: 113–118.

Fig. 30.1. Zur Natan, an open-air oil press consisting of a crushing basin (front, left) and cup-shaped depressions (background, right).

Fig. 30.2. Zur Natan, cup-shaped depressions used for pressing the olive mash.

31. Byzantine Wine Presses at Zur Natan

Etan Ayalon

In a detailed archaeological survey conducted over an area of approximately 3 km^2 on the low spur of Zur Natan in western Samaria, more than 100 wine presses were recorded (Ayalon, Charvit and Qidron 1990). At least 20 of these could be dated to the Byzantine period on the basis of their large dimensions, the finds discovered in them and their components (mosaic floors, mortices for fixed screw presses, types of plaster, etc.). The remarkably large number of these installations in such a limited geographical area reflects the extent of the wine industry that flourished in the region during that period. The wine presses extend over the entire spur that enjoys a rocky calcareous type of soil quite suitable for viticulture, and unlike the situation at other sites are not concentrated around the ancient village (H. Migdal, MR 1509/1832). Two of these wine presses that have certain unique characteristics will be described below.[1]

Wine Press D1

The installation was found south-west of the ancient village, adjacent to the Roman road leading to the port city of Apollonia and alongside the finest piece of land on the spur. It has not yet been completely unearthed, but the data required for understanding it are known. The wine press, with overall dimensions of about 9×10 m, consists of the following components (Figs. 1–2):

A large treading floor measuring approximately 6.0×6.4 m (38.4 m^2) was hewn out of the bed-rock and paved with white coarse mosaic. Most of the tesserae (except at the edges) were removed in ancient times, apparently for use in making lime, exposing the foundation of the floor that comprised fine stones and exceptionally hard mortar. From the treading floor the must flowed westward into the collecting vats through two channels cut into the bed-rock. The northern channel, 0.12 m wide and paved with coarse white mosaic, led to the larger of the two collecting vats; the southern channel, 0.35 m wide, to the smaller.

An arched niche hewn into the northern vertical rock wall of the treading floor served to anchor the end of a press beam. The niche was 1 m high, 0.55 m wide and 0.5 m deep. A large, heavy, rectangular stone was set into the rear of the niche on its narrow side to reinforce it. At the top of the niche the rock is cracked and broken. In the centre of the floor a large shallow circular pit originally 2.5 m in diameter and 0.35 m deep was carved into the bed-rock, possibly part of the beam press (see below). At a later stage the pit was deepened to 0.55 m but the extension had a smaller diameter – 1.75–1.95 m. A large round stone screw base 1.4 m in diameter and 0.55 m high was set into the pit. In its centre was carved a square mortice for installing a fixed wooden screw. As was common in this area (Frankel 1999: 141–142) two of the walls of the mortice broadened out at the bottom to lock the screw in place. The screw base was found moved out of place, apparently in an attempt to remove it for some secondary use.

On the southern edge of the treading floor an irregularly shaped trapezoidal compartment opening onto the floor was carved into the rock wall. It measured 1.0×2.3 m, with its central part carved deeper into the bed-rock. The compartment may have been used for storing grapes before treading, or perhaps it was a place where the beam weight (see below) was emplaced, since it is located opposite the beam anchoring niche. The eastern part of the wine press was not excavated, so it is not clear whether any grape storage compartments are located there, as they usually are in this type of wine press.

The main collecting vat is the northern one. Its overall dimensions are 3.0×2.7 m and 1.8 m deep, i.e. a volume of approximately 14.6 m^3. Its walls are plastered and there are coarse mosaic tesserae at the bottom of its western wall for reinforcement. Its floor, of similar mosaic construction, is mostly unexcavated, but it is reasonable to assume that it contains a sedimentation basin. There is a flight of five stairs adjacent to the western wall – four in a straight line, the fifth at a right angle to the others. The four lower steps are tiled in white mosaic. Construction was added to the vat's southern and western walls, hewn from the natural rock, to complete them. A shelf 0.25 m wide, on which the workers could stand or a wooden cover could be placed, was carved out along the eastern and northern edges of the vat. A stone roller, diameter 0.6 m, length 0.7 m, was found in the rubble that filled the collecting vat.

The smaller collecting vat was hewn out south of the main one, and there was apparently no connection between them – a familiar phenomenon in western Samaria. Its dimensions are 1.00×1.15 m and 0.5 m deep, volume only 0.575 m^3. It may have been used to collect the must produced by secondary pressing.

Outside the treading floor, opposite its north-west corner, a round compartment was discovered, 0.9 m in diameter and 0.55 m deep. It was plastered and floored with mosaic tiles laid in concentric circles. It may have been used for washing the feet before treading the grapes, for holding the marl sprinkled on the grapes, for preparing materials added to the must, etc. There may have been additional vats around the installation, as there are in two large wine presses in the area of H. Nasha (Site 33), south of Zur Natan.

This installation is unique in that no fewer than three distinct methods were used in it for secondary pressing of the rape (trodden grape pulp):

• A cylindrical stone rolled over the rape – a method known and practiced even today in the Mt. Hebron area (Frankel and Ayalon 1988 Fig. 56). Similar rollers were discovered in ancient wine presses at Mishmar Ha'Emeq (Safrai and Linn 1988 Fig. 37), in the Hassolelim-Sippori area in Lower Galilee and other places (see Amit and Baruch in this book).

• A beam press of the lever and weights type. The back end of the beam was fixed in the arched niche in the wall of the treading floor, sacks of rape were placed in the pit in the centre of the floor, and weights (that were not found) were apparently hung from the front end of the beam. Beam-anchoring niches have been found in many ancient wine presses (Frankel 1999: 145–146), including some at Zur Natan. This method continued to be used in the large Druze wine presses of the Golan Heights until recently (Dalman 1928–1942, IV: 354–355, B. 95–96).

• A fixed screw press of the type common in large Byzantine wine presses. It is reasonable to assume that in this installation the screw press replaced the beam press, either because of the introduction of the new technology into the area or because of the cracking and breaking of the upper part of the beam anchoring niche. It is not clear whether the roller was used in parallel with the more sophisticated pressing methods, for instance for pressing small quantities of rape, or perhaps on a smaller floor on the side, and not in the wine press itself.

The pottery found in the wine press indicates that it went out of use at the end of the Byzantine period.

Wine Press D46

This installation was found north-west of the ancient settlement, at the extremity of the excellent piece of land mentioned above. The installation (Figs. 3–4), unique in its dimensions (14×15 m) and ancillary apparatus, is part of a large group of installations that includes another large wine press, two open-air oil presses, three cisterns, tombs and more. A good paved road ran from here to the village, enabling transport by wagon of the huge output produced at the site. The wine press, first surveyed and partially published by Dar (1986: 149, Fig. 90), has now been completely unearthed (Neidinger, Matthews and Ayalon 1990: 18–21).

Some of the components of this enormous installation, most of which were hewn from bed-rock, were uprooted after it ceased functioning and were used for making lime. Two small lime kilns functioned here: one in the pit of the screw base in the treading floor, the other in the large collecting vat. The wine press comprises a large number of components, as follows:

• A large treading floor measuring 6.4×7.0 m, an area of 44.8 m² was originally paved with coarse white mosaic tesserae, most of which – except for those at the edges – were removed in antiquity. A pit was found in the centre of the floor measuring 1.65×1.45 m and 0.85 m deep. A stone base for a wooden screw was no doubt originally installed in it, but this has disappeared. At each of the two eastern corners a low, narrow shelf was carved out, the purpose of which is unclear. An irregularly-shaped compartment (1.65×2.80 m) opening onto the treading floor was hewn out of the eastern wall. The wall, 0.5 m wide, that once separated the treading floor from the two vats to its west, has been destroyed, so that we have no way of knowing the form and exact position of the pipes or channels through which the must flowed. The two ends of the wall have survived, however, so that it is clear where it stood. A rectangular depression in the rock floor, 0.45 m wide, apparently marks the continuation of the channel into the smaller collecting vat.

• Two small treading or pressing installations, found at the southern edge of the treading floor. Each includes a small round shallow upper platform 5 cm deep – the western 0.6 m in diameter, the eastern 0.7×0.8 m – each draining into an elongated receptacle with one semicircular end. The receptacles are open to the treading floor; their dimensions are 0.65×0.70 m, and their depth from the surface of the rock is 0.6 m. Their walls are plastered and the floors paved in white mosaic. At the front of each receptacle a partition was built of mortar and small stones; it was 0.2 m thick and its height was 0.15 cm, the maximum depth of liquid that the receptacle could hold. At the bottom of the partition a pipe 5 cm in diameter was installed, leading to the treading floor. Marks around the eastern receptacle indicate that it (and probably the second one as well) had been roofed by a constructed vault. These installations have parallels of different sizes in various places (Dar 1986: Fig. 42, Pl. 24; Frankel 1999: 139), all of which are too small to be used for storage. Apparently they were used to produce juices and other materials that were added to the must to improve and/or vary the flavor.

• A very large collecting vat, damaged in the past by the installation of a lime kiln inside it, measuring 3.6×2.9 m and 1.9 m deep, thus having a volume of nearly 20 m³. A small filtration basin, about 0.55×0.55 m, was hewn from the bed-rock between the treading floor and the collecting vat. Its floor was overlaid with white mosaic, of which only that in the north-eastern part has been preserved.

Four steps were cut into the south-west corner of the collecting vat. The walls of the vat are plastered and its floor covered with coarse white mosaic. In the floor of its eastern part a rectangular pit was hewn, 0.75×0.40 m and

0.35 m deep, with a floor of natural bed-rock into which a small depression had been carved. Alongside it, on a slightly higher level, there is a square pit 0.37×0.37 m, with a mosaic floor. Both these pits were apparently used as sedimentation basins for precipitating foreign matter. On the eastern and southern sides a narrow shelf 0.65 m wide had been carved around the perimeter of the vat, upon which the workers were stationed.

- A smaller collecting vat – 2.35×1.75 m, depth 1.7 m, volume ~7 m^3. Its walls were plastered and its floor paved with coarse white mosaic tiles. Five steps were hewn into the western wall – four in a straight line, the fifth at a right angle to the others. A rectangular sedimentation basin was carved into the floor of the vat at its north-east corner, measuring 0.7×0.4 m and 0.35 m deep, with a mosaic floor. Around the eastern, northern and western sides of the vat a shelf of varying width was hewn out of the rock.

In the wall separating the collecting vats two connecting outlets were cut, each 0.12 m in diameter, that slanted downward from the large vat to the small one, for runoff of surplus must. The western outlet is an open channel, the eastern a pipe.

- Two devices carved into the top of the wall on the north side of the smaller collecting vat: a round basin 0.5 m in diameter and 0.33 m deep, and a tiny expressing installation comprising a square pressing surface 0.2×0.2 m and 2 cm deep, slanting toward a depression 0.17 m in diameter and 0.13 m deep. These installations were apparently used to produce plant juices which were added to the must.

- Remnants of a rigid frame direct pressure screw press typical of those usually used in oil presses (Frankel 1999: 122–137). Similar presses have been found beside wine presses in the past (e.g. at another wine press at Zur Natan [Ayalon, Charvit and Qidron 1990: 97–100] and at a wine press in Lower Galilee [Shenhav 1984: 67, 123–124]). The remnants of the press were discovered west of the small collecting vat. In the bed-rock on each side of what had been the press-bed was carved a rectangular mortice for securing the frame of the press and beside them sloping approach channels. Centered between them remained the northern edge of the press-bed, that had been 0.65 m in diameter and encircled by a shallow groove, in which the must expressed from the rape collected and flowed to the collecting basin of the press. Only the bottom of the basin was preserved as a shallow depression on top of the shelf surrounding the collecting vat in front of the press-bed. This indicates that the press was in operation before the shelf was carved. It is possible that at first the wine press had only one collecting vat (the larger) and a direct pressure pressing installation. The second (smaller) vat was hewn at a later stage, causing the destruction of the direct pressure screw press. At about that time the fixed screw press, typical of Byzantine wine presses, was probably installed in the centre of the treading floor.

Two installations hewn out of the bed-rock south of the collecting vats were used for storing the grapes before they were trodden (Fig. 5). Each consisted of an irregularly-shaped shallow platform about 5 cm deep with a small gutter through which must, draining from the grapes stored on the surface, flowed into a relatively small collecting vat. The northern platform was 2.25×3.00 m (area 6.75 m^2), the southern about 2×2 m (4 m^2) in area. The collecting vats were hewn on a level approximately 0.5 m lower than the platforms. They were roughly cubical, approximately 0.6×0.6×0.6 m, with plastered walls. The floor of the northern one was tiled with white mosaic while that of the southern one was bed-rock. Alongside the northern pit and slightly above it on the vertical rock wall, a projection was left in the shape of a small rounded shelf, apparently for placing the vessel used for removing liquid from the vat. To the north two steps had been hewn out to facilitate descending into these small vats. A storage surface like these, where provisions had been made for collecting the choice must that drained from the stored, untrodden grapes, was called *mishtah shel ‘alim* – “surface of leaves” (Mishnah *Toh.* 10, 4; Tos. *Toh.* 11, 8).

Because of the destruction of parts of the wine press and the operation of lime kilns in it, no finds at all remained in the installation. However, its dimensions and its components – the plaster, the mosaics and the screw presses – have enabled dating it to the Byzantine period. Perhaps its large size and sophistication can be explained in the light of the history of the ancient village, as elucidated by the excavations (Ayalon 2008).

At the beginning of the Byzantine period, the village (that, according to the inscription found in its synagogue, was called *Antesion*) was settled by Samaritans. Apparently each family cultivated its own plot of ground and cut its own installations out of the bed-rock (see Site 30). That would explain why there are so many Byzantine-type wine presses scattered along the entire length of the spur. At the end of the 5th–beginning of the 6th cent. the village was damaged, apparently during the Samaritan revolts against the Byzantine empire, and a large Christian monastery was established there (Neidinger, Matthews and Ayalon 1994: 11–12). At this stage agricultural activity may have been organized under a single administration, and the need may have arisen for larger installations for producing agricultural products.

This is a rare example of the possibility of linking – with all due caution – technological development with ethnic and/or historical changes.

Notes

1. The wine presses were excavated by the author on behalf of the Eretz Israel Museum, Tel Aviv and the Texas Foundation for Archaeological and Historical Research, Houston.

Bibliography

Ayalon, E. 2008. "Migdal, Horbat". *NEAEHL* 5: 1950–1952.

Ayalon, E., Charvit, Y. and Qidron, A. 1990. "Grouped Installations from the Roman-Byzantine Period in the fields of Zur Natan". *Israel – People and Land, Eretz Israel Museum Yearbook* 5–6 (23–24): 93–120 (Hebrew).

Dalman, G. 1928–1942. *Arbeit und Sitte in Palästina*. 7 Vols. Gütersloh (rep. Hildesheim 1964).

Dar, S. 1986. *Landscape and Pattern. An Archaeological Survey of Samaria 800 B.C.E. – 636 C.E.* (Bar IS 308). Oxford.

Frankel R. 1999. *Wine and Oil Production in Antiquity in Israel and Other Mediterranean Countries*. Sheffield.

Frankel, R. and Ayalon, E. 1988. *Vines, Winepresses and Wine in Antiquity*. Tel Aviv (Hebrew).

Neidinger, W., Matthews, E. and Ayalon, E. 1990. *Preliminary Report on the Excavations at Zur Natan, 1989 and 1990 Seasons*. Houston.

Neidinger, W., Matthews, E. and Ayalon, E. 1994. *Report on TFAHR Excavations at Zur Natan, Israel*. Houston.

Safrai, Z. and Linn, M. 1988. "Excavations and Surveys in the Mishmar Ha-'Emeq Area". In: B. Mazar (ed.). *Geva, Archaeological Discoveries at Tell Abu-Shusha, Mishmar Ha-'Emeq*. Tel Aviv: 167–214 (Hebrew).

Shenhav, E. 1984. "Ya'ar Lavi Survey". *HA* 84: 18 (Hebrew).

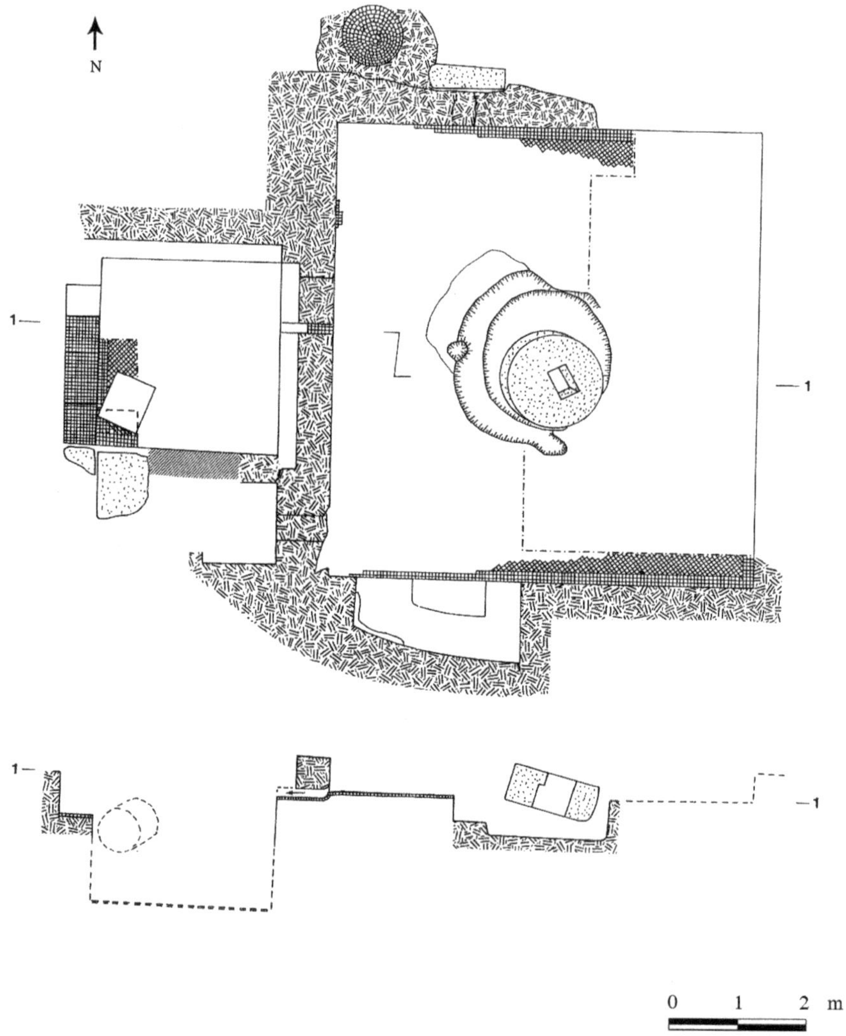

Fig. 31.1. Zur Natan, wine press D1, plan and cross-section.

Fig. 31.2. Zur Natan, wine press D1, view to the north.

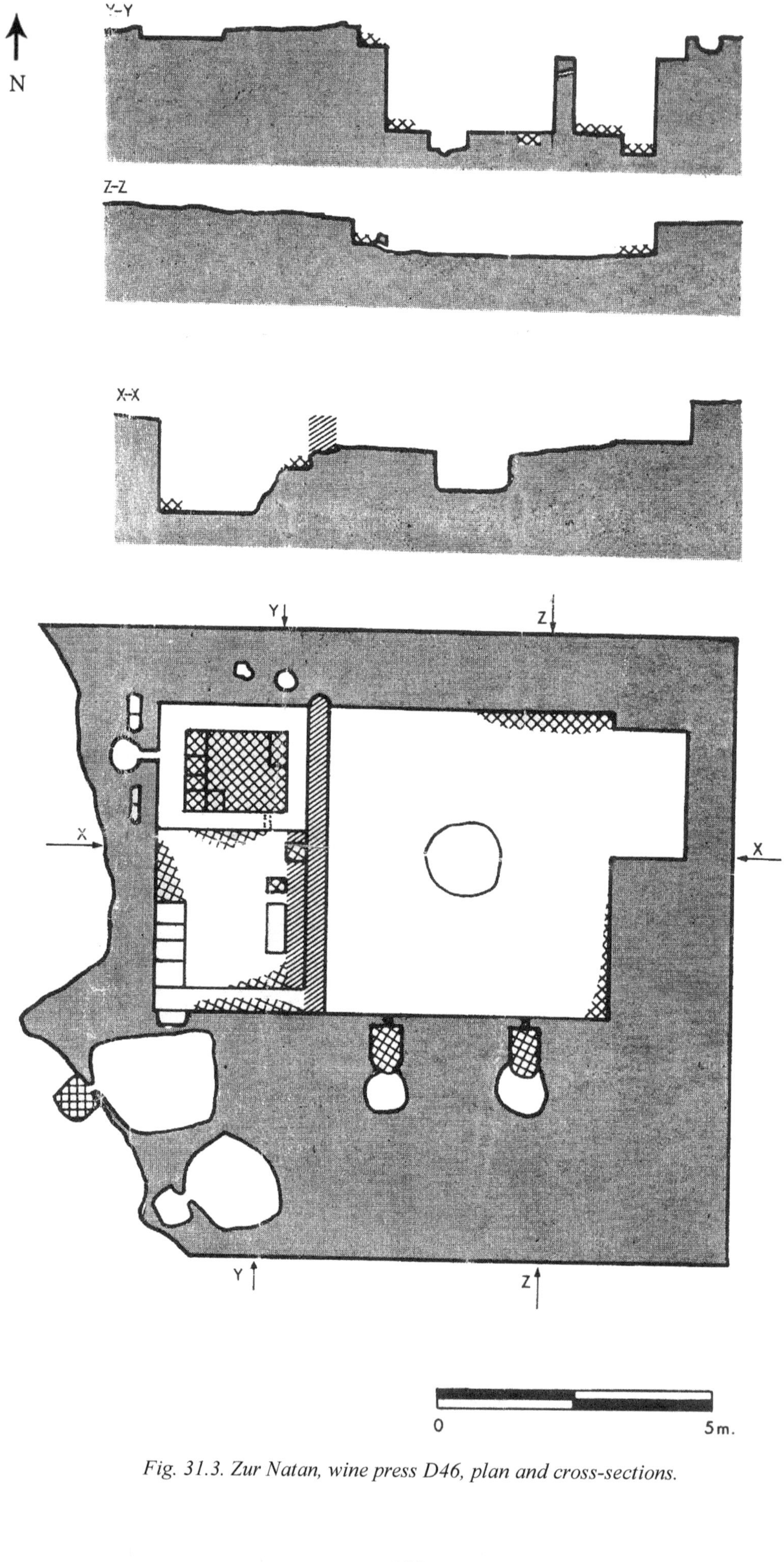

Fig. 31.3. Zur Natan, wine press D46, plan and cross-sections.

Fig. 31.4. Zur Natan, wine press D46, view to the south-east.

Fig. 31.5. Zur Natan, wine press D46, two installations for storing the grapes before they were trodden, view to the north-east.

32. Byzantine 'Industrial' Oil Presses at Zur Natan

Etan Ayalon and Yehoshua Dray

Within the confines of H. Migdal – the central site on the low spur of Zur Natan in western Samaria (MR 1510/1832) – two "industrial" oil presses from the Byzantine period were unearthed (Ayalon 2008).[1] These are quite different from the dozens of open-air oil presses found from one end of the spur to the other (see Site 30). The two large oil presses were located at the south-western part of the ancient village, near the ancient Samaritan synagogue.[2]

The Oil Press in Area B

This installation was erected to the north-west of the synagogue, together with two ritual baths (*miqva'oth*). It was severely damaged when a tower was built on top of it during the Early Arab period (7th cent. CE), but nevertheless sufficient remains to make it possible to comprehend its major components (Dray 1994):

• The Crushing chamber, located in the centre of the installation. It was apparently entered from the west, alongside one of the ritual baths (below). The dimensions of the room (L. 2087, 2092) were 4.8×5.8 m, i.e. area of 27.8 m^2. An opening in the northern wall led to one pressing chamber, and there is no doubt that a similar opening led to the second pressing chamber to the south.

The crushing chamber is floored with hard plaster, upon which remains a layer of ash containing olive pits. Fragments of the crushing basin and the crushing stone were found in the room. The basin, approximately 2.4 m in diameter, had a base about 0.5 m high; it was set upon a hard foundation 7 cm thick. Alongside the basin were found remains of a fibrous plant material that could have been remnants of rope or of a frail, as well as charred wooden beams that had been part of the roof and/or the installation.

• The northern pressing chamber (L. 2090, 2091) measures 2.2×7.3 m, i.e. area of 16 m^2. Its walls were plastered and its floor was made of polygonal slabs of hard limestone, well-fitted to each other. Integrated into the floor in secondary use was a *bodeda* – a small portable device for expressing oil or other liquids. On the floor a thick layer comprising clay and charred tree branches was found, apparently the remains of the roof. The pressing installation, discovered intact, included the following parts (east to west):

- A niche for anchoring the back end of the beam, built into a wall specially reinforced for the purpose. The niche is 0.55 m wide and 0.5 m deep; the original height could not be established.

- A central collecting vat, with a lid/press-bed consisting of a round stone slab 1.4 m in diameter with circular and radial grooves leading to a central opening. The expressed liquid flowed along the grooves and through the central opening into the collecting vat that was under the press-bed. The vat was not investigated because a later wall had been built above it.

- Beyond the pressing point was found a large cut stone (a lintel or doorsill in secondary use) placed across the line of the beam on a foundation of other large stones. In the two holes hewn into it, wooden poles were installed one on either side of the press beam, to stabilize it and keep it from moving sideways.

- The cylindrical screw weight (diameter 1.03 m, height 1.5 m and weighing approximately 3.45 tons) was found *in situ*. A socket was carved into its upper surface, in which the wooden screw rotated. In the socket were found remains of the screw, made from the wood of the common jujube (*Zizyphus jujuba*).[3] A pair of dovetail mortices were carved into the sides of the weight, into which were fitted planks of oak (*Quercus* sp.) which fixed the screw to the weight. The distance of the weight from the niche in the wall shows that the length of the beam was approximately 7.5 m.

• Only two components of the southern pressing chamber have survived: the screw weight and (outside the room) the press-bed. The weight is of the type described above, i.e. a cylindrical weight with screw socket and dovetail mortices. It is 1.05 m in diameter and 1.43 m high, therefore weighing approximately 3.45 tons. The press-bed is made of hard limestone; it is rectangular and measures 0.94×1.37 m. Around its perimeter is a circular groove terminating in a gutter in the centre of one of the long sides, indicating that this installation had a lateral collecting vat.

• Adjacent to the oil press, two ritual baths (*mikva'oth*, sing. *mikveh*) were found – a phenomenon known from other settlements populated by Samaritans who, like the Jews, punctiliously observed the ritual cleanness (Magen 1993). The upper part of both *mikva'oth* had been destroyed. The northern one (L. 2094) was built adjacent to the crushing chamber of the oil press. It is L-shaped. The base of the L measures 1.65 m across and 0.55 m wide, while the leg is 1.75 m long and 0.8 m across; the bath has been preserved to a depth of 1.25 m. Of the steps generally descending into a *mikveh*, only the high bottom one remains; its length was about two-thirds of the width of the base of the L. The inside of the ritual bath is coated with fine pink plaster.

The second *mikveh* (L. 2124), found south of the oil press, is rectangular and measures 1.65×1.75 m. Two steps built across its entire width were preserved in it, as was the mortar base of the one above them. The lower step and the floor of the *mikveh* were tiled with large mosaic tesserae in white, red and black – a rare phenomenon in *mikva'oth*. The inside walls are coated with fine pink plaster.

- Several rooms found adjacent to the oil press were probably used as storerooms.

The Oil Press in Area C

This installation (situated 25 m to the west of the press of Area B described above) was apparently very similar to it in form and dimensions. It suffered severe damage as a result of recent development activities (Neidinger, Matthews and Ayalon 1990: 13–17). Excavations in this location revealed the walls of several rooms and a water cistern. It is reasonable to assume that here too the central room was the crushing chamber – a broken crushing stone about 1 m in diameter was found in it. The northern pressing installation (L. 3015, 3021) was the only complete one (Figs. 1–3). Its internal measurements are 3.6×8.2 m and it is divided by a partition (WC5) into two unequal sections: a pressing room to the west and a room on the east for the weight.

- The pressing chamber was well-floored with slabs of hard limestone. Its entrance was from the south-east, where one of the jambs of the entryway was preserved. The rear wall (WC4) that held the niche (that was not preserved) for anchoring the beam was thicker and more solidly constructed than the other walls. The central collecting vat was integrated into the floor of the room. Its lid/press-bed measures 1.2×1.3 m and is 0.24 m thick. It is also equipped with circular and radial grooves leading to the central opening whose diameter is 0.33 m. The lid was mounted on top of a cylindrical collecting vat made of stone, with an inside diameter of 0.94 m and a depth of 0.6, and with a sump in its floor. In the sump, as in the entire installation, olive pits were found.

- In the partition separating the two parts of the room there was an opening to enable passage between them. On top of the northern section of the partition a large stone was mounted in which four sockets had been cut; poles were inserted into the sockets to stabilize the beam on both sides.

- In the small eastern room (L. 3021) the screw weight was found in a pit dug into the beaten-earth floor and lined with stones. The weight was 1.07 m in diameter and 1.55 m high, weighing 3.86 tons. A socket was cut into its top and two dovetail mortices in its sides for fastening the screw in place. In the weight pit were found charred pieces of a heavy beam that had been made from a Cedar of Lebanon (*Cedrus libani*), that had apparently been either a roof joist or the press beam of the installation. The length of the press beam was calculated as approximately 7.5 m (Fig. 4).

Discussion

The two oil presses described here are very similar in form and structure and are located close to each other. It is obvious that they were built together in an organized manner. In other sites as well, e.g. Qedumim (Site 34), several oil presses were discovered in the same village (Magen 1993). Their location on the south-western fringes of the site was close enough to make it easy for their operators to get to work and to guard the presses, yet far enough away so that this industrial enterprise did not affect the quality of life in the village.

The crushing mills here are no different from those in hundreds of other oil presses, but the central collecting vats are of a subtype that is not widely distributed. Similar installations have been discovered at, for example, Shuni (Site 25), Shiloh, Kh. 'Azzun and one hewn from bed-rock at Megaddim in the Carmel (Frankel 1999: 85). This distribution pattern raises the possibility that this type of collecting vat was developed by the Samaritans (Ayalon 1996).

One unusual feature of the presses in Area B is that in one the collection of the expressed liquid is central while in the other it is lateral.

Placing the weight in a pit contributed to its stability when it was lifted into the air and hung suspended from the screw under the beam end. This phenomenon is well-known from many sites. It also prevented the whole press from being too high above floor level.

In light of their location, form, parallels and stratigraphy it appears that these oil presses were built in the 4th or 5th cent. CE by the Samaritans, that explains the presence of the ritual baths in the eastern installation. At some stage, apparently in the 6th cent., the *mikva'oth* were abandoned and other buildings built in their place. This was apparently connected with the establishment of a large complex of buildings at the site – a monastery or estate – by the Christians. The eastern oil press (Area B), integrated into a new building, continued to operate (as shown by the artifacts found in it) until the 7th cent., when a tower was built above it. The Samaritan *mikva'oth*, however, went out of use and were covered by the later building. The western press (Area C), on the other hand, continued to be used until the 8th cent., when the entire site was abandoned. Near the end of its service the crushing chamber went out of use and small installations, not connected to oil production, were built in it.

Notes

1. A third oil press of this type may have operated at the north-west corner of the site, where a screw weight was discovered *in situ*; however, the area has not yet been excavated.
2. Excavations at the site were carried out on behalf of the Eretz Israel Museum, Tel Aviv and The Texas Foundation for Archaeological and Historical Research, Houston. They were directed by E. Ayalon, W. Neidinger and E. Matthews with the participation of Y. Dray and volunteers.
3. The wood remains were identified by Prof. Nili Liphschitz from the Institute of Archaeology, Tel Aviv University.

Bibliography

Ayalon, E. 1996. "The "Samaritan" Olive Oil Press". In: D. Eitam and M. Heltzer (eds.). Oli*ve Oil in Antiquity, Israel and Neighbouring Countries from the Neolithic to the Early Arab Period* (HANE/S VII). Padova: 149–156.

Ayalon, E. 2008. "Migdal, Horvat". NEAEHL 5: 1950–1952.

Dray, Y. 1994. "The Oil Presses of Zur Natan". In: W. Neidinger, E. Matthews and E. Ayalon. *Report on TFAHR Excavations at Zur Natan, Israel*. Houston: 14–15.

Frankel, R. 1999. *Wine and Oil Production in Antiquity in Israel and Other Mediterranean Countries*. Sheffield.

Magen, Y. 1993. "Qedumim – A Samaritan Site of the Roman-Byzantine Period". In: F. Manns and E. Alliata (eds.). *Early Christianity in Context, Monuments and Documents* (Studium Biblicum Franciscanum, Collectio Maior, 38). Jerusalem: 167–180.

Neidinger, W., Matthews, E. and Ayalon, E. 1990. *Preliminary Report on the Excavations at Zur Natan, 1989 and 1990 Seasons*. Houston.

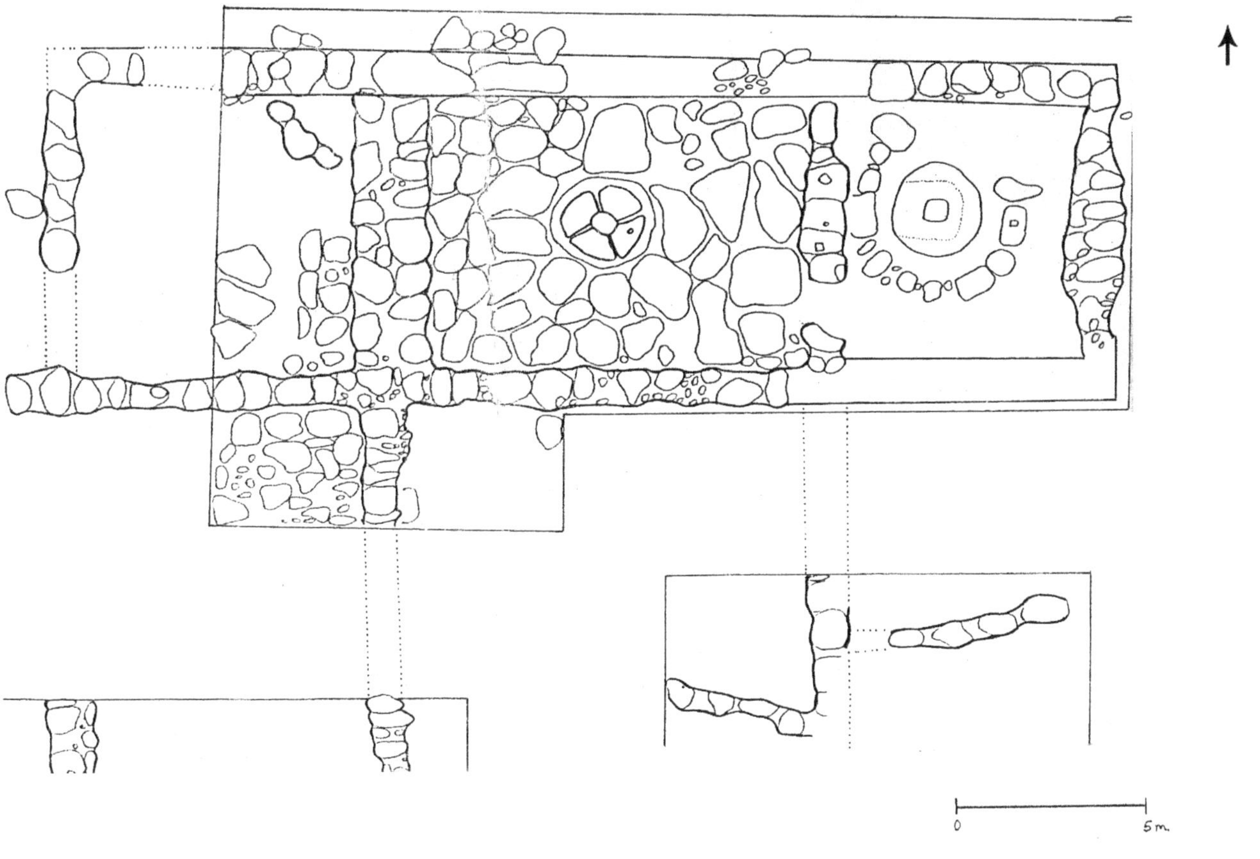

Fig. 32.1. Zur Natan, oil press in Area C, plan of the northern pressing installation.

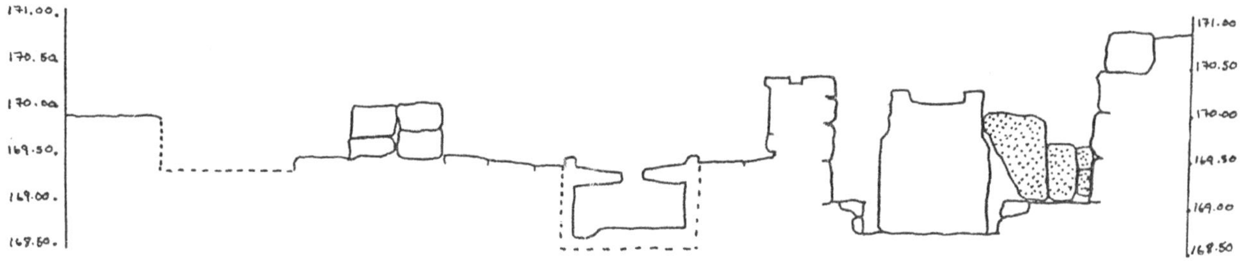

Fig. 32.2. Zur Natan, oil press in Area C, section of the northern pressing installation.

Fig. 32.3. Zur Natan, oil press in Area C, the northern pressing installation, view to the west.

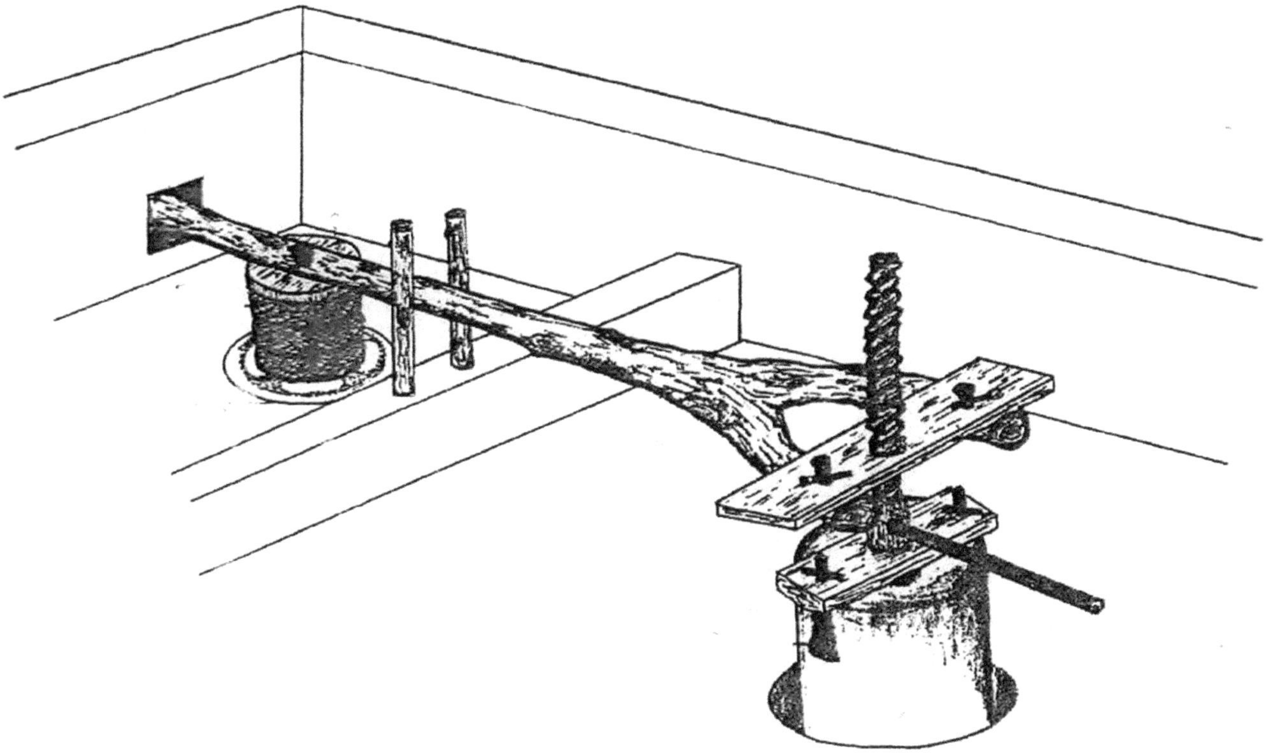

Fig. 32.4. Zur Natan, oil press in Area C, reconstruction of the northern pressing installation.

33. A Byzantine Wine Press at H. Nasha

Etan Ayalon

H. Nasha is a rural settlement situated on the western fringes of the Samarian Hills, at the edge of Kokhav Yair (MR 1502/1807). A small-scale salvage excavation has recently been carried out at the site (Mettens 2004). A detailed survey of the settlement carried out by the author revealed three oil presses within its confines and a few wine presses at its periphery. Pottery finds prove that the site was settled from the Middle Bronze Age II until the Early Arab period. The wine press under discussion here was discovered at the western edge of the site and was cleared by the author in 1989 (Ayalon 1997).

The large complex installation, measuring 13×16 m (208 m^2), comprises the following parts (Fig. 1):

- Two or three storage areas for holding grapes before treading. The inside measurements of the eastern one (L. 3) are 3.8×5.1 m (19.4 m^2). It is floored in coarse white mosaic and surrounded by a low wall. The western storage area (L. 4), measuring approximately 4×4 m (16 m^2) overall, is made partly of a large rock smoothed flat and partly of a mosaic floor surrounded by a low wall. It appears that there was a similar storage area south of the treading floor, on a level 1.7 m above it (L. 5). There are many white tesserae scattered in this area but because of its proximity to the surface it was evidently damaged and subsequently destroyed.

- A large treading floor (6.5×6.8 m [44.2 m^2]) of which only the eastern half (L. 1) has been cleared. Its mosaic floor has only been preserved in a few places. It was laid on a weak under layer of small stones, a fact that explains its inferior state of preservation. A stone screw base is built into the centre of the floor and in it has been hewn a mortice for installing a fixed wooden screw (Fig. 2). The base is elliptical, measuring 1.9×1.6 m. The central dovetail mortice, with the southern part being wider than the northern, is 0.5 m deep. In the wider part, two of the sides of the mortice broaden out toward the bottom so that the screw can be fixed in place. A screw mortice of this shape is unusual in this region in which the mortices are almost always square (Frankel 1999: 140–145).

- A small square filtration basin (L. 8) measuring 0.9×0.9 m and 0.5 m in depth (approximately 0.4 m^3), built adjacent to the north-east corner of the treading floor. A shallow channel 0.3 m wide, paved in mosaic, leads from the treading floor to the filtration basin, cutting through the wall that separates it from the treading floor. The basin is plastered and floored in white mosaic. From its bottom a ceramic pipe 7 cm in diameter leads to the collecting vat situated on its western side. As both vats were built in a large pit that had been dug out previously and as the filtration vat was narrower than the collecting vat the cavity left to its north was filled with stones.

- A large rectangular collecting vat measuring 2.8×3.2 m and 1.6 m in depth, the volume being 14.3 m^3. Its western wall was hewn out of the rock and the vat was originally wider there; however, the heavy layer of mortar and plaster that was applied (apparently to give the vat a more regular shape) narrowed it somewhat. The vat is well plastered in two layers: a base layer of coarse gray plaster reinforced with fragments of ribbed pottery, and a finer, smoother outer layer. The bottom of the vat is paved with large white mosaic tesserae. Three stairs were installed in the north-east corner; fragments of mosaic remain on the topmost step. In the north-west corner was found a large, round sedimentation basin with a mosaic floor. The corners of the collecting vat were reinforced by small rounded mortar additions, and the entire vat was supported on the north side towards the slope by a solid retaining wall. In the south-west corner of the vat, on the wall separating it from the treading floor, a gutter was installed, about 0.3 m wide and floored with mosaic, that enabled the flow of must from the treading floor directly into the collecting vat, in addition to the indirect route via the filtration basin.

- Four small compartments built around the wine press. The two round northern ones are contiguous with the corners of the treading floor. Of these, the western one (L. 7) is the larger, diameter 1.3 m, depth 0.8 m. The eastern one (L. 9) is 0.9 m in diameter and 0.85 m deep; it has mosaic around its edge, on the wall. Both are plastered and paved with white mosaic comprising relatively small tesserae in concentric circles.

The two southern compartments that are at a slight distance from the treading floor are semi-circular in shape. The eastern one (L. 10) measures 1.2×1.2 m and is 0.8 m in depth. Of the western one, only a corner remains. Both are plastered, with mosaic floors.

It is not clear what purpose was served by the compartments (Frankel 1999: 139). They could have been used for washing the feet before treading; to store marl that was sprinkled on the grapes to clarify the must; to produce and/or prepare materials to be added to the must to improve its taste; or for other purposes. One hypothesis that arose – that they were used as holding tanks for the first must expressed – is unlikely as no pipes or channels were found connecting them to any part of the wine press. Similar compartments were common in this region: they are known from a large wine press at H. Hanut, west of Kokhav Yair, from at least two wine presses from Zur Natan to the north (Site 31) and from Kefar Sirkin (Sidi, Amit and 'Ad 2003) and Rishon le-Zion (Site 42) to the south.

Because of its shape, dimensions and components (plaster, mosaics and screw) the wine press is dated to the Byzantine period. On the basis of pottery found on the treading floor and in the vats and basins – some of it of the "Mafjar" type – it can be assumed that like many others of its kind, this wine press went out of use at the beginning of the Early Arab period (Ayalon 1997). A small portable stone expression device (Hebrew: *bodeda*) was found in the collecting vat; it was apparently used for expressing small quantities of liquid, perhaps for improving the must.

It should be noted that in this wine press, a measurement of 0.45 m appears repeatedly. Apparently it was a standard measurement in use in the area at that time, possibly a short cubit (Hebrew: *amah*).

Bibliography

Ayalon, E. 1997. "The End of the Ancient Wine Production in the Central Coastal Plain". In: Y. Friedman, Z. Safrai and J. Schwartz (eds.). *Hikrei Eretz, Studies in the History of the Land of Israel Dedicated to Prof. Yehuda Feliks*. Ramat Gan: 149–166 (Hebrew).

Frankel R. 1999. *Wine and Oil Production in Antiquity in Israel and Other Mediterranean Countries*. Sheffield.

Mettens, A. 2004. "Horbat Nashé". *HA-ESI* 116: 29*-33*.

Sidi, N., Amit, D. and 'Ad, U. 2003. "Two Winepresses from Kefar Sirkin and Mazor". *'Atiqot* 44: 253–266.

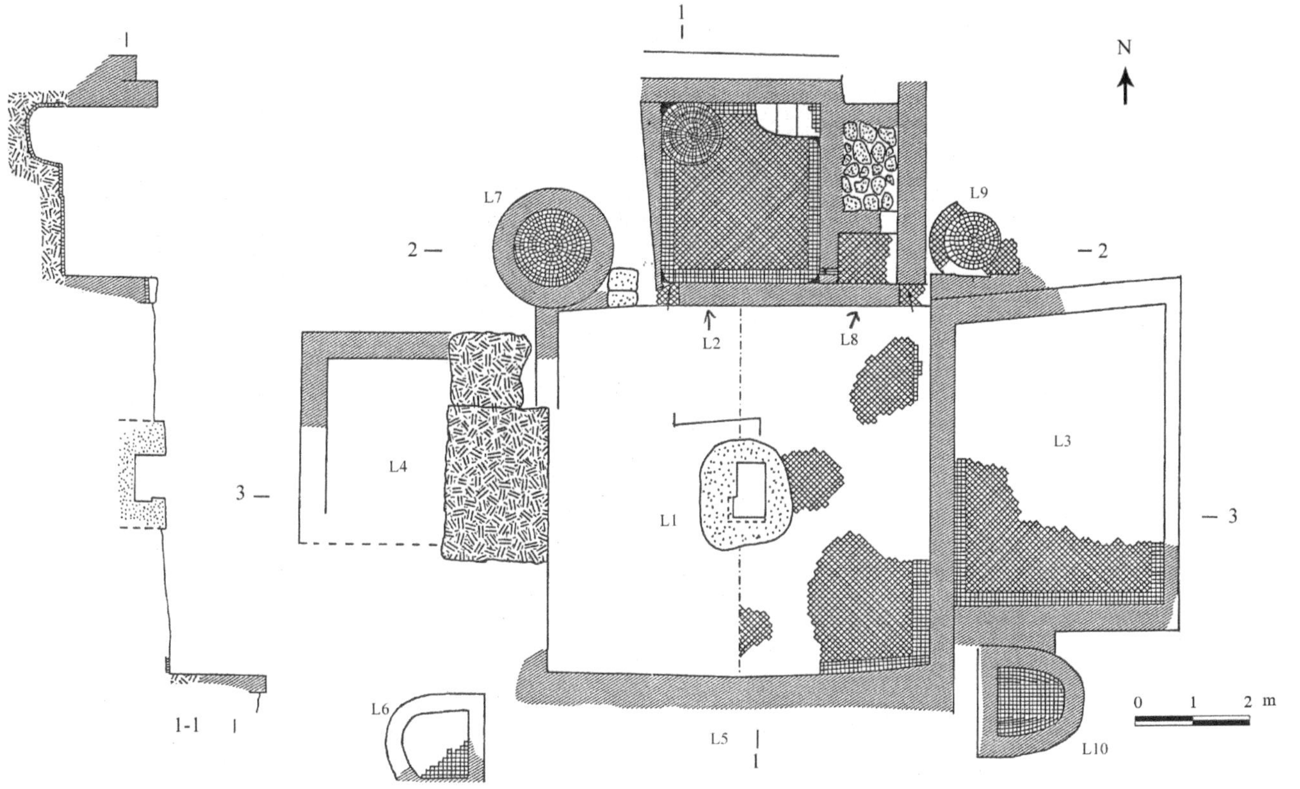

Fig. 33.1. Nasha, plan and section of the wine press.

Fig. 33.2. Nasha, the screw base, looking north.

34. Oil Presses at Qedumim

Etan Ayalon, Rafael Frankel and Amos Kloner

The Site

Qedumim is a Samaritan site (MR 165/179) about 10 km west of Shechem/Nablus that was excavated since 1979 by Dr. Izchak Magen, the Archaeological Staff Officer of the Judaea and Samaria Command (Magen 1982; 1993).

The excavations revealed burials at the site from the Early Bronze Age I and Middle Bronze Age II and some sherds from the Iron Age II, but actual settlement started only in the Persian period. There was evidence that the site was occupied also in the Hellenistic period but there was no settlement in the 1st cent. BCE. Occupation of the site was renewed in the Roman period at the end of the 1st cent. CE and continued in the Byzantine period. There was again a break in settlement at the end of the 5th cent. lasting till the end of the 6th cent., when once more settlement was renewed lasting till the Early Arab period and the Middle Ages.

The Oil Presses

Three Oil press complexes were excavated at the site: a twin lever press from the Roman period in Area C, a lever and screw press from the end of the 5th-beginning of the 6th cent. in Area F further to the south and various elements probably from more than one press in Area M.

In Area C the crushing mill was destroyed in recent building activities. The press room (8.0×6.5 m – 52 m^2) was paved with stones and its walls well plastered (Fig. 1). In it were two parallel beam presses. The rectangular beam niches are unusually large reaching the floor (Fig. 2). On both sides of each press-bed were two plain piers that served to support the frails and the beam. The press-beds were covered by later construction but were apparently round and were connected by bores to rectangular collecting vats with stone covers in which there were small round apertures. The weights were not found. Three plastered compartments found in the room probably served to separate the oil from the watery lees. In Area C a free standing collecting vat was found, with circular and radial grooves on the wide rim (Fig. 3). This was probably used to produce "first oil". Frails of olive mash were probably placed on the vat between crushing and pressing. Amongst the finds connected to the press were bone shovels that were probably used to load olive mash or residue (compare Mishnah *Shab.* 8, 6), storage jars and many oil lamps. The press room was destroyed by fire in the end of the 5th cent.-beginning of the 6th cent. and abandoned. A ritual bath (*mikveh*) found near the oil press probably served its Samaritan workers.

In the press of Area F (Fig. 4), the beam niche was not found but all the other components were found *in situ*. The crushing basin is situated 1.5 m north of the press-bed and has a flat, worn crushing surface and sunken socket. The press-bed and collecting vat are identical to those of Area C but lack plain piers. The screw weight is in a built round pit 2.5 m west of the press-bed. The screw weight is of an unusual type: as well as a central socket in which the screw revolved it has both two lateral dovetail mortices like those of the Samaria weight and two internal mortices like those of the Kasfa weight. A *mikveh* was found near this oil press also. The finds show that this complex was built in the end of the 5th cent.-beginning of the 6th cent. and abandoned in the 7th-8th cent.

In Area M there is a crushing basin similar to that in Area F but with a slightly raised socket. A broken crushing stone was found nearby. In another room there is a screw weight of the Samaria type but rectangular in shape with rounded corners instead of the more usual cylindrical shape. In yet another room there are two small beam weights with bores in the shape of a reversed T. The weights of the presses in area C were probably of this type. This complex also included a *mikveh*, storage rooms and a stable where the customers' animals were kept during work. The oil press operated in the end of the 6th and the 7th cent.

Conclusions

The oil presses of Qedumim are unique in three aspects that were both first noticed at this site:

- The concentration of no less than three large and sophisticated oil presses in a small part of a modest village is impressive and testifies to the importance of oil production in this mountainous area during the Byzantine period.

- The fact that those industrial complexes included ritual baths (*mikva'ot*) reflects the importance of ritual cleanness in the oil production process at this Samaritan village.

- Of particular interest is that while in both presses C and F the collection is lateral, C has plain piers on either side of the press-bed while F does not.

Bibliography

Magen, I. 1982. *The Archaeological Discoveries at Qedumim – Samaria, Qedem Museum*. Jerusalem.

Magen, Y. 1993. "Qedumim – A Samaritan Site of the Roman-Byzantine Period". In: F. Manns and E. Alliata (eds.). *Early Christianity in Context, Monuments and Documents* (Studium Biblicum Franciscanum, Collectio Maior, 38). Jerusalem: 167–180.

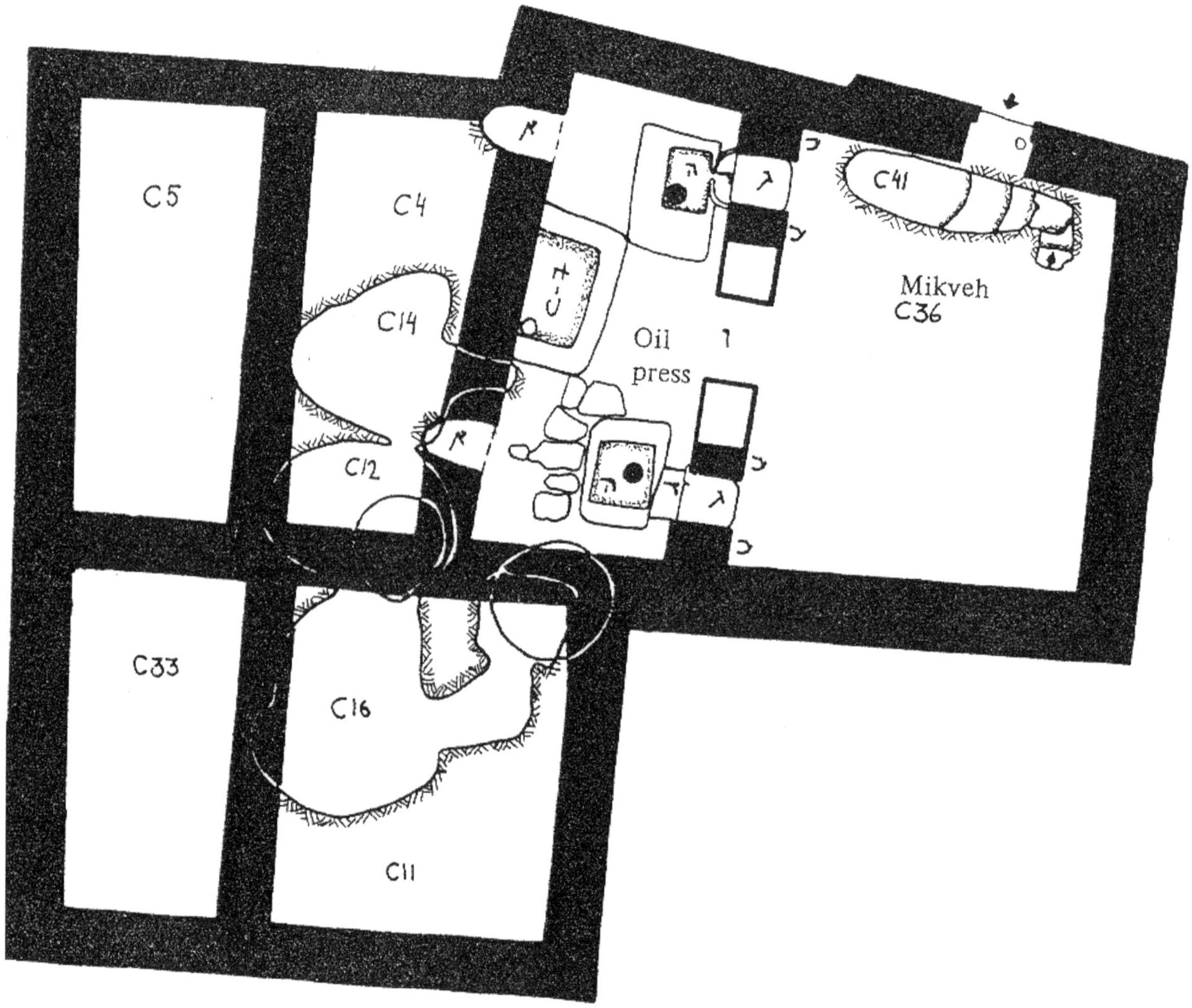

Building I

Fig. 34.1. Qedumim, plan of the oil press in Area C.

Fig. 34.2. Qedumim, oil press in Area C, part of the pressing installation.

Fig. 34.3. Qedumim, oil press implements in Area C.

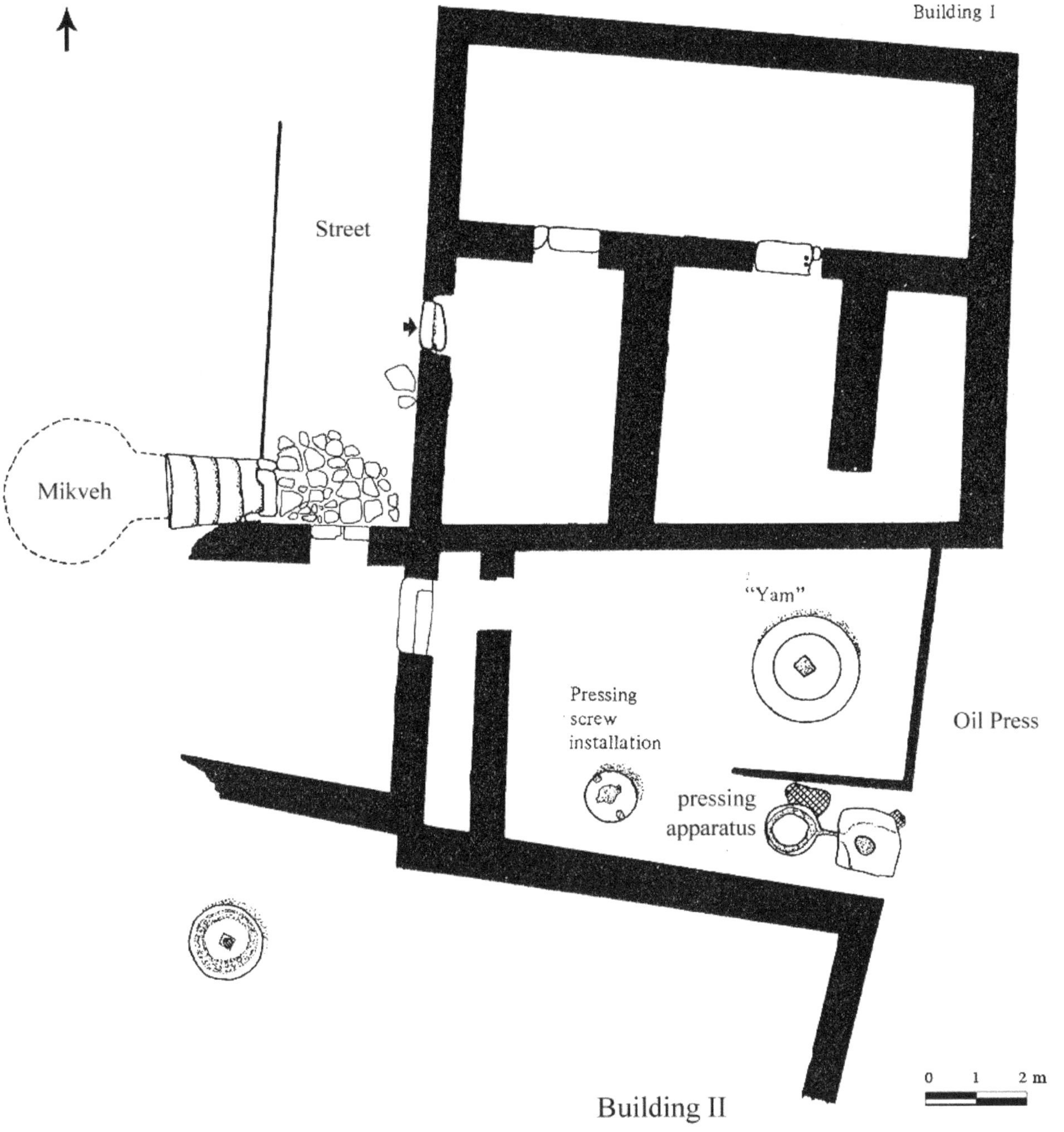

Fig. 34.4. Qedumim, plan of oil press in Area F.

35. A Large Hellenistic Wine Press at Tel Michal

Ze'ev Herzog

The Site

Tel Michal is located on a sandstone ridge overlooking the beach at Herzliya (MR 1310/1742). Remnants of the settlement are scattered over five small hills, the principal one of which is a high tell, some 30 m above sea level and covering an area of about 3 dunams. Excavations on the tell revealed remains from all the periods during which the site was settled, from the Middle Bronze Age IIb to the Early Arab period. On the northern hill there was a settlement that existed mainly during the Persian period.

The site was first surveyed in 1922 by Jacob Ory. From 1958 to 1960 the north-eastern hillock of the mound was excavated by Nahman Avigad on behalf of the Department of Antiquities. From 1977 to 1980 the Archaeological Institute of Tel Aviv University undertook comprehensive excavation of the tell, under the direction of the author.

The Wine Press

The wine press was in use in Strata V-IV (Figs. 1–2). It was built on the northern hill at a time when the high tell was occupied by a fort and houses. It was constructed on top of the remains of buildings from the Persian period. Some of the earlier walls were reused in the framework around the wine press, but most of them were razed and their stones reused as an additional protective zone beyond the surrounding outer wall, perhaps to reduce the amount of sand and dust in the vicinity.

The reuse of segments of earlier walls may have caused the discrepancy in orientation that exists between the outer irregular surrounding wall and the fine rectangular wine press. The stone pavement around the press varies in width from 1 m on its west side to 2.75 m on its north.

The plastered wine press (6.05×9.20 m) is sunk about 0.3 m lower than the stone pavement that surrounds it. Its components consist of Treading Floor 556 (5.25×6.05 m – 31.8 m^2) separated by Wall 1191 from the two deep vats to its west. Although the plastered area of the wine press is almost a perfect rectangle, this wall is not parallel to the eastern and western walls, and the vats are not completely regular in shape.

Vat 551B that was filled via a pipe from Floor 556 has rounded corners; it measures 1.55×1.40 m, depth 1.24 m, i.e., total (as opposed to usable – see below) volume of about 2.6 m^3. A second pipe connects this vat with the adjacent Vat 551A that is much larger, measuring 2.0×1.9 m, depth 1.9 m, i.e. total capacity of about 7 m^3. Three steps lead down into it. Both vats had a bowl-like depression in the bottom for collecting the last drops of the precious liquid.

The surfaces of the wine press were plastered with several coats of a lime composition with a fine outer surface. The wine press probably had no roof since no supporting pillars were found inside and the surrounding walls were too far apart, too low and too poorly constructed to bear the weight of a roof.

As mentioned above, the wine press was in use in Strata V and IV (3rd-2nd cent. BCE), but in the latter stratum some alterations were made north of Floor 556. Whereas in Stratum V the entire treading floor was utilized (and Compartment 571 was attached to it at the north), it seems that at some time the northern side of Floor 556 sank below the level of the pipe that led into Vat 551B (as the result of an earthquake?), rendering this part of the treading floor unusable. The problem was solved by decreasing the size of the floor and constructing new compartments on it. These were made of ashlars laid on the original (sunken) floor and plastered on their southern sides (Walls 1192 and 1195). However, most of Wall 1192 was plundered during a later period.

In reconstructing the wine press (Fig. 3) the author has assumed that it was entered from the east. Grapes brought from the vineyard were piled on the stone pavement to await pressing, eventually being spread on the treading floor and trodden by barefoot workers. The must drained into Vat 551B.

The pipe connecting Vat 551B with Vat 551A is 1 m above the floor level of the former. Therefore Vat 551B has a usable volume of approximately 2.1 m^3, and has to fill up to this volume before the must begins to spill over into Vat 551A. It is not clear whether the must was permitted to ferment in that vat or was removed and fermented in jars. The stairs in this vat were used by workers when they climbed in and out to fill the jars with must or wine. Above the top step a triangular foothold was cut into the wall to help the workers climb out.

A channel 0.25 m wide running across the pavement and outside under the southern outer wall was most likely built to drain off the water used for washing out the vats before processing a new batch of grapes.

The purpose of the compartments on the northern side of the treading floor is unknown. They may have been used for storing the grapes before treading, or they may have been installations for secondary pressing of the rape after the initial treading.

The dimensions of the wine press are impressive. This is one of the earliest examples in the Land of Israel of a "public"-sized wine press – a type that was to become fairly common during later periods. The produce of the wine press in excess of that required by the occupants of the local fort and its attendant village was almost certainly destined for marketing.

Bibliography

Herzog, Z., Rapp, G. Jr. and Negbi, O. (eds.). 1989. *Excavations at Tel Michal, Israel*. Minneapolis and Tel Aviv: 168–173.

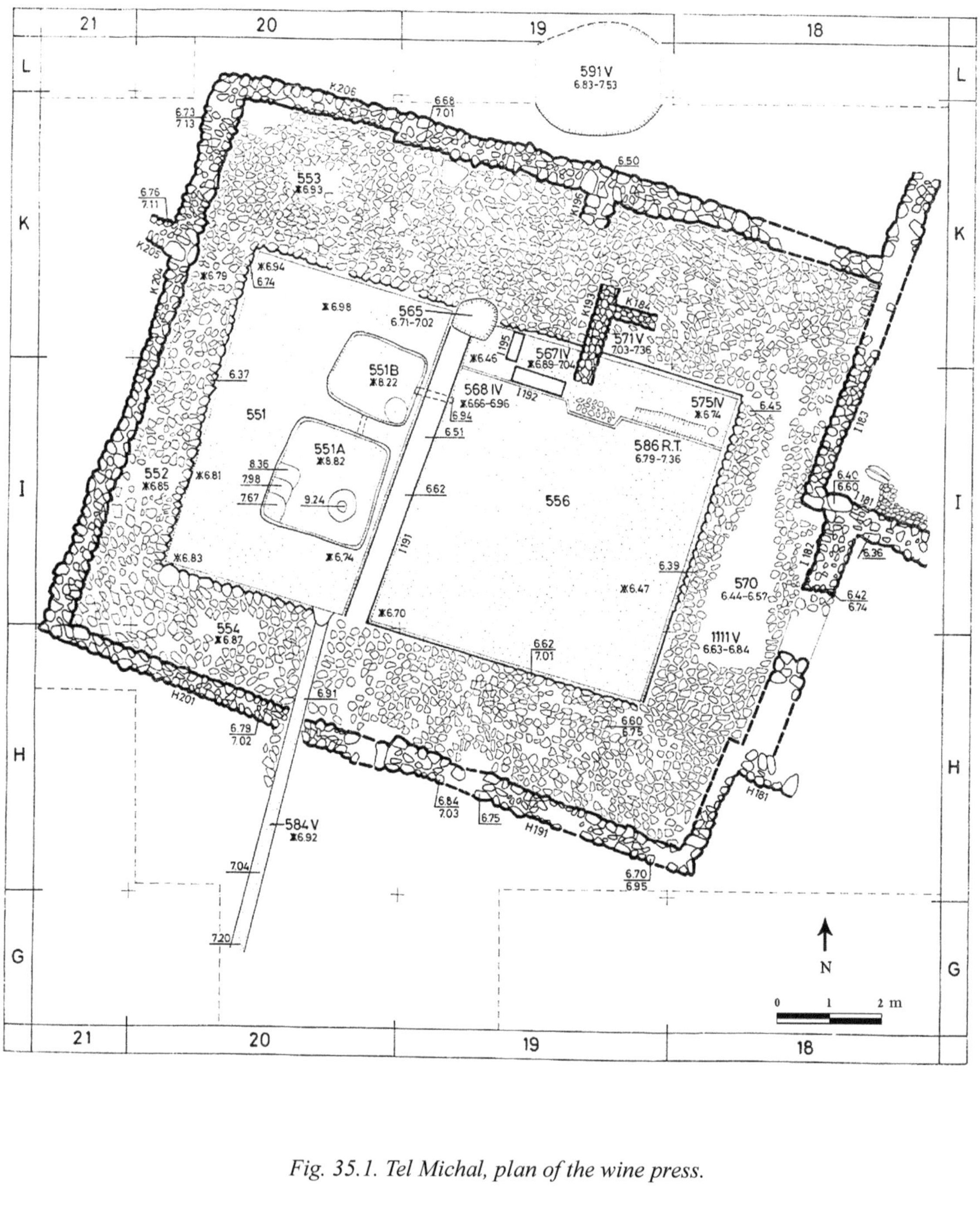

Fig. 35.1. Tel Michal, plan of the wine press.

Fig. 35.2. Tel Michal, the wine press after reconstruction, view to the south-east.

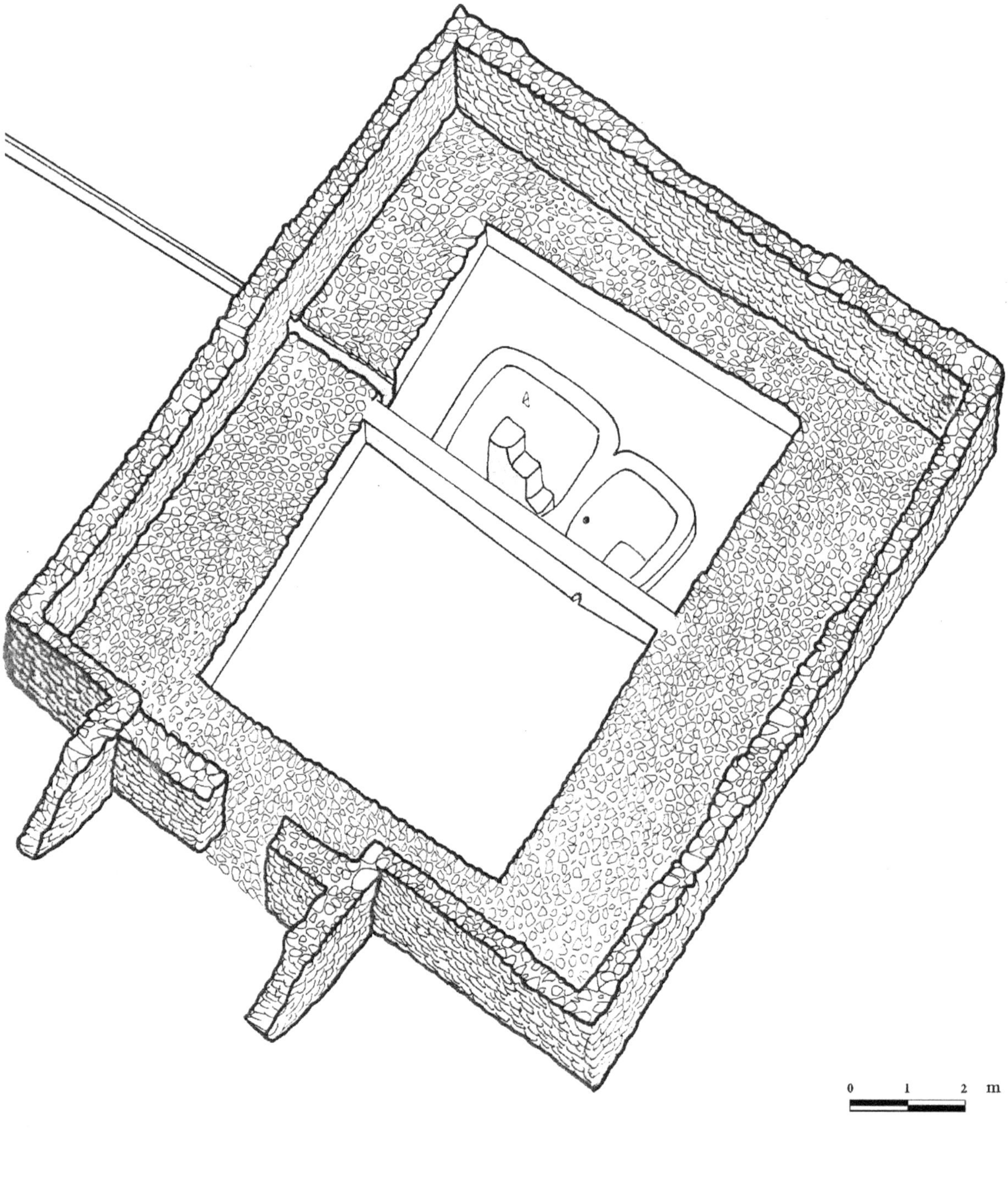

Fig. 35.3. Tel Michal, isometric view of the wine press.

36 Roman and Byzantine Wine Presses at Tell Qasile

Etan Ayalon

The Site

Tell Qasile (MR 1308/1676) is located in northern Tel Aviv, adjacent to the Yarkon River and within the campus of the Eretz Israel Museum. The site was settled almost continuously from the Iron Age through the Middle Ages. From the settlement that occupied the site during the Roman-Byzantine periods excavations have revealed large buildings and a bath on the tell itself, and a Samaritan synagogue, another bath, burial caves, wine presses and pottery kilns in surrounding areas (Ayalon 1988). Several wine presses were discovered on and around the site, but this article deals with two pairs only – one pair Roman, one pair Byzantine.

The Roman Wine Presses

Two adjacent wine presses were discovered in 1982 to the north-east of the tell. They were excavated by the author (Ayalon 1984) and preserved *in situ*. In total the two installations measure 12×13 m (156 m^2). These are large wine presses, each of which undoubtedly served several growers or one large estate. Each includes (Figs. 1–3):

- A small compartment for storing grapes before treading, that of one wine press measures 0.95×1.20 m (1.1 m^2), and the other 1.15×1.50 m (1.7 m^2).

- A large rectangular treading floor measuring 5.0×5.5 m (27.5 m^2). In both cases there is a small, deep vat in the treading floor, one round 0.6 m in diameter and 0.9 m deep (volume 0.25 m^3), the other rectangular measuring 0.6×0.8 m and 1.15 m deep (volume 0.55 m^3). These small vats doubtless served to collect the must produced during secondary pressing.

- A round sedimentation basin, that of one wine press measures 1.00×1.15 m and 0.8 m deep (volume 0.92 m^3), and the other 1.3×1.4 m and 0.7 m deep (volume 1.27 m^3).

- A large rectangular collecting vat, one measuring 2.20×2.25 m and 1.75 m deep with a volume of 8.6 m^3, the other 2×2 m and 1.5 m deep holding 6 m^3, with a second sedimentation basin in its floor.

The must flowed through gutters from the treading floor to the sedimentation vat and thence to the collecting vat. Steps were built in the north-east corner of the northern collecting vat to facilitate descent into it.

South of the sedimentation basin of the northern wine press, a large rectangular vat was hewn with no gutter, steps or sedimentation basin. It measured 1.25×1.85 m and was 1.55 m deep, i.e. a volume of 3.6 m^3. It was probably used to store grapes, to produce the fine "first must" from grapes before treading or to ferment must that had been expressed in the wine press itself.

The treading floors and especially the vats were built with a certain degree of carelessness, with lines and angles that were not straight. On the other hand, the walls and floors were punctiliously plastered. Because of the damage done to the plastered surfaces of the treading floors by intensive treading, they were repaired from time to time with a new layer of plaster. On each of the treading floors, six layers of plaster were discovered, one above the other. On the northern floor, on the earliest layer of plaster, a mosaic floor was discovered made of large, coarse white tesserae cut "out of square". Apparently an attempt was made here – one of the first of its kind in the Holy Land – to use a mosaic floor (that is stronger than a plaster floor) on the treading floor of a wine press. But the innovation was unsuccessful and afterward layers of plaster were again used.

Another interesting innovation in these wine presses are the small vats in the treading floors. The vat found in the centre of the northern one is surrounded by a recessed frame, lower than the floor, in which apparently a perforated board was placed. The vat adjacent to the edge of the southern treading floor is encircled by stones upon which a similar board could have been placed. A sack containing the rape from previously trodden grapes was evidently placed on the board and pressed to extract the remaining must that was collected in the small vats. Separation of this relatively inferior must from the primary must expressed by treading was a usual practice. No evidence remains of the type of installation used for the second pressing; from the absence of signs indicating emplacement of a screw (see below) it appears that a beam (lever) and stone weights were used. The niches for anchoring the back end of the beam may possibly have been built on the northern wall of the northern vat and the southern wall of the southern vat; in both these places signs remain of the walls having been reinforced with stones. Yet another (and simpler) possibility exists: that pressing was accomplished by heavy stones being laid on top of the sacks of rape.

Shards of ribbed Roman pottery were set into the plaster used on the walls and floors. In the debris that filled the vats, particularly in the northern wine press, were found ceramic and glass vessels from the 2nd-3rd cent. CE which were thrown there after the wine presses were abandoned. That, and the type of plaster used in the installations enable dating the construction of these wine presses to approximately the 1st or 2nd cent. CE and their abandonment to the 3rd cent.

The Byzantine Wine Presses

The two wine presses, placed 6 m apart, were discovered at the top of the hill to the east of the tell. The northern one was excavated in 1963 by the late Dr. Ya'akov Kaplan (1993: 1456). The southern one was excavated in 1987 by the author and his colleagues (Chidiosan, Ayalon and Yosef 1990: 23–36). Both were preserved *in situ*. From the similarity in form and components it can be deduced that the two wine presses were built at about the same time (Fig. 4).

These installations are also large: the overall dimensions of the northern one are 5.9×10.2 m (60 m^2) while those of the southern one are 7.9×10.7 m (84 m^2). Each includes:

- Several semi-circular storage compartments (in the southern wine press 1.2×1.5 m [1.8 m^2]) constructed around the treading floor and in some cases connected to it by a narrow pipe.

- A large treading floor. The northern one measures 5.25×5.40 m (28.3 m^2) while the southern one is an exact square, 5.3×5.3 m (28.1 m^2). In the centre of each floor is a large square stone screw base measuring 0.80×0.85 m in the northern wine press and 1.02×1.02 m in the southern. Each has a mortice for installing a fixed wooden screw that widens at the bottom on two adjacent sides as is usual in wine presses in the coastal plain (Frankel 1999: 141–142).

- In each, a pipe in the wall of the treading floor leads to a filtration basin measuring approximately 1.15×1.35 m. The southern one is 0.4 m in depth and the northern 0.66 m.

- A large collecting vat to which a pipe leads from the filtration basin. In the southern wine press it measures 2.2×2.3 m and 1.1 m in depth – a volume of 5.6 m^3. The dimensions of the northern one are 2.57×2.10 m and 1.5 m deep – volume 8 m^3. Each collecting vat has stairs for descent into it and a sedimentation basin in its floor.

The walls of both wine presses are well plastered and all floors tiled in coarse white mosaic, carefully laid on four under layers, from the bottom up: packed red sand, mortar with stones in it, very hard mortar containing earth and small stones, and hard mortar mixed with organic matter.

There was one find worthy of special mention. At the bottom of each screw mortice there is an opening from which a lead pipe led to the collecting vat. In the northern wine press the pipe was 5 cm in diameter and 3.1 m long, and in the southern one 8 cm in diameter and 2.85 m long (Fig. 5). The pipes were carefully set on beds of hard mortar and covered over with stone slabs, upon which was overlaid the infrastructure of the mosaic-tiled treading floor. Apparently this special effort to drain the mortice was meant to prevent rotting of the screw base due to the ingress of must or water; these liquids could also have caused swelling of the wooden screw, thus preventing or hindering rotation of the nut. It should be noted that the contact between the acidic must and the lead pipe undoubtedly resulted in lead poisoning of those who drank the wine (Ayalon and Rosen 2007).

In accordance with descriptions of wooden screws in Byzantine mosaic floors from Trans-Jordan and Lebanon (Roll and Ayalon 1981) a screw fashioned of oak was reconstructed by Yehoshua Dray and installed in the northern wine press (Fig. 6). It is presently on display for the enlightenment of visitors to the Eretz Israel Museum, Tel Aviv.

In light of the form of the wine presses and of the characteristics of their components (the screw, the use of mosaic, the type of plaster and the pieces of ribbed pottery mixed with it) they were dated to the Byzantine period (Fig. 7). At one stage the southern installation went out of use and a structure was built over it. The pottery found in the vats of the wine press and in the foundations of the later building date the abandonment of the installation to the beginning of the Early Arab period – the 8th-9th cent. CE. With this, it joined the long list of large wine presses in the Holy Land that were abandoned following the Muslim occupation (Ayalon 1997).

Bibliography

Ayalon, E. 1984. "Two Wine Presses from the Roman Period at the Haaretz Museum Center". *Tel Aviv* 11: 173–182.

Ayalon, E. 1988. "A Byzantine Road and Other Discoveries in the Vicinity of Tell Qasile". *Israel – People and Land, Eretz Israel Museum Yearbook* 4 (22): 9–34 (Hebrew).

Ayalon, E. 1997. "The End of the Ancient Wine Production in the Central Coastal Plain". In: Y. Friedman, Z. Safrai and J. Schwartz (eds.). *Hikrei Eretz, Studies in the History of the Land of Israel Dedicated to Prof. Yehuda Feliks*. Ramat Gan: 149–166 (Hebrew).

Ayalon, E. and Rosen, B. 2007. "The Wines of the Land of Israel in the Roman-Byzantine Period – Sweet but Perhaps Poisonous". In: A. Baruch (ed.). *Jerusalem and Eretz Israel* 4–5. Ramat Gan: 135–156 (Hebrew).

Chidiosan, N., Ayalon, E. and Yosef, A. 1990. "Two Byzantine Winepresses in the Eretz Israel Museum, Tel Aviv". *Israel – People and Land, Eretz Israel Museum Yearbook* 5–6 (23–24): 23–36 (Hebrew).

Frankel R. 1999. *Wine and Oil Production in Antiquity in Israel and Other Mediterranean Countries*. Sheffield.

Kaplan, J. 1993. "Tel Aviv". *NEAEHL* IV: 1451–1457.

Roll, I. and Ayalon, E. 1981. "Two Large Wine Presses in the Red Soil Regions of Israel". *PEQ* 13: 111–125.

Fig. 36.1. Tell Qasile, the Roman wine presses, view to the north.

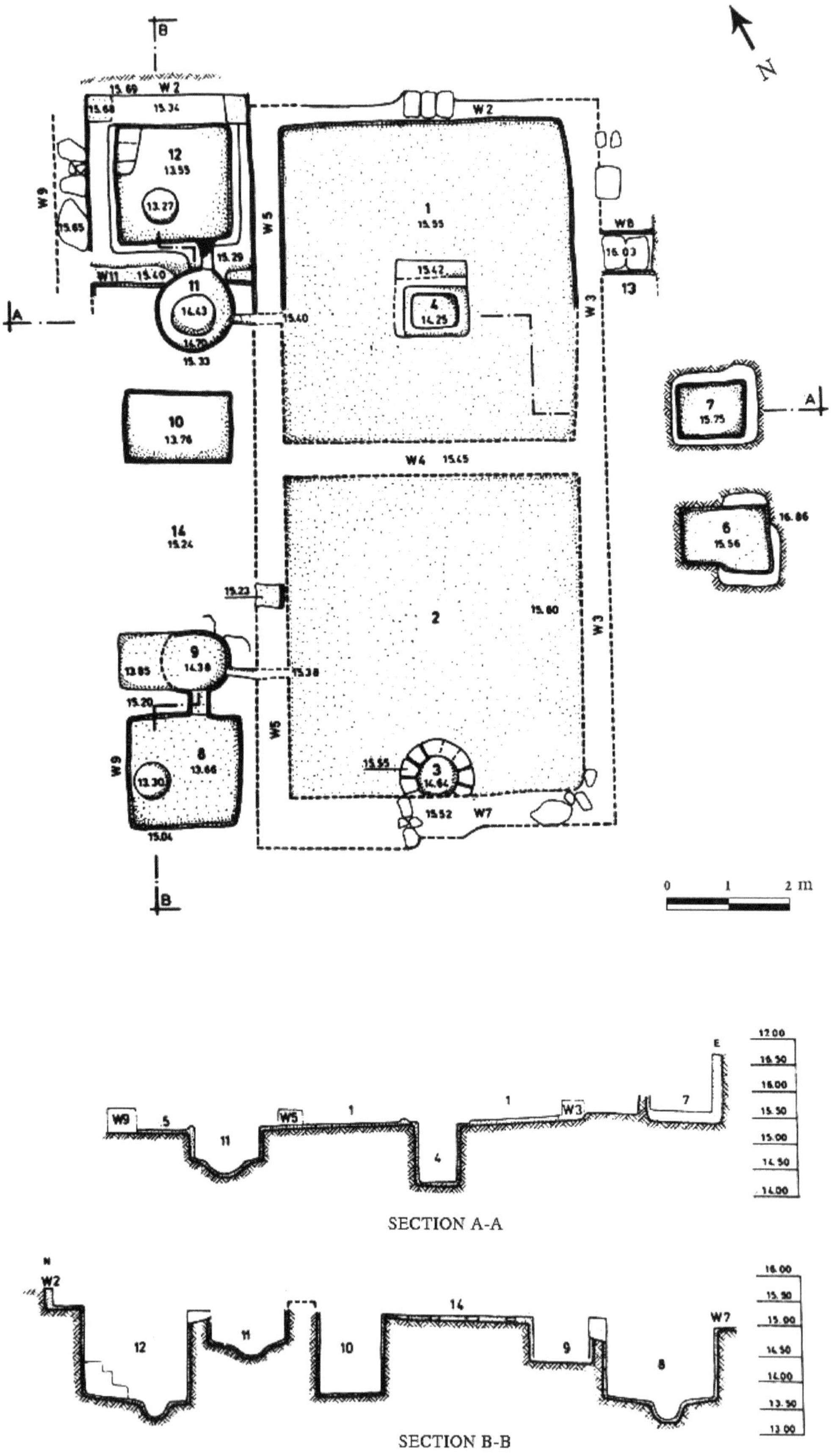

Fig. 36.2. Tell Qasile, plan and sections of the Roman wine presses.

Fig. 36.3. Tell Qasile, the northern Roman wine press, view to the east.

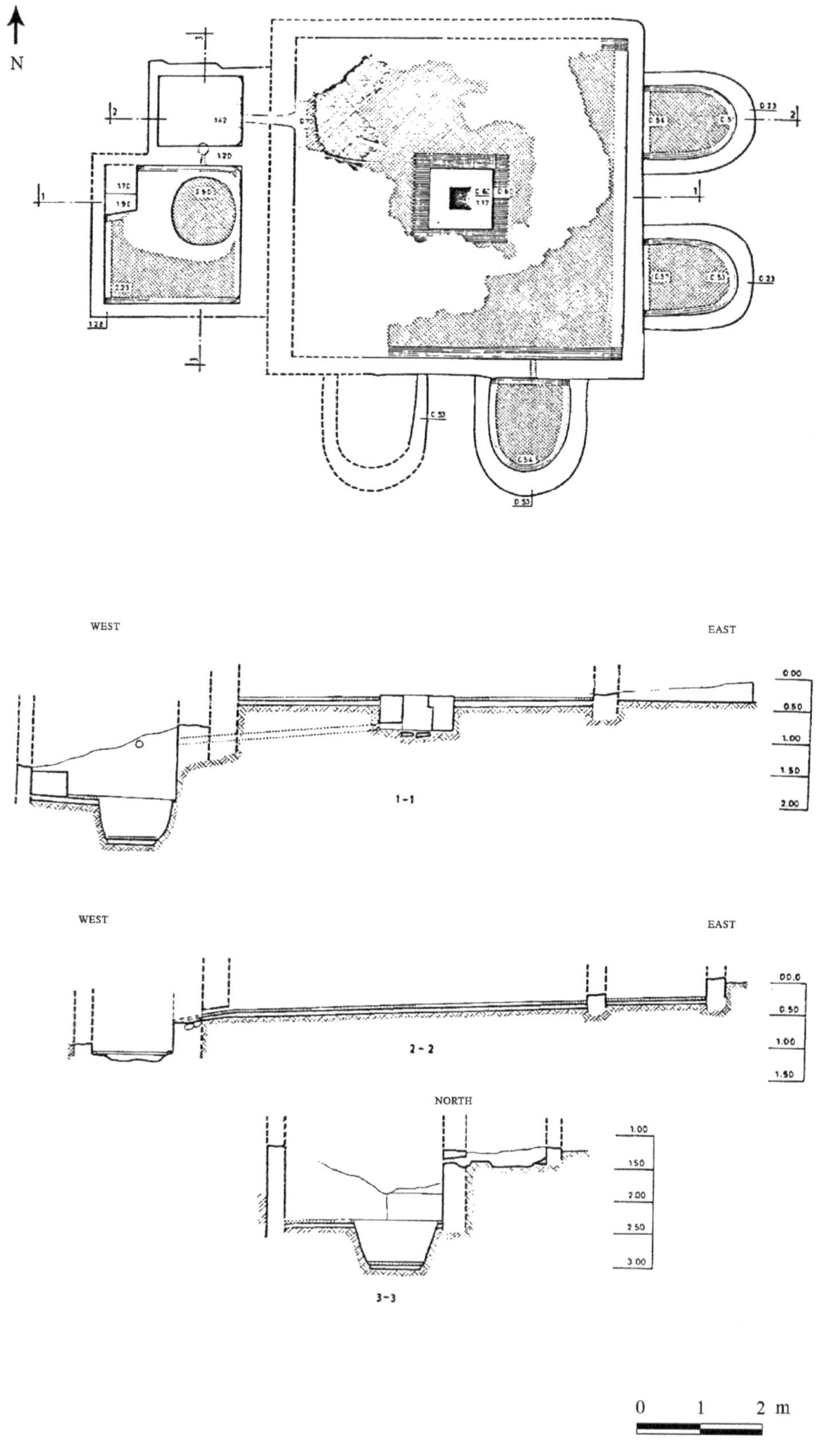

Fig. 36.4. Tell Qasile, plan and sections of the southern Byzantine wine press.

Fig. 36.5. Tell Qasile, lead pipe in situ *in the southern Byzantine wine press.*

Fig. 36.6. Tell Qasile, reconstructed wooden screw press in the northern Byzantine wine press.

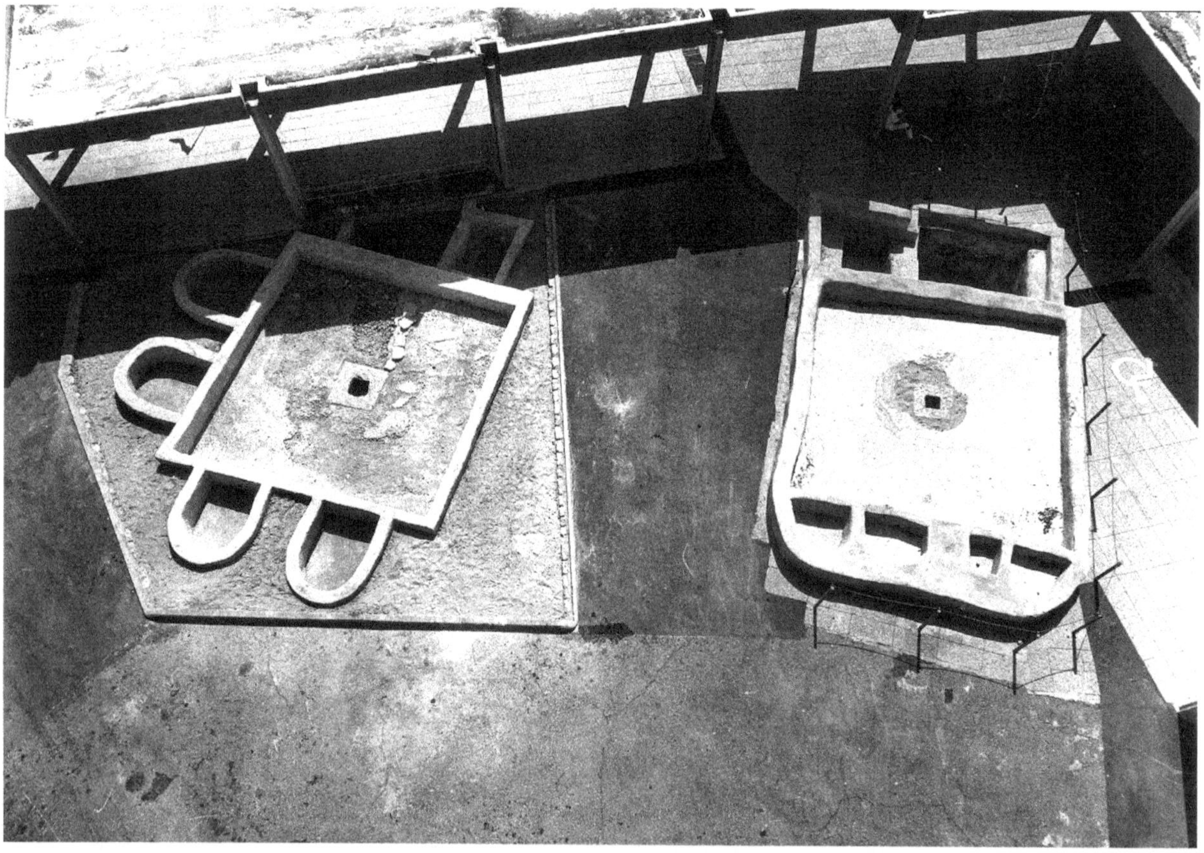

Fig. 36.7. Tell Qasile, the two Byzantine wine presses after reconstruction, view to the west.

37 Wine Presses at Mulabbis (Petah Tiqwa)

Shlomo Gudovitch

The Site

Between 1994 and 2003 several salvage excavations were carried out at Mulabbis (MR 138/166), in the industrial area of Petah Tiqwa. They revealed a settlement that was occupied from the late Persian till the Ottoman period. Both buildings and installations were uncovered as well as a variety of small finds.

In the centre of the site were found two oil presses dating to the Early Arab period and to their north an oil press from the Byzantine period was uncovered. Near it the press-bed of an oil press was also found not *in situ*. In the western part of the site two wine presses from the Byzantine period were excavated the earlier of which was built on a building from the Roman period. The later wine press is almost identical to the earlier and was built adjacent to it to the south. The earlier wine press apparently continued to function together with the later after the latter was added. In the southern part of the site the remains of a complex wine press were excavated. It was dated to the end of the Byzantine period. These three wine presses which were excavated by the author on behalf of the Israel Antiquities Authority will be described below.

The Two Western Wine Presses (Fig. 1)

The wine presses were built on the western slope of the hill in sandy red soil that does not crack in the dry summer. The area was first leveled so that the slope would be suitable for the treading floors. At the same time the pits in which the vats were to be built were probably also dug.

The Early Wine Press

In square D/7 a treading floor was uncovered (L. 10037; dimensions 5×5.5 m., area 27.5 m^2). The floor was paved in mosaic made up of large white tesserae (22–25 per square decimeter). The treading floor was surrounded by a wall 0.6 m wide built in masonry consisting of field stones and large pebbles and grey mortar. All the walls, except the north-east corner, had been robbed leaving only the foundations. On the remaining corner traces of the plaster that covered the inner face of the walls of the treading floor could be seen. At the end of the excavation the treading floor was removed thus making it possible to understand the manner in which it was constructed. The foundation consisted of three layers of stones of diminishing size: the bottom layer flat field stones and pebbles 9–12 cm in diameter, the second pebbles 8–10 cm in diameter and the third pebbles 5 cm in diameter. Above each layer of stones was a layer of mortar light grey in color, containing lime and ashes and 2.5–3 cm thick.

Above this foundation there was an additional layer of mortar, lighter in color on which the mosaic floor was laid. The floor and foundation were in all more than 30 cm thick. In all the mosaic floors of this installation except for that of the treading floor the tesserae were laid in straight lines parallel to the surrounding walls, while in the treading floor they were laid diagonally to the surrounding walls.

The area to the west of the treading floor was also paved in mosaic (L.10024; c 22 tesserae to the square decimeter). Originally it was surrounded by walls similar to those surrounding the treading floor but only the eastern (W 108) and the northern (W 106) remain the others having been destroyed by development operations carried out in the area before the excavation took place. The northern wall (W 106) was built on the remains of a building from the Roman period (L. 10031). In the north-eastern corner of square D/6 a sieving vat was uncovered (L. 10039, dimensions 1×1.2 m, depth 0.72 m; vol. 0.86 m^3; Fig. 2). Its floor was of white mosaic and the walls were coated in white hydraulic plaster. The must flowed from the treading floor to the sieving vat in a channel built of rubble and pebbles and coated in similar plaster, and from the sieving vat to the collecting vat (L. 10038) through a clay pipe 9 cm in diameter.

The collecting vat was to the west of the sieving vat (dimensions 2.3×2.3 m, depth 1.6 m, volume 8.5 m^3, effective volume up to the outlet of the connecting pipe 4.7 m^3). Its walls were coated in three layers of plaster, the bottom one gray and including ribbed shards and the other two of white hydraulic plaster. The floor was paved in white mosaic (32 tesserae per square decimeter). In the centre of the floor was a round sump (total diameter 1 m with an inner sump 0.75 m in diameter, depth 0.5 m) and in the south-western corner two plastered steps were built to facilitate descent and ascent to and from the vat.

The Late Wine Press

To the south of and parallel to the two vats of the early wine press another sieving vat and another collecting vat were uncovered. These are later than those of the early as shown by the fact that the southern wall of the early collecting vat (L. 10038), after its outer face had been plastered, served also as the northern wall of the late collecting vat

(L. 10044). This wall separating the two collecting vats was 0.5 m. wide.

The late treading floor, the area of which we presume to be c. 34 m^2 was apparently to the south-east of the two vats and south of the treading floor of the early wine press. Remains of an area paved with field stones and pebbles were found (L. 10018) that were apparently the foundation of the treading floor and wall 110 a small section of which remains was apparently the eastern wall of the floor. The must flowed northwards to the sieving vat in a built channel with a mosaic floor (L. 10032) that is placed between the vat and wall 108 (the western wall of the early treading floor). The sieving vat (L. 10041) has a mosaic floor and plastered walls (1×1.1 m, depth 0.58 m, volume 0.64 m^3). The must flowed through a ceramic pipe (diameter 9 cm) to the collecting vat (2.8×2.5 m, depth 1.51 m, vol. 10.5 m^2, effective volume 7 m^3). In the collecting vat there were two plastered steps and in its floor near the northern wall there is a sump (diameter 1 m depth 0.25 m) in the centre of which is a deeper round depression (diameter 0.75 m, additional depth 0.35 m). In most places the walls of the vat were coated in two layers of plaster, in the first layer the plaster is grey with ribbed shards and above it there is a layer of pink mortar containing many red brown small grits, apparently crushed shards. In the north-eastern corner, however, five layers of grey plaster were discerned of slightly varying hues apparently the result of repeated repairs over a long period. The floor was paved in white mosaic (32 tesserae per square decimeter) that was laid diagonally except for three rows around the edges that were parallel to the walls and the steps and around the sump. The mosaic to the east of the collecting vat (L. 10042) was arranged in alternating triangles.

After the wine press ceased to function the sump was widened to a diameter of 1.8 m and 0.40 m depth. The widened part of the sump was not paved and in it there was a burnt layer containing animal bones and pottery from the Ottoman period.

The early wine press almost certainly continued in use after the late wine press was added, as may be concluded from the fact that all the later components were built alongside the early ones, even though in an unusual manner. The two wine presses were probably abandoned when the southern complex wine press was built but when the latter was destroyed (as a result of the Persian or Muslim conquests in the 7th cent. that led to a damage to vineyards and a decline in yields) the later of the two western wine presses was restored. At this time the collecting vat of the early wine press was blocked up and in the fill (L. 10029) were found a large quantity of fragments of jars, some fragments of glass, a complete bottle from the Umayyad period and a coin dated 491–518 CE. Above the southern section of this fill up to the northern edge of the late (southern) collecting vat a mosaic floor was laid. The foundation layers of the floor were 0.25 m thick and consisted of a layer of ribbed jar shards, then of a layer of medium sized pebbles laid in grey mortar and finally a layer of mortar containing lime, ashes and straw on which the mosaic was laid. The floor (L. 10036, 1.10×0.65 m) was flush with the northern edge of the late collecting vat, joined the previous mosaic floor to the east and west and to the north was over the southern part of the early collecting vat. In this floor the tesserae were apparently laid in a pattern consisting of a series of semicircles, perhaps five, of which two remain.

Although the two wine presses are similar in form and built at the same level there are considerable differences in the way they were constructed. In the treading floor of the early wine press the tesserae were laid diagonally while in the other mosaic floors of this wine press the tesserae were laid in lines parallel to the walls. In the floors of the late wine press most of the tesserae were laid diagonally with a border of two or three lines laid parallel to the walls (the exception is the sieving vat [L. 10041] in which the whole floor was laid in lines parallel to the walls). There were similar borders around the upper mosaic floors which were divided into separate sections.

The Southern Complex Wine Press (Fig. 3)

In squares P-Q/11–12 a series of white mosaic pavements were first uncovered (L871–16–22 tesserae per square decimeter). Above the floors a burnt oily black layer was found. Amongst the finds in this layer were a fragment of bowl of African Red Slip Ware dated to the 4th cent. CE, a fragment of a bowl from the Byzantine period, a coin from the 6th cent. and a metal pendant in the form of a coin. Above the burnt layer was a layer of alluvium in which mixed shards were found.

There were three upper trapezoidal floors (L. 501 [21 m^2], L. 503 [c. 24 m^2] and L. 504 [24m^2]) and between them were partitions built of field stones and mortar that included shards some of which were ribbed. The partitions were 0.25–0.35 m wide and survived to a height of only a few cm (Fig. 4).

In each of the three floors there was a small opening that led down to the compartment that was below the floor. The only one to survive complete was bordered by four small oblong stones and measured 7–8 cm in diameter (L. 501, P.13; Fig. 5). The floors probably sloped down towards the openings.

The northern face of the wall that encloses the floors to the north (W305 in squares P/11–13) was cleared to a depth of 2.5 m (till level 29.92 m) without reaching the bottom of the foundations. The wall was built partly of roughly hewn stones and partly of field stones and of hard grey mortar. The lower section of the wall was coated in grey plaster while the upper section, till 0.65 m from the top which was probably that exposed above ground was coated in pink plaster typical of the Byzantine period. In the pink plaster small shells could be discerned as well as many small brown-pink insertions presumably crushed shards.

The wall (W357) to the west of the western floor (L. 504) is apparently the outer wall of the wine press in this direction. It was found largely destroyed especially at its southern end.

In squares Q/12–14, the area to the south of floors 501, 503 and 504, the excavation was deepened to level 30.53 m where the main treading floor of the wine press was uncovered (L. 521, 527, 528, 529). It was paved in mosaic similar to that of the upper floors. To the north and east of the central treading floor four compartments were uncovered (Figs. 4, 6). They were roofed in barrel vaults and all were built below upper mosaic floors, from west to east: compartment 526 under floor 504, compartment 533 under floor 501 and compartment 532 under floor 503. The floor of compartment 531 which is to the east of the treading floor has only partly survived. To the east of the treading floor and south of compartment 531 a section of a wall running from west to east was uncovered (W105, L. 10051). This was clearly the southern wall of another compartment but as the area further south was not excavated we cannot be certain if this is the outer wall of the wine press in this area or whether there were additional upper floors and compartments further south. (In our reconstruction we have included an additional floor and an additional compartment).

The walls of the compartments are made up of medium sized and large field stones and hard grey mortar and coated in white plaster. The walls up to the base of the vault are coated in two layers of hydraulic plaster and the vaults in one layer of plaster containing lime. The openings that connect the upper floors to the compartments are in the centre of the vault and close to the back wall of the compartment.

The compartments are separated from the central treading floor by partitions that were 0.6 m high (the height of the walls up to the beginning of the vault). The inner face of the partitions was coated in hydraulic plaster and the outer, in the direction of the treading floor was covered in plaster containing lime. At the bottom and in the centre of each partition was a pipe that connected the compartment to the treading floor. The tesserae of the mosaic floors of the compartments were comparatively small (30–32 tesserae per square decimeter) and were laid diagonally except those along the partition which were laid parallel to it. The tesserae adjacent to the partition were larger than the others and of a similar size to those of the upper floors (for dimensions of compartments including the connecting openings see Table 1).

Square S/ 13

Shortage of time did not allow for the excavation of most of the central treading floor so that it was not possible to ascertain whether there was a mortice for a screw press in the centre of the floor. As the northern edge of the treading floor was 9.2 m long and on the assumption that the treading floor was square an excavation area was opened 10 m to the south of the wall in order to discover what was to the south of the treading floor. At this point was found a section, c.1 m long, of a wall (W 366, width 0.63 m) that was aligned east to west and that was clearly the southern wall of the treading floor. To the north of the wall the mosaic floor of the treading floor was uncovered. The tesserae along the wall, i.e. the border of the floor, were laid in lines parallel to the wall. The slope of the treading floor was 2.2%. The must flowed to the sieving vat through a channel of which only the bottom survived.

The sieving vat was almost square (L. 584, 1.2×1.3 m depth 0.72 m vol. 1.12 m^3), paved in mosaic and its walls were coated in plaster. Two leaden pipes led from the southern corners of the vat one to each of the two round collecting vats. These were only partly excavated but they were probably identical and enough of the western vat was uncovered to determine its dimensions and character (Square T/13–14; L. 541; diameter 3.8 m depth 1.68 m vol. 19 m^3; effective vol. [below entrance point of lead pipe 0.74 m below top] 11 m^3). In the centre of the floor, which is paved in mosaic similar to that of the upper floors, is a round sump 1 m in diameter and 0.43 m deep and in the centre of the sump is an additional small round depression

Table 1. Dimentions of compartments and openings

Compartment	Length in m	Width in m	Height in m	Height of base of vault in m	Dimensions of opening in m	Volume in m^3
L. 526	2.75	1.6	1.25	0.5	l. 13, w. 4	2.2
L. 533	2.7	1.3	1.25	0.6	l. 24, w. 5, h. 15	2.1
L. 532	2.6	1.5	1.2	0.58	l. 26, w. 5; h. 14	2.28
L. 531	2.2	1.35	1.14	0.45	l. 11, w. 4	1.34
L. 100051	2.2	1.3	1.14?	?	?	1.2?
					Total	9.12?

(diameter 0.4 m, depth not excavated). Remains of a mosaic floor were found south of the collecting vats suggesting that all the area around both was paved in this manner.

During the excavation many fragments of tiles were found suggesting that at least part of the installation was roofed. Shards from the Early Arab period were found on a thin layer of soil above the treading floor thus providing a date for the abandonment of the wine press.

Discussion

At Mulabbis three wine presses were excavated that were in use from the early Byzantine till the Early Arab period. In the two early wine presses there was no evidence for either auxiliary floors or compartments or for devices for secondary pressing. Both are built in a usual plan known also from earlier periods and the differences between them are in the way the mosaic floor was laid and in the direction of the treading floor. In the early wine press except for the treading floor the tesserae were laid in lines parallel to the walls while in the later wine press (there the treading floor has not survived and therefore we do not know how its floor was laid) the tesserae are laid in diagonal lines in the centre with a border of two to three rows parallel to the walls. The floor of the southern complex wine press was laid in a similar manner.

In the three wine presses there was a difference between the size of the tesserae in the parts where the must remains for a short time or did not stand at all (the treading floors and the areas around the vats), where there were 22–25 tesserae per square decimeter and the size of the tesserae in the vats where there were 30–32 tesserae per decimeter. The tesserae in the latter were smaller in order to make the floor more impermeable to liquids.

Although the traditional type of wine press, such as the two northern wine presses at Mulabbis originated much earlier than the complex wine presses, such as the southern wine press at Mulabbis which appeared no earlier than the Byzantine period, nevertheless the early type continued in use beside the later. It is possible that some of those of the traditional type that were found near Byzantine sites were built in earlier periods but were improved later, for example by adding a mosaic floor. The type and size of wine press used was probably dependant on the size of the vineyard. The owner of a small plot of ground could almost certainly not afford to buy an expensive screw press and probably used simpler methods to press out the last must, such as laying stones on sacks of rape (Frankel 1999: 146–147) while more elaborate presses were used in wine presses in large estates or monasteries.

The southern complex wine press is of a type of large wine presses known only in the Byzantine period. It was built on a moderate southern slope. It consisted of a central treading floor flanked on at least two sides, the northern and eastern and possibly also the western side, by compartments above which were upper auxiliary floors. In the centre of the treading floor there was probably a mortice to secure a screw press for secondary pressing of the rape (Ibid.: 140–141) but this area was not excavated. To the south of the treading floor there was a sieving vat through which the must flowed through lead pipes to the two round collecting vats where the must fermented and clarified and the solids were amassed in the sump at the bottom.

The parts of the wine press that were cleared in the excavation allow for it to be reconstructed in two different ways:

A. A wine press with six fermentation compartments arranged like a capital L on two sides of the central treading floor, the northern and eastern.

B. A wine press with nine fermentation compartments arranged like an inverted U on three sides of the treading floor – west, north and east.

The wine press operated in the following manner. The upper auxiliary floors served as upper treading floors. On each floor piles of grapes were laid and as a result of their weight and some minimal treading the first must was produced which flowed through the openings in the floor to the compartment below. Compartments that are usually placed around the treading floor in smaller wine presses and there serve to store the grapes have, in the press at Mulabbis, been improved to receive the first must. The fact that there are no sumps in the compartments and the small size of the openings in the upper floors show that the rape, the skins and stalks of the grapes did not reach the compartments. After the must stopped flowing the grapes were transferred to the main treading floor and fresh batches of grapes were put on the upper floors until the compartments filled to the required depth of c. 0.50 m and then it fermented in the compartments to produce high quality wine.

The grapes were then trodden again on the central treading floor and the must flowed via the sieving vat to one of the collecting vats. The rape was then pressed in the screw press that probably stood in the centre of the treading floor (Hadas, in press). When the treading was completed and the treading floor cleared the pipes of the fermentation compartments were opened and the wine allowed to flow to the second vacant collecting vat and then poured into storage jars.

Attempts to asses the capacity of the wine presses must be based on the quantity of grapes that could be trodden on the treading floors and the volume of the collecting vats. In the earlier of the western wine presses the area of the treading floor is 27.5 m^2 and the effective volume of the collecting vat 4.5 m^3. It would have needed more than seven tons of grapes to fill the collecting vat. In the later of the western wine presses the treading floor has not survived but the volume of the collecting vat is 7 m^3 and it would have needed c. 11 tons of grapes to fill the vat.

The presumed volume of the five fermentation compartments that were excavated was c. 9 m³ and it would have needed 15 tons of grapes to fill them. The ratio between the area of the upper floors and the volume of the fermentation compartments shows that to fill the compartments it would have needed 100 litres of must from each square metre of floor. Therefore there would have to have been 160 kg of grapes on each square metre, a pile that would have been c. 0.5 m high. The walls between the upper floors are sufficiently strong (c. 0.30 m wide) to support piles of grapes this high. The fact that the foundation layers of the upper floors are 0.3 m thick, much thicker than is usual, would also have made them sufficiently strong to support the weight involved. The weight of grapes on 25 square metres of floor space would have been 3.2 tons which would have produced 2 m³ of must sufficient to fill one compartment. In order to fill one of the round collecting vats it would have been necessary to tread on the central treading floor (area 85 m²) a pile of grapes 0.6 m high and weighing 13.5 tons. This would have produced 8.2 m³ of must sufficient to fill the collecting vat.

In all 40 tons of grapes would have been needed for one batch of grapes for all three wine presses and if the wine presses were used five times in a season the total quantity of grapes would have been more than 200 tons. The more uniform the vineyards are the more liable they are to suffer from disease and pests even in modern agricultural conditions. That is one of the reasons that in my opinion the three wine presses did not function together.

It is not clear why the upper floor area was divided into separate floors each of which was connected to one compartment. Was it to keep the yield belonging to different people separate? Or perhaps to keep grapes of different varieties apart? Or perhaps to produce different types of wine by adding different ingredients and spices? In my opinion the separation into several floors also contributed to work efficiency as shown by the fact that the volume of one collecting vat sufficed to contain the must from all the compartments.

Although the collecting vats vary in depth the effective depth, up to the point where the pipe or channel through which the must flowed entered the vat, was c. one metre, the depth at which the fermentation process is at its best.

Conclusions

At Mulabbis three wine presses were excavated that demonstrate the development and improvement that took place in wine production in the Byzantine period making it a central element in the economy of the settlement.

The early western wine press that was built on a building from the Roman period was comparatively small and consisted only of basic components. Later, perhaps as a result of increased yields because of the cultivation of a larger area or perhaps because of improved cultivation methods, a second wine press was built next to the first and almost identical to it. There were however two differences between the two wine presses: [1] the mosaic floors were laid in a different manner; [2] The relationship between the treading floor and the other parts of the wine press. The treading floor of the later wine press was built to the south because the treading floor of the early wine press was to the east, and the fact that the position of the early treading floor was taken into account when building the later shows clearly that the two wine presses functioned together.

Later apparently as a result of a further expansion of the area of the vineyards the southern complex wine press was built. At this stage the two western wine presses were abandoned. In the 7th cent., after the Persian or the Arab conquest the site and its agricultural land were damaged and the complex wine press was destroyed.

After a period of time some of the vineyards were rehabilitated and at the same time the southern of the two western wine presses was restored and came into use again. In order to widen the floor around the collecting vat (L. 10036) the collecting vat of the northern and earlier of the two western wine presses (L. 10038) was filled and a floor built above it. In the fill in the vat the finds were typical of the Umayyad period and included a complete glass bottle thus showing that the renewal of the wine press took place at this time or slightly later. At one point this installation ceased to serve as a wine press and was changed to serve some other purpose. The sump in the collecting vat was widened resulting in the destruction of part of the mosaic floor making the installation unfit for use as a wine press. It is not clear when this change took place but the fact that shards from the Ottoman period were found in the lower layers of the fill suggests that it served in its new function for a long period.

Editors' Notes

1. As regards the method in which the southern complex wine press functioned, it is more probable that the first must that flowed through the small openings in the auxiliary floors to the compartments below was collected directly into jars. The compartments would then have been working areas where the jars stood before and after they were filled. The partition between the treading floor and the compartments was primarily the back wall of the treading floor. Its being only 0.60 m high made it easy to step over it. The pipes that connected the compartments to the treading floor probably served both to drain any must that was spilt in the compartments and to facilitate the washing of the compartments.
To allow the must to flow from the compartments over the large treading floor to one of the collecting vats seems a very unlikely procedure for many reasons. We will mention some. Firstly, the general method of work in a complex wine press such as the southern one at Mulabbis was one that allowed the wine press to be used continuously. One batch of grapes stood in the sun on the auxiliary floors while the previous batch was trodden on the treading floor, the must flowing into one collecting vat while the must from the batch before that was fermenting in the second vat. Thus all the time at one end wine was poured into jars and at the other fresh grapes were brought from the vineyards. For the must to flow from the compartments to one of the collecting vats the treading floor would have had to be cleared and cleaned and one vat to remain empty and this would have interrupted the whole work procedure. Secondly to allow the precious must to flow over the treading floor would appear to be a very inefficient and wasteful method. The third and main reason against the procedure suggested here is that if the must had already fermented in the compartments, why transfer it to another collecting vat and not directly

to storage jars? (See also note 2 to Introduction).

2. A possible explanation for there being auxiliary floors on only two of the three possible sides (there are other examples of this arrangement) is that it makes it possible sometimes to approach the central treading floor directly from one direction and therefore makes it possible to bring grapes to the floor without them first going via the upper auxiliary floors. This would allow for making another type of wine.

Bibliography

Hadas, A. In press. "Methods of Wine Production in Ancient Wine Presses". In: S. Gudovitz, "The Excavations at Mulabbis". *'Atiqot*.

Frankel, R. 1999. *Wine and Oil Production in Antiquity in Israel and Other Mediterranean Countries*. Sheffield.

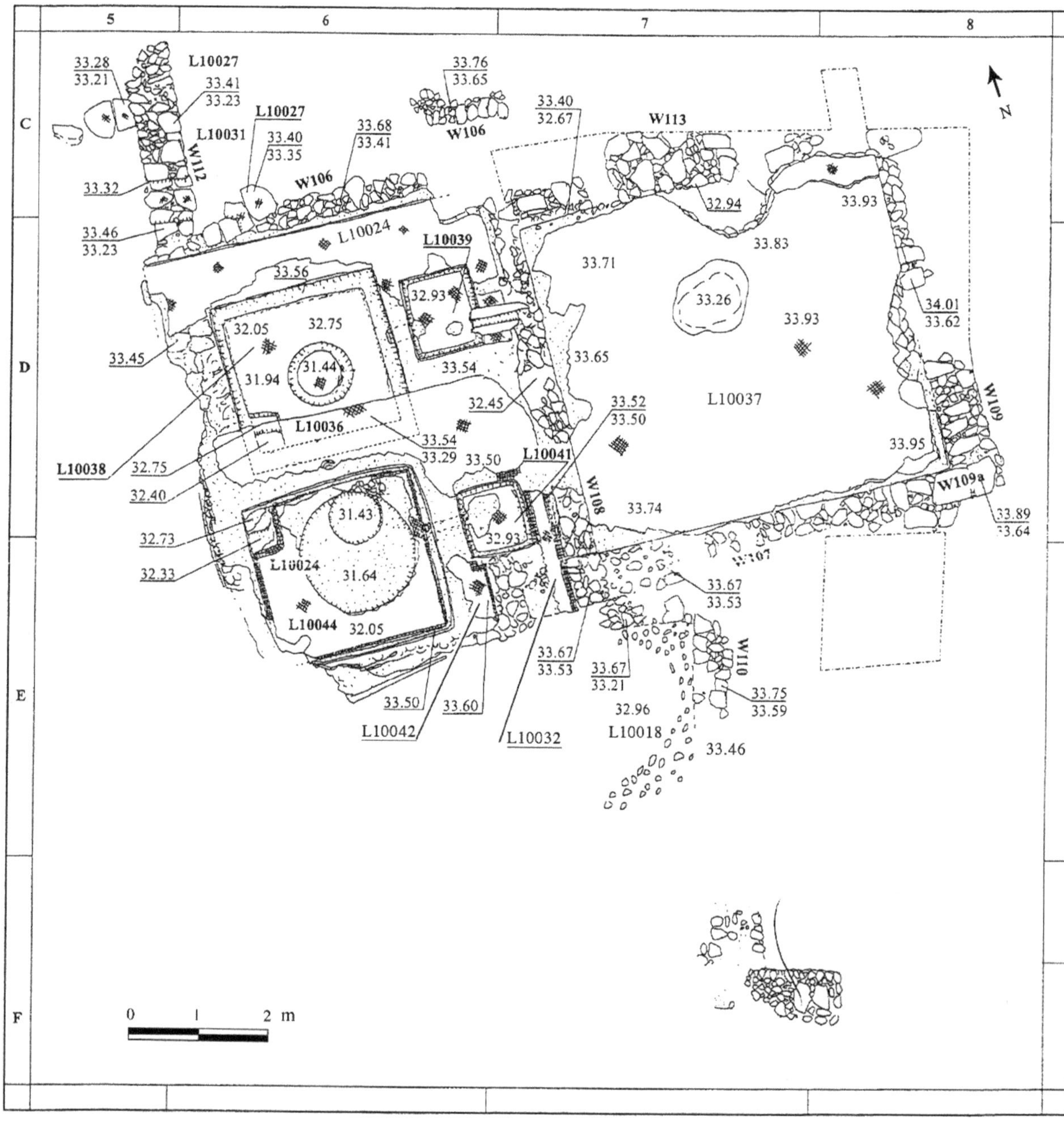

Fig. 37.1. The two western wine presses: plan.

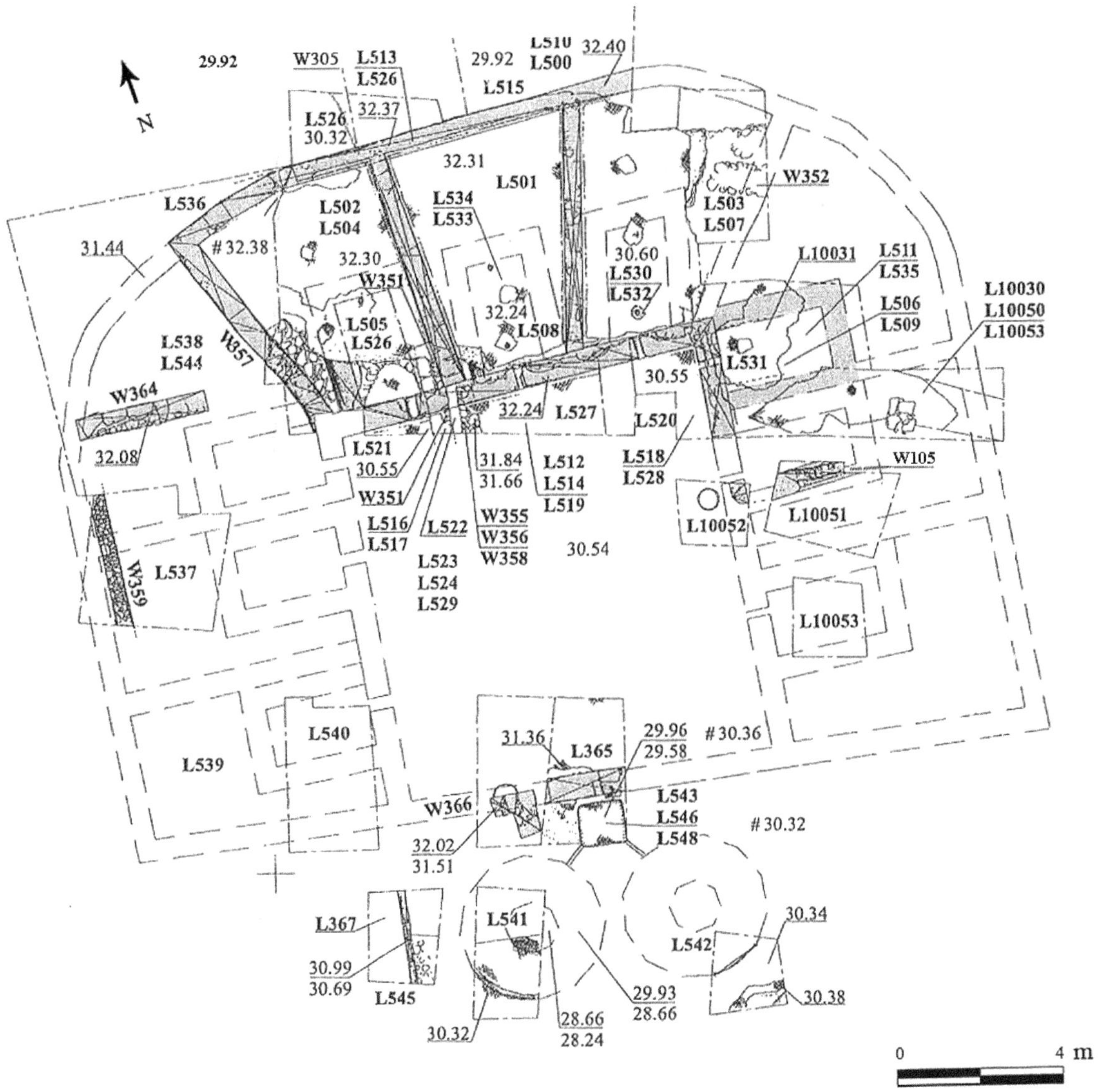

Fig. 37.2. The complex southern wine press: plan.

Fig. 37.3. The early western wine press: the border of the treading floor (front), sieving vat (L. 10039; centre) and collecting vat (back), looking west.

Fig. 37.4. The complex wine press: the upper treading floors with vaulted compartments beneath them, with the border of the central treading floor (left), looking north-west.

Fig. 37.5. The complex wine press: the drainage opening in the upper treading floor (L. 501, P/13), looking west.

Fig. 37.6. The complex wine press: a complete vaulted compartment beneath the upper treading floor.

38. Wine and Oil Presses from H. Mazor and the Surrounding Region

David Amit

The Site

During the years 1993–1996 extensive archaeological surveys and excavations were carried out in preparation for building the new town of El'ad in the western foothills of the Samarian hills (Amit *et al.* 1998: 57–66; 1999: 55*-64*; Sidi, Amit and 'Ad 2003: 258–261). The main site investigated was H. Mazor (MR 1455/1618). This was shown to have been a farmstead in the Roman period that developed into a large well planned village in the Byzantine period. There were few finds from the Early Arab period and from the Middle Ages but the site was resettled in the Ottoman period. About 500 m to the north-west of H. Mazor another farmstead was excavated that was shown to have been occupied during the Hellenistic and Early Roman periods (2nd cent. BCE-2nd cent. CE). At these two sites and in the surrounding region in all nine agricultural installations were found: six wine presses, two dated to the Hellenistic period and four to the Byzantine and the remains of three oil presses, one Hellenistic and two Byzantine.

The Wine Presses

1. Area S7 (Fig. 1)

This wine press is located at the north-eastern end of the north-western farmstead. It is partly cut in bed-rock and partly built and plastered in gray plaster. It consists of a square treading floor (4.5×4.5 m, area 20.25 m^2) connected by a channel to a square intermediate vat (0.75×0.75 m, depth 1.08 m) that is connected by a bore close to the bottom to a square collecting vat (1.5×1.5 m, depth 2 m, volume 4.5 m^3). A channel connects the treading floor to the collecting vat thus making it possible to allow the must to flow either directly to that vat or via the intermediate vat. The plan of the wine press is similar to but does not conform exactly to the "four rectangle plan" (Frankel 1999: 149–150) (see wine press P below) as the two vats are not placed exactly side by side. The typical grey plaster dates the construction of the press to the Hellenistic period but the finds suggest that it was still in use in the Byzantine period.

2. Area P (Fig. 2)

The wine press is located approximately 400 m east of the north-western farmstead and 200 m north-west of the centre of H. Mazor, and was probably close to the vineyards. It is cut in bed-rock and plastered in gray plaster that includes tiny stone grits. Similarly to press No. 1 it consists of an almost square treading floor (3.7×3.8 m, area 14 m^2, height of walls 0.2–0.5 m) connected by a bore to a square intermediate vat (0.7×0.7 m, depth 0.4 m). This vat is connected by a bore 10 cm above the floor to a square collecting vat (1.3×1.3 m, depth 1.7 m, volume 2.9 m^3). In both vats there are small sumps to aid in removing the last remaining liquid. Similarly to wine press No. 1 all the components are square and in this case the plan conforms exactly to the "four rectangle plan". The typical plaster dates this press also to the Hellenistic period.

3. Area H2 (Fig. 3)

This installation is situated at the western edge of H. Mazor and also consists of three almost square components making up a "four rectangle plan", but it has other components not found in Nos. 1 & 2. The treading floor (6.6–5.7×6.0 m, area 38 m^2, height of walls 0.3 m) is connected by channels both to the intermediate vat (0.8×0.9 m, depth 0.5 m) and to the collecting vat (2×2 m, depth 1.4 m, volume 5.6 m^3). The two vats are connected by a bore 0.2 m above the floor of the intermediate vat and in the floors of both there are small sumps. There are two steps in the collecting vat. The floors of both are paved in white mosaic and the walls covered in thick plaster. Near the centre of the treading floor there is a rectangular stone block (1.35×1.20 m) in which a central dovetail mortice was cut, that was the base of a fixed screw press and was connected to the collecting vat by a covered channel. The treading floor was paved in stone slabs but originally these probably served as the base for a mosaic floor. This wine press in its final phase is from the Byzantine period but probably the mosaic floors and the screw base with connecting channel were later additions and originally – perhaps in the Hellenistic period – the treading floor was the bed-rock that was probably plastered.

4. Area H4 (Fig. 4)

The wine press is located 150 m south-west of No. 3 on the edge of the village probably near the vineyard. Wine presses 3 and 4 are both of the Byzantine period but they differ considerably. Similarly to No. 3 the treading floor was almost square (6.2×6.4 m, area 40 m^2) and was originally paved in coarse white mosaic. In the centre of the floor there was a base for a fixed screw press, whose mortice was connected to the collecting vat by a ceramic pipe. The actual screw base is missing but the depression remaining in the floor (0.8×1.7 m) shows that it was rectangular. In the south-eastern corner was a square plastered depression that measured 0.5×0.7 m and was 10 cm deep.

The first elements that appear in No. 4 and not in No. 3 are semi-circular compartments placed around the treading floor (0.80–1.15×0.8–0.9 m, depth 0.4 m, volume 256–415 litres). Seven were found but there were probably eight – three on each lateral side and two along the side opposite the vats. They were plastered, had mosaic floors and were connected to the treading floor by bores.

A ceramic pipe connected the treading floor to a square intermediate vat (1.25×1.25 m, 1 m deep) that was almost certainly connected to the collecting vat although this channel was not found. The collecting vat is the second element that differs from those of all the other complex wine presses at the site. It was not square but round (diameter 2.5 m, depth 1.5 m, volume 7.3 m^3) and in the centre of its floor was a shallow sump (diameter 0.85 m, depth 0.3 m).

This wine press with its round collecting vat and semicircular compartments is of a type that can be regarded as a local subtype of the "four rectangle plan", several examples of which have been found in the vicinity (Sidi, Amit and 'Ad 2003: 262–263, Fig. 1).

5. Area M2 (Fig. 5)

The wine press is located close to the centre of H. Mazor and is different in several aspects from the other wine presses at the site. It actually consists of two wine presses placed at a right angle one to the other. The main one (5A) was made up of an almost square treading floor (L. 268; 5.4×6.0 m, area 32.4 m^2) connected by a narrow channel to a square collecting vat (L. 272; 2.2×2.2 m, depth 1.6 m, volume 7.7 m^3).

In the centre of the treading floor there is a rectangular stone screw base with a central dovetail mortice connected to the collecting vat by a lead pipe. There were six auxiliary floors associated with and about 1 m higher than the treading floor measuring in total c. 24 m^2. Three were to the north: L. 701 and 702 adjacent to the main floor and the third, L. 279 north of L. 702 and connected to it by a channel. Three were adjacent to the southern side of the treading floor: L. 704, 700 & 262. The must from these floors flowed through a bore to rock-cut semi-circular compartments with vaulted ceilings and mosaic floors. The outlets of these compartments consisted of lead pipes. The outlets of the compartments of the two northern floors that were adjacent to the treading floor opened above the treading floor but the three southern auxiliary floors sloped southwards away from the treading floor and their compartments and outlets were to the south.

The treading floor of the smaller winery (5B) was to the east of the northern auxiliary floors of the larger one. It was also almost square (3.7×4.0 m, area 14.8 m^2) and was connected by a channel to a rectangular collecting vat (1.5×1.1 m, depth 1.1 m, volume 1.8 m^3) that was next to the collecting vat of 5A. In this case also there was in the centre of the treading floor a rectangular screw base with a central dovetail mortice connected to the collecting vat by a lead pipe.

This press is also to be dated to the Byzantine period.

6. Area U 100

This small simple rock-cut wine press is an integral part of the north-western farmstead and consists of a trapezoidal treading floor (2.75×2.25/1.75 m, area 5.56 m^2) that is connected by a channel to a round collecting vat (diameter 1.1 m, depth 1.1 m, volume 1.1 m^3). The farmstead and the wine press in it are dated to the Byzantine period.

Wine Presses – Conclusions

The wine presses of H. Mazor have a great deal in common. In most cases the treading floors and vats are exact squares or almost so. Three (Nos. 1, 2 & 3) are in the "four rectangle plan" and a fourth (4) is a local variant of this plan. The four complex Byzantine wine presses (3, 4, 5A, 5B) were all equipped with fixed screw presses, the screw bases being rectangular and connected by pipes to the collecting vats. In the three cases in which the screw base was found the mortice was a central dovetail one, a type found in the region but more common in Upper Galilee and Phoenicia (Frankel 1999: 144–145).

Two of the wine presses (4 & 5A) had additional floors and compartments around the treading floor that were almost certainly connected to treatment of the grapes before treading on the main floor although the slight difference in the character of these ancillary installations hint at different techniques. Of particular interest are the southern auxiliary floors of No. 5A that show clearly that the "first must" was collected separately. It should be noted that presses 5A & 5B lack intermediate vats. Is this connected perhaps to the presence of auxiliary treading floors? The relationship between presses 5A & 5B also raises questions of production practices.

Table 1. Dimensions of wine presses

No.	Period	Floor m^2	Vat m^3
1	Hel	20.25	4.5
2	Hel	14	2.9
3	Byz	38	5.6
4	Byz	40	7.3
5A	Byz	32.4	7.7
5B	Byz	14.8	1.8
6	Byz	5.56	1.1

The Oil Presses

7. Area S 2–3 L. 40

The oil press was in one of the rooms in the western building of the north-western farmhouse. The olive crushing mill was not found. The press room measured 5×8 m and originally contained two lever and weights presses from the Hellenistic period. In the northern press four weights with reversed T bores were found, three *in situ* and a fourth nearby. 2 m to the east stood a central collecting vat. Next to it there was an additional vat to which the oil from the central vat apparently flowed. Although the niche in which the beam was anchored was not found the foundation of the wall remains and the length of the beam can be calculated as 6.5 m. Of the southern press all that remained are the depressions in which the central and additional vat originally stood.

8. Area H7

This oil press was found in a building located between H. Mazor and the north-western farmstead – perhaps part of a separate site. The press-bed was a slightly sloping square rock-cut surface (1.1×1.1 m), the expressed liquid flowing to a free standing square collecting vat (internal measurements 0.6×0.6 m, depth 0.45 m, volume 0.16 m^3, thickness of the wall 0.7 m) that was emplaced in an especially prepared depression in the rock. In the bottom of the vat was a small sump. The vat probably originally had a cover but this was not found. 1 m from the collecting vat was a weights pit in the shape of the sole of a shoe, probably intended to contain two weights. If this is so the diameter of the larger would have been about 1 m and of the smaller about 0.8 m. Between the collecting vat and the weights pit there were two small holes in which upright poles were secured that served to guide the beam. Neither the beam anchoring niche nor the wall of which it was part were found.

9. Area MI (Fig. 6)

This oil press was found in a building 50 m south of the centre of H. Mazor (area N). It was very disturbed but a square collecting vat and a weights pit in the shape of the sole of a shoe were found, both identical to those of No. 8. In addition a cylindrical screw weight was found with central socket and internal dovetail mortices (diameter 1.25 m, height 1.3 m). This type is not common, is found in the centre of the country and probably originated in Syria. An example was recorded at Kh. Kasfa, 6.5 km to the north-west of H. Mazor (Frankel 1999: 114–115).

Bibliography

Amit, D. *et al.* 1998. "Mazor – 1993–1996". *ESI* 18: 57–66.

Amit, D. *et al.* 1999. "Mazor (El'ad) 1996 – 1997". *HA-ESI* 109: 55*-64*.

Frankel R. 1999. *Wine and Oil Production in Antiquity in Israel and Other Mediterranean Countries*. Sheffield.

Sidi, N., Amit, D. and 'Ad, U. 2003. "Two Winepresses from Kefar Sirkin and Mazor". *'Atiqot* 44: 253–266.

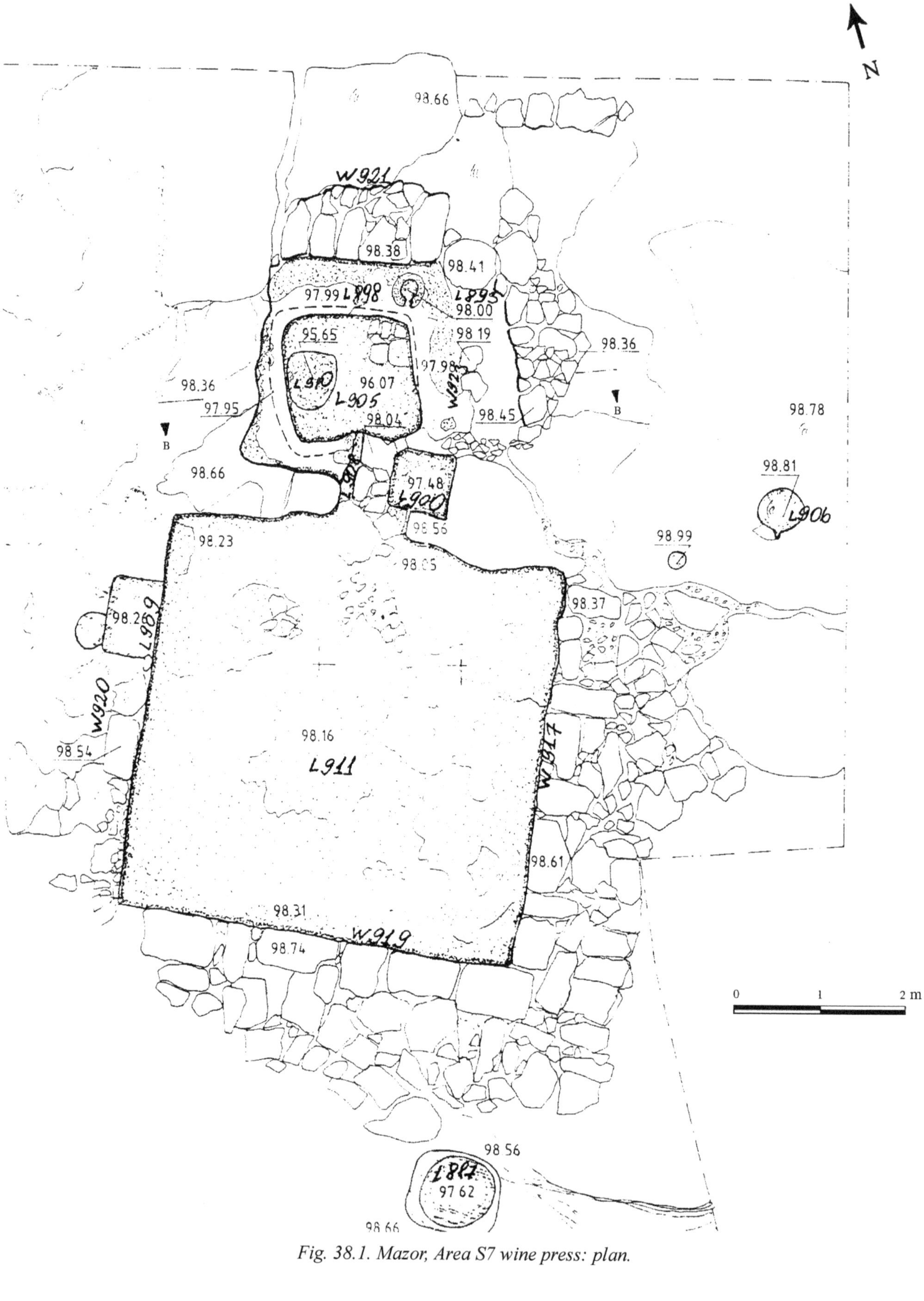

Fig. 38.1. Mazor, Area S7 wine press: plan.

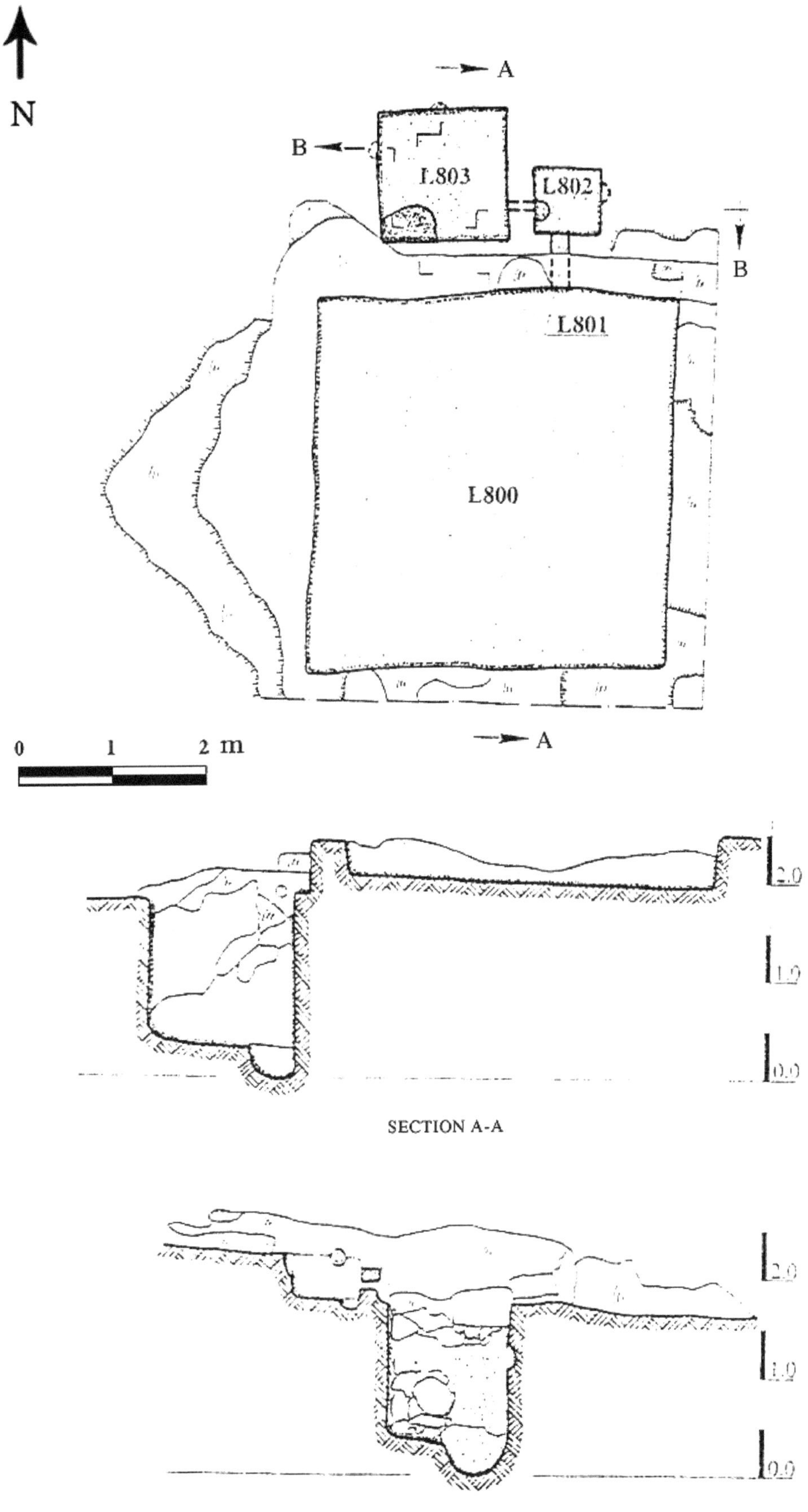

Fig. 38.2. Mazor, Area P wine press: plan and sections.

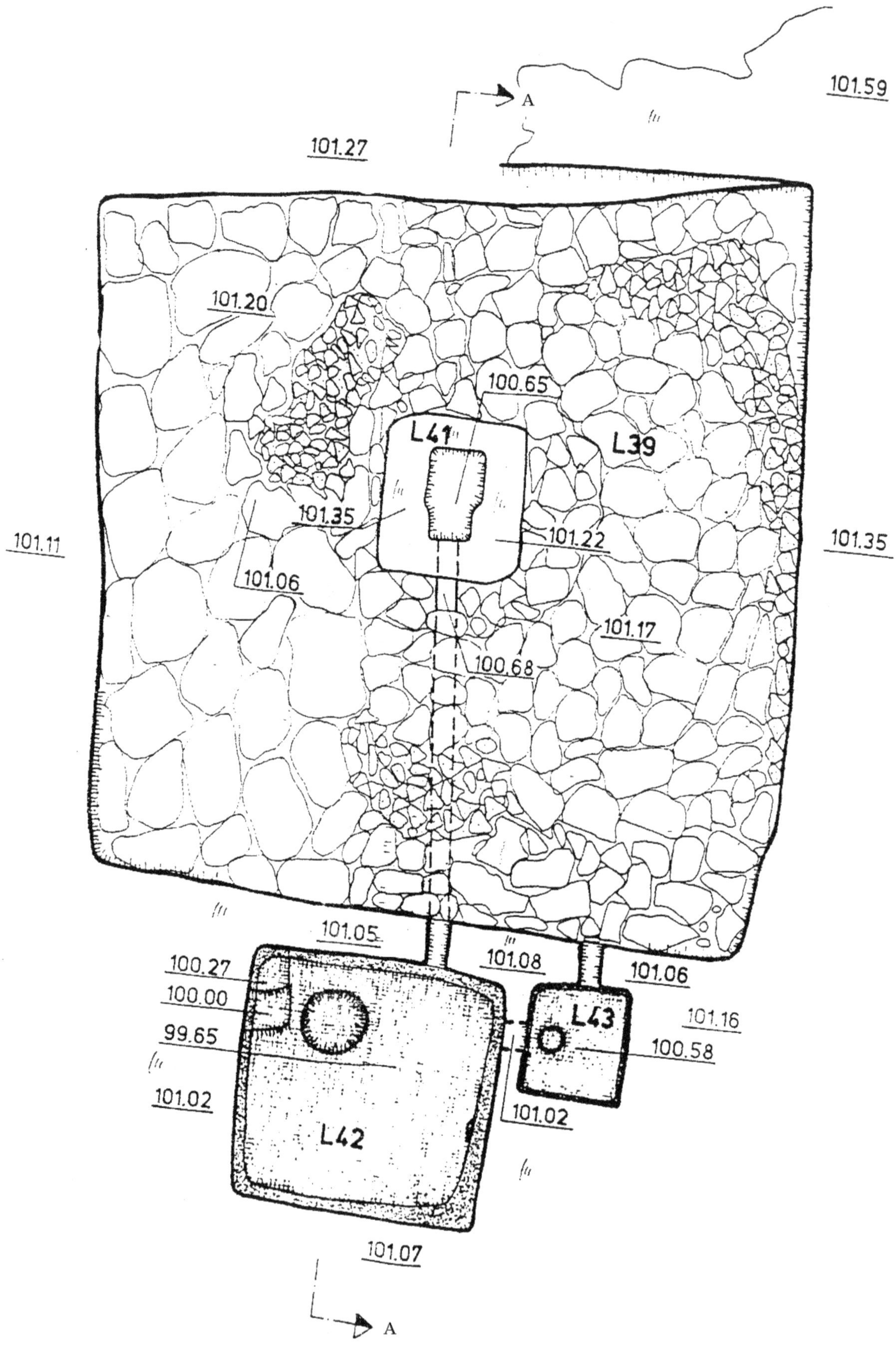

Fig. 38.3. Mazor, Area H2 wine press: plan and sections.

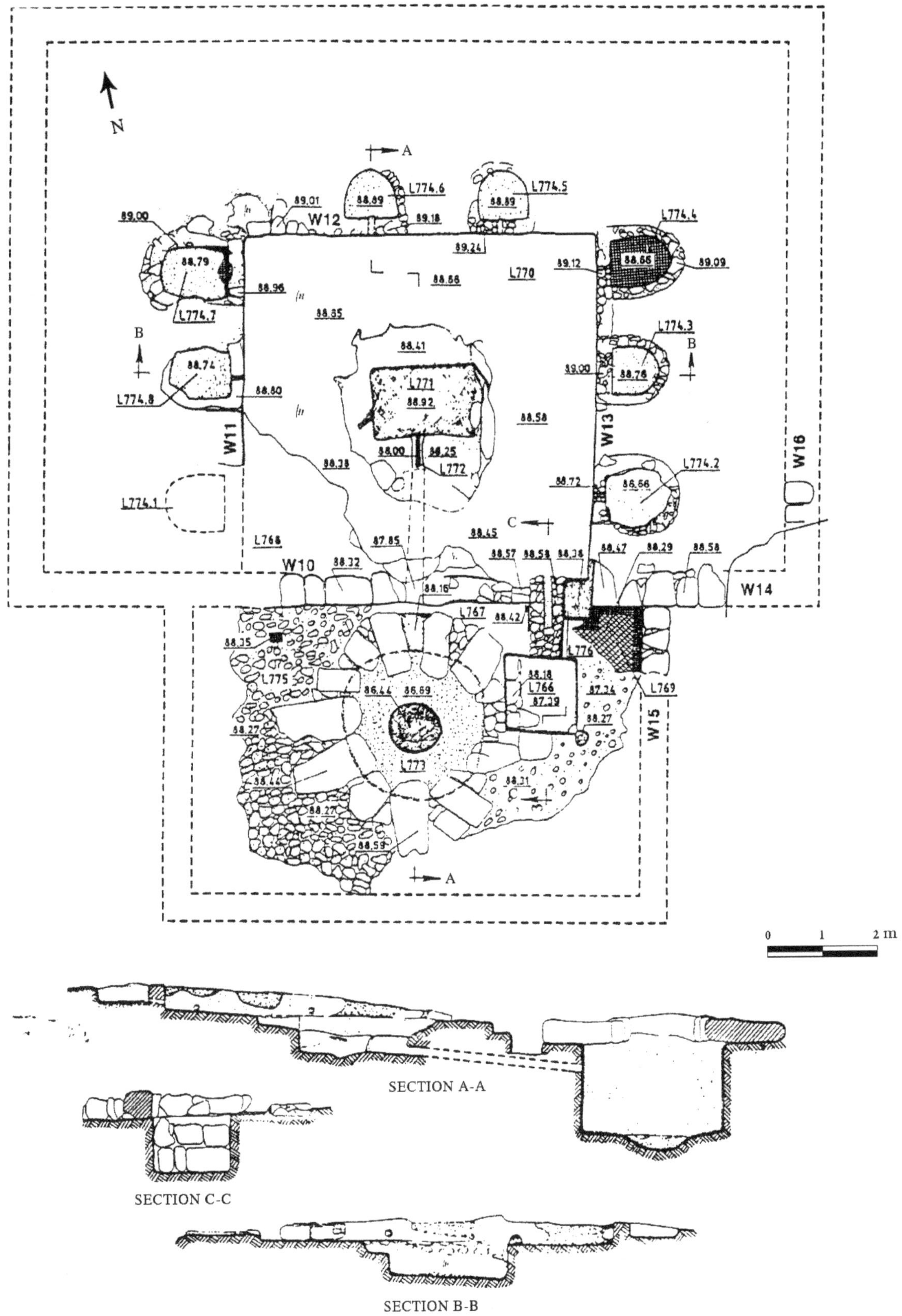

Fig. 38.4. Mazor, Area H4 wine press: plan.

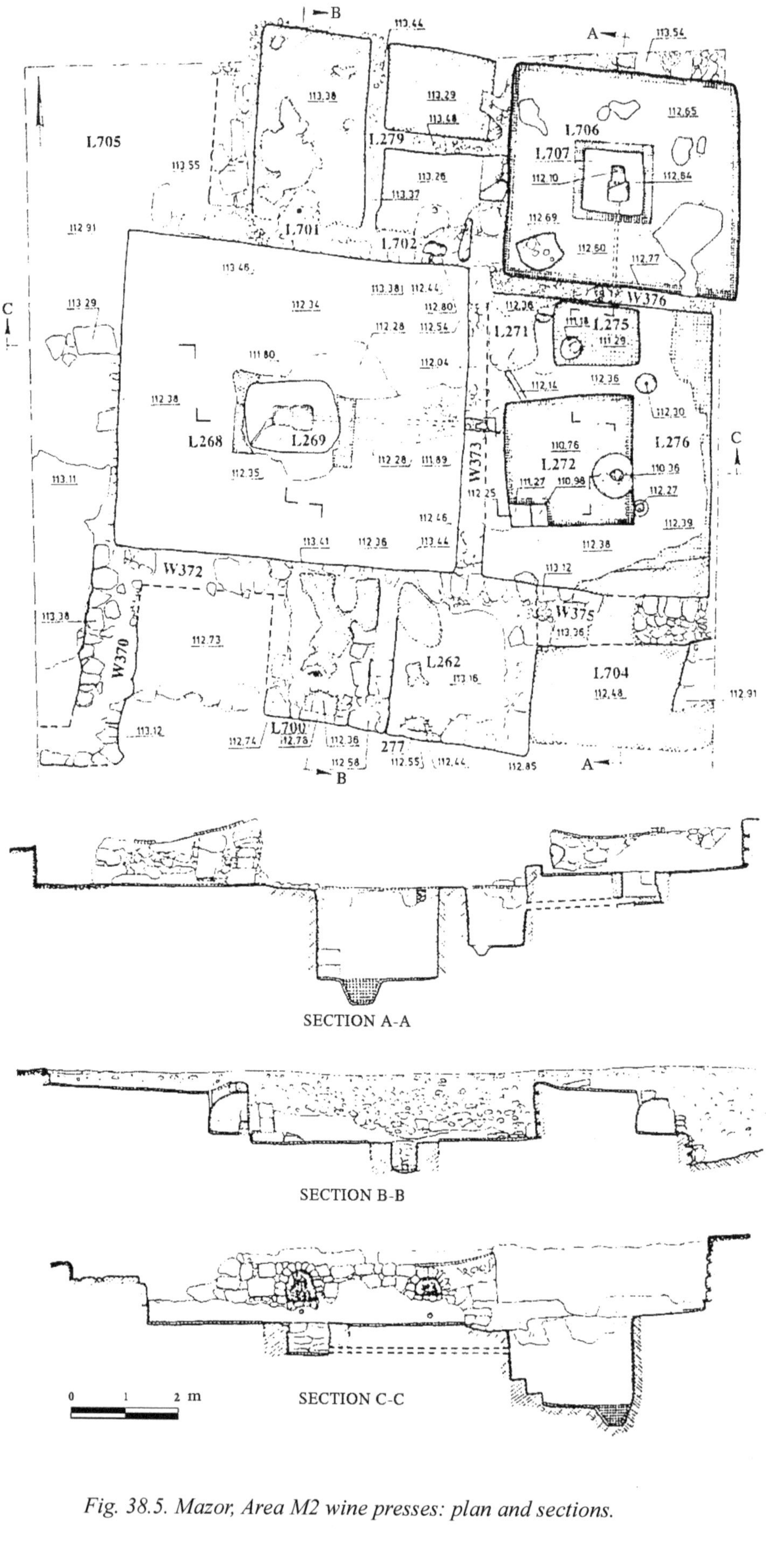

Fig. 38.5. Mazor, Area M2 wine presses: plan and sections.

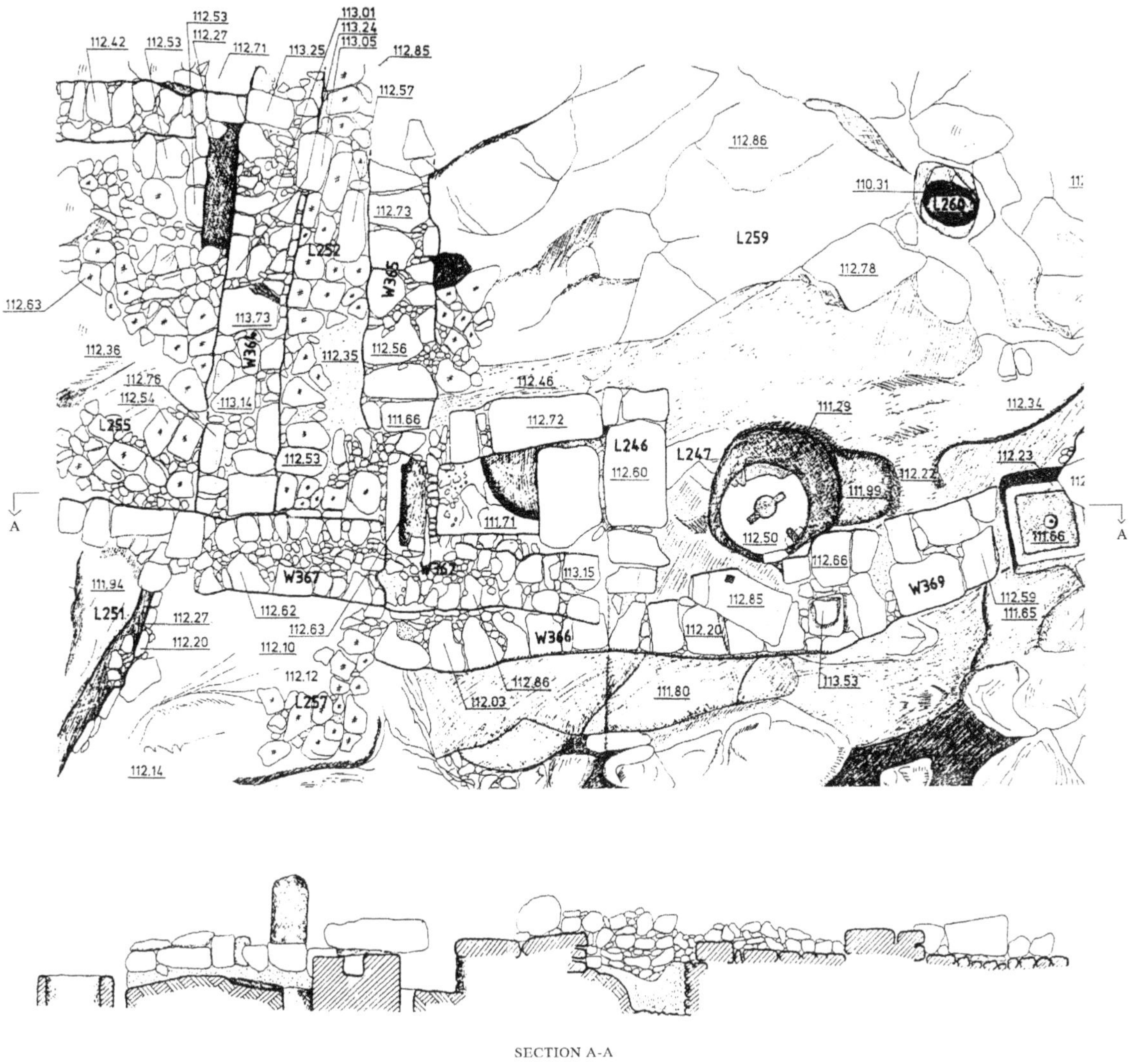

Fig. 38.6. Mazor, Area MI oil press: plan and section.

39. Treading Installations at Kh. Hablata (Hallamish/Neveh Suf)

Ofer Sion and Zohar 'Amar

The Site

Kh. Hablata is in the centre of the present settlement Halamish (Neveh Suf) in the Samarian Hills (MR 1621/1571). The rock formation at the site is of limestone and dolomite. The Roman road from Antipatris to Gophna and Jerusalem passed close to it. The site was surveyed in the past by Guérin (1880, 5: 105) and in 1941 by Husseini of the Mandatory Department of Antiquities of Palestine.[1] In 1991 it was surveyed by the authors and wine presses and other installations were cleared (Sion and 'Amar 1997: 205–207).

The Installations

These unusual installations (Fig. 1) were hewn on two adjacent rock surfaces, Nos. 3 and 4, that are separated by a channel 1–2 m wide. The southern surface No. 3 is 0.3 m higher than the northern surface No. 4.

On surface 3, 14 rock-cut round treading installations were found all very similar in character (Fig. 2). The average diameter of the treading surfaces is 2.3 m and the depth 10 cm. The treading surfaces of installations 1–11 are concave while those of installations 12–14 are flat and are surrounded by a shallow channel. The small collecting vats are also round with a maximum diameter of 0.75 m and a maximum depth of 0.4 m. Installations 12 and 13 are connected by a channel.

On surface 4 there were two square wine presses (Nos. 15 & 16) and two installations (Nos. 17 & 18) similar to those on surface 3 that had either not been completed or had been damaged in the past.

The pottery found in the installations is dated to the Byzantine period.

Installations Nos. 15 & 16 are regular wine presses. The unusual round floors Nos. 1–14 & 17–18 were probably used to store the grapes before the treading and the fine must squeezed by their own weight was collected in the vats (*Mishtah shel 'alim* - Mishnah *Toh.* 10.4, 5; see Frankel 1999: 188). The relatively small volume of the collecting vats, however, hints to the possibility that other fruits were squeezed on those floors, as known also from pre-industrial installations (Dalman 1928–1942, IV: 241).

Notes

1. Archives of the Israel Antiquities Authority: site number 8270/0.

Bibliography

Dalman, G. 1928–1942. *Arbeit und Sitte in Palästina.* 7 Vols. Gütersloh (repr. Hildesheim 1964).

Frankel, R. 1999. *Wine and Oil Production in Antiquity in Israel and Other Mediterranean Countries*. Sheffield.

Guérin, V. 1880. *Description géographique, historique et archéologique de la Palestine.* 7 Vols. Paris (repr. Amsterdam 1969).

Sion, O. and 'Amar, Z. 1997. "Treading Installations at Khirbet Hablâtâ (Hallamish)". *'Atiqot* 32: 205–207 (Hebrew).

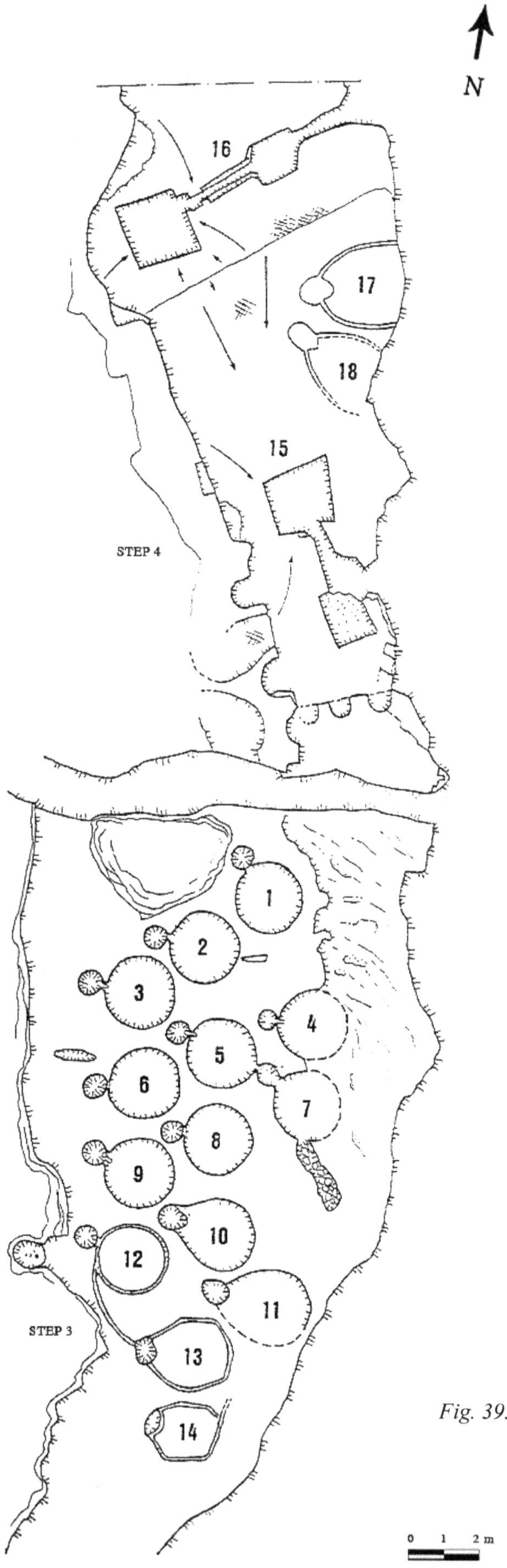

Fig. 39.1. Hablata, plan of the treading installations.

Fig. 39.2. Hablata, the round treading installations, view to the south-east.

40. A Byzantine Wine Press at Wadi Hermiya

Yehiel Zelinger

The wine press was unearthed by the author in 1995 at Wadi Hermiya in the Samarian hills, adjacent to the Ram'allah bypass road (MR 1736/1556). The components of the installation were hewn from the natural bed-rock, leveled and straightened, and floored with coarse white mosaic tesserae on a sub-stratum of clay and pebbles. Its western section was destroyed in the past as the result of agricultural cultivation. The wine press consists of the following components (Figs. 1–2):

- A treading floor on the south measuring approximately 5.0×5.3 m – 26.5 m². Its northern wall was the bed-rock, carved to a depth of 0.6 m and filled in where required with stones of different sizes. On the east and west it was bounded by cut stones set on their narrow sides, 0.30–0.45 m high. They were preserved only in the northern part of the area. The western wall was not preserved. Those parts of the wall that remained were covered with a thick layer of white plaster mixed with crushed pottery fragments. Along the edges of the mosaic floor were three rows of tesserae laid parallel to the lines of the walls; the remaining tesserae were set in diagonal rows.

- In the centre of the treading floor was a stone base for fixing the screw of a press. The base is rectangular (0.9×1.2 m) and has a closed dovetail mortice for holding the screw, measuring 0.35×0.60 m and 0.65 m deep. This is apparently the *tapu'ah* (the apple) mentioned in Talmudic sources (Frankel 1999: 187).

- A plastered channel (or perhaps originally a pipe – the destruction of the wall precludes a definite decision on the issue), 0.5×0.1 m and 5 cm deep, leads from the treading floor to the collecting vat. Where the channel leaves the treading area, a flat stone has been set into the floor, apparently to reinforce the connection. It, too, is surrounded by three or four rows of tesserae parallel to its sides.

- A single collecting vat with no filtration mechanism was installed opposite the western part of the treading floor. The vat measures 2×2 m and 1.7 m deep, thus its volume is 6.8 m³. In its south-east corner was found a sedimentation basin 0.6 m in both diameter and depth, i.e. a volume of 170 litres. The total volume is therefore 6.97 m³. The walls of the vat are heavily coated with white plaster and the floor is white mosaic. On the east, north, and apparently on the west also it was surrounded by an area floored in white mosaic, measuring 3.5–4.5×5.0 m. Its northern side is irregular, partly following the contour of the natural rock. This is apparently an area where those working around the vat could stand.

- North of the paved area around the collecting vat is an irregularly shaped area of rock, well straightened and leveled, with maximum dimensions of 2.7×3.5 m. Its purpose is not clear, and since its western part has been demolished it cannot be established whether some kind of vat existed alongside it.

- Three small round built vats were unearthed on the edges of the treading floor – two on the eastern side (Fig. 3) and one at the eastern end of the southern side. It is possible that such vats also existed on the southern and western sides, but this part of the wine press was demolished. The average inner diameter of the vats is 0.4 m, their walls are 5 cm thick and they have been preserved to a maximum depth of 0.3 m. The vats are floored with white mosaic tiles and are covered on the outside with similar tesserae – a rather exceptional phenomenon apparently carried out to reinforce the wall. On the inside the wall is covered by plaster reinforced with ribbed pieces of pottery. No openings were found in the bottoms of the vats for draining liquids, so apparently they were used for storing and preparing ancillary materials to be added to the must, e.g. marl. Similar vats were found, for instance, in Zur Natan (Site 31) and H. Nasha (Site 33).

In light of its form, its components and its technology and according to the pieces of pottery embedded in the plaster, the wine press can be dated to the Byzantine period.

Bibliography

Frankel R. 1999. *Wine and Oil Production in Antiquity in Israel and Other Mediterranean Countries*. Sheffield.

Hirschfeld, Y. 1983. "Ancient Wine Presses in the Park of Aijalon". *IEJ* 33: 207–218.

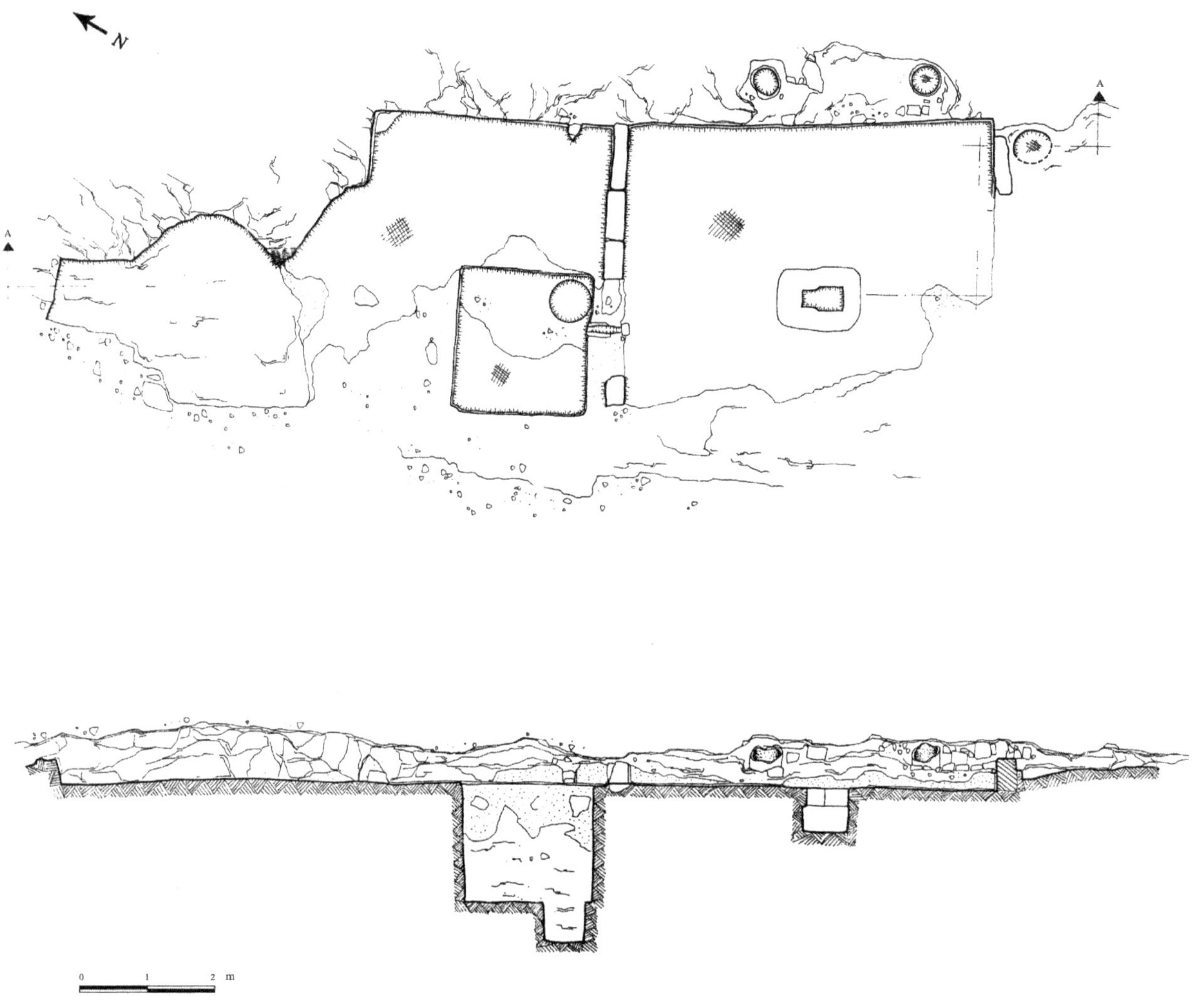

Fig. 40.1. Hermiya, plan and cross-section of the wine press.

Fig. 40.2. Hermiya, overall view of the wine press to the south-east.

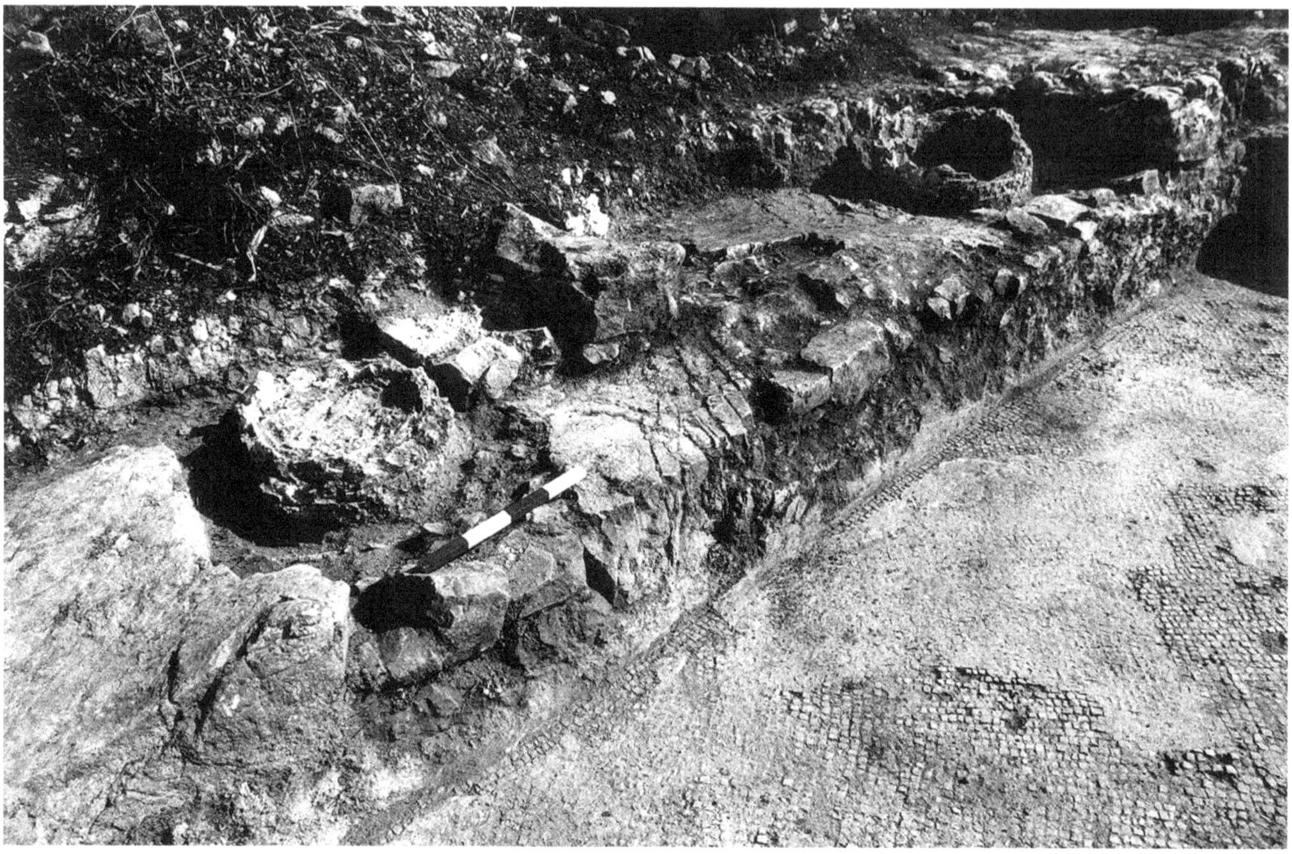

Fig. 40.3. Hermiya, two plastered vats at the edges of the treading floor, view to the south-east.

41. A Byzantine Oil Press at Jifna

Yehiel Zelinger

In a salvage excavation at Jifna in the Samarian Hills (MR 1703/1525), situated about 25 km north of Jerusalem and identified with the town of Gophna from the Second Temple period (Zelinger 2000),[1] a large water reservoir was unearthed in which an elaborate oil press had been installed (Figs. 1–2). The reservoir – 13×12 m and 5 m deep – was lined with fine white plaster to prepare it for storing water. A flight of stairs, of which 12 were preserved, had been hewn out of its western wall, enabling descent into the reservoir. The piers of arches preserved in its southern wall and the remains of bases on the floor show that the reservoir had a covering supported by constructed arches. This roof was probably constructed of light materials (wood, thatch, etc.) since no stones from a vault were found in the rubble within the reservoir.

At a certain stage the reservoir was no longer used to store water, and inside it a complex oil press was constructed. The plant consisted of five presses and systems for storing the olives and collecting the expressed liquid. The crushing mill was not preserved; it probably stood in the empty northern end of the reservoir, between the stairs and the presses, close to the compartments that had been built to store the olives. It was almost certainly a stone mill of the conventional type whose parts were stolen after the oil press was abandoned.

In the complex, two types of presses were recorded both of which were operated by means of direct pressure exerted by wooden screws:

1. One single screw cross press hewn into bed-rock (Frankel 1994: 70–71);

2. Four rigid frame screw presses with free standing rectangular press-beds in which open mortices had been hewn (Ibid.: 71–73).

A cross press was carved into the bed-rock in the wall at the south-east corner of the reservoir (Fig. 3). The complete press is 3.25 m high, part of it below the floor of the cistern. It is shaped roughly like a cross and includes, from the top down, the following components:

- A vertical niche comprising the upright of the cross, roughly square in cross-section, in which the upper part of the screw rotated. Its width and depth were 0.25–0.30 m.

- A horizontal niche representing the transom of the cross, 1.4 m long, to accommodate the "nut" – a plank with a threaded hole in which the screw rotated thus raising or lowering it.

- A larger cavity in which the frails were piled one above the other.

- At the bottom of the frail space was hewn the circular press-bed with central opening around which were a circular and radial grooves. The expressed liquid flowed along them into the collecting vat through the central opening. The diameter of the press-bed was 0.8 m.

- A central cylindrical collecting vat, hewn into the rock below floor level, with a sedimentation basin at the bottom. The vat was plastered and measured 0.4 m in diameter and 0.6 m deep, thus having a volume of approximately 75 litres. The olive oil and watery lees were drawn off with a small jar or ladle through the same opening by which they had flowed into the vat.

The cross press was common mainly in the Jerusalem and Benjamin areas and was probably specially adapted for cave installation (Frankel 1999: 130). This makes it difficult to establish a date for its use, since in caves we generally cannot refer to excavated assemblages with their concomitant finds. An installation like this was recently excavated in the "Cave of the Horses" at Bet Guvrin and dated to the Byzantine period.[2] At Jifna also, in light of the accompanying finds (below), the installation should probably be dated to that period.

Of the second type – the rigid frame direct-pressure screw press – four screw press bases were found, three of them *in situ* (Figs. 1–2 above). These presses were free standing and had no architectural connection to the walls of the reservoir. Each of the press bases has a circular groove for draining the liquid, terminating in a channel leading to the edge of the base and two open mortices on the sides (Figs. 4–5) for installing the press piers. Each base measures approximately 1×2 m and the diameter of the round press-bed measures 0.8 m in two instances and 0.5 m in a third. The press-bases were hewn and dressed elsewhere, brought to the reservoir and installed in depressions carved into its floor. The piers were set into the side mortices; above them was fastened a plank with a threaded hole in it, in which the screw rotated. The piers did not survive; if they were made of wood, it decomposed and if they were hewn from stone, they were probably stolen for secondary use.

Adjacent to each press-bed, under the end of the channel,

a container was placed to collect the expressed liquid. In the eastern press this container was a pottery vessel set into a pit hewn into the floor; it measured 0.6 m in diameter and 0.75 m deep – i.e. a total volume of 212 litres. In the central press the liquid flowed directly into a rectangular vat cut into the floor. It had a shallow sedimentation basin in its floor and measured 0.6×0.7 m and 0.45 m deep. Its volume was therefore about 190 litres. There was a ledge cut into the rim of the vat on all four sides to accommodate a cover of stone or wood (Fig. 4). A hewn pipe 0.6 m long and a few centimeters in diameter connected the bottom of this vat with the cylindrical collecting vat of the western press that was also carved into the floor. It also had a sedimentation basin and measured 0.55 m in both diameter and depth, i.e. a volume of 130 litres. The connection between the two vats – a rare phenomenon in ancient oil presses – enabled two possible activities: production of large quantities of expressed liquid from either of the presses or separation of the oil by allowing the watery lees at the bottom of one vat to flow through the pipe into the other. Two additional circular vats or pits, 0.45 and 0.7 m in diameter were carved in the floor adjacent to the presses, but had no direct connection with the collection systems. They may have been used for holding the jars used to remove the oil from the vats.

In the north-east corner of the pool raised areas were found with floors that sloped toward and drained into collection vats with pottery vessels inside them. Apparently, olive mash was placed on these surfaces for a short period of time so that the finest type of oil that leaked out of it before pressing would drain into the vessels.

Niches were carved into the eastern and southern walls of the reservoir, probably for the placement of lamps.

When the oil press was abandoned, many of its portable parts were stolen and over the years the reservoir became filled with earth and garbage from the surrounding area.

The pottery that has been collected is from two periods:

- The Early Roman period, probably when the reservoir was hewn from bed-rock, the type of plaster used suggesting a similar date.
- The Late Roman-Early Byzantine period, probably when the oil press was in operation.

The chronological relationship between the pressing installations is unclear: Did the cross press precede the frame presses or *vice versa*? Did all the frame presses work simultaneously or did one replace another that had gone out of service?

A Byzantine oil press with a similar arrangement of three direct pressure rigid-frame screw presses placed around a crushing mill was discovered, for example, at Tel Safsafot (Frankel 1988–1989; Site 22).

The Jifna region is compatible with olive growing and both ancient and traditional olive oil presses are known from the surrounding area (Frankel, Avitsur and Ayalon 1994:17).

Notes

1. The excavation was carried out in 1995 on behalf of the Staff Officer for Archaeology in Judea and Samaria, under the leadership of the author and with the assistance of Ibrahim Abu-'Amar, Archaeological supervisor of the Ram'allah district. Also assisting were Mendel Kahn and Alex Opokin (draftsmen), Miriam Menokian (artist) and Shlomi Amami (photographer).
2. Many thanks to Professor Amos Kloner for permitting me to mention this as yet unpublished information.

Bibliography

Frankel, R. 1988–1989. "An Oil Press at Tel Safsafot". *Tel Aviv* 15/16: 77–91.

Frankel, R. 1994. "Ancient Oil Mills and Presses". In: R. Frankel, S. Avitsur and E. Ayalon. *History and Technology of Olive Oil in the Holy Land.* Arlington and Tel Aviv: 19–89.

Frankel R. 1999. *Wine and Oil Production in Antiquity in Israel and Other Mediterranean Countries.* Sheffield.

Zelinger, Y. 2000. "Gophna during the Second Temple Period". In: Y. Eshel (ed.). *Judea and Samaria Studies* 10: 103–112 (Hebrew).

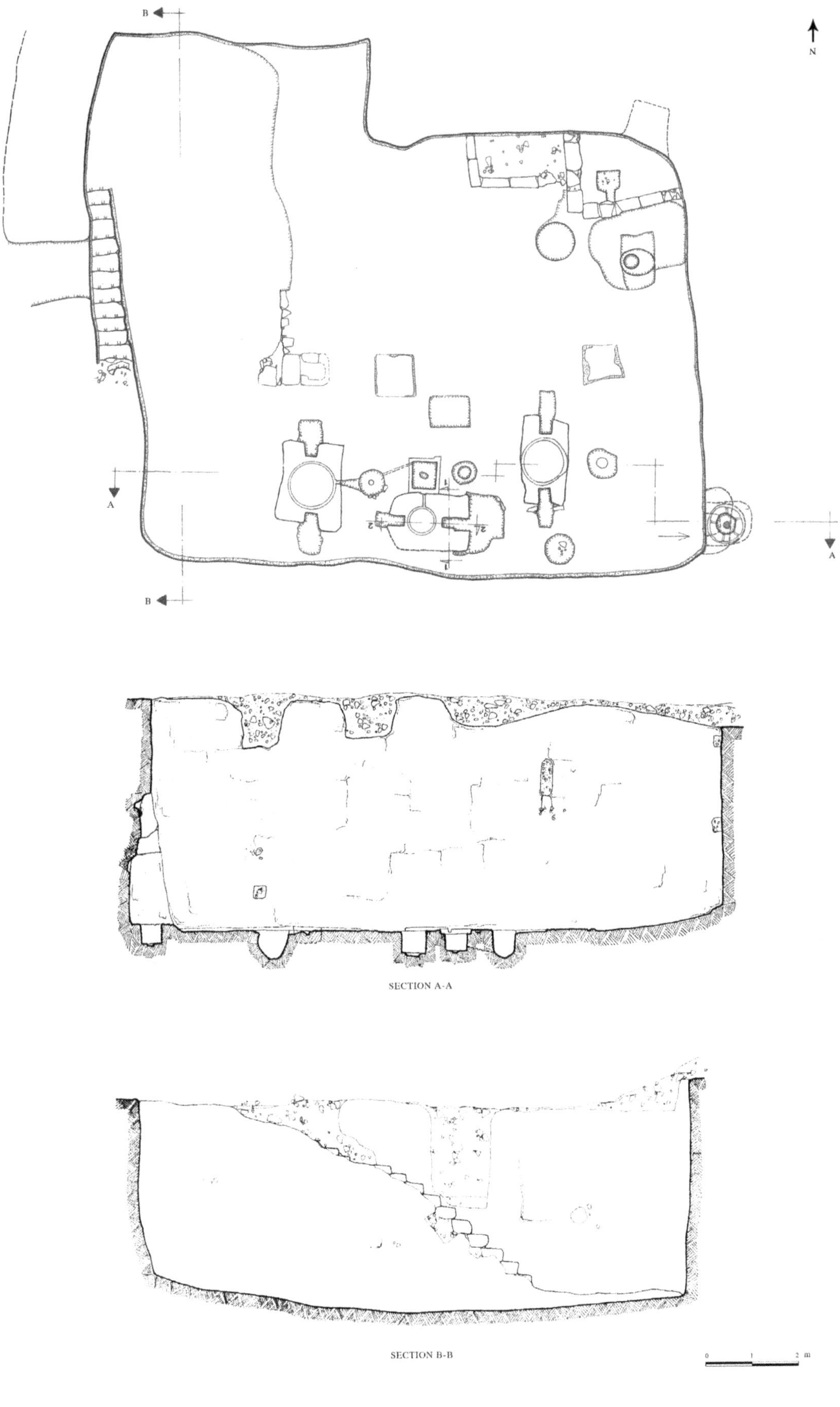

Fig. 41.1. Jifna, plan and sections of the water reservoir and the oil presses.

Fig. 41.2. Jifna, the reservoir and the oil presses, looking south-east.

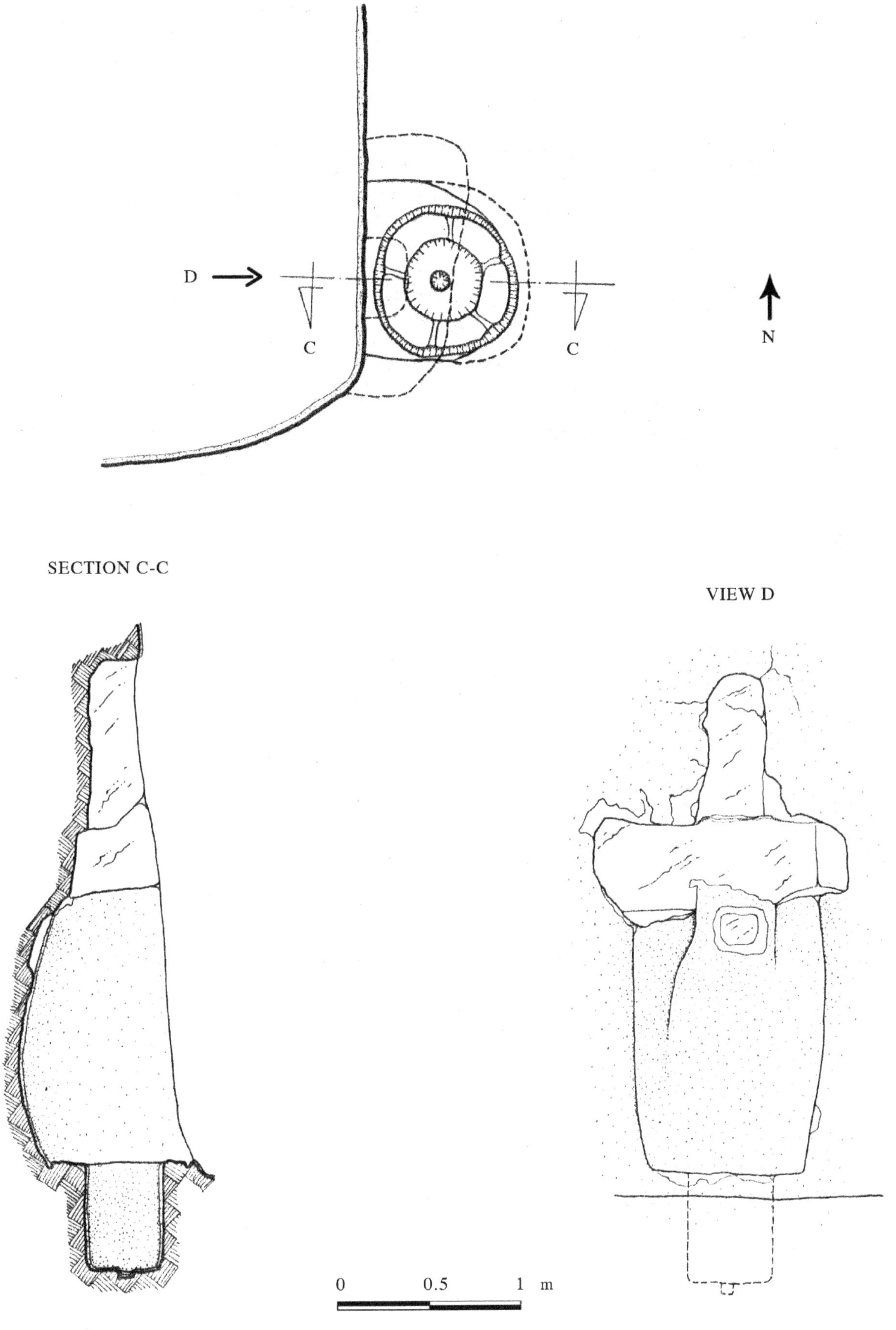

Fig. 41.3. Jifna, plan and sections of the cross press.

Fig. 41.4. Jifna, press-bed and collecting vat of the central rigid frame direct pressure screw press, looking north.

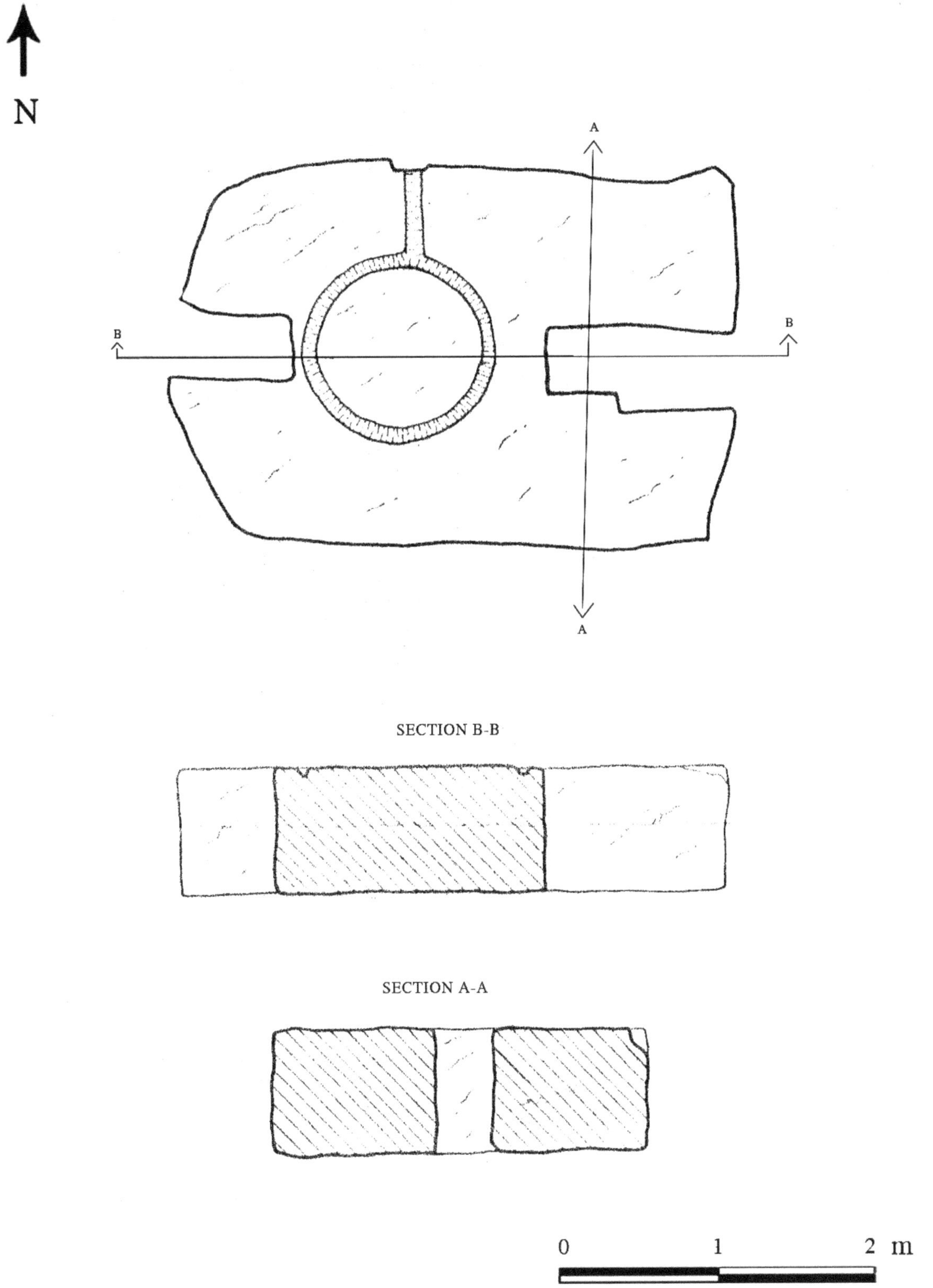

Fig. 41.5. Jifna, drawing of the press-bed of the central direct pressure screw press.

42. A Complex of Two Wine Presses at Rishon Le-Zion

Eran Shavit

The Site

The wine presses are part of a Byzantine site buried under more than 6 m of sand. It is located to the north-east of Ha-Qore (Palmahim) junction (MR 1302/1524) on both sides of the Shprinzak Street near its junction with Rothchild Street.

The site was mainly occupied from the 4th cent. until the end of the Byzantine period. The earliest date possible for the building of Wine Press I is the middle of the 5th cent., as coins predating this period (Theodosius II/Valentinian III) were found on the floors of a structure unearthed underneath and south of its vat area. This structure was covered with soil when the wine press was built.

The latest numismatic evidence at the site are two coins from the time of Justinian I (AD 529–533) that were found on the floor of a building, perhaps a storehouse unearthed in Area C near wine press I, together with broken storage jars. The site was abandoned in the 6th cent., used for a short time as a garbage disposal in the Umayyad period and then covered with sand dunes.

A cross in the mosaic floor of the collecting vat of wine press I and pig bones found in several places suggest that the inhabitants of the site in the Byzantine period were Christians. The lack of pig bones in the later garbage deposit testify to non-Christian population.

The two wine presses and some remains of other structures were excavated by the author in 1991 on behalf of the Israel Antiquities Authority.

Wine Press I

This wine press is located at the west side of Shprinzak Street. It consists of three square paved areas, arranged from west to east (Fig. 1). First the treading floor (3–4) will be described, then the vats (5–6) and only then the eastern area (1–2).

The Treading Floor (3)

The treading floor (4.3×4.3 m, area 18.5 m^2) is paved in white mosaic. It is surrounded on all sides by 0.3 m wide walls, plastered on the inside with pink plaster. The walls were preserved to a height of no more than 0.2 m. In the centre of the floor a round stone was embedded (4). The mosaic tesserae form a margin of three parallel lines around its perimeter as well as along the walls. The block, 1.2 m in diameter, has a rectangular mortice (0.25×0.30 m, 0.7 m deep) in its centre. The hole widens on two of its sides, in order to accommodate a wooden screw. This type is common in the region (Frankel 1999 Map 35). The mortice does not penetrate through the whole width of the stone, so this width could not be checked in the excavation.

A clay pipe, some 0.1 m in diameter, was laid under the treading floor. It was made of five pipes each 0.45–0.50 m long, that were laid into a mortar and stones foundation. Today the pipe starts in the natural red loam soil 0.15 m to the east of the stone screw base and with no connection to it, so no liquid could enter it. Its lower end, leading into the filtration vat (5), was blocked at a later stage by a stone and covered with plaster. It seems that the pipe belongs to an earlier stage of use, in which the existing screw base had not yet been put in place[1] (see editors' note).

The Vat Area (5–6)

This rectangular section was separated from the treading floor by a wall 0.4 m wide. It was paved in white mosaic and enclosed within a wall of which, however, only part remains. The treading floor was connected to the filtration vat (5) by a plastered stone channel that was incorporated into the dividing wall. It was cut in a *kurkar* (sandstone) block 0.4 m long and 0.26 m wide. The outlet of the subterranean pipe was 0.1 m below that of the channel. The filtration vat (1.1×1.2 m, 0.55 m deep) had a white mosaic floor and its walls were covered in pink plaster.

A pipe 0.6 m long and 0.1 m in diameter that left the filtration vat close to its floor flowed into the collecting vat (6) 0.75 m below its rim. This circular vat (diameter 2.75 m, depth 2.50 m, total volume 14.8 m^3) had a mosaic floor in which a marble slab (0.25×0.40 m) was inlaid immediately below the outlet of the pipe so that the flow of must would not damage the floor (Fig. 2). A somewhat similar arrangement was found in the wine press at nearby Kh. Duran (Site 45). In the centre of the vat is a circular sump 0.3 m deep and gradually diminishing in diameter from 1.1 to 0.6 m. The mosaic floor of the sump was laid in the form of a cross – a rather unusual phenomenon (Fig. 3). On the floor, attached to the south wall was a *kurkar* stone 0.45 m high that served as a step. The plaster that covered the wall of the collecting vat consisted of four layers, first small stones, then grey plaster, then a layer of sherds and finally on the surface pink plaster.

The Western Area (1–2)

This section is divided by a 0.3 m wide east-west wall into two parts (1). The southern, larger part (3.5×3.5 m) is paved partly in pebble stones and partly in white mosaic. The mosaic tesserae were laid, as in all other mosaic floors of both installations, in several straight lines along the edges of the floor and diagonally in its inner part. This floor went through several changes and repairs. The northern part (3.5×1.9 m) is paved in beaten earth. Both parts can be explained as an accessory area, probably for temporary storage of grapes before the treading process.

Two round compartments, plastered and paved with white mosaic, were found on either side of the western area. The southern one (2) is 0.9 m in diameter and 0.4 m deep, while the northern one, only partially preserved, is 0.7 m in diameter. Each one is surrounded by a rectangular floor (2.0×1.5 and 1.5×1.7 m respectively) that protrudes from the general rectangular shape of the western area (1). Similar compartments have been found in other wine presses in the region such as H. Nasha (Site 33) and Kefar Sirkin (Sidi, Amit and 'Ad 2003). They might have been used for mixing some liquid to be added to the must or just for washing the feet before treading the grapes. No indicative sediment was found on the floors of these compartments.

After the installation was abandoned its collecting vat was used as a dump and filled with debris, sand and a large quantity of bones of donkeys, camels and cattle originating from a nearby Early Arab site.

Wine press I was reconstructed *in situ* by the Conservation Department of the Israel Antiquities Authority and Yehoshua Dray reconstructed the wooden screw press, for the first time in Israel. Recently two other wine presses found in the region, of the Iron Age and the Hellenistic period, have been transferred to this site, creating together a "Wine-Industry Park".

Wine Press II

This installation is located on the eastern side of Shprinzak Street, to the east of wine press I and in a 90° angle to it, i.e. its parts were built in a general south-north line. Today it is covered again with soil.

This wine press is smaller and less symmetrical than the one described above and generally built in a less professional manner (Figs. 4–6). The arrangement of its various components and the shape of the filtration vat (2) are different, and techniques such as plastered floors or floors made of vertical pottery sherds are found only here. It also seems that this wine press was used for a longer period of time: unlike wine press I, it underwent many changes, and at least five different stages can be discerned. The main components of the installation – the treading floor (6) and the vats (1–2) – will be described first in order to make its understanding easier, and only then the western area will follow.

The Treading Floor

The square treading floor (6) measures 4×4 m (16 m^2). Its south-eastern corner was not excavated. Three phases could be defined in it, and in all of them it was paved with white mosaic.

The lowest, earliest floor (A) probably covered a much smaller area than the later ones. The must flew from it towards the vats through a plastered channel whose remains were found adjacent to the surviving part of the floor, underneath the later channel (see below). It crossed the dividing wall between the treading floor and the vat area.

The middle floor (B) was laid 0.1 m above the first one. Only few of its tesserae remained next to the northern wall due to a re-plastering of the wall at this stage, that covered the row of cubes closest to it. A new channel was installed at this phase in the dividing wall, on top of the previous one. It was made from a single stone, 0.45 m long and 0.2 m wide, quite similar to that of wine press I described above.

The upper, latest floor (C) was laid 0.15 m above the middle one. Only two rows of mosaic tesserae survived next to and parallel to the eastern wall, while in most of the area the mortar foundation of the floor remained. In the centre was found a screw base (0.25 m thick) that was probably an olive crushing stone in secondary use. This round stone is 1.1 m in diameter and has a 0.3×0.3 m hole in its centre in which the screw was installed. The hole runs through the entire stone. The screw base was embedded with its top at the level of the upper floor, making it possible to date it to the latest phase. A clay pipe that led under the floor from the general direction of the screw base towards the filtration vat (see below) is not connected to the hole of the base. This fact strengthens the impression that this base was laid here in the latest stage while the pipe belongs to one of the earlier phases.

The Vat Area

The vat area (1–2) has an asymmetrical trapezoidal shape with walls 3.0–4.5 m long (Fig. 7), that are only partly preserved. It is paved with white mosaic that, as usual here, have three straight lines of tesserae along the edges. The remains of an earlier floor are preserved in some places along the edges, testifying that the walls bordering the earlier area were later replaced or widened. Unlike in wine press I, here both the filtration vat and the collecting vat are round and together with the screw base in the treading floor (6) are arranged in a straight south-north line, a fact that shows some pre-planning.

The filtration vat (2), to which the later stone channel leads, is 0.9 m in diameter and 0.6 m deep. Its walls and floor are plastered. A later layer of plaster covers the mosaic floor at its perimeter, replacing missing mosaic tessearae. A similar phenomenon can also be found in the collecting vat. The opening of a clay pipe that originated somewhere in the centre of the treading floor (but not in the hole of the later screw base) and passed under it was found in the wall of the filtration vat (unmarked in the section). It was blocked in the later phase. A clay pipe 0.55 m long leaves the filtration vat close to its floor and leads diagonally into the collecting vat.

The collecting vat (1) is 2.4 m in diameter and 1.5 m deep, i.e. a volume of c. 6.8 m^3. The floor of the vat is in mosaic and in its centre there is a sump 0.5–0.7 m in diameter and 0.3 m deep. The outlet of the pipe coming from the filtration vat is about 1 m above the floor level.

Vat 3, of which only a small part survived in the south-western corner of the vat area clearly belongs to an earlier phase of the wine press. It was later covered by the western and southern walls of the vat area and by Vat 4. However, the fact that the border of the mosaic floor of the vat area skirts Vat 3 proves that Vat 3 operated for some time contemporarily with Vats 1 and 2. Maybe this was the stage in which the small wine press in the western area (see below) was used alongside the main installation.

The Western Area

This section, of which only the eastern part was unearthed, is situated at right angles to the other two described above. Two major phases can be discerned one early and one late.

In the early phase the remains of two floors were found in the eastern edge of this area. In the north a mosaic floor was found made of small tesserae and pottery sherds vertically embedded in mortar (D), in which the upper part of a basalt hourglass mill was embedded in secondary use, probably as a small base for a wooden screw – a very unusual phenomenon. This floor might have been connected to Vat 3 mentioned above, but functioned as a separated unit, with no connection to the main installation. South of this floor are the remains of another floor paved with pottery sherds vertically embedded in a layer of mortar (E; compare to a similar floor in the nearby H. Duran [Site 45] wine press). Both floors were laid in the same level and were later covered by the upper mosaic pavement of this area.

In the later phase the whole western part was covered by a white mosaic (F) 0.25–0.30 m higher than the early floors and 0.2 m higher than the uppermost floor in the treading floor (6). No remains were found of the walls that bordered this later floor. It is unknown whether at this phase the western area was part of wine press II or formed a section of another installation that continued westwards, beyond the border of the excavation.

Compartment 5, situated south of this mosaic floor is round, 1 m in diameter and 0.7 m deep. Its walls are coated with pink plaster that has been repaired at least three times, and its floor is paved with white mosaic. Most of the top of the wall of the compartment was in the form of a step or ledge 0.2 m wide. There is no pipe or channel leading into the compartment.

Compartment 4, at the northern edge of this area, has a diameter of 0.7 m and is 0.5 m deep. Its wall is coated from the inside and outside with a thick layer of plaster that also covers the compartment's floor. A plastered, poorly preserved channel led into it from the south-east (part of the channel's floor is shown in the plan (Fig. 4) adjacent to the rim of the compartment), possibly originating in the corner of the upper treading floor (6).

Summary

The earliest stage in wine press II was completely different from the later one. From this phase we found three small floors – two (A and D) of mosaic tesserae (one with a tiny base for a screw and pottery sherds) and the third (E) made only of pottery sherds. The channel of only one of these floors was found, in a different location and level than the later gutter. The original vats that belong to this phase are probably buried under the later vat area. It seems that Vat 3, that was blocked later, is one of the early vats. The remains of this phase are so scanty that it is almost impossible to reconstruct the shape of this installation.

In the main later stage (with several sub phases) wine press II is easier to understand due to its better preservation. Its main components (Treading Floor 6 and Vats 1 and 2) are built in one line. Compartment 4 belongs to this press, similar in shape and function to the round compartments in wine press I. The later remnants in the western area (Floor F and Compartment 5, that are not connected) probably also belong to this phase but not necessarily to wine press II.

Notes

1. It may therefore be suggested that the pipe is a remnant of a different, earlier pressing system. The logic behind an underground pipe like that can be understood only if the must was pressed at a level lower than the later treading floor and could not, therefore, flow to the vat through the upper gutter. A central vat, into which the rape were collected after the first treading for a secondary pressing may be an optional suggestion.

Editors' Notes

1. Another possibility is that the pipe in the treading floor of wine press I originally started from the bottom of the mortice in the screw base, as known from other wine presses in the region, such as Tell Qasile [Site 36], a base that belonged to the earlier phase and was later replaced by the one found in the excavation.
2. Amos Kloner is of the opinion that the fact that the tesserae in the mosaic floor of the collecting vat are arranged in a pattern similar to a cross is only a convenient way of laying the floor and should not be regarded as evidence that this is a Christian wine press.

Bibliography

Frankel, R. 1999. *Wine and Oil Production in Antiquity in Israel and Other Mediterranean Countries*. Sheffield.

Shavit, E. 1995. "Rishon LeZiyyon". *ESI* 13: 57.

Sidi, N., Amit, D. and 'Ad, U. 2003. "Two Winepresses from Kefar Sirkin and Mazor". *'Atiqot* 44: 253–266.

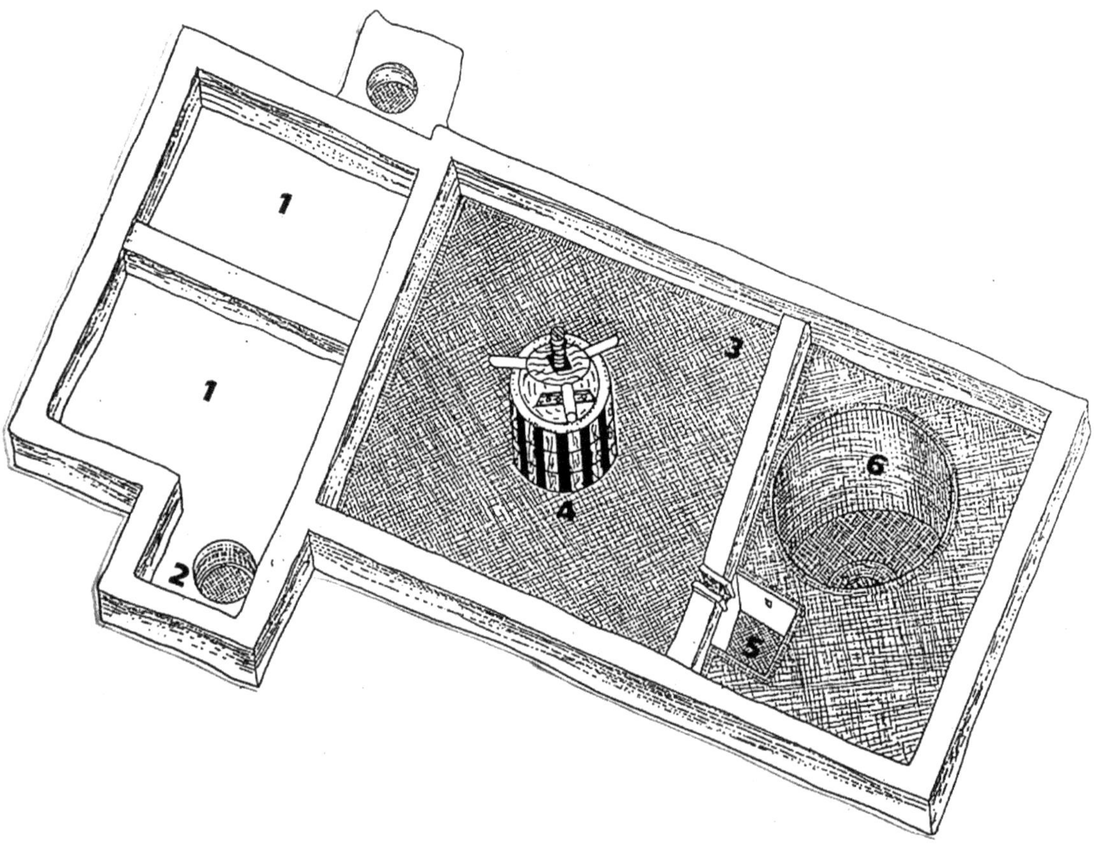

Fig. 42.1. Rishon Le-Zion, isometric view of wine press I, looking north.

Fig. 42.2. Rishon Le-Zion, wine press I, the collecting vat, looking south; note the flagstone embedded in the floor.

Fig. 42.3. Rishon Le-Zion, wine press I, the sump in the collecting vat, paved in mosaic layed in the form of a cross.

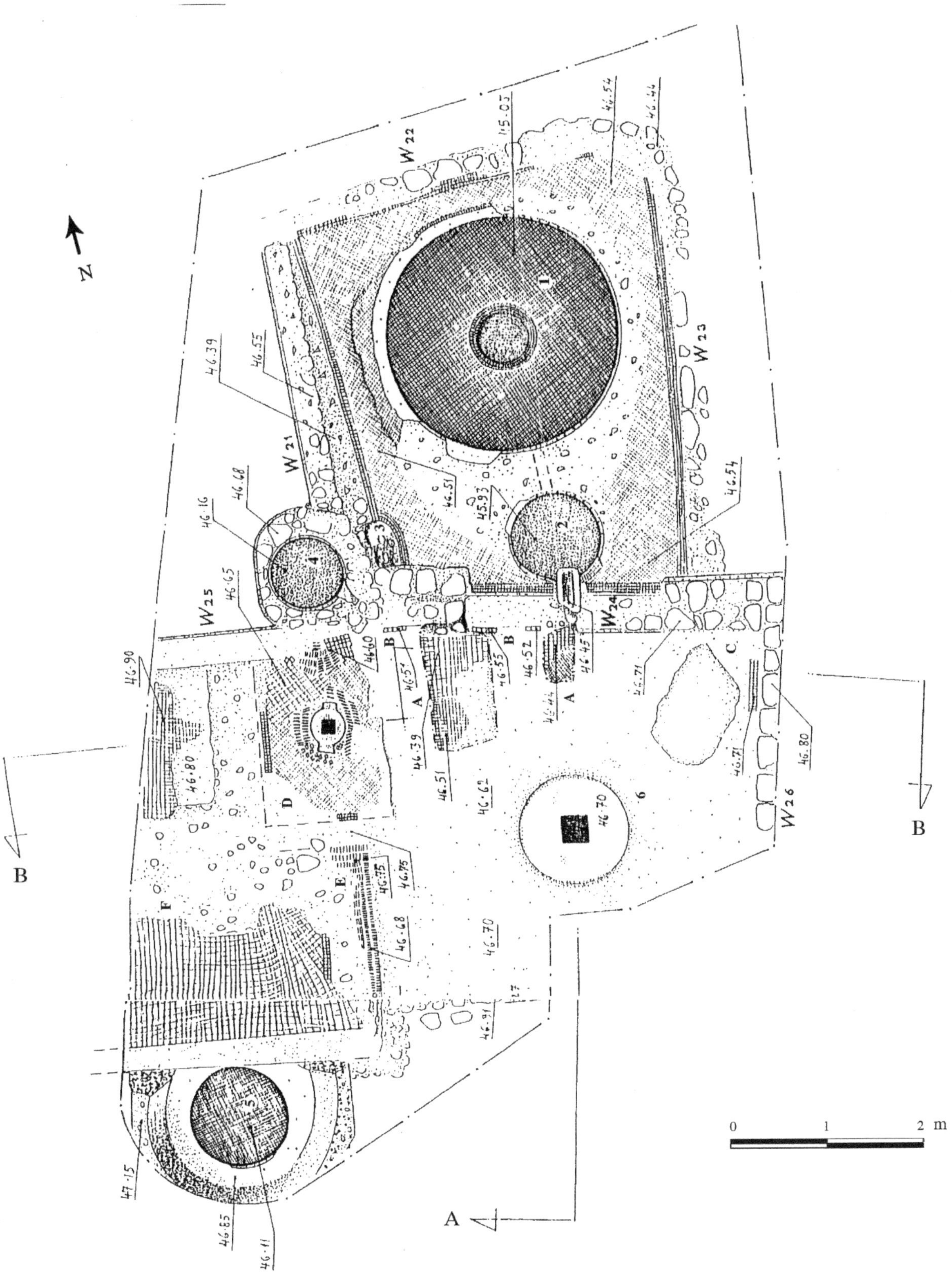

Fig. 42.4. Rishon Le-Zion, wine press II, plan.

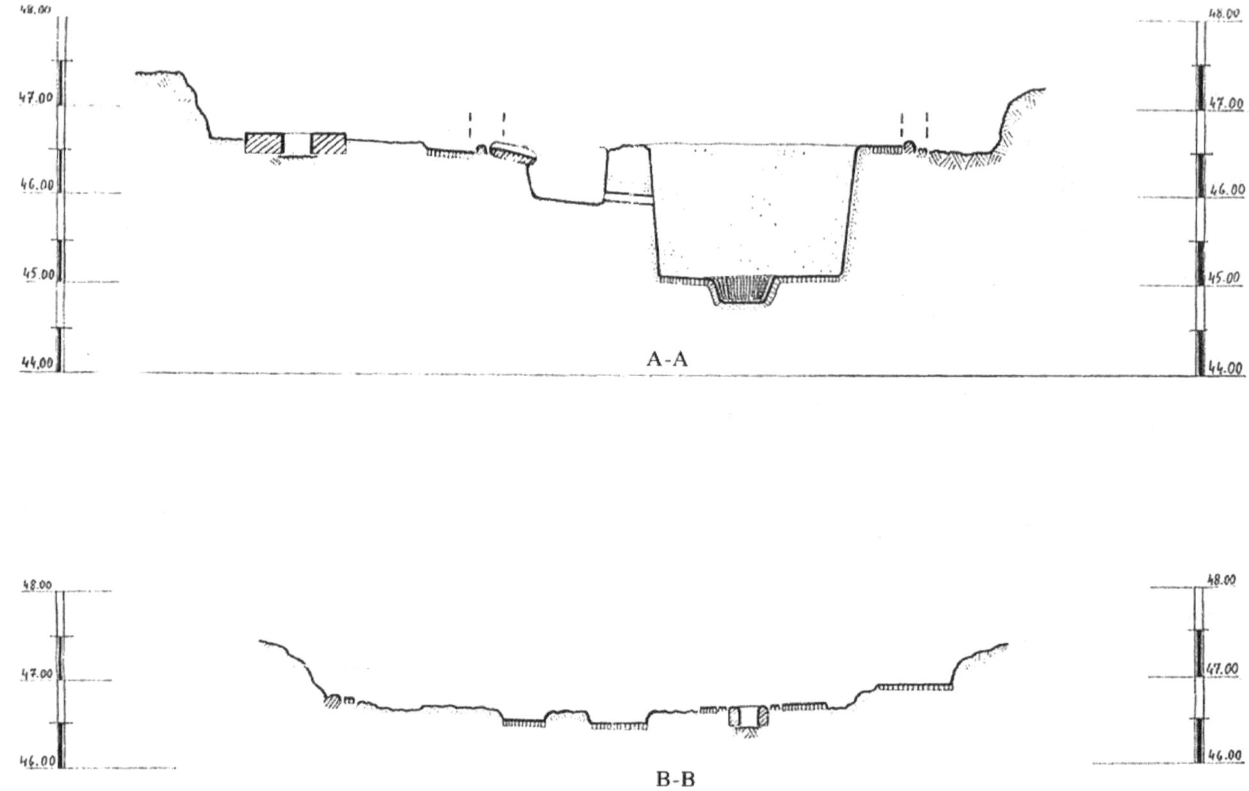

Fig. 42.5. Rishon Le-Zion, wine press II, sections.

Fig. 42.6. Rishon Le-Zion, wine press II, view to the north.

Fig. 42.7. Rishon Le-Zion, wine press II, the vat area, looking west.

43. Wine and Oil Presses at H. Hermeshit (Ne'ot Kedumim)

Zvi Greenhut and Michal Yron-Lubin

The Site

H. Hermeshit (MR 1476/1508) is situated within Ne'ot Kedumim, The Biblical Landscape Reserve in Israel, approximately 6.5 km east of Lod. The site is located on a long north-south ridge, 206 m above sea level at its highest point. The ancient settlement was concentrated on the north side of the ridge. Its core covered c. 60 dunams and together with its installations and fields c. 250 dunams.

The site was first surveyed and described by the Survey of Western Palestine (Conder and Kitchener 1881–1883, II: 335). It was surveyed again within the framework of the survey of the Map of Lod by the survey team of Tel Aviv University (Gophna and Beit-Arieh 1997: 97–98, Site 215) and knowledge of the site was up-dated following the excavations. Since 1988 a detailed survey of the site and its surroundings as well as archaeological excavations have been conducted.[1]

H. Hermeshit was first occupied during the Early Roman period, although coins from the Hasmonean period found at the site may indicate an even earlier date (Ariel 1998: 1*-4*). The Jewish settlement included domestic areas, ritual baths, wine presses, cisterns and a hiding complex dated on the basis of its shape to the time of the Bar Kochba Revolt.[2] It is difficult to establish the size of the settlement during this period due to the heavy building operations, secondary use of building stones and fills and leveling conducted during the Byzantine period. Most of the architecture and finds from the site date to the late Byzantine period – 6th-7th cent. CE, although this settlement was founded even earlier, during the 3rd-4th cent. on the settlement from the Early Roman period (Greenhut 1998: 151; Ariel 1998: 1*). A settlement also existed here during the Umayyad period (Greenhut 1998: 135; Ariel 1998: 1*) and perhaps, according to the numismatic finds (Ariel in preparation) also during the Mameluke period as an isolated farmstead.

The Wine Presses

In all 13 wine presses were discovered at H. Hermeshit.[3] Of these two – Nos. 49a and 400 – were large and complex, comprising in addition to the usual treading floor and collecting vat also various other components and devices. In wine presses 65 and 58 the collecting vats had floors of white mosaic. No. 58 differed from most of the other wine presses in that it had a filtration vat in addition to the collecting vat, while Nos. 16, 22, 37a, 63, 64, 65, 65a, 83, 91, and 108 were of the simple type having only treading floor and collecting vat.

Complex Wine Presses

Wine Press 49a (Fig. 1) (Greenhut 1998: 147–151) is located on the north-eastern slope of the hill within the area of the ancient settlement. The treading floor (4.5×4.5 m, area 40.25 m^2, depth 0.6 m) was cut in the rock. Its walls were covered in plaster the make-up of which included shards of the Byzantine period, and its floor was paved in white mosaic that consisted of five concentric strips each in the form of a square one within the other.[4] The floor reflects several signs of repair showing that it was used for a long period. An almost square stone block (1.1×1.2 m) was found in the centre of the floor and in its centre a square mortice (0.4×0.4 m) that widened at the bottom. The wooden screw of a screw press was secured in the mortice with the aid of wedges. The mortice was of a type characteristic of the coastal plain in the Byzantine period (Frankel 1999: 141–142). It was connected to the intermediate vat to the east by an underground lead pipe.

Adjacent to each of the sides of the treading floor there are two semi-circular compartments 0.6–1.0 m in diameter. Their floors are of mosaic, their walls plastered and they are connected to the treading floor by narrow bores. Next to the south-western compartment traces of mosaic were found suggesting that there were additional floors above all the compartments.

To the east of the treading floor, but separated from it by a low stone partition, was a plastered floor in which were located both the intermediate vat (0.75×0.75 m, depth 0.7 m, volume 0.38 m^3) and the collecting vat (1.55×1.45 m, depth 1.7 m, volume 3.82 m^3). Both vats have a circular depression in their floors in which the dregs could settle to facilitate collecting the final drops. The collecting vat was paved in mosaic. To the east of the wine press and close to the collecting vat was found a cup-mark surrounded by a shallow square frame upon which a cover was laid. Many bricks were found all over the installation that might indicate of the existence of high walls around it.

Wine Press 400 (Fig. 2) is located north-west of the building remains in Area A, at a distance of 32 m from wine press 49A. The plan of those two wine presses is somewhat similar although the former is smaller than the latter and also lacks compartments around the treading floor. Wine press 400 consists of a plastered treading floor (L. 401; 3.70×3.95 m [area 14.6 m^2], depth 0.25–0.75 m), a filtration vat and a collecting vat hewn into the rock to the west of the floor. The two vats, connected by a small slanting channel, were plastered and paved with mosaic. The northern filtration vat (0.9×0.8 m, depth 0.7 m, volume 0.5 m^3) was connected to the treading floor by a channel. The southern collecting vat, the larger of the two (L. 402; 1.1×1.5 m, depth 1.4 m, volume 2.3 m^3) is coated with a thick layer of plaster, and in its north-east corner a step was hewn. To the west of this pit a plastered floor (1.7×2.9 m) was found paved with small stones.

The central mortice for the screw (0.35×0.45 m, depth 0.2 m) was hewn in bed-rock in the centre of the treading floor and was blocked at a later stage. The area around the mortice was slightly raised above the level of the floor, creating a press base in the shape of a rectangle with rounded corners (1.1×1.2 m). It appears that at first there was an installation for pressing the rape with one fixed screw (Frankel 1999: 140ff). The rape was heaped around the screw and enclosed within a movable frame, and then the pressing board was forced down the screw by revolving the nut and pressed the skins. The expressed juice flowed across the treading floor to the small (northern) filtration vat. The fixed screw press went out of use and the screw mortice was blocked up. Pressing of the skins and seeds was now carried out by a beam whose end was anchored in a niche in the southern wall of the wine press while the other end was tied down to a tethering ring hewn into the bed-rock. The rape was placed in baskets and the juice ran down the sloping surface of the treading floor into the small collecting vat.

Wine Press 58 (Fig. 3). This installation is located on the eastern slope of H. Hermeshit and it appears that it served the vineyards located east and south-east of the site. The wine press was cut in bed-rock although the south-west corner of the treading floor was built up because the rock outcrop was too small. It consisted of three elements, all almost square:

- A treading floor that was originally probably paved with white mosaic (4.6×4.2 m, area 19.3 m^2), with a circular depression for collecting the refuse in the north-east corner.

- Intermediate vat (0.75×0.75 m, depth 0.7 m, volume 394 litres) that was connected to the treading floor by a channel and had a circular depression in its floor.

- Collecting vat (1.3×1.5 m, depth 1.25 m, volume 2.4 m^3) that was paved with white mosaic and also had a circular depression in its floor.

Simple Wine Presses

Not all the wine presses were cleared and in those that were the collecting vat was not always excavated, but the information gathered shows clearly the character of these installations. They varied both in shape and size but were all rock-cut although in No. 65 the floor of the collecting vat was paved in mosaic and the walls plastered. In some both treading floor and collecting vat were irregular in shape (e.g. Nos. 65, 65a, 91) while in others both elements were square, rectangular or trapezoid (Nos. 16, 22, 63). The simple wine presses were all much smaller than the three more complex ones.

The Types of Wine Press and their Location

All the wine presses except the small irregular simple ones have in common that all their main components – treading floors, intermediate vats and collecting vats – are straight sided, square, rectangular or trapezoid. Some of them (16, 49a, 63) have circular depression next to them, that might have been used for storing material associated with the wine manufacturing process. However, Nos. 400 and 49a had mortices for a fixed screw press while the latter had also semi-circular compartments. In No. 400 the screw press was later replaced by a lever press.

Wine presses 49A and 400 are located in the centre of the original settlement, on the north-eastern side near the dwellings (Areas A & B), as opposed to the simple wine presses hewn into the lower slopes, mainly to the east or north-east and in the heart of the vineyards. It seems that wine presses 400 and 49A received the grape harvest from vineyards situated further away on agricultural terraces at a distance of c. 200–300 m.

The Date of the Wine Presses

Wine presses with fixed screw presses were widespread during the 5th-6th cent. CE in the Judean Shephelah (Frankel 1999: Map 34). It is more difficult to find parallels for wine presses that have a beam press for pressing the rape although Frankel (Ibid.: 145–146) mentions a number of examples.

At the site of H. Hermeshit c. 10 simple wine presses have been found consisting only of a treading floor and one collecting vat. Nos. 400 and 49A are improved wineries and it can be assumed that they were in use contemporaneously or at different times during the late Byzantine period. According to the ceramic and numismatic finds it can be determined that the abandonment of wine press 49a was at the beginning of the Early Arab period (Greenhut 1998: 151). At some time during the late Byzantine period the method of pressing the rape in wine press 400 was converted from the single fixed screw method to that of the anchored beam press.

At H. Hermeshit there are at least two cases of wine presses (Nos. 16, 65) that went out of use at the Byzantine period, probably during its earlier part (Ibid.). One simple wine press (No. 64) went out of use during the Early Roman period, when a ritual bath (*mikveh*) was quarried into its collecting vat. We assume that the system of simple wine presses is earlier than the complex wineries and dates to the early Byzantine period at the latest. Nos. 400 and 49A represent a centralized agricultural organization that differs from the economic system connected to the smaller simple wine presses in which each individual vineyard owner was responsible for the agricultural labor and wine production himself.

The Oil Presses

Three very similar lever and screw oil presses were excavated at H. Hermeshit: No. 2 to the north-west of the site, L. 562 & 558 in area G in the northern part of the settlement and L. 446 & 448 in area A at its southern portion. In addition the pier of a grooved pier screw press was also found in area G (No. 29a in the survey).

Oil Press 2 (Fig. 4)

The oil press is located in the south-eastern corner of a large building (Greenhut 1998: 128–143). A raised circular plinth cut in the rock was found in the centre of a big room encircled by four pillar bases. This might have served as the base for the crushing basin part of which, broken, was found a few dozen meters west of the building. The pressing area was located in the east part of the room. The beam anchoring niche was in the southern wall that was at this point 1.5 m wide. The press-bed was rock-cut and surrounded by a circular groove (diameter 0.65 m) that led by a gutter to the corner of a square collecting vat (1×1 m, depth 0.5 m, volume 0.5 m^3) that had a semi-circular depression in its floor, beneath the gutter outlet. There were two screw weights of the Samaria type (Frankel 1999: 111–113), both 1 m in diameter; the southern, nearer to the press-bed, was smaller (0.9 m high, weight 1.8 tons) while the northern was bigger (1.3 m high, weight 2.5 tons).[5] On the stone partition between the press-bed and the weights were two small holes clearly for posts that stabilized the beam. The beam was 7 m long. On the walls on both sides of the press-bed were found a few pairs of recesses and some single ones. Those might have served for inserting poles that supported upright wooden boards that supported the olive frails laid on the press-bed.

Oil Press in Area A (Figs. 5–6)

This installation (L. 446, 448) is almost identical to oil press No. 2. The beam anchoring niche was in a wall 1.9 m thick, the press-bed was surrounded by a circular groove (diameter 0.7 m) that led to the corner of a rectangular collecting vat (0.65×1.70 m, depth 0.65 m, volume 0.7 m^3). There were round pits (diameter 1.2 & 1.6 m, depth 0.3 & 0.5 m) for two screw weights but only one, of a Samaria type, was found *in situ* (diameter 0.85 m, height 1.15 m, weight 1.8 ton). In this press too there were two holes for posts that stabilized the beam on the partition between the press-bed and the weights. Here too the beam length was 7 m. On the walls on both sides of the press-bed were found a few pairs of recesses and some single ones as in oil press No. 2.

Oil Press in Area G (Fig. 7)

This installation (L. 562, 558) was also a lever and screw press in which the beam anchoring niche was in a thick wall (1.8 m) with two Samaria type screw weights (diameter of both 0.9 m, height 1.4/1.25 m, weight 2.25/2 tons respectively) in round weight pits (diameter c. 1.3 m, depth 0.8 m). The press-bed and vat were both free standing. The diameter of the circular groove on the press-bed was 0.85 m. The collecting vat was round (diameter 0.7 m, depth 0.78 m), this being the only significant difference between this press and the other two. There were also holes for posts that stabilized the beam, in this case on a large stone slab, one an actual hole and the other a niche on the edge of the slab. The beam length was 7.5 m.

The Grooved Pier Press (Area G, Site 29a, Fig. 8)

This type of direct pressure screw press known also as the "Judean Pier Press" is typical of the region (See Frankel 1999: 126–127; Kloner, in this book). It was not found *in situ* and so it is difficult to determine if it was used at the same time with the lever and screw press from area G, preceded it or replaced it.

Oil Presses – Conclusions

The oil presses are dated to the Byzantine period, probably to its later stage. The great similarity between the three presses at H. Hermeshit shows a clearly defined local technical tradition. The most unusual characteristic of these installations is the use of two screw weights together in one press, as well as the use of the small recesses in the walls beside the press-bed in all three complexes. The use of two or three beam weights in one lever and weights press is usual, and the use of a screw weight and a beam weight together is also known (e.g. H. Karkara [Site 2] and H. 'Ovesh in western Galilee and at Umm el-'Amad in southern Lebanon [Frankel 1999: 171]). But the use of two screw weights together in one press is very rare and appears only in this area of the Shephelah.[6]

Notes

1. The research at the site was conducted on behalf of the Israel Antiquities Authority and financed by Ne'ot Kedumim. Seven seasons of excavation have been carried out. The 1988–1989 excavation seasons and survey were directed by Z. Greenhut and the 1990 season by Z. Greenhut and M. Yron-Lubin (Greenhut 1998). Since 1991 the project has been directed by M. Yron-Lubin (Yron-Lubin 1995: 83–85; 1996a: 66–68; 1999: 46*-49*; in preparation; Yron-Lubin and Sebah 2000).

2. The hiding complex was surveyed by the staff of the Hiding-Complex Survey, headed by Y. Tepper.

3. Two additional simple wine presses were discovered recently at the western and eastern boundaries of the site.

4. The treading floor of the wine press at Hulda – 15 km to the south – was paved in a similar manner (Kloner 1986: 198–199; Site 49).

5. A general phenomenon in all the oil presses at H. Hermeshit is the arrangement of the screw weights relating to their size: the smaller screw weight is always closer to the press-bed.

6. The only parallel of which we know for this kind of press comes from H. Tinshement, located in this region and dated also to the late Byzantine period (Dahari 1998: 68). Very recently another olive press of this kind

has been exposed at Ben Shemen interchange, in the very same region, though it comprises of two different types of screw weights in one device (Our thanks are to Y. Zelinger for this information).

Bibliography

Ariel, D.T. 1998. "Coins from the Survey and Excavations at Horvat Hermeshit (1988–1990)". *'Atiqot* 34: 1*-4*.

Ariel, D.T. In preparation. "Coins from the Survey and Excavations at Horvat Hermeshit".

Conder, C.R. and Kitchener, H.H. 1881–1883. *The Survey of Western Palestine*. 3 Vols. London.

Dahari, U. 1998. "Horbat Tinshemet, Church of St. Bacchus". *ESI* 18: 67–68.

Frankel R. 1999. *Wine and Oil Production in Antiquity in Israel and Other Mediterranean Countries*. Sheffield.

Gophna, R. and Beit-Arieh, Y. 1997. *Archaeological Survey of Israel, Map of Lod (80)*. Jerusalem.

Greenhut, Z. 1998. "Horvat Hermeshit (1988–1990)". *'Atiqot* 34: 121–172 (Hebrew).

Kloner, A. 1986. "The Structure at Hulda and its Use as a Wine Press". In: A. Kasher, A. Oppenheimer and U. Rappaport (eds.). *Man and Land in Eretz-Israel in Antiquity*. Jerusalem: 197–208 (Hebrew).

Yron-Lubin, M. 1995. "Horvat Hermeshit (Ne'ot Qedumim)". *ESI* 14: 83–85.

Yron-Lubin, M. 1996a. "Horvat Hermeshit (Ne'ot Qedumim) – 1993". *ESI* 15: 66–68.

Yron-Lubin, M. 1996b. *Settlements in the Northern Shephelah in the Early Bronze Age, Iron Age II, and the Byzantine Period: Site Cachment, Carrying Capacity and Landscape Archaeology*. Unpublished M.A. Thesis, Tel Aviv University (Hebrew).

Yron-Lubin, M. 1999. "Horbat Hermeshit (Ne'ot Qedumim) – 1995". *ESI* 19: 46*-49*.

Yron-Lubin, M. In preparation. *Four Seasons of Excavation in Horvat Hermeshit* (Hebrew).

Yron-Lubin, M. and Sebah, A. 2000. "Horbat Hermeshit (Ne'ot Qedumim) – Preservation and Restoration". *ESI* 20: 69*-73*.

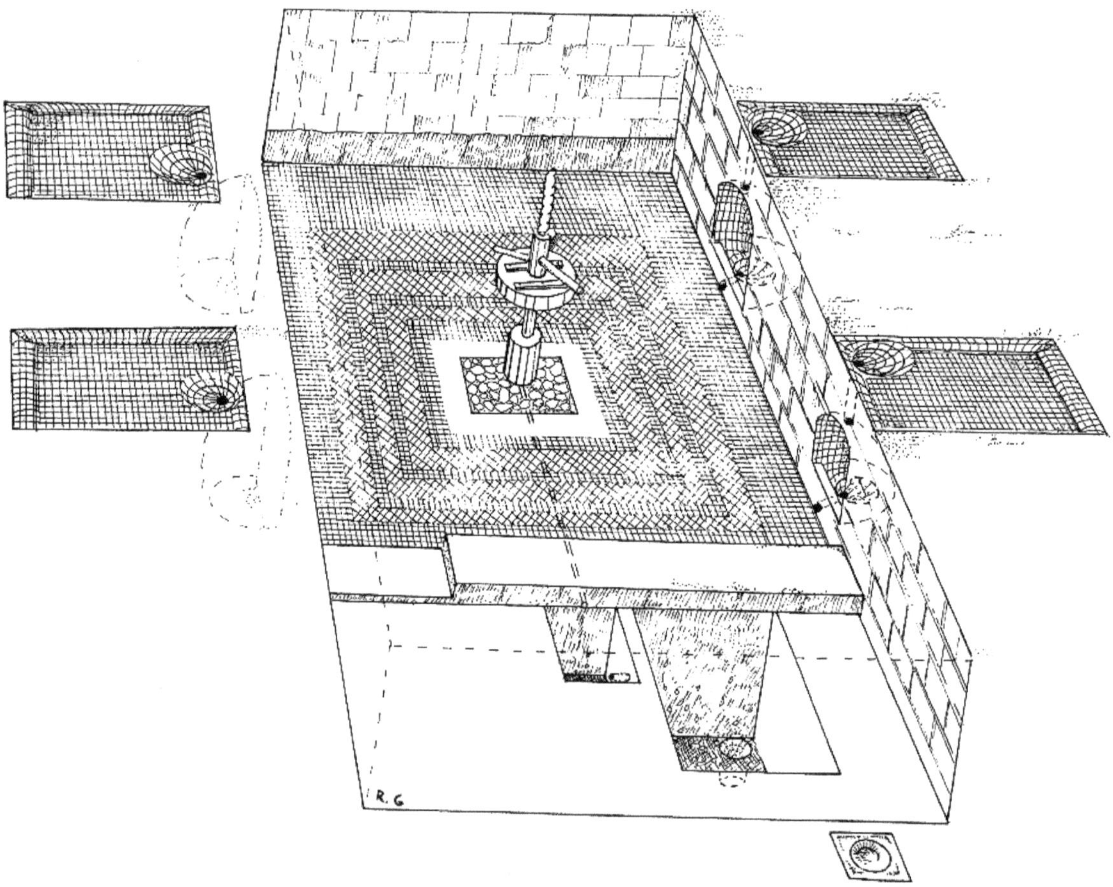

Fig. 43.1. Hermeshit, isometric reconstruction of wine press 49a, looking west.

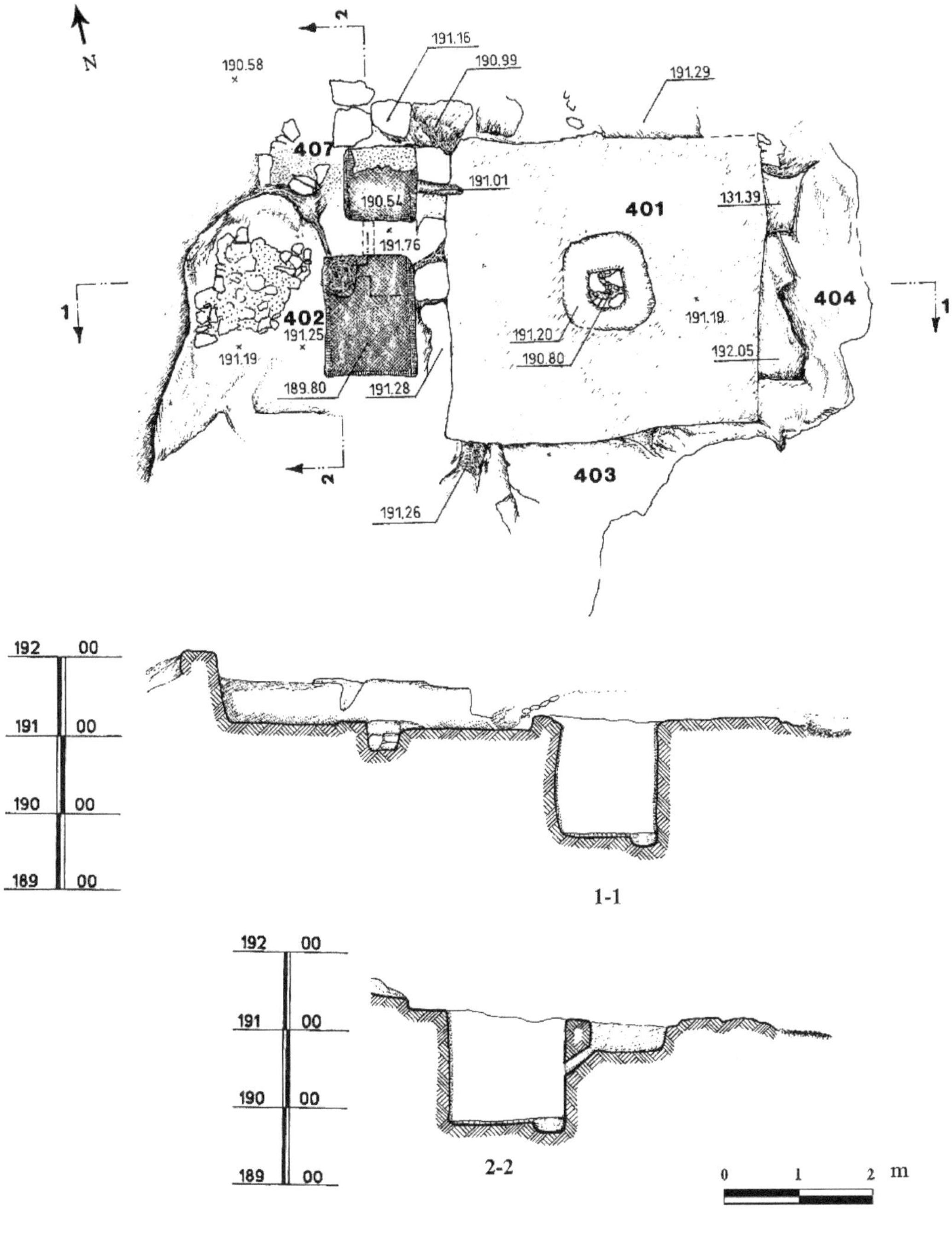

Fig. 43.2. Hermeshit, plan and sections of wine press 400.

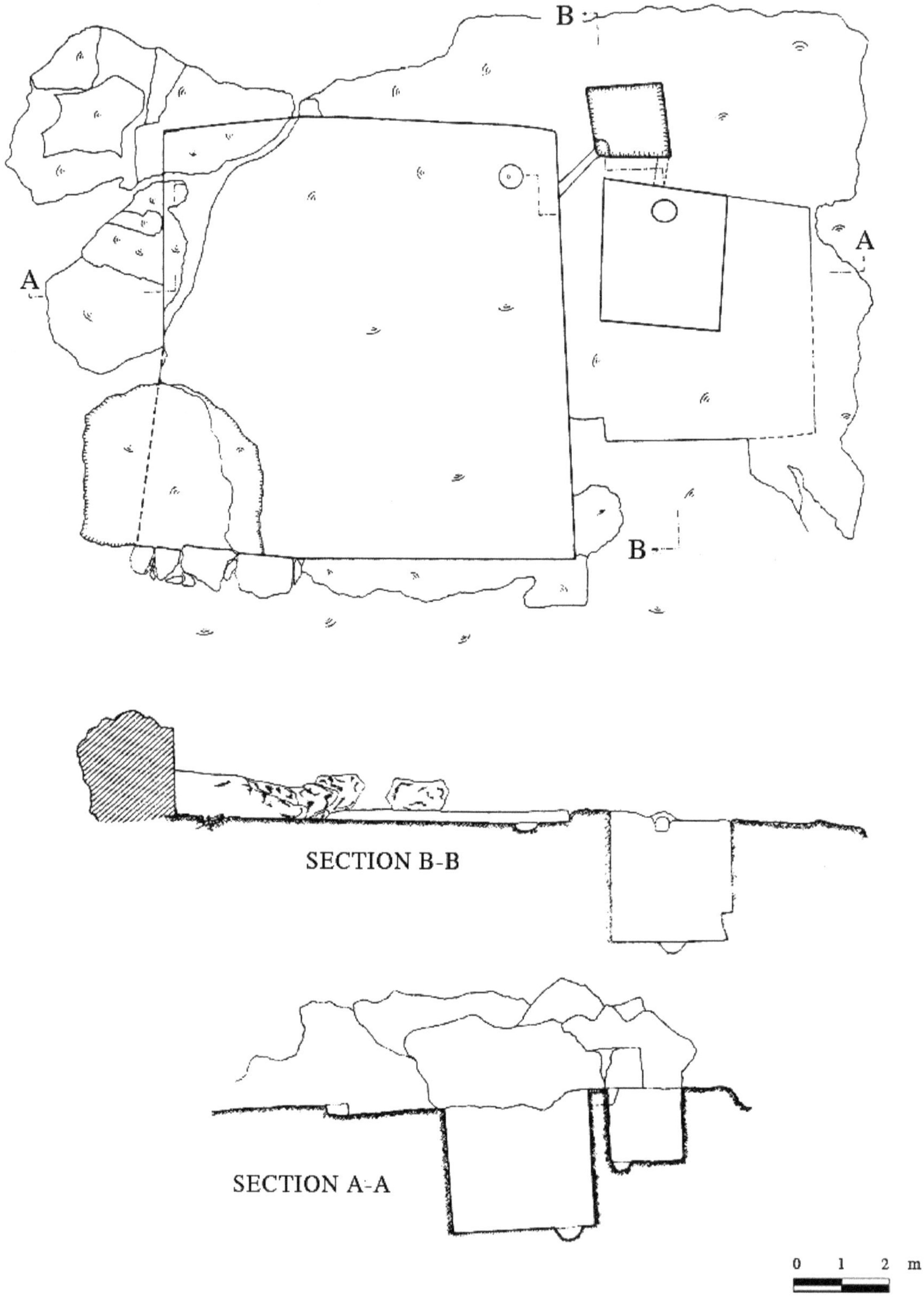

Fig. 43.3. Hermeshit, plan and sections of wine press 58.

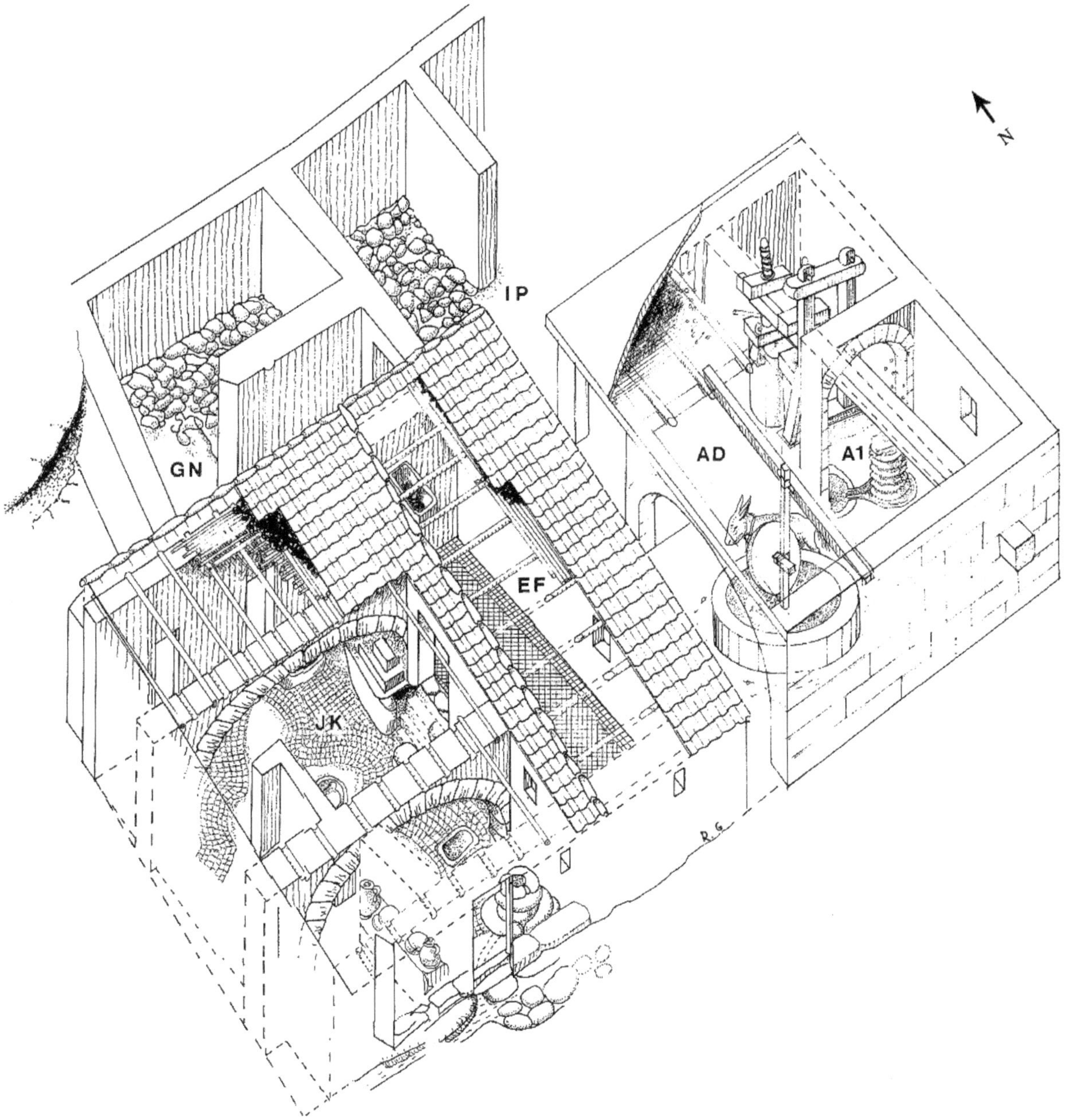

Fig. 43.4. Hermeshit, isometric reconstruction of oil press 2, looking north-east.

Fig. 43.5. Hermeshit, oil press in area A, looking east.

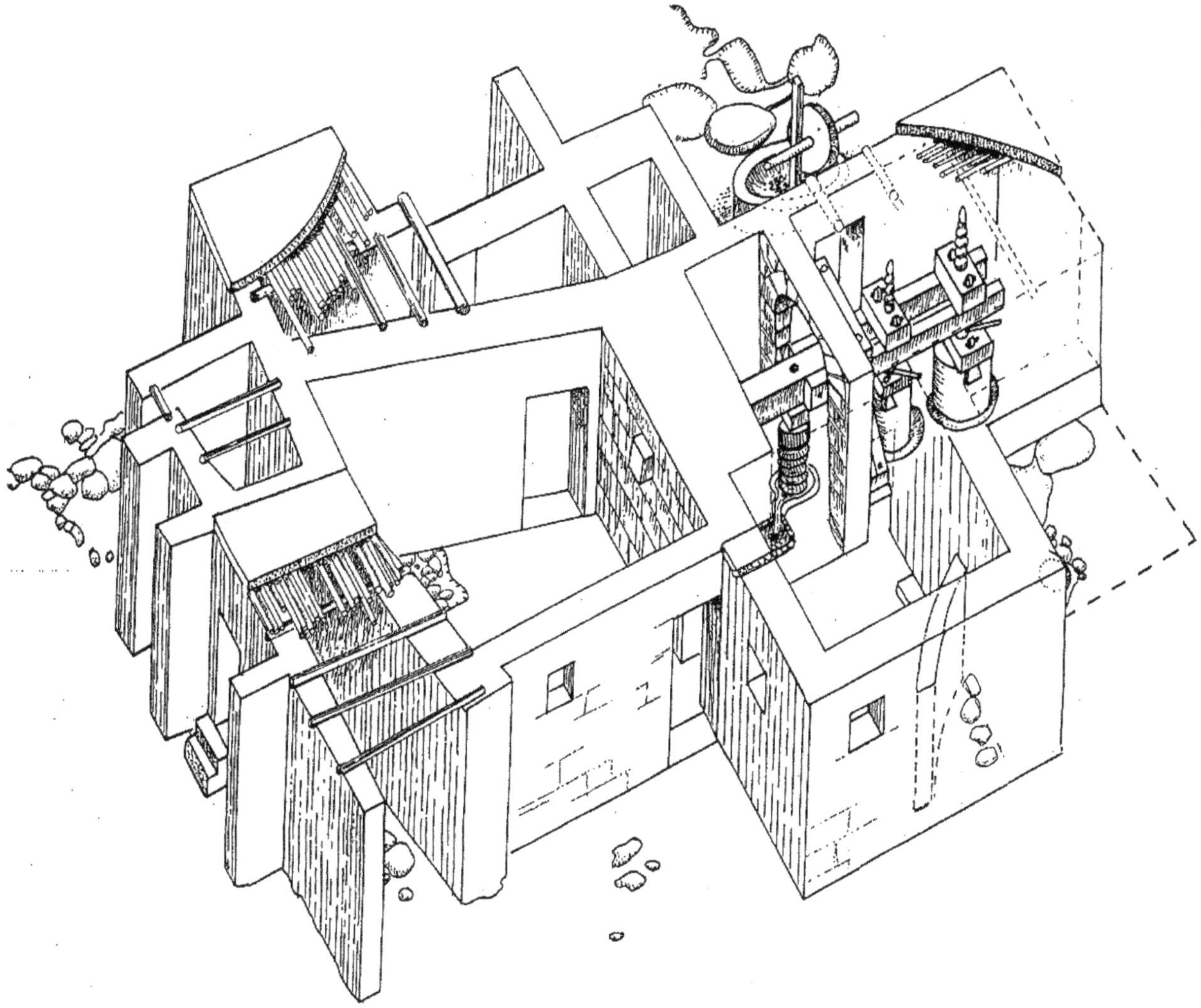

Fig. 43.6. Hermeshit, isometric reconstruction of the oil press in area A, looking north.

Fig. 43.7. Hermeshit, oil press in area G, looking south-west.

Fig. 43.8. Hermeshit, grooved pier in area G, looking south-east.

44. Byzantine Wine and Oil Presses at Mevo' Modi'im

Emanuel Eisenberg and Ruth Ovadiah

The Site

The site is located at the top of a rocky spur 219 m above sea level (MR 1489/1490). The existence there of a Christian edifice had been known since 1964, when a segment of colored mosaic floor was discovered, adorned with a cross (*HA* 13 [1965]: 10–11). During the course of development work at Mevo' Modi'im, rescue excavations were carried out at the site on behalf of the Israel Antiquities Authority by E. Eisenberg in September-October 1976 (Eisenberg and Ovadiah 1998: 1*-19*). During the excavations it became clear that the building was a monastery (Fig. 1) extending over an area of approximately 700 m^2 and comprising a central courtyard, a chapel and additional rooms arranged around the courtyard. Three cisterns and a wine press hewn from bed-rock were found adjacent to the monastery.

The Wine Press

The wine press is situated 13 m west of the monastery (Fig. 2). It comprises the following elements: a treading floor (122), a filtration basin (123), a collecting vat (124), and an additional small vat (125).

The various parts of the wine press are hewn from bed-rock. The treading floor is lower than the surface of the rock by approximately 0.5 m; its eastern, southern and western walls are hewn from the bed-rock, while its northern wall is built from a course of uncut stones. The treading floor measures 3.5–4×3.7–4 m (14 m^2). Patches of coarse white mosaic tesserae were found on the rock floor. At its northern edge two holes have been carved out, apparently sockets for wooden posts. A third, smaller hole was found about 1.25 m east of the treading floor. A broken stone drum used as a base for a fixed screw press, 1.55 m in diameter and 0.47 m high, was found *in situ* in the northern part of the treading floor. Because it was found broken and slightly raised it may be deduced that it was broken during an attempt to remove it from its place. In the centre of the drum is a square mortice measuring 0.35×0.35 m for a wooden screw. Two adjacent walls of the socket are wider at the bottom to fix the screw in place (Frankel 1994: 75–77). A cubical cavity had been cut in the rock floor below the square mortice and from the cavity extends a channel 0.15 m wide and deep that leads to the filtration basin (compare: Ibid.: Fig. 84A).

The filtration basin measures 0.55×0.50 m and 0.3 m deep. A short pipe cut through the base of its wall connects the bottom of the basin with the collecting vat. The collecting vat is a cube 2×2×2 m (volume 8 m^3) with a mosaic floor and plastered walls. There is a sedimentation basin in the centre of the floor; in the northern corner a small triangular step was left during cutting to facilitate ingress and egress. The perimeter of the vat has been squared off and finished, partly in plaster and partly in white mosaic, to a width of 0.4 m. The mosaic sections are surrounded by a row of stones.

A third vat (125), 0.5×0.9 m and 0.9 m deep (volume 0.4 m^3), is located beside the filtration basin. It is connected to the collecting vat by two openings – one at the bottom, the second 0.32 m from the top. While the purpose of this vat is not clear, it may have been used to add spices and supplemental materials to the wine to improve its taste, or perhaps to store the grapes before treading and collect the "first must" that was expressed by their own weight and that was thought of as the raw material for the finest wines.

The wine press has been dated, as has the monastery to the late Byzantine period (second half of the 6th cent. CE).

The Oil Press

The installation was an integral part of the monastery, situated in a large hall (113) to the east of the chapel. Its northern wall continues the northern wall of the chapel eastwards. The oil press hall is rectangular; its inner dimensions are 11.2×9.2 m (103 m^2).

The walls of the oil press are built of dressed limestone blocks on the outside while the inner face has been straightened with small stones covered with a layer of plaster. The walls are about 0.7 m thick and some of them have been preserved to the height of several courses. Part of the southern wall was hewn from bed-rock as was part of the floor. The remaining part, that continued eastward, was floored with stone slabs

The press has four entrances with stone doorsills found *in situ*. Two are located in the western wall: The northern of the two leads into the narthex (106) while the southern – twice as wide, 1.6 m in width – opens onto the central courtyard (107). The third entryway, situated in the southern wall, leads to a long hall (115) that was not excavated and may have served as a storeroom. The fourth is in the eastern outer wall and apparently leads to the outside courtyard. The western entrances have steps hewn from the bed-rock that descend to the floor of the oil press that was built on a lower level.

The oil press was divided into two unequal parts:

The crushing wing in the southern part takes up about two-thirds of its area and measures 7.5×9.2 m. The wing was used for storing and crushing olives and storing the olive mash. Three compartments were built against the wall in the north-west corner, measuring 0.7×0.8 m, 0.5×0.7 m and 0.4×0.7 m respectively. They were possibly used for storing olives. In the south-west corner there is a shallow rectangular recess cut into the rock, ending at the south-west entrance. Two bores, open from side to side, have been cut into its wall, possibly for tethering beasts of burden that brought the olives to the oil press. About 2.5 m east of the western wall a crushing basin of hard limestone was found *in situ*. It is 2 m in diameter and 0.6 m high with a raised rim 10 cm wide and about 2 cm high. There is a rectangular socket in its centre that was used to anchor the pivot upon which the axle of the crushing stone was mounted. Two broken pieces of crushing stones, also made of hard limestone, were found adjacent to the crushing basin. (Today an intact crushing stone stands upon the crushing basin. The residents of the settlement say it was found in the area of the monastery).

Two stone basins found south of the crushing installation were apparently used for storing the olive mash. One is rectangular in shape with two rounded corners. Its inside measurements are 0.80×0.75 m and 0.72 m deep. There is a hole in its edge. The second is a sarcophagus apparently brought here to fulfill a secondary use as a container. It was found between two of the three pillars that supported the ceiling (below).

The pressing wing is in the northern part of the oil press. A partition constructed of a single course of cut stones separates this section from the southern crushing wing. On one of these stone blocks a round depression, diameter 0.3 m, has been carved out, possibly for standing a pottery vessel. The dimensions of the room are 3.0×9.2 m. The floor of its western part is lower than that of the rest of the oil press by approximately 0.55 m and it is entirely covered with well trimmed rectangular stone slabs, closely fitted and laid on hewn bed-rock. The paved floor is slightly inclined with the gradient directed toward the centre of the southern wall, to the collecting vat (see below).

The pressing installation includes the following components:

- A wall 1.1 m thick was built adjacent to the eastern wall of the chapel (101). It was constructed of dressed stones and integrated into it was a pair of large slotted blocks for installing the beam. The space between the blocks left a niche 0.85 m wide and 1 m deep in the centre of the wall. The bottom of the niche is 1 m above floor level. The upper parts of the slotted blocks were broken at the places where there were horizontal openings, rectangular in cross-section. The opening in the northern block went all the way through but that in the southern one was only carved to a depth of 0.15 m.

- Integrated into the eastern part of the floor is a circular press-bed of hard limestone, diameter 1.2 m. There is a circular groove at its edge, ranging in depth from 0.12 to 0.21 m, ending in a channel leading southward to the collecting vat. A number of grooves were cut into the press-bed to facilitate the flow of expressed liquid into the circular groove.

- The rectangular collecting vat measures 1.20×0.65 m and depth 0.6 m (volume 0.46 m^3). It is cut from the bed-rock near the southern wall and its sides are plastered. Its mosaic floor has a rounded depression in its centre into which is set the bottom of a ceramic jar that served as a sump. Two small depressions were carved into the stone flooring, connected by a shallow little groove that also leads to the collecting vat.

- To the east of the slab floor and the press-bed three shallow rectangular sockets had been carved into the bed-rock. The central one is the largest of the three and measures 0.40×0.53 m. It appears that in the two outer pits, posts were installed for stabilizing the press beam while a forked prop that was used to raise the beam when it was not in use was probably emplaced in the central pit.

- On the eastern side of the pressing wing an irregularly-shaped pit, 1.0–1.3 m across and 0.6 m deep was hewn out of the rock below floor level. It is located along the extension of a straight line determined by the beam niche and the press-bed and probably held the screw weight (below).

Roofing: In the centre of the oil press, two square plinths were found located 3.8 m apart. The base of a third pillar was found along the same line southward, attached to the wall of the storehouse (115). Another pillar 0.8 m high, together with the springer of an arch, was found attached to the eastern wall of the chapel (101), and adjacent to it was found a fallen voussoir. These data indicate that the oil press was roofed, as was customary in such installations since their operations were carried out during the rainy season. It can be surmised that the ceiling was supported on a series of arches; judging by the distance between bases it can be calculated that the ceiling was at least 3.2 m high. Since nothing was found in the debris to indicate the nature of the roofing it may be inferred that it was made of perishable materials.

Functioning: In oil presses of the Byzantine period it was customary to allot two thirds of the area for crushing and one third for pressing.[1] In this installation a lever and screw press (Frankel 1994: 50–53) was used for expressing the oil. The slotted blocks forming the niche were designed to anchor the beam by means of a rod inserted through the holes in the blocks. The centre of the press-bed integrated into the floor was about 3.7 m from the back of the niche.

Further along that line posts intended to stabilize the beam were set into two of the sockets carved into the surface of the bed-rock floor. A cylindrical screw weight with a central socket and two external dovetail mortices (Ibid.: 53, Type 1) was found on the floor of the pulpit in the chapel. It is virtually certain that this weight was originally situated in the pit at the end of the beam. The screw weight is relatively small (diameter and height both measure 0.6 m, thus weighing about 470 kg) and it may be assumed that additional weights (utilizing the pit) were used in the pressing process,[2] although they were not found in the excavation. In order to exert maximum pressure on the frails, the press beam had to be long enough so that its end almost touched the eastern wall, since the press-bed had been installed relatively far from the fulcrum (Ibid.: 35–36, Fig. 26). In other oil presses of the same type the length of the beam was about 8 m (compare Frankel 1985: 111), a measurement that could have been compatible with the length of the beam here.

The original oil press has been dated, like the monastery itself, to the late Byzantine period (second half of the 6th cent. CE). During the Early Arab period it was moved to the chapel, in which were discovered the screw weight and an infrastructure constructed of stones upon which the crushing basin rested. The installation operated in that location during the 8th and 9th cent. CE.

Notes

1. Compare, for example, to oil presses discovered at Qedumim in Sectors M and F, in which the area of the crushing wing was double that of the pressing wing (Magen 1993; Site 34).
2. An oil press operating with the lever and screw method but with the aid of additional weights was discovered at H. Karkara (Frankel 1985; Site 2).

Bibliography

Eisenberg, E. and Ovadiah, R. 1998. "A Byzantine Monastery at Mevo Modi'im". *'Atiqot* 36: 1*-19* (Hebrew).

Frankel, R. 1985. "Western Galilee, Oil Presses". *ESI* 4: 110–114.

Frankel, R. 1994. "Ancient Oil Mills and Presses in the Land of Israel". In: R. Frankel, S. Avitsur and E. Ayalon. *History and Technology of Olive Oil in the Holy Land.* Arlington and Tel Aviv: 19–89.

HA 13 (1965): 10–11: "Mosaic Pavement Discovered at Modi'im" (Hebrew).

Magen, Y. 1993. "Qedumim – A Samaritan Site of the Roman-Byzantine Period". In: F. Manns and E. Alliata (eds.). *Early Christianity in Context, Monuments and Documents* (Studium Biblicum Franciscanum, Collectio Maior, 38). Jerusalem: 167–180.

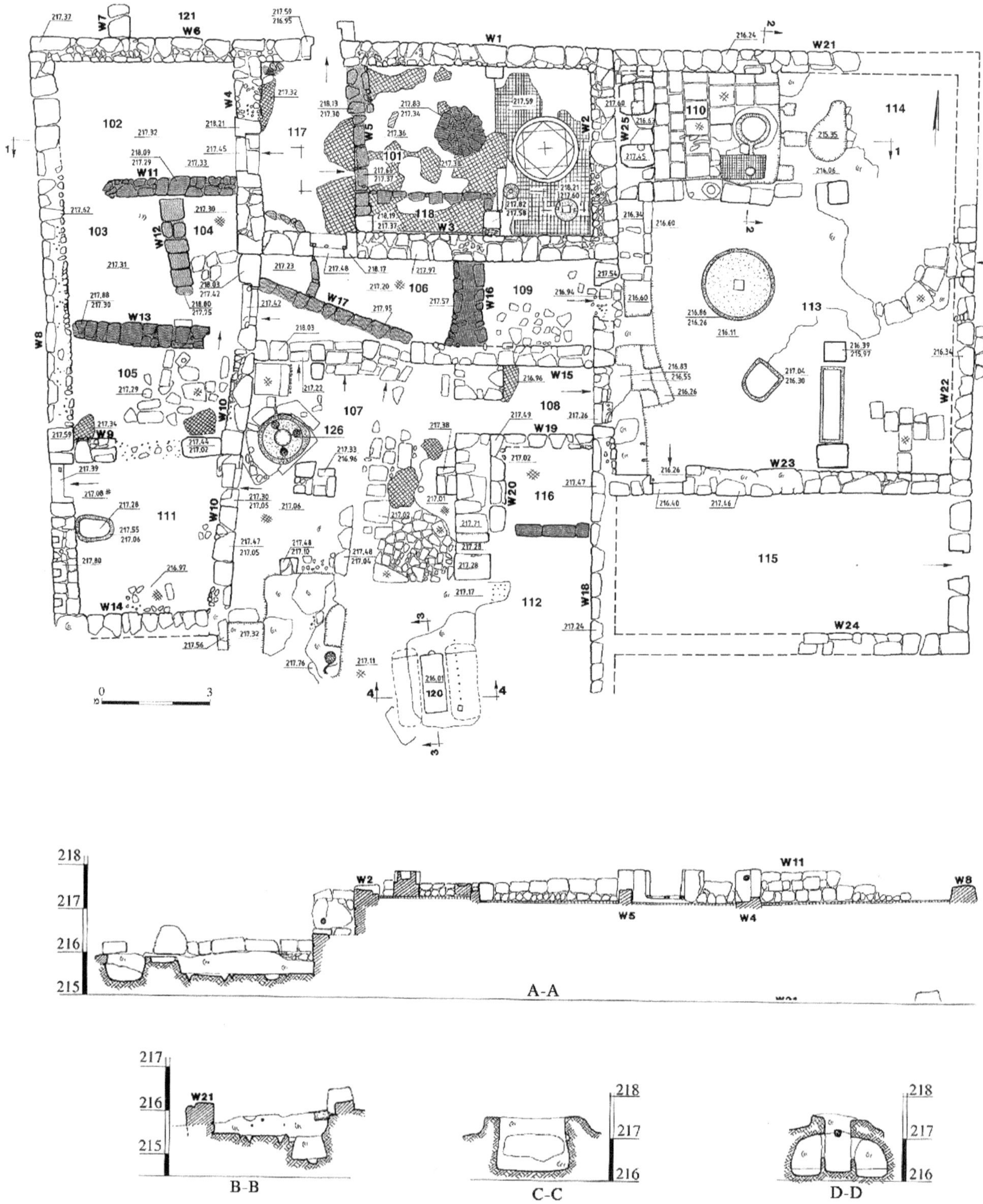

Fig. 44.1. Modi'im, plan and sections of the monastery.

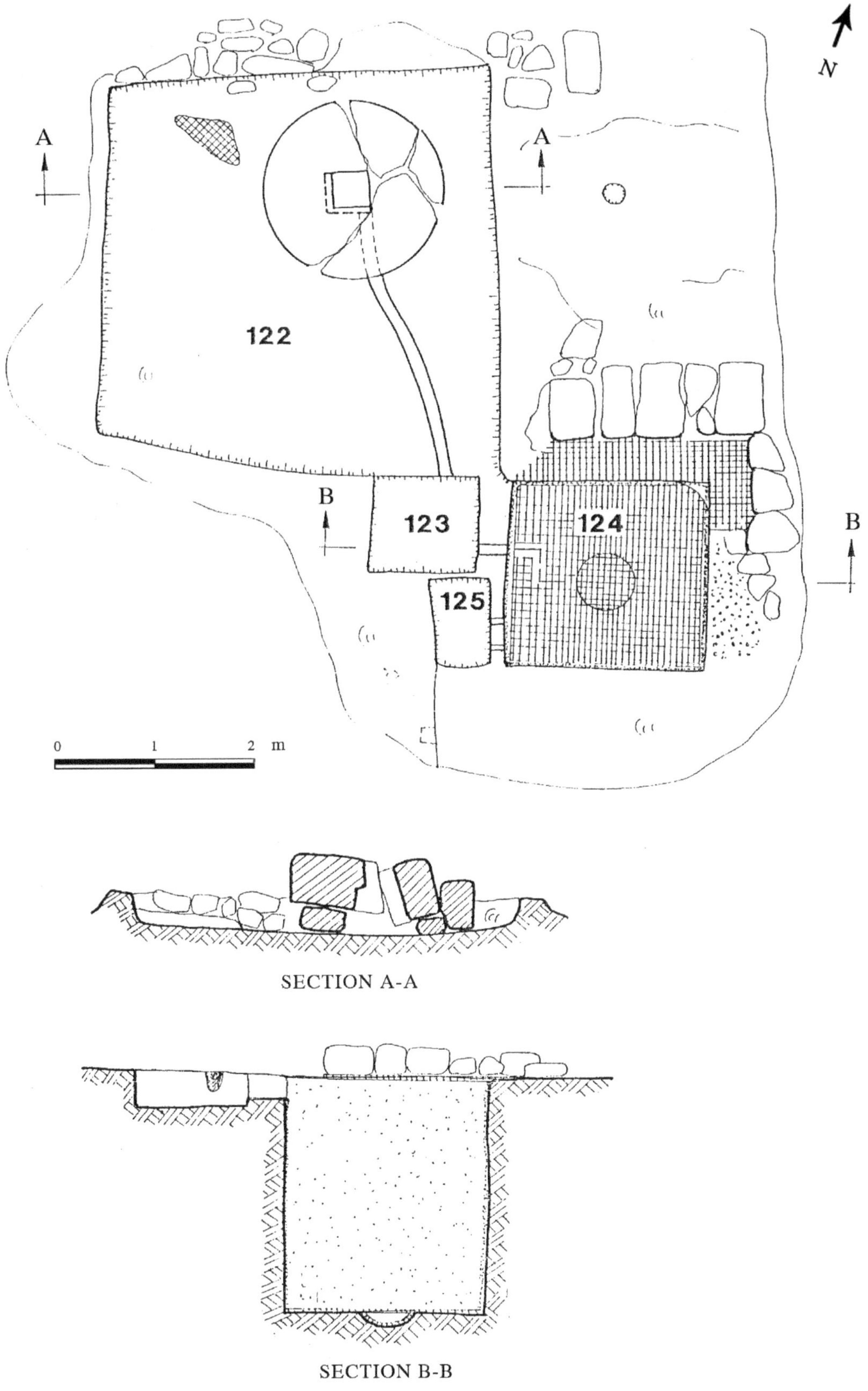

Fig. 44.2. Modi'im, plan and sections of the wine press.

45. A Byzantine Wine Press at Kh. Duran (Rehovot)

Israel Roll and Etan Ayalon

The Site

Kh. Duran is located within Rehovot in the southern coastal plain (MR 1324/1448). It is a rural site that existed from the Hellenistic to the Early Arab periods. In the past, buildings have been found there, as have installations such as an oil press and a tomb (Sussman 1969; Bushnino 2007; Nadelman 2000; Kogan-Zehavi 2000). The wine press was unearthed at the southern edge of the site.[1] Part of it was not excavated and another part destroyed by urban development operations, but those remnants that survived and the symmetry of the components enable reconstruction of the missing parts.

The large wine press (Figs. 1–4) measures 11.1×12.5 m (138 m^2). Its walls were carefully plastered with several layers of strong gray plaster. The walls of the vats were finished with a layer of finer pink plaster. The floors were tiled with coarse white mosaic, repaired in a few places with smooth, flat stones or pottery shards placed vertically in a matrix of mortar. The treading floor and storage compartments were constructed at ground level, while the vats were built within a large cavity hewn out of the bedrock. The wine press comprises the following elements:

- Six identical compartments in which grapes were probably stored before treading and/or where the "first must" was produced. These were installed on three sides of the treading floor – the north, west and south. Their dimensions are 1.7×1.2 m, depth 0.6 m. Narrow bores enabled the "first must" to drain onto the treading floor. At least one of the compartments was roofed by a vault, known also from other sites (Frankel 1999: 139).

- Seven or eight platforms on which the workers could stand or set the baskets of grapes, installed level with the top of the compartment walls – one alongside each compartment.

- A large treading floor measuring 5.5×6.0 m (33 m^2), into the centre of which a round stone screw base 1.6 m in diameter and 0.8 m high had been incorporated. In the centre of the stone base a mortice had been chiseled out to accommodate a fixed wooden screw, with two of its sides wider at the bottom than the top to secure the screw in position. A small gutter leading to the outside was cut into the bottom of the mortice, and perhaps in this wine press – as in those at Tell Qasile (Site 36) – there was once a pipe leading from the screw mortice to one of the vats. The must trodden from the grapes on the treading floor flowed through a stone channel into the filtration vat.

- The filtration vat is situated between two collecting vats in a symmetrical arrangement along the fourth, eastern wall of the treading floor. The vat measures 2.3×0.8 m, depth 0.5 m. Its floor is made of pottery shards embedded in a matrix of mortar. A large stone was set in the floor under the outlet of the channel leading into the vat to prevent the flow of must from eroding the vat floor. A ceramic vessel has been set into the centre of the floor as a sedimentation basin. When excavated, it was full of a white material – apparently marl that had been sprinkled on the grapes to absorb dirt and clarify the must (See Site 24). This practice still persists among the Arabs of the Mt. Hebron area (Frankel 1999: 42). After passing through the filtration vat the must flowed through ceramic pipes to the two collecting vats.

- Two large collecting vats, the southern one of which has been almost completely destroyed. The dimensions of the northern vat are 2.30×2.07 m, depth 1.35 m – a volume of ~6 m^3. A step has been installed on the floor under the outlet of the pipe to prevent the flow of must from eroding the vat floor. In the centre of the floor a double arrangement has been installed for sedimentation: a round cavity, in the bottom of which was sunk a ceramic vessel 0.47 m deep. Flush with the top of the vats and placed around them to north-east and south there was a platform floored with mosaic that was probably used by the workers. The discovery of two collecting vats confirms the "public" nature of the wine press, since they enabled various growers to produce must and have it flow into one or the other of the vats, or to tread grapes of different varieties and to keep their musts separate. Another possibility is that while one batch was fermenting in one vat the next could be prepared and flow into the second.

On the basis of its form and components, and particularly because of the pieces of ribbed pottery incorporated into the plaster of the walls and floors, the installation was dated to the Byzantine period. Like many other wine presses in the Land of Israel, it went out of service at the beginning of the Early Arab period, 8th-9th cent. CE (Ayalon 1997). Attesting to this is a rich pottery assemblage unearthed in a structure dating to that period that was erected over the treading floor, and mainly in the collecting vat, that became a trash pit. The wine press at Rehovot was the first to be excavated in the region that exhibited the abandonment of a large Byzantine wine press following the Muslim conquest of the 7th cent. The same phenomenon has since been noted at other sites. The desertion of these installations can be explained by the loss of export markets for wine, as well as the Muslim religious prohibition against its production and consumption. It appears that the breakdown of wine production was a major factor in the settlement crisis that befell the country during the Early Arab period.

Notes

1. The installation was unearthed during a rescue excavation (1975–1976) headed by the authors on behalf of the Israel Antiquities Authority and the Department of Classical Studies at Tel Aviv University.

Bibliography

Ayalon, E. 1997. "The End of the Ancient Wine Production in the Central Coastal Plain". In: Y. Friedman, Z. Safrai and J. Schwartz (eds.). *Hikrei Eretz, Studies in the History of the Land of Israel Dedicated to Prof. Yehuda Feliks*. Ramat Gan: 149–166 (Hebrew).

Buchenino, A. 2007. "Building Remains and Industrial Installations from the Early Islamic Period at Kh. Deiran, Rehovot". *'Atiqot* 56: 119–144 (Hebrew); 84*-85* (English Summary).

Frankel R. 1999. *Wine and Oil Production in Antiquity in Israel and Other Mediterranean Countries*. Sheffield.

Kogan-Zehavi, E. 2000. "Rehovot, Khirbet Deiran". *HA-ESI* 111: 59*.

Nadelman, Y. 2000. "Rehovot, Kh. Deiran (A)". *ESI* 20: 93*.

Roll, I. and Ayalon, E. 1981. "Two Large Wine Presses in the Red Soil Regions of Israel". *PEQ* 113: 111–125.

Sussman, V. 1969. "An Ancient Burial Cave at Rehovot". *'Atiqot* 5: 69–71 (Hebrew).

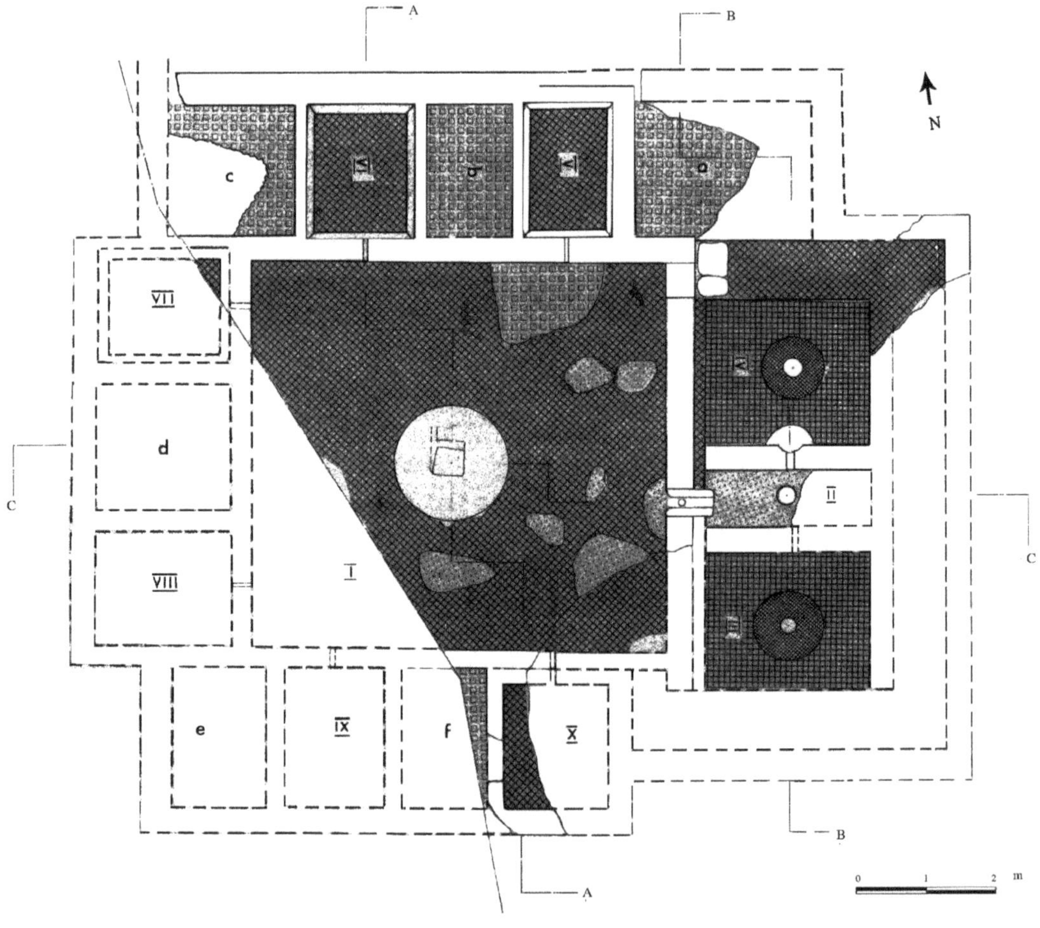

Fig. 45.1. Rehovot, plan of the wine press.

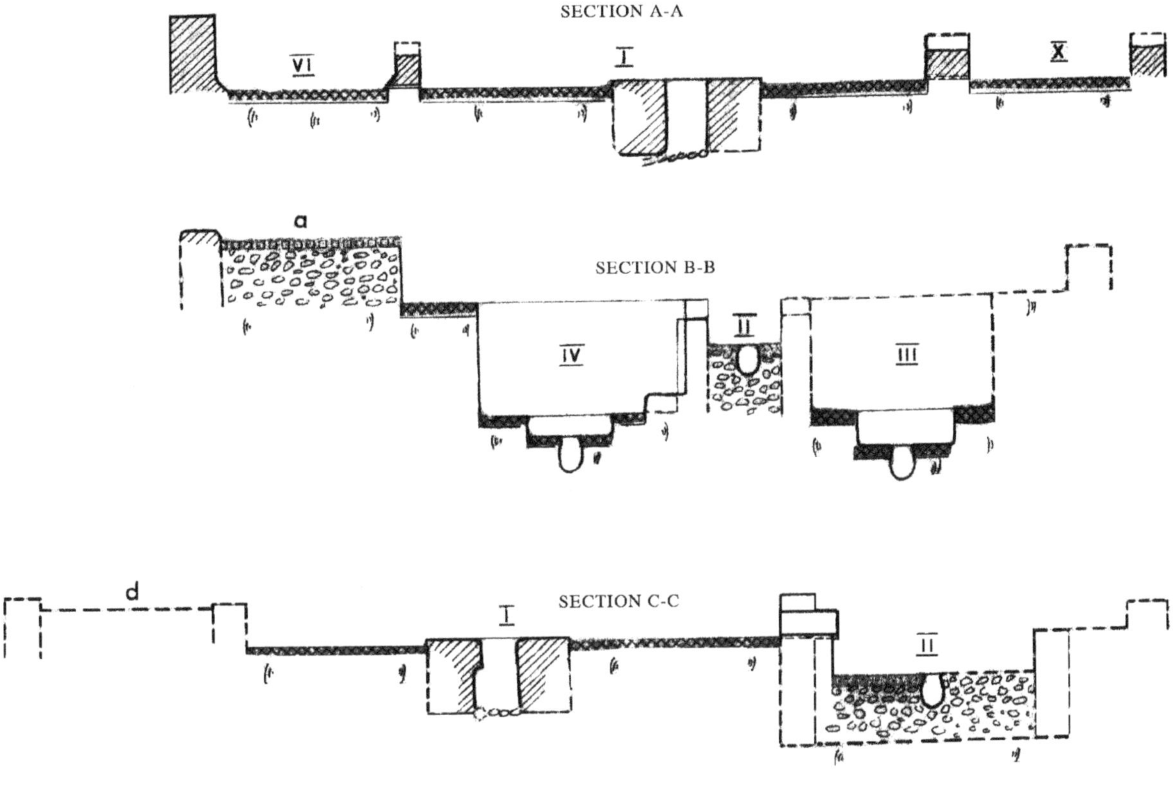

Fig. 45.2. Rehovot, sections of the wine press.

Fig. 45.3. Rehovot, the wine press, looking east.

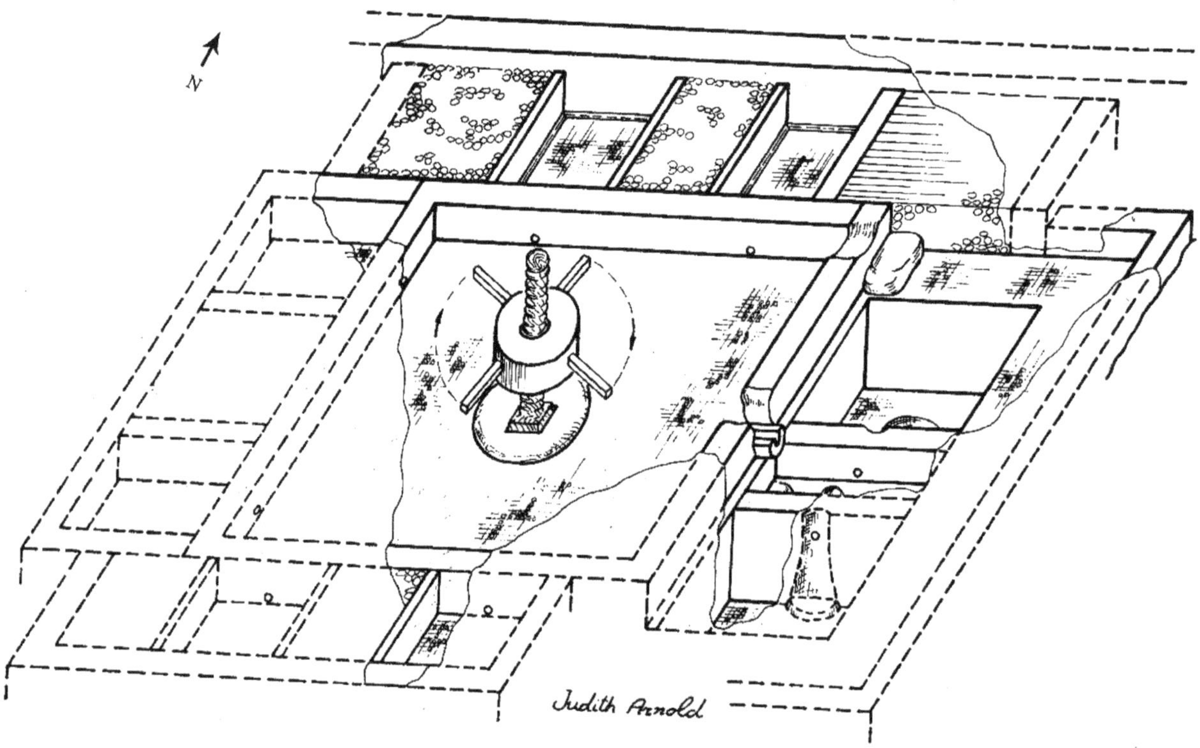

Fig. 45.4. Rehovot, isometric view of the wine press.

46. Qalandiya[1]

Etan Ayalon, Rafael Frankel and Amos Kloner

This installation (MR 169/141) was excavated in 1931 and published by Baramki (1933) as a Byzantine bath but there is no doubt that it was a twin wine press. It was, however, not found complete and some of the details that are important to understand how it functioned were omitted from the report. Nevertheless it includes some unique elements and has, therefore, been included in this book (Fig. 1).

In the western wine press, rooms A, B and C, Baramki discerned two stages. In the first stage rooms B and C were undivided and comprised one square room (9.00×8.85 m) floored in mosaic. The collecting vat was placed about 1 m from the southern wall. It was rectangular, had steps at east and west ends and apparently measured 6.00×1.65 m, depth 1.98 m, i.e. volume of 19.6 m^3. The exact dimensions of the treading floor are not clear but its southern edge is probably below the later wall that divided "rooms" B from C. The two pottery pipes mentioned by Baramki (1933: 107) almost certainly connected the treading floor of the first stage to the vat. In the second stage the treading floor and vat were separated by a wall and the floor (now 8.85×5.45 m = 48.2 m^2) was paved with stone slabs. The vat (a) was reduced and made shallower, now measuring 2.25×1.65 m, depth 1.7 m, volume of 6.3m^3. It included steps and a sump. The round "basin" to the east of the vat was probably an intermediate vat although in the plan it is not connected either to the floor or to the vat whereas there is a duct connecting the floor directly to the vat. A possible explanation is that in a third stage the round settling vat no longer functioned.

In the walls of the treading floor there were four compartments ("stoups") – a typical element of wine presses in the region. Three of these were semi-elliptical, measuring approximately 0.5×0.5 m, depth 0.6 m, and one rectangular. They appear not to have been connected to the treading floor. An unusual feature of these compartments is a ledge to accommodate a lid (Ibid., Pl. 38 Fig.1). Another unusual feature in the western wine press are the two funnel-shaped apertures on the edge of the vat that are connected to it by a sloping bore. The one to the south of the vat is beautifully made in mosaic and was probably already part of the installation in the first stage. The second to the east of the vat must have been added in the second stage, this explaining why it was made in concrete.[2] These funnels probably served to add a liquid to the must in the vat, using the funnels apparently allowing for greater control of quantities.

Room A to the north, also paved in mosaic, was at a higher level than the treading floor B. It was probably an auxiliary treading floor – another common feature in wine presses in the region – although there is no evidence as to how A and B were connected. Baramki mentions a furnace in room A. This again could have been a later addition.

The second wine press (D, E) is to the east. The treading floor E (4.1×5.5 m = 22.5 m^2) was first paved in mosaics and then with plaster. There were two vats to the east. The larger one (b) was in the north-east corner and had steps and a sump. The size of the vat is not completely clear (Ibid., Fig 95, Pl. 40 Fig. 2) but presuming that the steps started at the edge of the vat it measured 2.50×1.35 m, depth 1.25 m, volume of 4.2 m^3. There is another smaller vat (c) in the east wall (1×1 m, depth 0.93 m, volume of 0.93 m^3) with only a foot hole cut in the wall as a step. It was perhaps an afterthought as it cuts both into the floor and into the wall. In this winery too there were two semi-elliptical compartments (stoups), and room D was also probably an auxiliary treading floor. The purpose of the round depression to the east (Ibid., Pl. 41 Fig 2) is not clear.

In the report it is suggested that a round structure to the north (Ibid., plan Fig. 95, section y-y in Fig. 96 and Pl. 41 Fig. 1) is a furnace, and as stated above another furnace is mentioned in room A. Baramki connected these furnaces to the function of the installation as a bath. If, however, we are correct in interpreting it as a wine press these furnaces could be connected with the production of boiled down wine, the *sapa, defrutum* or *carenum* of classical literature and the *yayin mebushal* of Talmudic literature. Alternatively the large furnace to the north could have been a pottery kiln such as is often found attached to large wine press, although there is no trace in the plan or the photograph of the usual central protrusion that supported the upper part of the kiln, on which the pottery vessels were placed.

Notes

1. Dimensions taken from plans in Baramki 1933.
2. Both funnels are described in the text p. 105 and appear on Pl. 79 Fig. 1; neither appear on the plan but both appear on the sections, Fig. 96 – the mosaic funnel on section y-y and the concrete one on section x-x.

Bibliography

Baramki, D.C. 1933. "A Byzantine Bath at Qalandia". *QDAP* 2: 105–109.

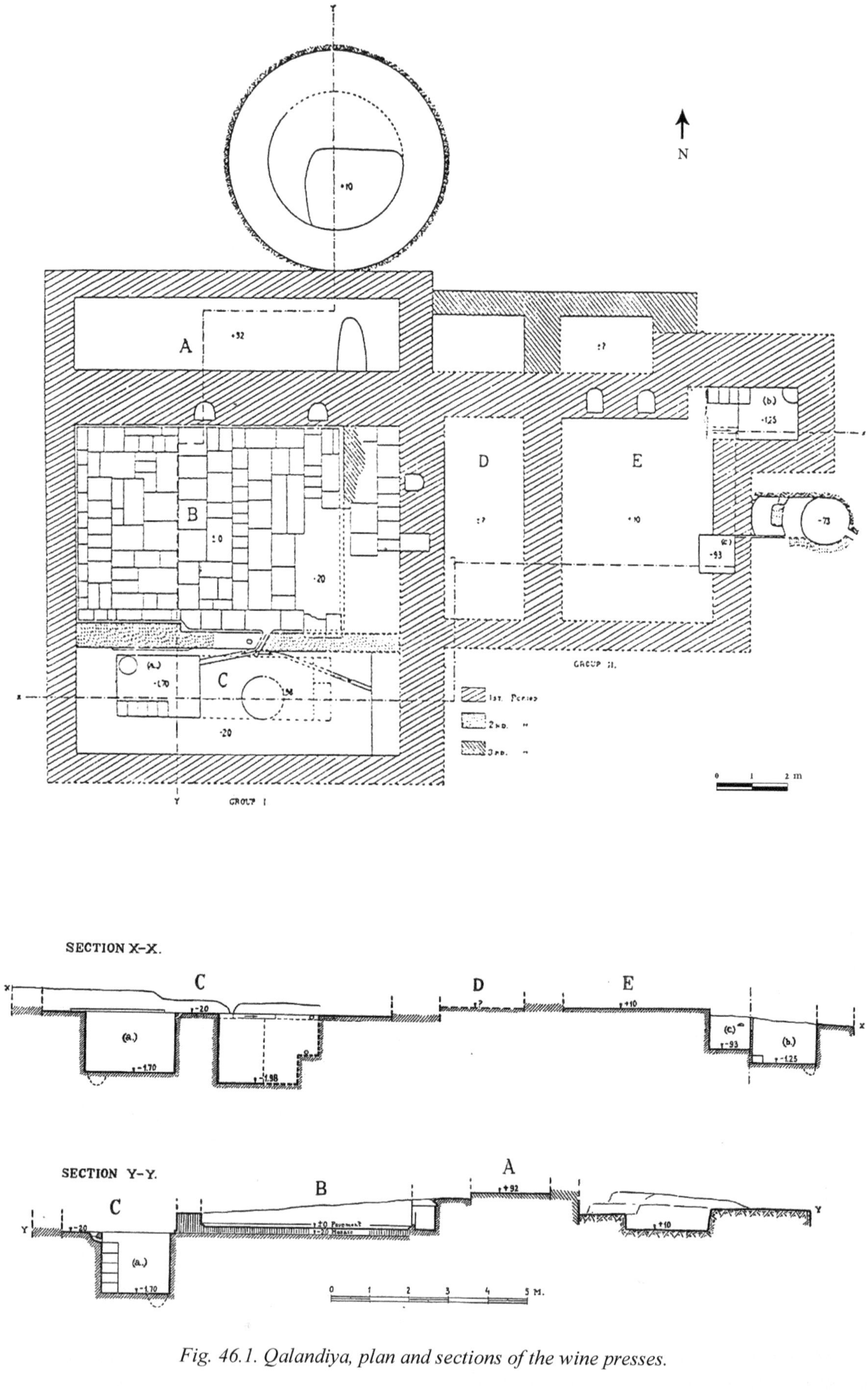

Fig. 46.1. Qalandiya, plan and sections of the wine presses.

47. Two Wine Presses at Ras et-Tawil (Pisgat Ze'ev) in North Jerusalem

Shimon Gibson

The Site

The hill of Ras et-Tawil (present-day Pisgat Ze'ev, a suburb of Jerusalem) is located close to Tell el-Ful (identified as Gibeah of Saul), north of Jerusalem (MR 1729/1376). The remains of walls, cisterns and a wine press were already observed on the summit of the site in the 19th cent. by the Survey of Western Palestine expedition (Conder and Kitchener 1881–1883, III: 124). The site was also mentioned by other scholars, notably Féderlin (1906: 268) and Abel (1938: 260). In 1942 the site was examined by Baramki of the Department of Antiquities, but his short report was not published.[1] For some reason, Baramki was unable to locate the wine press at the site during his visit.

The site was thoroughly investigated during an archaeological survey of the region in 1981, and later that year it was excavated (Gibson 1985–86).[2] These excavations uncovered the remains of a *coenobium*-type monastery on top of the hill that had a chapel, a complex of rooms and courtyards. Nearby were revealed a bread oven, a cistern, and two wine presses, one large and one small. The buildings and installations were entirely surrounded by an enclosure wall of stone that also extended around a system of agricultural terraces on the northern slope. Pottery and coins found at the site indicated that the monastery was established during the Byzantine period and continued to exist until it was eventually abandoned or destroyed during the Abbasid period, towards the end of the 8th cent. The ruins became a source of building stone that was quarried there from medieval times onwards.

The Wine Presses

The Large Wine Press

This installation is located some 15 m to the north-west of the monastery building complex. It is rock-cut and partly built and comprises a rectangular treading floor, a basin, a vat and a storage chamber (Figs. 1–2). The western half of the installation was visible before the excavations began. This is apparently the wine press seen by the SWP explorers. Late Ottoman and modern artifacts were found in the eroded *terra rossa* fills covering the installation.

The treading floor (L. 40, 5.70×5.95 m, area c. 34 m^2) is surrounded by walls, rock-cut in their lower parts and built of large field stones. The southern wall has a thickness of 2 m. It is assumed that the entrance to the structure was in the west wall but it has not been preserved. The walls were coated with a hard pinkish plaster (2 cm thick) containing crushed potsherd inclusions, against a backing of small flat stones (averaging 0.1×0.2 m). The plaster was covered with a layer of paint of a deep wine-red color; brushstrokes are still clearly visible on the surface.

The rock-cut floor was originally paved with coarse white tesserae and two patches of tesserae are preserved in the south-east corner and along the east wall. The mosaic floor was laid before the plastering of the walls was undertaken. The floor has an average of 30 tesserae to the square decimetre. The tesserae used were oblong in shape (2×3 cm) and were set vertically into the plaster bedding of the floor (similarly-shaped tesserae are known from a Byzantine chapel and structure at Masada [Yadin 1965: 112, 114]). The bedding of the floor consists of a layer of light grey plaster (4 cm thick) containing crushed chalk inclusions with a few crushed potsherd inclusions, above a rock-cut surface. The plaster bedding is preserved on the eastern side of the floor only. The fills of soil that were cleared away from the eastern part of the floor did not contain pottery.

In the centre of the floor is a circular rock-cut basin, 1.15 m in diameter and 0.15 m deep. In its centre is a small rectangular mortice (0.45×0.40 m, 0.33 m deep) with concave sides (7 cm) to the east and south. This mortice served as the socket for the stone base of an upright screw press (Fig. 3). The stone base was not found but it was probably circular and its top surface would have been more or less flush with the surface of the mosaic floor (cf. Frankel 1999: 140 ff.). The wooden screw upright would have been anchored in place by this stone base. Suitable downward pressure was applied with a wooden nut that was rotated with the help of handles inserted into holes located in its side.

Immediately north of the treading floor is a rock-cut basin (L. 41, 1.65×1.50 m and 0.55 m deep). There are no signs of plaster visible. On the southern and eastern sides of this basin is a rock-cut bench (0.15–0.23 m wide). The function of this basin is uncertain; it may have been used for storing the grapes prior to their being pressed. The fills from within this basin contained potsherds and roof-tile fragments from the Byzantine period and a few sherds of medieval or Ottoman handmade vessels.

The collecting vat is located west of the treading floor and a rock-cut channel (0.15 m in width and depth) links the two. The vat (L. 42, 2.1×1.9 m and 1.23 m deep, volume of

4.9 m^3) was found full of rubble and soil containing mixed Byzantine and modern potsherds as well as an iron ring and bullet cartridges. Only the southern half of the vat was excavated. There are no signs of plaster on its walls.

To the east of the treading floor is an oblong sunken chamber (L. 44, 2.10×0.85 m and 1.3 m deep). The southern and eastern walls were rock-cut whereas the other two walls were built of two courses of squared blocks of limestone above un-coursed small stones (the north wall had a width of 0.65 m). The walls and rock-cut floor of this chamber were coated with two layers of white plaster (2 cm thick) containing crushed potsherd inclusions. At the western end of this chamber was a plastered step (0.15 m in width and height) with protruding triangular shaped stones (0.15 m high) at both ends of the step in the corners. This step and the flanking stones may have served as the base for a wooden ladder. It seems probable that this chamber was used as a cellar for storage purposes. Potsherds and roof-tile fragments from the Byzantine period and a few animal bones (unidentified) were found on the floor of this chamber.

Excavations were undertaken to the north of this chamber to investigate the possible existence of additional sunken chambers but none were found. The fills excavated in this area (L. 43) revealed potsherds, roof-tile fragments, a marble fragment and glass window pane fragments from the Byzantine and Early Arab periods. A few medieval or Ottoman potsherds and an iron peg or hook of modern date were also found.

The Small Wine Press

This installation (L. 45) was located on the western slope of the hill not far from the main cistern used by the monastery. It was also cleared during the excavations at the site. The finds from the soil covering the wine press consisted of potsherds of Byzantine or Early Arab date and one glazed sherd of medieval date.

The press consists of a treading floor, a niche in the east wall, a cup-mark and a vat (Fig. 4). It is completely rock-cut and there are no signs that it was enclosed within a structure like the large wine press described above. There were also no signs of plaster on the walls and floor of this installation.

The treading floor is rectangular with rounded corners (1.45×1.60 m – area 2.3 m^2) and slopes down to the west. The rock-cut niche for a pressing beam is located in the east wall (0.30×0.25 m, 0.35 m deep). The far side of the niche is slightly concave. The pressing beam was hinged by two sockets cut into the sides of the niche. The northern socket measures 20×7 cm and the southern socket 10×7 cm; both have a height of 10 cm. The niche is open from the top which means that when the pressing beam was inserted a large stone must have been placed on top to keep it in position. The beam probably had a length of about 2 m. It may have been weighted down with pierced stones but none were found. A cup-mark is located at the western end of the treading floor (0.13 m in diameter and 6 cm deep). Numerous beam presses of this type were investigated during a survey in the vicinity, some of them dating from the Early Roman and Byzantine periods (Gibson and Edelstein 1985: 147–149).

The vat is rectangular with rounded corners (1.00×0.86 m and 0.55 m deep, volume c. 500 litres) and has a rounded sump cut into the floor in the north-western corner (0.35 m in diameter and 0.13 m deep) that was probably used to collect the settled residues of the grape juice.

Summary

The large wine press was clearly the main installation that served the wine-producing needs of the monastic community at Ras et-Tawil and most of the wine was probably sold by the monks in the markets of nearby Jerusalem. Numerous examples of installations of this type are known dating from the Byzantine period (Hirschfeld 1983; Gibson and Edelstein 1985: 149).

The vineyards were cultivated on the agricultural terraces located on the northern slope of Ras et-Tawil (an area of about 15 dunams) and at harvest the grapes were brought to the large installation where they were probably stored temporarily in the basin to the north of the press before eventually being trodden on the mosaic-paved floor. This was followed by the pressing of the grape residue with the use of a central fixed screw press. The grape juice was collected within the vat and then transferred to jars and some of it was probably stored within the plastered oblong chamber located immediately east of the press.

The small beam press installation was clearly used only for pressing small quantities of grapes. Additional wine presses were located during the survey in 1981 beyond the limits of the monastic settlement and these may also date from the Byzantine period (Gibson 1995: 266–267).

Notes

1. IAA Archives.
2. The excavation, directed by Shimon Gibson and Amos Kloner, was a salvage operation for the Israel Department of Antiquities (licence No. 1053). The team consisted of O. Shmilshuk, M. Sedgwick, M. Isaacs, B. Alpert, A. Caplan, N. Nagel and M. Kreiner. Surveying was undertaken by D. Huli and photography by T. Wachs. The drawings published here were prepared by S. Gibson.

Bibliography

Abel, F.M. 1938. *Géographie de la Palestine* Vol. 2. Paris.

Conder, C.R. and Kitchener, H.H. 1881–1883. *The Survey of Western Palestine*. 3 Vols. London.

Féderlin, L. 1906. "À propos d'Isaie X 29–31". *RB* 15: 266–273.

Frankel R. 1999. *Wine and Oil Production in Antiquity in Israel and Other Mediterranean Countries*. Sheffield.

Gibson, S. 1985–86. "Ras et-Tawil: A Byzantine Monastery North of Jerusalem". *BAIAS* 5: 69–73.

Gibson, S. 1995. *Landscape Archaeology and Ancient Agricultural Field Systems in Palestine*. Unpublished Ph.D. Thesis. University College of London. London.

Gibson, S. and Edelstein, G. 1985. "Investigating Jerusalem's Rural Landscape". *Levant* 17: 139–155.

Hirschfeld, Y. 1983. "Ancient Wine Presses in the Park of Aijalon". *IEJ* 33: 207–218.

Yadin, Y. 1965. *Masada: First Season of Excavations, 1963–64*. Jerusalem.

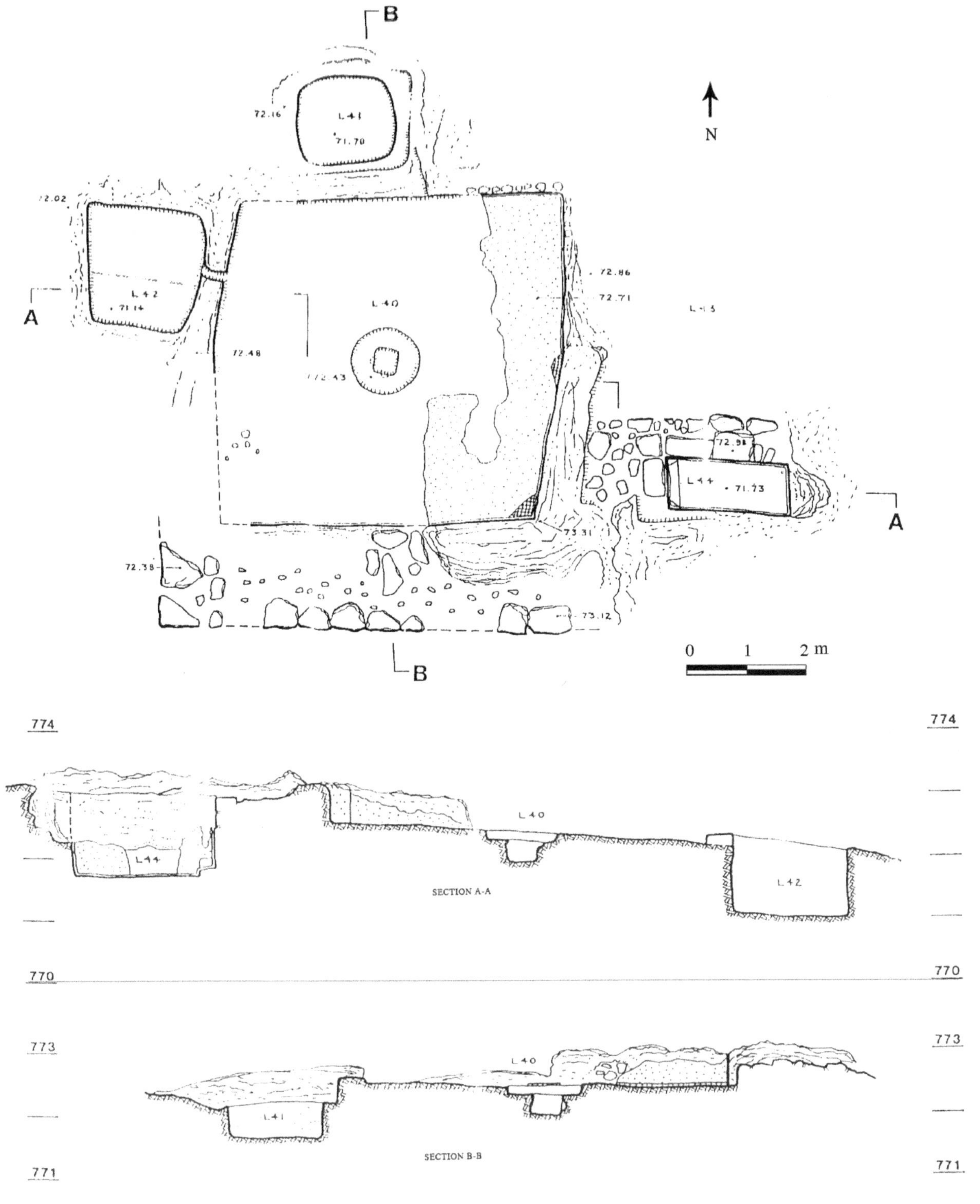

Fig. 47.1. Ras et-Tawil, plan and sections of large wine press.

Fig. 47.2. Ras et-Tawil, large wine press, looking east.

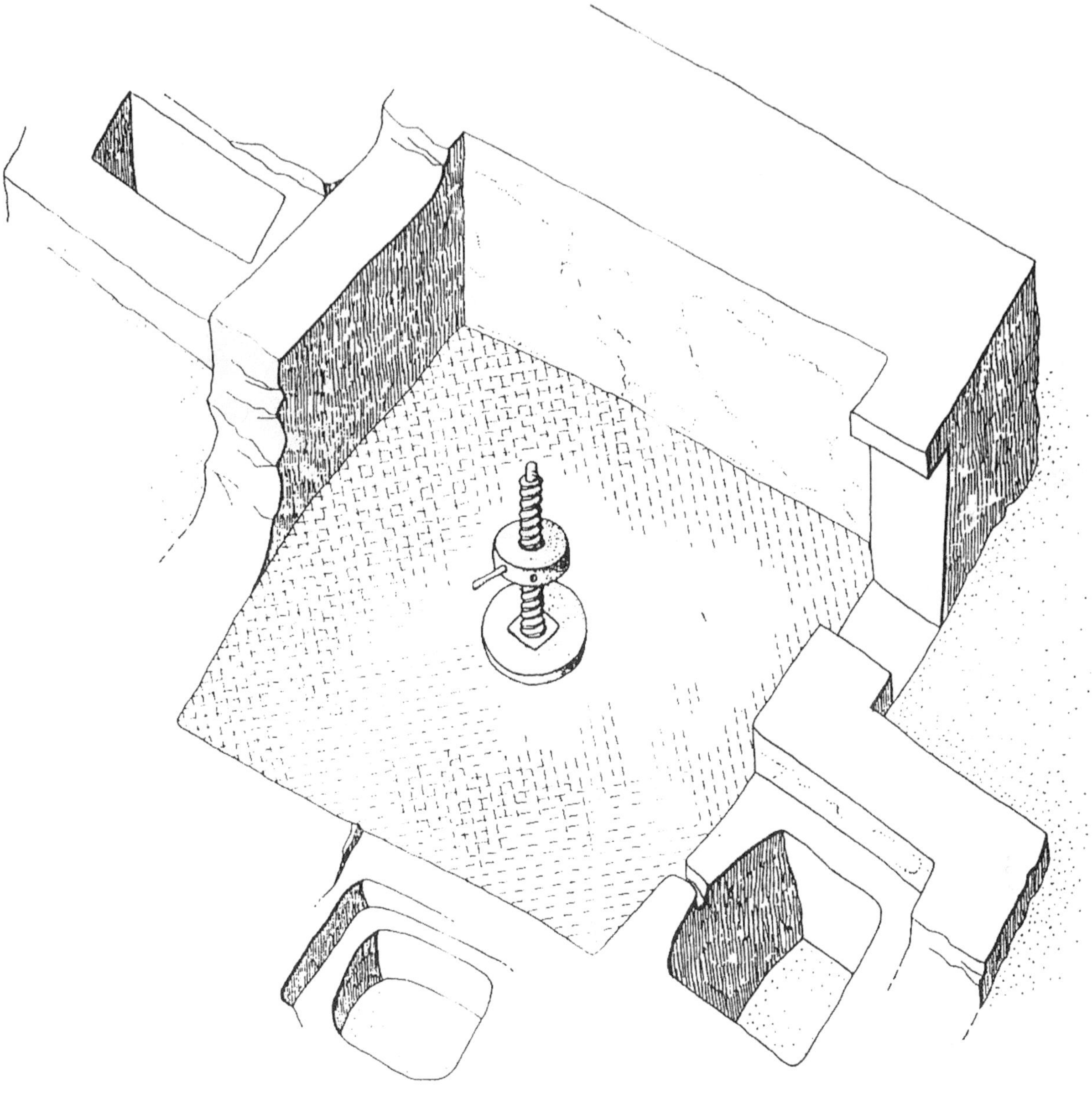

Fig. 47.3. Ras et-Tawil, reconstruction of large wine press.

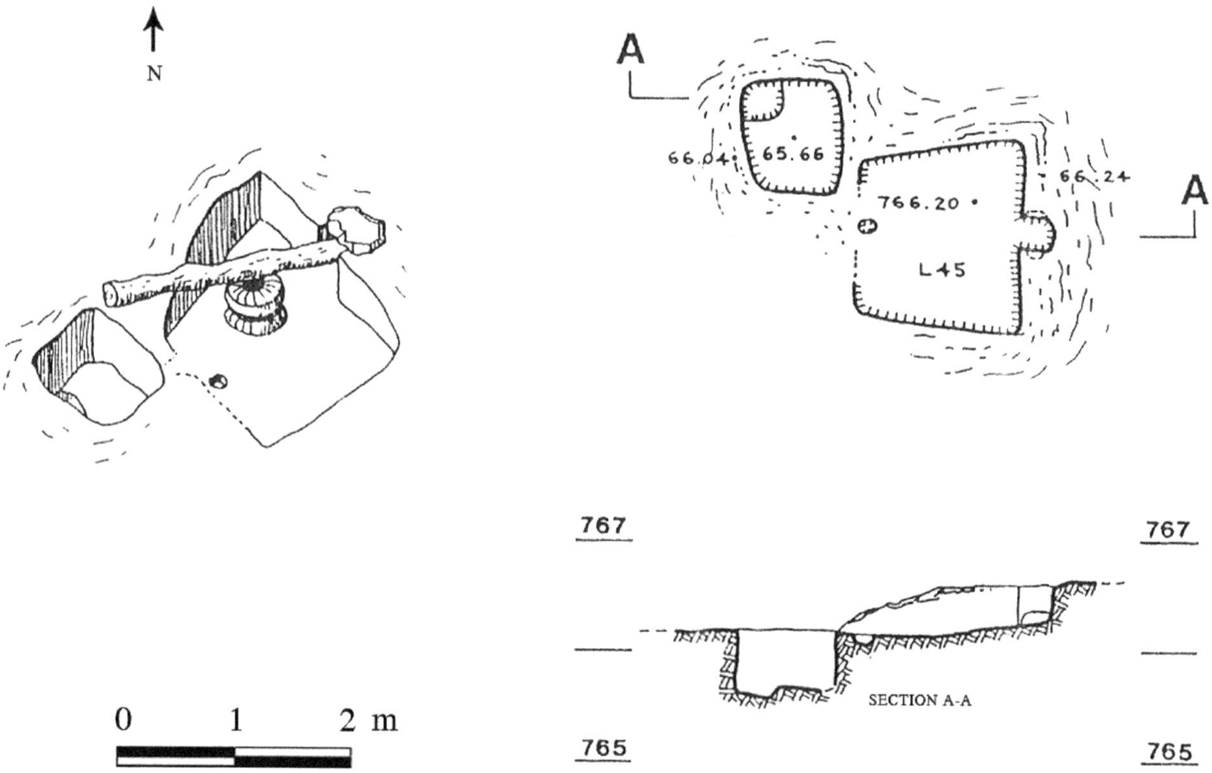

Fig. 47.4. Ras et-Tawil, plan, section and reconstruction of small wine press.

48. Oil and Wine Presses at Ras Abu Ma'aruf (Pisgat Ze'ev East 'A)

Jon Seligman

The Site

Ras Abu Ma'aruf (MR 173/137) is a hill located between the villages of Hizme, 'Anata and Shu'afat, some 900 m east of Tell el-Ful (Biblical Gibeah) and 4.4 km north-west of Jerusalem's Damascus Gate.[1] The area is today semi-arid but the quantity of agricultural installations in the immediate vicinity would point to rich agriculture settlement in the Roman and Byzantine periods (Gibson 1985–6 [Site 47]; Avner, Azmon and Sabariego 1992 [Site 50]; Shukron and Savariego 1994).

Two large sites were excavated including oil and wine presses on the south-west edge of the hill and a wine press west of the summit (Seligman 1994, 1999). These sites, that are both similar and contemporary, will, after description, be jointly discussed.

The first site, area 4, consists of an oil press built into a man-made cave and a building complex on the rock surface above, including storerooms and a large wine press.

The Oil Press (Figs. 1–4)

The cave was originally a *mikveh* (ritual bath) of the Herodian period. During the Byzantine period the cave was extended to measure 6.0×5.3 m. Two hewn staircases connected it to the structure above.

In the former immersion area of the *mikveh* the collecting vat and press-bed were constructed. This consisted of a large ceramic bowl in the shape of a truncated cone set into the rubble and supported by it. A crushed lime floor sealed the unit at general floor level. The collecting vat was 0.57 m deep and had a capacity of 190 litres. At its bottom was a sump used to collect the last drops of liquid after pressing. The press-bed was made of two stone slabs (1.10×0.55 m) cut to fit above the vat. A round channel and two cross channels, at the point where the slabs joined, helped the expressed liquid flow into the vat. In the centre was a square inset in which was placed the perforated wooden board used to cover the mouth of the vat and on which the frails were placed during pressing. Around the press-bed were discovered a number of olive pits.

To the south of this installation was cut a round pit in which was placed the screw weight, measuring 1.03 m in height and 1 m in diameter, set in the hole 0.66 m below floor level. The weight had two external dovetailed mortices and a cross slot added on the west side to strengthen the hold of the tenon wedges. On the weight's top surface was a round socket in which the screw base could turn.

At the back (north) of the cave was a small natural but enlarged cavern with a gap left in the wall dividing it from the cave. The gap was set 1.7 m above the level of the press-bed. In this gap (fulcrum) the press beam was anchored.

In the ceiling of the cave were two holes on the line of the beam. Two uprights inserted here had once secured the beam in place during operation by preventing lateral movement.

On the western part of the floor was a raised, curved area with a square hole above it in the ceiling. We suggest that this was the base for the crushing basin, a crucial installation now missing, the socket used to fix the central pivot in position. A broken crushing stone was found in the fill of the cave. It was small (0.7×0.4 m) and had a square hole in which the wooden axle was attached.

This oil press, in common with many others, was built into a cave. The reasons for housing the installation in a closed structure are to protect the valuable equipment while not in use, to allow the process to continue in all weather conditions as harvesting is at the start of the rains and to enable internal illumination to permit work 24 hours a-day during the short pressing season (Elitzur 1986: 179). In addition the strength provided by the natural rock niche could withstand the large forces produced by the leverage of the installation.

Crushing of the olives was conducted in a crushing basin, that was absent though its location was evident due to the socket in the roof of the cave and the rock-cut base on the floor. A single, slightly tapered, crushing stone was discovered in the debris of the cave.

The press operated on the principle of lever and screw that is comprehensively discussed by Frankel (1984: 150–167; 1999: 107 ff). In short the beam was anchored in the niche at the back of the cave. This beam was probably forked at the opposite end, a nut fixed between the forks. A wooden screw, thread in the nut, turned within a socket in the stone screw weight below to which it was attached. When the screw was turned the beam lowered and at a certain point lifted the weight off the floor. The weight had two dovetailed external mortices, one strengthened by a cross slot. Wooden tenon wedges in the slots fixed in mortices in a board on top of the weight, into which the screw was fixed so that it would turn without being released.

Pressing of the olive pulp was achieved by placing the beam on a pressing board below which were piled the frails containing the pulp. These baskets were set on the press-bed. In the centre of the press-bed is a square inset over the opening to the vat. A perforated wooden board or slats were inserted within this inset via which dripped the oil and watery lees into the collecting vat. After processing the frails and board were removed and the separated floating oil was removed with a dipper juglet.

This type of installation was common in the Byzantine period, especially in the north and centre of the country (Frankel 1984: 156; 1999: 112)[2] and weights with external mortices are well known.[3] The central, rather then lateral, collecting vat is a regional feature common through various periods in Judea and the Jerusalem area.[4] Similar collecting facilities with a lever and screw press were found at Kh. Siyar el-Ghanam (Corbo 1955: 36–37), H. Beit Loya (Frankel, Patrich and Tsafrir 1990; Site 67), Shu'afat (Site 51) and Kh. el-Quseir (Frankel 1992: 49–59; Site 8).

The Area 4 Wine Press (Figs. 5–6)

Two clear phases of construction seem to be evident in the wine press. The first may belong to the Herodian period while the second is clearly set both by the finds and the very nature of the installation to the Byzantine period.

Of the first phase of the wine press is the almost square central treading floor, 6.3×5.7 m (c. 36 m^2). It was cut into the bed-rock so that the back wall was quarried 2 m into the rock, its upper part then built on a leveled area on the hewn edge, some mortar stains here providing evidence of a built superstructure. A small stone was set into the south-western part of the floor and a niche had been cut into the bed-rock wall to the north-east some 0.5 m from the floor. In this niche a beam was anchored to press frails of rape set on the central stone and weighted down by a stone rested, when not in use, on the stone surface to the south-west.[5]

Also clearly of the first phase of use of the wine press are the collecting vats located south-west of the main treading floor. The north-eastern vat was small (0.85×0.85 m and 0.8 m deep, volume 0.58 m^2), coated with a hydraulic plaster and paved with plain white mosaic. A ceramic pipe embedded in the wall allowed the must to flow from the main floor into the vat. Another ceramic pipe on the bottom of the opposite side led to the next larger vat to its south-west. This vat measured 1.9×2.3 m and was 2 m deep (volume c. 8.7 m^3). Five steps were cut on its south-eastern side down to the bottom that was paved with plain white mosaic. In the southern corner of the floor was a deep sump (diameter 0.7 m, depth 0.55 m), also covered with mosaic and presumably used to gather the last drops of must. The walls were covered with a hydraulic plaster using flat stones as a base. While this was most certainly a collecting vat the other vat is more difficult to understand. The fact that the outlet pipe was set at its base suggests that it was a filtration vat, the filter probably consisting of bunches of uprooted brush such as Thorny Burnet.

The second phase of use enabled production on a larger scale and at a greater degree of complexity. The central treading floor continued in use. The floor was paved with white tesserae. In the centre of the floor, again bordered by the mosaic, was a round stone (diameter 1.3 m) with a square mortice, widened to one side (0.45×0.45 m and 0.4 m deep).[6] The mortice would indicate the use of a single fixed screw press to squeeze the trodden rape and that the weighted beam and thus the niche were no longer functional. The walls to the south-east and north-west were cut in part from the bed-rock and built up in parts to form the arched compartments under the auxiliary treading floors.[7] Bed-rock pillars in the centre of the wall between the main area and each auxiliary treading floor, in addition to further pillars at the south-western end of walls may show that the treading floor was roofed with three arches that would have carried the wooden roof beams to cover the area. The fact that large quantities of charcoal were found in the debris here and on the floors of the auxiliary installations strengthens this supposition. The walls were all covered with a white lime plaster.

The four auxiliary treading floors were all similar, though their sizes were variable. Each was paved with a plain white mosaic. Each had a compartment under the floor with a duct leading from the floor and a further gutter leading through the wall to the central floor. The sides were plastered and the floors covered with plain white mosaic.

On either side of the vats were approximately equally sized rooms (both around 2.8×2.8 m). Both were paved with a white mosaic laid square to the plastered walls and had a closed dovetail mortice stone set in the centre of the floor, that would have held a single fixed screw to press the rape.[8] A ceramic pipe led through the wall of the south-eastern floor to a separate collecting vat (1.1×1.1 m, 1.4 m deep, volume 1.7 m^3). It too had a sump in one corner. Triangular niches in the walls formed footholds to allow access to the unit. The opposite floor had no such vat, but the angle of the floor showed that the must gathered here would have poured into the adjacent central vat.

The Area 13 Wine Press (Figs. 7–8)

Area 13, around 100 m east of the former site, consists of a large pool and an adjacent complex wine press, similar to that described above. The installation clearly has a number of phases that are difficult to date individually due to the nature of the construction.

Phase I

The central treading area measures 5.3×5.6 m (area 29.7 m^2). White tesserae formed the floor. In the centre of the room was a large square stone block (1.1×1.3 m) bordered by three rows of mosaic.[9] This had been either sunk into a hole in the floor or had been carved during the original leveling of the bed-rock. In the centre of the stone was a mortice into which the screw, used during the pressing process, was fitted. The mortice measured 0.40×0.38 m, was 0.4 m deep and widened at the bottom on two adjacent sides. On the southern part of the block was a huge perforated rock that had probably been used as a weight during the pressing.

In the northern part of the room, against the eastern and western walls that stand to a height of 0.7 m, are the bases of two opposing pilasters that supported the roof. These pilasters seem to be later then the mosaic and probably sit upon it.

To both the east and west were two supplementary treading floors respectively. These were situated 1 m higher than the central treading floor. These floors underwent major changes in later phases as shown by various intact features such as surrounding walls. Cut under each floor was a small vaulted compartment, built in two sections, the bottom carved from the bed-rock and the upper arched part built of small stones and mortar. All the compartments were plastered and had plain white mosaic floors. The floors were connected to the compartments via a duct, the compartments connected in turn to the treading floor via holes in the walls. This allowed must to flow from component to component.

South of the room of the main floor were two collecting vats. The eastern L-shaped vat measures 2.0×1.9 m and 1.5 m deep, a volume of 5.8 m^3. A ceramic pipe connects the vat to the main floor through the wall. Only part of the floor is covered with a mosaic, the rest plastered, along with the walls. A sump lined with mosaic was built on the southern side of the floor. In the south-western corner are two niches in the wall that were used to help climb out of the pool. On the edge of the vat, on either side of the same corner are depressions carved in the rock in which jars can be laid, channels connecting back into the vat to collect spillages.

The western vat measures 2.4×1.2 m and 1.64 m deep, with a volume of 4.7 m^3. It probably shows two building phases, the eastern part being originally round and later enlarged westwards, the extension being rectangular with three steps built into it. The floor was a mosaic, speckled in black and red with a small sump emphasized with an arrangement of tesserae around it. The walls and steps were plastered. No pipe or other feature connected this vat to the wine press or to the other vat, though its function seems clear.

Phase II

In this phase the north-western auxiliary treading floor was renovated (Fig. 9). It was paved with mosaics different from all the other mosaic floors as it has large tesserae (9 per 10×10 cm) that indicate a later renovation. On the eastern side of the floor is a hole framed in mosaic that led to a compartment below. Internally this compartment is plastered, though the former mosaic floor did not survive. The must flowed through this compartment to a collecting vat built on the main treading floor of the previous phase. This vat was constructed north of the western pilaster. It is made up of a small mosaic paved area (0.73×0.84 m) preserved from the original mosaic of the treading floor and surrounded by a low wall. The floor has a shallow sump on one edge.

On the central floor along the southern and eastern walls were placed two rows of stones with signs of burning between them. A similar arrangement of stones with signs of burning between them is found along the eastern wall. A possible explanation for these cauldron stands is presented below.

North of the wine press is an elongated room measuring 3×10 m. The floor was leveled rock. At the eastern end of the room is a large stone slab with smaller ones laid in a line to its south. This slab covered a pit with a settling depression in its bottom measuring 1.9×1.5 m and 1.8 m deep. The slab fitted exactly into an inset on the rock edge. The other stones covered a channel leading towards the wine press. The use of both the channel and the pit is enigmatic, though it seems to be connected to the wine press.

Operation of the Wine Presses

The two complex wine presses are of the same type and represent a form known from a number of other sites in the Byzantine period. Between them are various ancillary installations that characterize a large cross-section of those present during the period under discussion.

The grapes brought from the field were placed on the auxiliary floors and probably left there for around three days to allow the static pressure of the bunches to produce the "first must". This poured through the ducts in the floors into the vaulted collecting compartments underneath. The holes to the main treading floors were blocked to allow the must to collect. This was probably collected separately to produce wine of the highest quality (*prototropum mustum*). If these floors were roofed, as we suggest, then this would not follow the usual practice of letting the grapes stand directly under the sun, a procedure usually followed to produce *vinum dulce* the sweet wine (Roll and Ayalon 1981: 119; Frankel and Ayalon 1988: 42; Frankel 1999: 42, 139).

The grapes were then shoveled onto the main treading floor to be pressed under-foot, the must flowing into the vat, in area 4 via a settling tank to remove any solids. The juice would then be left for 3–4 days in the vat for initial fermentation and then transferred into jars for slow secondary fermentation in a cool dark place.

The remains left after the treading were collected for further processing in one of three ways to produce a low quality wine (*mustum tortivum, circumsicium*) (Roll and Ayalon 1981: 122; Frankel 1999: 42):

1. Area 4 wine press, phase I: The trodden grape remains were collected in frails and placed on the press-bed, a pressing board placed above surmounted by a wooden beam anchored in the niche in the northern wall and weighted by a stone attached by rope to its opposite end. The location of the weight is marked by a stone circle bordered by the mosaic in the south-west part of the floor (Frankel 1984: 191; 1999: 145–146).

2. Phase II: A square mortice, widened to one side, was cut in the centre of the round press-bed and a single screw fixed into it, jammed in by wooden pegs. A heavy rope was snaked around the stone keeping the rape in position. The nut, above the pressing board, was turned down the screw squeezing the grape remains, the juice dripping through the rope to flow into the collecting vat (Frankel 1984: 194; 1999: 140).

3. Area 13 and the ancillary floors of Area 4, phase II: The process was similar to that described above but in place of the rope a square wooden frame was erected around the screw, in its square or rectangular base, the rape placed within and then pressed (Frankel 1984: 194; 1999: 140).

The two collecting vats found in area 13 allowed the press to be worked while the other vat was full with fermenting must.[10] This raised efficiency and also explains the need for multiple auxiliary floors as various phases of the pressing process could be conducted simultaneously. As these are most probably not single family wineries but rather industrial scale units the multiple floors also allowed the separation of varieties of grapes, or those belonging to various growers. Assuming the wine presses belonged to a single owner, the need for two adjacent installations can also be similarly explained, as while one waits for initial fermentation within the vats the main treading floors and the screw presses are effectively out of action. Two such units would allow continuous processing during the busy harvesting season.

Among the numerous examples of complex wine presses, only a few similar installations with auxiliary floors and compartments have been published. Contemporary examples were found at e-Tur (Bethphage; Saller 1961; Site 52), Bir el-Katt (Corbo 1955: 118–119), Kh. el-Mekhayyat (Saller 1941: 193–195), Emmaus (Hirschfeld 1983: 211–218), a so-called 'bath' found in the excavations of the Jerusalem YMCA (Iliffe 1935: 76–77), Qirya and the Valley of the Cross in Jerusalem (Rahmani 1991; Site 54), a yet unpublished press in Shu'afat (Site 51) and Kh. Jalame (Weinberg 1988: 11–15; Site 20).

The total volume of the collecting vats, excluding the capacity of the compartments, is 17.5 m^3. This allowed production on a scale that is obviously not of a family nature. Rather it shows operation, management and probably marketing on an industrial level. The large area around devoted to viticulture and the fact that other wine presses in the area could also process further quantities point to the existence of large vineyard estates.

After the area 13 wine press went out of use, probably in the 7th cent. CE, one auxiliary floor was renovated as noted above. As there was no longer the need to produce large quantities of wine, due to the spread of Islam, it is possible that small amounts of grapes were pressed to boil up to form *dibbes*, a grape syrup used as a sweetener. The boiling of the syrup may have been conducted in cauldrons set on the oven stones placed on the main treading floor in this phase.[11]

Discussion

Three clear phases were represented in the various areas, phase II being the central phase of the site, then in its fullest phase of activity. This activity is well evidenced by both the architecture and other finds.

Phase I

During this phase, that is dated by the *mikva'ot* to the end of the Second Temple period, Ras Abu Ma'aruf was a fortified farmhouse inhabited by Jews. It included, we suggest, the first phase of the wine press in area 4 (lever and weights press). Similarly to other farmhouses of this type (Qalandiya [Magen 1984; 2004; Site 46]; Shu'afat [Onn and Rapuano 1995; Site 51]) the site was probably abandoned with the fall of Jerusalem in 70 CE.

Phase II

Both the pottery and glass that are overwhelmingly dated to the 4th-7th cent. CE, and the typology of the presses place us in the Byzantine period. The site was in Jerusalem's agricultural hinterland. Wine and oil were produced here in industrial quantities for secular and probably ritual use for the city's inhabitants and the monks in the nearby monasteries at Ras et-Tawil (Gibson 1985–6; Site 47) and Deir Ghazali (Avner, Azmon and Savariego 1992; Avner 2000; Site 50). The site would have been abandoned or even burned, in common with contemporary installations, with the growing influence of Islam in the 7th cent. that drastically reduced the demand for wine (Hirschfeld 1983: 218; Frankel and Ayalon 1988: 76; Ayalon 1997).

Phase III

Limited quantities of finds and the renovated floor in area 13 relate to this phase. The site was probably used seasonally to operate the *dibbes* press, as no remains of more extensive inhabitation were present. Based on the material finds, the Umayyad occupation of the site does not seem to extend beyond the mid-8th-early 9th cent.

Notes

1. The excavation (Permit 85/90), conducted during the summer of 1990,

was directed by the author with the assistance of May Gordon under the auspices of the Israel Antiquities Authority.

2. Screw and lever presses are common in this period in all areas except Judea where this stage in oil press development is usually skipped and only the next technological improvement on the lever and weight press – the direct frame press – is found (Frankel 1996).

3. Screw base weight with socket and two external dovetail mortices (the Samaria type – 6.2.1.1, 56 parallels given: Frankel 1984, I: 156, II: 52–53; 1996: 111–113). This type is especially common in the Sharon and Samaria as well as in many foreign countries. The type probably originates in Italy, where it is found from the 1st cent. BCE, the earliest appearance in Israel being the Byzantine period (Ibid.: 172–173).

4. Regionality in oil press development is fully treated by Frankel (1984; 1987; 1996; 1999; Frankel, Patrich and Tsafrir 1990).

5. Frankel 1984: 191, beam lever press, type 8.5; 1999: 145–146, Ts8.5, 12 parallels given.

6. Frankel 1984: 187–188; 1999: 141–143, hewn, round base stone for single fixed screw press, square mortice widening in one/two direction/s, type 8.1.5.1/2, Ts81–1/2. This type of installation with a round or square base stone has a distribution centered in Samaria.

7. Frankel 1984: 183; 1999: 139, auxiliary treading floors with collecting vats forming compartments of the main floor, type 8.0.1.2, T803, five parallels given. Rahmani notes that these compartments do not usually face west to shield the contents from the west wind (Rahmani 1991: 105), that from personal experience blows strongly every day from around eleven o'clock. It must be stated, however, that two of our eight compartments indeed face westward making this supposition erroneous, at least at Ras Abu Ma'aruf. Two others face north-west.

8. Frankel 1984: 189; 1999: 144–145, rectangular stone base for a single fixed screw press with a dovetail mortice, type 8.3, T831, 60 parallels given. This installation type characterizes the Upper Galilee, though a concentration (6) is found in the Jerusalem area. Floor B at e-Tur (Bethphage) is the only parallel known of an auxiliary floor with a base for a single fixed screw press (Saller 1961: 198; Site 52).

9. Frankel 1984: 187–188; 1999: 142, hewn, square stone base for a single fixed screw press, square mortice widening in one direction, type 8.1.1.1, Ts81–1.

10. The two parallel vats, as in area 13, are part of an architectural plan called the 'four rectangle plan', type 9.1, T91, of which ours is a sub-type 9.1.3, T913 (Frankel 1999: 149). In area 4 the settling tank and the collecting vat are on a single axis, our wine press following sub-type 9.6.2, T962 (Frankel 1984: 200; 1999: 152–153).

11. See the similar suggestion for the Qirya and Valley of the Cross installations (Rahmani 1991: 106; Site 54).

Bibliography

Avner, R. 2000. "Deir Ghazali: A Byzantine Monastery Northeast of Jerusalem". *'Atiqot* 40: 25*-52* (Hebrew).

Avner, R., Azmon, S. and Savariego, A. 1992. "Jerusalem, Pisgat Zeev (Deir Ghazala)". *ESI* 10: 128–130.

Ayalon, E. 1997. "The End of the Ancient Wine Production in the Central Coastal Plain". In: Y. Friedman, Z. Safrai and J. Schwartz (eds.). *Hikrei Eretz, Studies in the History of the Land of Israel Dedicated to Prof. Yehuda Feliks*. Ramat Gan: 149–166 (Hebrew).

Corbo, V. 1955. *Gli Scavi di Kh. Siyar el-Ghanam*. Jerusalem.

Elizur, Y. 1986. "Oil Press Caves – "Oil Press Built into the Rock"". *NZ* 13: 179–190 (Hebrew).

Frankel, R. 1984. *The History of the Processing of Oil and Wine in Galilee in the Period of the Bible, the Mishna and the Talmud*. Unpublished Ph.D. Thesis. Tel Aviv University (Hebrew).

Frankel, R. 1992. "Some Oil Presses from Western Galilee". *BASOR* 286: 39–71.

Frankel, R. 1996. "Oil Presses in Western Galilee and Judaea - A Comparison". In: D. Eitam and M. Heltzer (eds). *Olive Oil in Antiquity, Israel and Neighbouring Countries from the Neolithic to the Early Arab Period*. Padova: 197–218.

Frankel, R. 1999. *Wine and Oil Production in Antiquity in Israel and Other Mediterranean Countries*. Sheffield.

Frankel, R. and Ayalon, E. 1988. *Vines, Wine Presses and Wine in Antiquity*. Tel Aviv (Hebrew).

Frankel, R., Patrich, J. and Tsafrir, Y. 1990. "The Oil Press at Horvath Beit Loya". In: G.C. Bottini, L. Di Segni and E. Alliata (eds.). *Christian Archaeology in the Holy Land. New Discoveries*. Jerusalem: 287–300.

Gibson, S. 1985–6. "Ras et-Tawil: Byzantine Monastery North of Jerusalem". *BAIAS* : 69–73.

Hirschfeld, Y. 1983. "Ancient Wine Presses in the Park of Aijalon". *IEJ* 33: 207–218.

Iliffe, J.H. 1935. "Cemeteries and a 'Monastery' at the YMCA, Jerusalem". *QDAP* 4: 70–80.

Magen, Y. 1984. "Kalandia – A Vineyard Farm and Winery of Second Temple Times". *Qadmoniot* 17 (66–67): 61–71. (Hebrew).

Magen, Y. 2004. "Kalandia - A Second Temple-period Viticulture and Wine-manufacturing Agricultural Settlement". In: Y. Magen *et al. The Land of Benjamin*. Jerusalem: 29–144.

Onn, A. and Rapuano, Y. 1995. "Jerusalem, Kh. er-Ras". *ESI* 13: 71.

Rahmani, L.Y. 1991. "Two Byzantine Winepresses in Jerusalem". *'Atiqot* 20: 95–110.

Roll, I. and Ayalon, E. 1981. "Two Large Wine Presses in the Red Soil Regions of Israel". *PEQ* 130: 111–125.

Saller, S. 1941. *The Memorial of Moses on Mount Nebo*. Jerusalem.

Saller, S. 1961. "The Archaeological Setting of the Shrine of Bethphage". *LA* 11: 194–208.

Seligman, J. 1994. "Jerusalem, Pisgat Ze'ev (East 'A)". *ESI* 12: 52–54.

Seligman, J. 1999. "Agricultural Complexes at Ras Abu Ma'aruf (Pisgat Ze'ev East 'A)". *'Atiqot* 38: 137–170.

Shukron, E. and Savariego, A. 1994. "Jerusalem, Pisgat

Ze'ev (Villa Quarter) ". *ESI* 12: 56–58.

Weinberg, S.S. 1988. "The Buildings and Installations". In: G. Davidson Weinberg (ed.). *Excavations at Jalame, Site of a Glass Factory in Late Roman Palestine*. Columbia: 5–23.

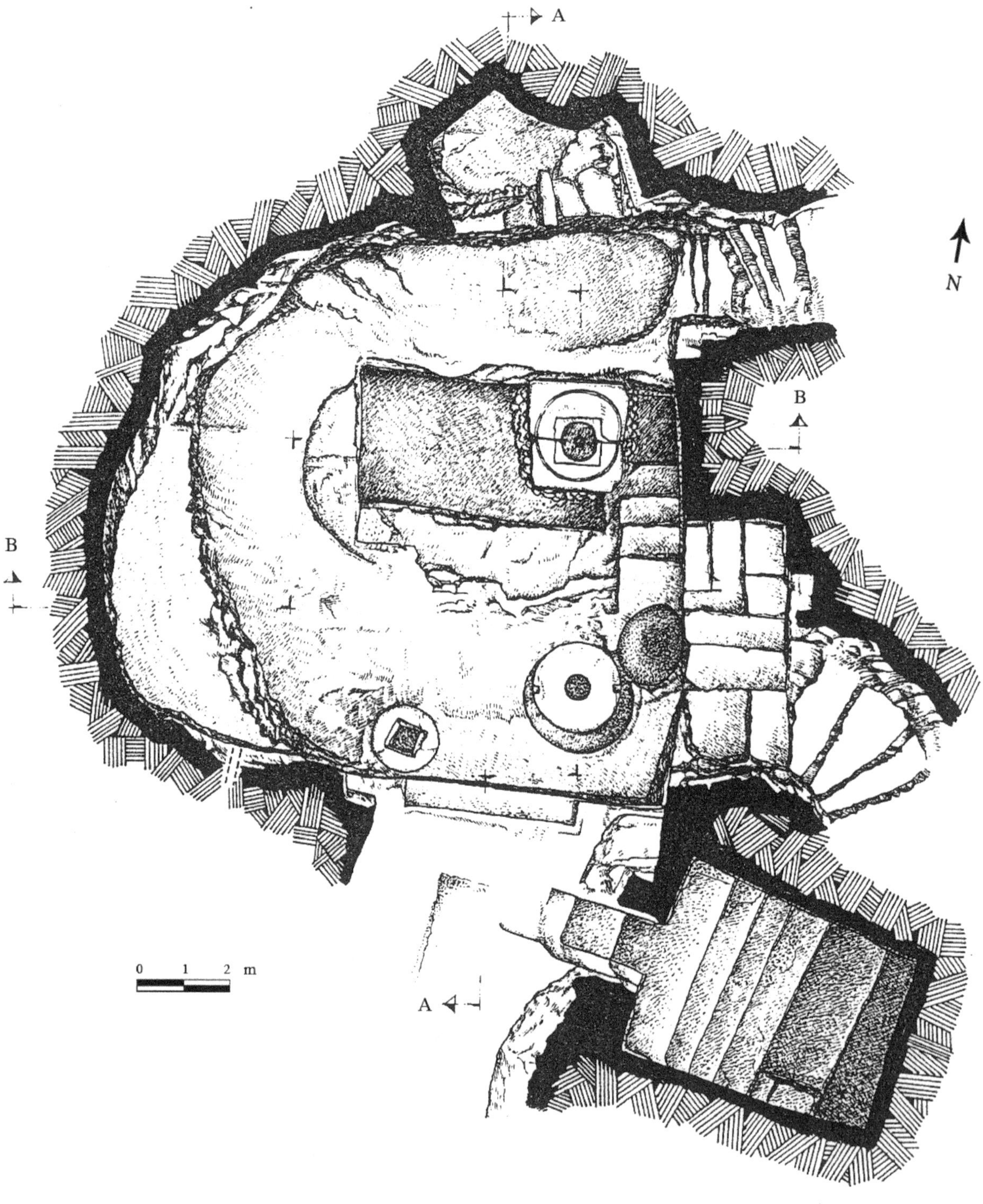

Fig. 48.1. Ras Abu Ma'aruf, plan of the oil press.

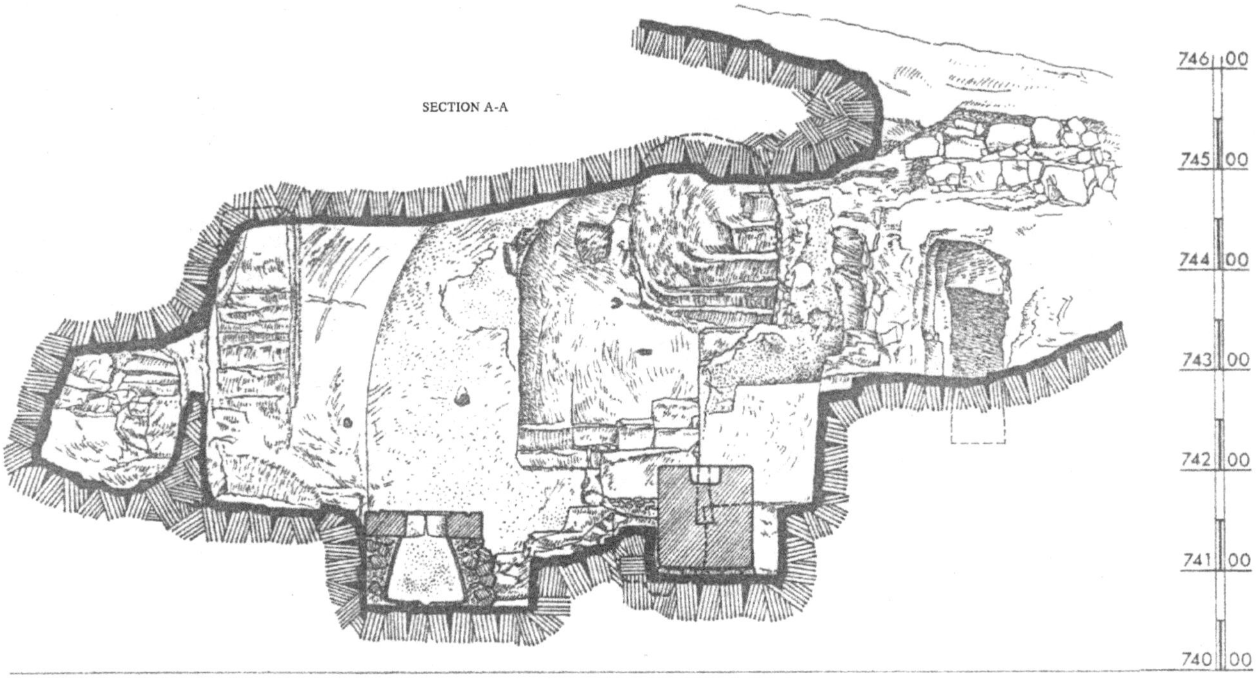

Fig. 48.2. Ras Abu Ma'aruf, section of the oil press.

Fig. 48.3. Ras Abu Ma'aruf, the oil press, view to the east.

Fig. 48.4. Ras Abu Ma'aruf, lever and screw press, view to the north.

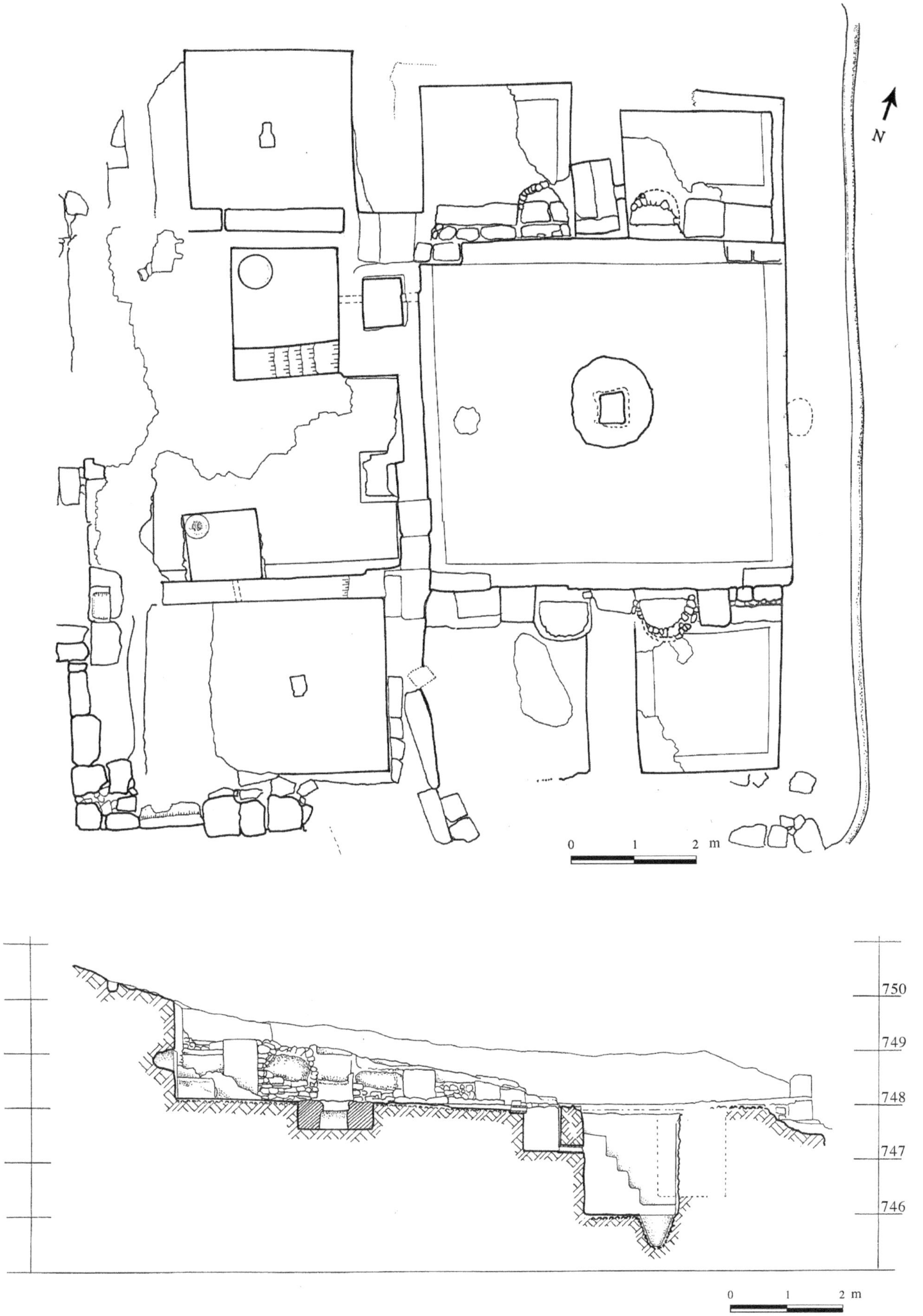

Fig. 48.5. Ras Abu Ma'aruf, plan and section of area 4 wine press.

Fig. 48.6. Ras Abu Ma'aruf, area 4 wine press, view to the east.

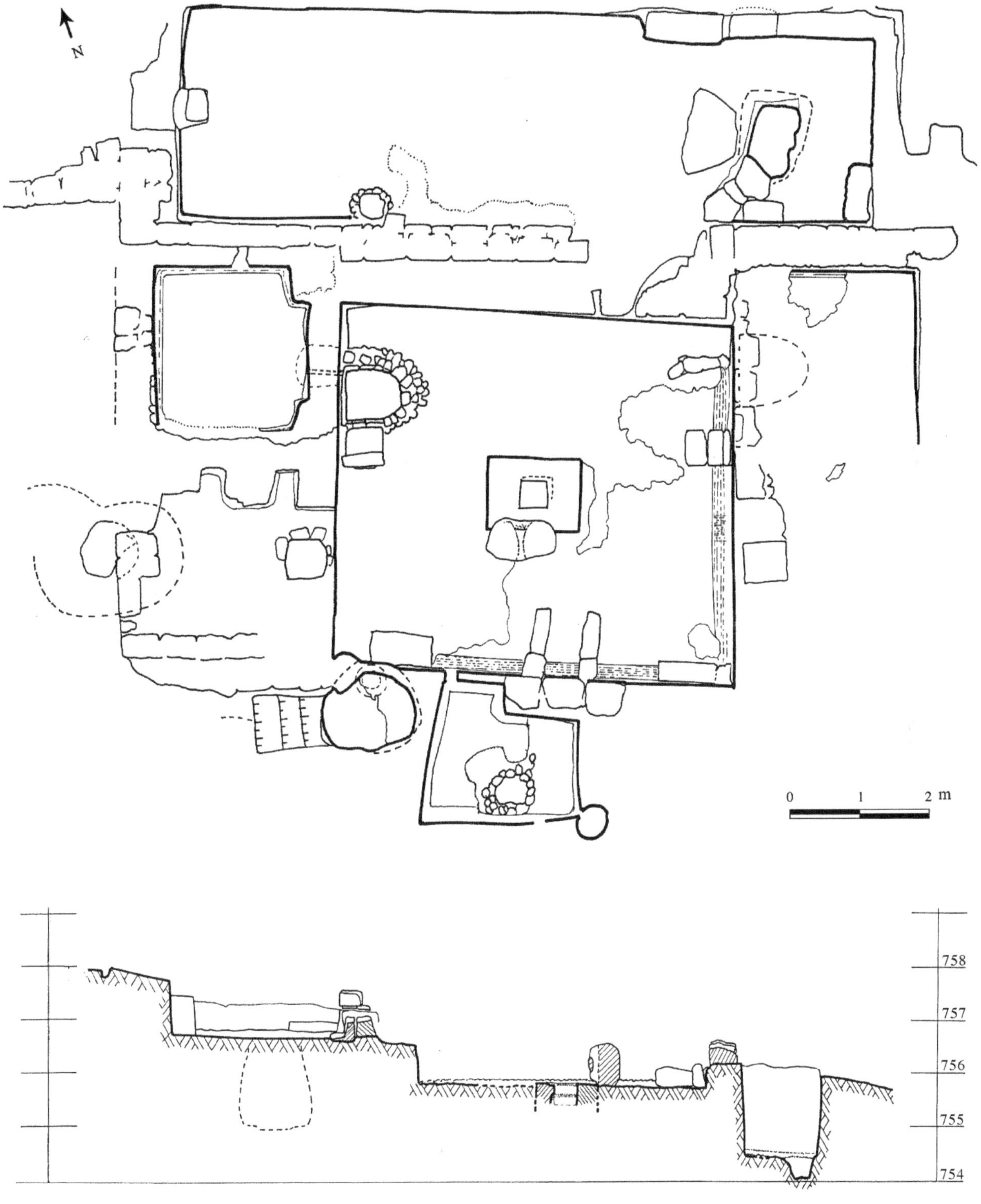

Fig. 48.7. Ras Abu Ma'aruf, plan and section of area 13 wine press.

Fig. 48.8. Ras Abu Ma'aruf, area 13 wine press, view to the south.

Fig. 48.9. Ras Abu Ma'aruf, auxiliary pressing floor and collecting vat, view to the west.

49. Hulda[1]

Amos Kloner

An installation with a mosaic floor was excavated at Hulda in 1953 (MR 140/136; Fig. 1). Because of the Jewish ritual items and Greek inscriptions depicted in the floor the building was interpreted as a synagogue (Avi-Yonah 1960: 57–60). Later studies have shown this was a Byzantine wine press constructed and used by Jews or Samaritans and the ritual items represented in the floor indicated the wine was being produced in accordance with Jewish religious customs.

The structure is rectangular with an estimated outer length north-south of at least 13.6 m and width east-west of 8.6 m. The interior dimensions are 10.4×5.5 m; the walls were c. 1.6 m thick (Fig. 2).[2] The entrance to the building is in the south wall; a parallel opening is presumed to have been located in the 0.3 m wide dividing wall between the southern and northern parts of the building. Only a few interior sections of the walls around the mosaic floor and some linear traces of walls remain; none of the outer faces of the walls were preserved.

The northern area which served as the treading floor measures 5.76 m east-west and 5.53 m north-south (area c. 32 m^2). The mosaic floor is made of large white tesserae in a design of three nested squares, the outlines formed by darker tesserae. The central area of the treading floor is destroyed and only a gaping hole remains, presumably the original location of the screw press used to extract the last juice from the rape.

Six compartments each measuring 1.0×0.6 m, 0.5–0.6 m deep and spaced up to 1.5 m apart were built into the walls of the northern area (Fig. 3). The compartment walls are plastered and the floors paved with tesserae. In one of the corners of each compartment facing the center of the building is a small shallow bowl-shaped depression for collecting liquid from the floor. The partitions between the compartments and the central treading floor are thin and built of small stones that were added after the main wall around the floor, made of much larger stones, had been built. Only the stones on the inner side of the eastern wall remain; those of the outer wall are robbed out. Most of the stones of the northern and western walls were also robbed out; the compartments can be identified by their preserved mosaic floors.

The southern area measures 5.61 m east-west and 4.92 m north-south and is also paved with white tesserae. A square vat 1.2×1.2 m and 0.79 m deep paved with tesserae was built into the floor in its north-eastern part. It has a small shallow bowl-shaped recess in the center. Its walls are covered with high-quality plaster tempered with many potsherds, most of them from Late Roman and Byzantine amphorae.

A slightly inclined lead pipe 0.46 m long with an internal diameter of 5.4 cm connects the square vat and the round one (below). The square vat served as a settling basin; it appears that the channel leading from it to the collecting vat served as part of a filtering process that ensured that solid ingredients were removed from the liquid.

A round collecting vat 2.48 m in diameter and c. 1.34 m deep (volume c. 6.4 m^3) is located west of the square basin (Fig. 4). Three small semicircular steps, the upper two 0.23 m high, lead into the vat at its eastern part. The floor of the pool is paved with a mosaic and its walls are plastered with the same material as the small vat. Since the plaster and potsherds are integral to the installation, these features date the wine press to no later than the 5th century CE.

Two ornamental panels decorate the mosaic floor south of the vats. One, a rectangle measuring 0.92×0.50 m south of the round vat, is composed of small colored tesserae depicting a seven-branched *menorah*, a *shofar*, an incense shovel, an *etrog* and a *lulav* with a three-line Greek inscription between the *lulav* and *menorah* (Fig. 5):

Blessing to the people

A second inscription of six lines in a square within a lopsided round medallion c. 1.17 m in diameter is located east of the first panel:

Good luck to

Eustochios

and Hesychios

and Evagrios

the founders

The inscription is in black tesserae and the donors' names in red tesserae.

Wine press mosaics were usually composed only of white tessarae, but a few exceptions such as the installation at H. Massah near Kefar Tabor (*HA* 65–66 [1978]: 5) are known. The use of symbols known mainly from synagogue ornamentation in the treading floor of the Hulda wine press is a unique phenomenon; the inscriptions also attest to the special status of this installation.

The closest parallels for the winery at Hulda are Late Roman and mainly Byzantine wine presses found in various parts of Israel. These can be divided into two types:

1. Wine presses consisting of 4–6 m^2 enclosed surfaces usually paved with stone slabs (a few have mosaics), such as those from the Negev (Mazor 1981; Site 70). Installations with recesses for collecting grape juice at the sides of the central treading floor were also found at Emmaus (Hirschfeld 1983), Mt. Nebo (Saller 1941, I: 193–194, II Pl. 77: 2), on the eastern slope of the Mt. of Olives and in the two wine presses in the Qirya compound and the nearby Valley of the Cross (Rahmani 1991; Site 54). Two large wine presses of this type were recently published from Kh. Yajuz in Jordan (Khalil and al-Nammary 2000).

2. Wine presses with small cells 1–1.7 m long and 0.6–1.2 m wide and paved with mosaics, adjacent to the main treading floor and separated from it by a low partition.

The 6th century CE installation at Kh. Duran has six cells, two each on three sides of the treading floor (Roll and Ayalon 1981; Site 45). An installation at Qalandiya that has been described as a Byzantine bath (Baramki 1933; Site 46) is in fact a sophisticated wine press. The main treading Floor B was paved with stone slabs and its wall contains three niches similar to the cells at Hulda; two niches in Room E and the basins in Rooms C and E are also similar to those at Hulda. The wine press at Eshtamoa in the southern Hebron Mountains has similar small cells paved with mosaics (D. Amit, pers. comm.). These wine presses may be described in the Tosefta (*Oha.* 15.7a): "And what is the courtyard of the tomb? That is the floor (הגת) to the midst of which the caves open". Note that the Hebrew word הגת , translated here as "the floor" means actually "the wine press". It is possible that the cells were connected to smaller treading floors that would have been used in the first phase of the manufacture of the wine, when the liquid expressed from the grapes was left with the rape in order to obtain a redder color and to begin the fermentation process (Dray 2003; pers. comm.). No evidence supporting this theory is reported from the above-mentioned sites nor is there any indication of such at Hulda.[3] A close examination of the cells at Hulda also does not reveal any connection that would enable liquids to flow from the cells to the central treading floor.

Grape clusters were piled in the cells prior to their pressing, to the tops of the walls. The cells probably separated produce from different vineyard owners rather than different grape varieties. The grape clusters were either tipped out of wicker baskets or the full baskets were stacked in the cells. The clusters were transferred to the treading floor either with a shovel-like tool or large pottery jar fragments. Grapes were no doubt crushed in the process and the resulting juice ran into the small bowl-like depression at the bottom of each cell.

The must probably flowed down the gentle slope to the square vat, somehow passing through the partition between the two floors to clarify the liquid, an important stage in wine production, in the course of which solid residues remained on the treading floor.

To keep spoilage to a minimum the grapes had to be processed as quickly as possible. The main factor affecting efficiency in the production process would have been the maximum exploitation of the treading floor, in which a layer of grape clusters some 0.2 m high spread evenly over the floor would be crushed underfoot. The partition around the treading floor had to be low enough to allow for a relatively rapid transfer of the clusters from the cells onto the floor. It is reasonable to assume that the outer partition of the cells was also low enough for the convenient unloading of the grapes brought from the vineyard.

Wine production would have proceeded as follows:

- The grape clusters were collected from the vineyard into the cells;
- They were spread evenly over the central treading floor;
- The grapes were trodden underfoot;
- The liquid and solids were left on the treading floor for two days, beginning the fermentation process and the wine acquiring its red color;
- The liquids were diverted into the settling and collecting vats;
- The rape remaining on the treading floor was pressed in the screw press;
- The solid residual material was cleared and deposited outside the building;
- The clarified liquid was removed from the collecting vat into storage jars.

The wine produced at the Hulda wine press was either made by Jews or Samaritans, or at least under very strict Jewish supervision, an important aspect of Jewish religious customs and practices of the period. The Mishnah stresses the prohibition against using wine made by gentiles. Tractate *Av. Zar.* 4.8–12 enumerates the rulings in this regard, including activities such as treading and measuring done at the wine press. It appears that the sages intended their strictures to prevent the involvement of, among others, practitioners of foreign cults. This subject is also discussed in the Jerusalem Talmud (*Av. Zar*. 2.4, 41b; 5.4, 44d, among others) and is evidence of the great importance attached to this matter by the Amoraim. The wine press at Hulda dated to the 5th century CE provides significant archaeological evidence for the careful observance by the Jewish or Samaritan population of the rulings connected with the production of ritually pure wine at this time.

Notes

1. This article is a condensed version of Kloner 2006. The original article was published in Hebrew (Kloner 1986).
2. Avi-Yonah (1970: 106) gives different dimensions — 12.2×7.0 m — and gives the length of the building as 12.7 m. Huttenmeister (Huttenmeister and Reeg 1977, 1: 177) uses Avi-Yonah's dimensions. Cf. also Chiat 1982: 209–210, citing Avi-Yonah. The dimensions presented here are based on new measurements taken at the site by the author and students in the mid-1980s. The plan is the one published by Avi-Yonah. Figs.1, 3–8 are courtesy of the Israel Antiquities Authority.
3. Cf. also Conder and Kitchener 1881–1883, II: 330 for one of many examples from Samaria, as well as Ibid.: 278, "Judaea: Khirbet el'Arab in the Marissa Region of the Shefelah" described as a "pile of stones". The "*List of Monuments and Historical Sites*" (*Yalqut HaPirsumim* 1091, 1964: 1474 [Hebrew]) indicates that the latter site contains "remains of a prayer niche with three apses". This is actually a wine press with three small cells of the type found at Hulda.

Bibliography

Avi-Yonah, M. 1960. "Hulda". In: *Louis M. Rabinowitz Fund Bulletin* 3: 57–60.

Avi-Yonah, M. 1970. "Synagogues: Hulda". In: B. Mazar (ed.). *The Encyclopedia of Archaeological Excavations in Eretz-Israel*: 106 (Hebrew).

Baramki, D.C. 1933. "A Byzantine Bath at Qalandia". *QDAP* 2: 105–109.

Chiat, M.J.S. 1982. *Handbook of Synagogue Architecture* (Brown Judaic Studies 29). Chico, California.

Conder, C.R. and Kitchener, H.H. 1881–1883. *The Survey of Western Palestine*. 3 Vols. London.

Danby, H. 1933. *The Mishnah*. London.

Dray, Y. 2003. "Wine Production in the Improved Byzantine Wine Press". In: Y. Eshel (ed.). *Judea and Samaria Research Studies* 12: 219–228 (Hebrew).

Frankel, R. 1999. *Wine and Oil Production in Antiquity in Israel and Other Mediterranean Countries*. Sheffield.

HA 65–66 (1978): 5–6: "H. Massah" (Hebrew).

Hirschfeld, Y. 1983. "Ancient Wine Presses in the Park of Aijalon". *IEJ* 33: 207–218.

Huttenmeister, F. and Reeg, G. 1977. *Die Antiken Synagogen in Israel* I. Weisbaden.

Khalil, L.A. and al-Nammary, F.M. 2000. "Two Large Wine Presses at Khirbet Yajuz, Jordan". *BASOR* 318: 41–57.

Kloner, A. 1986. "The Structure at Hulda and Its Use as a Wine Press". In: A. Kasher, A. Oppenheimer and U. Rappaport (eds.). *Man and Land in Eretz-Israel in Antiquity*. Jerusalem: 197–208 (Hebrew).

Kloner, A. 2006. "The Structure and Installations at Hulda and Their Function as a Jewish Winepress". In: E.M. Maeir and P. de Miroschedji (eds.). *"I Will Speak the Riddle of Ancient Times", Archaeological and Historical Studies in Honor of Amihai Mazar*. Winona Lake: 853–864.

Mazor, G. 1981. "The Wine-Presses of the Negev". *Qadmoniot* 14: 51–60 (Hebrew).

Mishnah (See Danby 1933).

Rahmani, L.Y. 1991. "Two Byzantine Winepresses in Jerusalem". *'Atiqot* 20: 95–110.

Roll, I. and Ayalon, E. 1981. "Two Large Wine Presses in the Red Soil Regions of Israel". *PEQ* 113: 109–125.

Saller, S.J. 1941. *The Memorial of Moses on Mount Nebo*. Jerusalem.

Saller, S.J. 1961. "The Archaeological Setting of the Shrine of Bethphage". *LA* 11: 194–208.

Tosefta (trans. J. Neusner). Hoboken, N.J., 1977.

Yalqut HaPirsumim ("List of Monuments and Historical Sites") 1091, Jerusalem 1964 (Hebrew).

Fig. 49.1. Hulda, the structure with the mosaic floor, looking south.

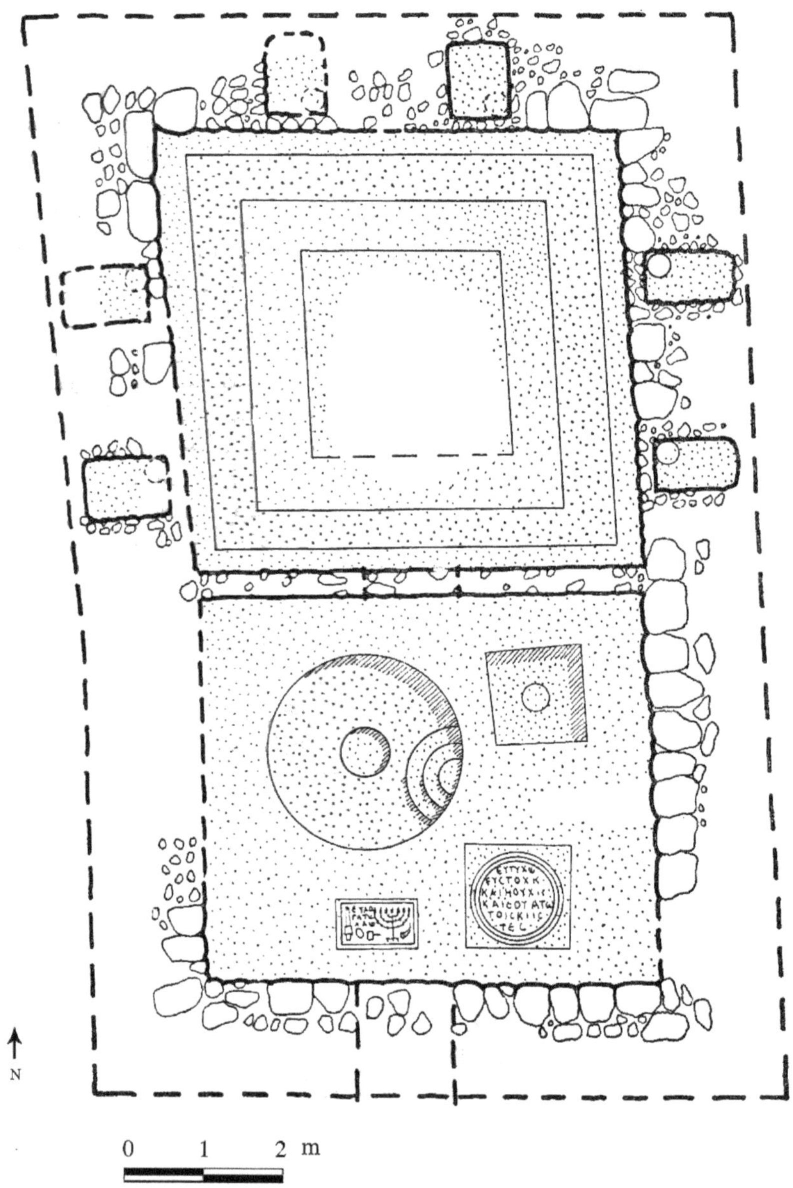

Fig. 49.2. Hulda, plan of the structure and mosaic floor.

Fig. 49.3. Hulda, cell for storing grapes in the eastern wall, looking west.

Fig. 49.4. Hulda, round collecting vat with steps, looking north-east.

Fig. 49.5. Hulda, western mosaic panel, looking north.

50. Deir Ghazali (Pisgat Ze'ev) – Jerusalem

Rina Avner

The Site

Pisgat Ze'ev, a suburb of modern Jerusalem, lies approximately 5 km to the north-east of ancient Jerusalem. The Arabic name of the site "Deir Ghazali" (Monastery of the Gazelles) and the crosses that decorate the installations (see below) hint at its being the remains of a monastery although it could also be a *villa rustica*. The site (MR 1739/1363) is located on a moderate eastern slope overlooking the Judean Desert, on a junction of three local ancient roads. The site and the roads were discovered in an archaeological survey carried out by S. Gibson in 1985 (Gibson and Edelstein 1985: 150). The monastery was excavated in 1990 (Avner 2000).[1]

Two kinds of agricultural installations were found in the monastery complex (Fig. 1): two wine presses in the southern section (rooms 9–10) and an oil press in the northern section (room 4).

The Wine Presses

Both wine presses are in the "four rectangle plan" as defined by Frankel (1999: 149–150), each consisting of a rectangular treading floor, a rectangular settling vat and a rectangular collecting vat so arranged as to fit into a fourth rectangle. Each of these units was paved with white mosaics although there is a difference in the stone size and density. Pillar bases along the walls and in the centre of the rooms prove that they were roofed.

The treading floor of the wine press in room 10 is not preserved but the channel shows it was in its southern part, while the vats are in the north-western section. The rectangular settling vat (1.1×0.5 m, depth 0.38 m) has a hole (diameter 5 cm) in the western wall that continues as a channel under the floor and leads to the collecting vat (1.75×1.50 m, depth 1.38 m, volume 3.6 m^3).[2] In its south-western corner is a sump in the floor and in the north-eastern corner some recesses in the wall allowed descent into the vat. Both vats are floored with white mosaic (100 stones per square decimetre).

The wine press in room 9 (Fig. 2) is the larger of the two and also the more complex. In the western wall (W41) there are two small semi-oval compartments (0.8×0.7 m, depth 0.8 m; 0.85×0.70 m, depth 0.85 m) the floors of which are paved with white mosaic (density: 100 stones per square decimetre) and the walls covered with grey plaster. Remains of a lead pipe were found that connected the northern compartment to the treading floor. The compartments apparently served to produce "first must" that dripped from the grapes left in the sun before treading. The treading floor (6.4×5.6 m, area 35.8 m^2) is paved with mosaic (density: 4.5–7 stones per square decimetre). The settling vat (0.6–0.9×0.7–0.8 m, depth 0.48 m) is located near the north-western corner of the treading floor. Its floor is white mosaic (100 stones per square decimetre) and the walls are coated with gray plaster. A channel built of stones connected the settling vat to the collecting vat (2.4×2.2 m, depth 1.22 m, volume 6.4 m^3) that is to its north-west. A sump (diameter 0.5 m, depth 5 cm) is in the mosaic floor (100 stones per square decimetre) in the north-western corner of the vat. Small recesses in the north-eastern corner allowed descent into the vat. On the rim of the northern, eastern and southern sides of the collecting vat there are small depressions in the mosaic pavement that drain into the vat via channels. Storage jars were probably placed in the depressions for filling and must that spilt drained back into the vat.

North of the treading floor a *kasfa* screw weight (diameter 0.6 m, height 0.74 m; see Frankel 1999: 114) stood in a rectangular pit, showing that there had been a lever and screw press here. The niche in which the beam was anchored would have been in the southern wall but has not survived. This installation was used to press the rape that remained after treading from both wine presses. A door in W15 connected the two wine presses allowing the rape to be moved from one to the other for pressing.

In the centre of the treading floor the tesserae formed a medallion containing a cross. The southern wall of the treading floor in room 10 (W44) and the northern and southern walls of the collecting vat were decorated with plastered crosses of different types. In the cross on W44 the arms were shorter than the foot while in the two in the collection vat the four arms were equal, in the northern triangular.

The Oil Press

The oil press in room 4 was severely damaged when the monastery was destroyed in the 7th cent. (Fig. 3). The room is floored with white mosaic (9 stones per square decimeter). A crushing basin was found *in situ* in its centre. A crushing stone was used to block the passage between two piers and another was found close to the north-western corner of the room. The collecting vat (diameter 0.65 m, depth 0.55 m, volume 176 litres) is preserved intact in the eastern side of the room. It has a stone cover equipped with a sunken shelf around the opening for a wooden perforated board, on which the frails were laid. Another broken stone vat was found in secondary use in the north-eastern corner. Two typical grooved piers of a "Judean press" were incorporated into the eastern wall of the room (W13) to thicken it. About 1 m west of the collecting vat there is a stone weight, the bore of which was in the form of a reversed T. Near it was found, not *in situ*, a stone weight of the Samaria type (Frankel 1999: 111–113) for a lever and screw press.

An unusual stone was found close to the north-western pier. It is cylindrical in form, has one lateral channel and three drilled holes on one of the round faces (Fig. 4). No parallels were found for this stone nor have we any suggestions as to its function (see editors' note below). A cross decorates one of the stones in wall W13.

Dating

These agricultural installations belong to the first phase of the monastery. In the second phase they went out of use and partly dismantled or built upon. Both wine presses were turned into cisterns. According to the finds the complex was established in the late 5th or 6th cent. and the first phase came to its end in the late 6th or 7th cent. CE.

Notes

1. The excavation was carried out by the author assisted by A. Savariego and S. Asmon on behalf of the Israel Antiquities Authority.
2. The depth is such according to the published section (Avner 2000 Plan 1); while the text defined it by mistake as 3.38 m.

Editors' Note

The stone with three holes is probably a beam weight of a lever and weights press and the holes are probably for metal clamps to which the rope was attached. Screw weights with holes for clamps are found both in southern France and Anatolia (Frankel 1999: 119; Ahmet 2001: Fig. 8) and one example from this country was found at 'Ain el-Jedide (Site 55). In Anatolia, however, there are also beam weights with holes for clamps (Balut 2005: 209, Fig. 16).

Bibliography

Ahmet, K. 2001. "A Middle Byzantine Olive Press Room at Aphrodisias". *Anatolian Studies* 51: 159–167.

Avner, R. 2000. "Deir Ghazali: A Byzantine Monastery Northeast of Jerusalem". *'Atiqot* 40: 25*-52* (Hebrew).

Balut, S. 2005. "Two Unique Olive Oil Presses from the Border Region of Lycia-Pamphylia-Pisidia". *Adalya* 8: 191–219 (Turkish with English summary).

Frankel, R. 1999. *Wine and Oil Production in Antiquity in Israel and Other Mediterranean Countries*. Sheffield.

Gibson, S. and Edelstein, G. 1985. "Investigating Jerusalem's Rural Landscape". *Levant* 17: 139–155.

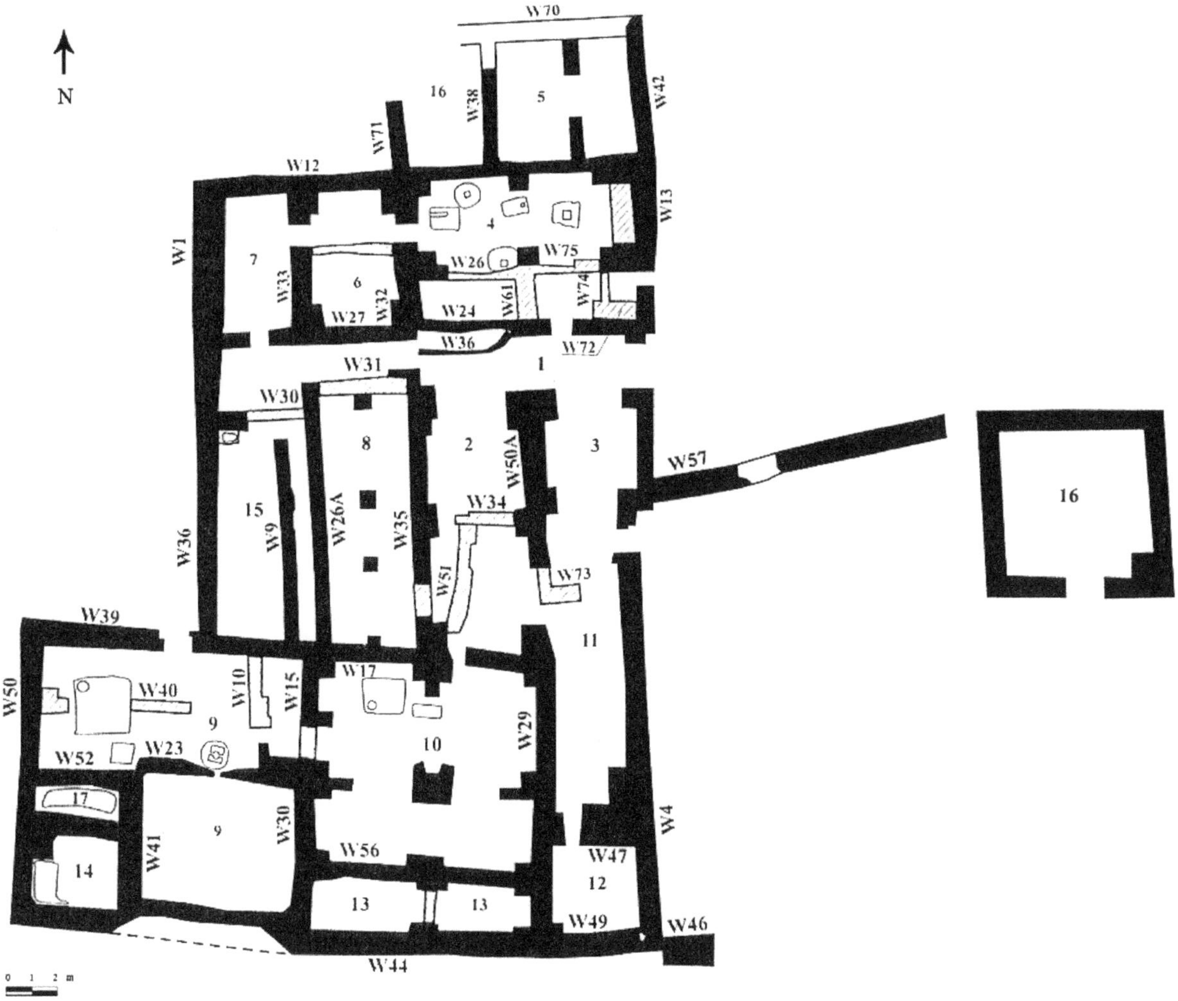

Fig. 50.1. Deir Ghazali, general plan of the monastery.

Fig. 50.2. Deir Ghazali, wine press in Room 9, looking south-east.

Fig. 50.3. Deir Ghazali, oil press in Room 4, looking east.

Fig. 50.4. Deir Ghazali, cylindrical stone with three holes and a lateral channel.

51. Byzantine Wine and Oil Presses at Shu'afat

Jon Seligman

The Site

The site at Shu'afat, known as Kh. er-Ras (MR 171/135), is located 3.5 km north-west of the Damascus Gate of Jerusalem on a ridge between Nahal Sorek (Sorek Stream) and Nahal Tzofim (Tzofim Stream). Excavations were conducted here in 1991 by A. Onn in a large number of sub-sites of various periods, prior to development[1] (Onn and Rapuano 1995). The agricultural installations exposed at Shu'afat fit into the ever expanding picture of the agricultural hinterland of Jerusalem in general and in the Byzantine period in particular. Contemporary sites around the city almost inevitably uncover wine and/or oil presses, showing the central position these two commodities held, both nutritionally and economically.[2]

The presses exposed here were found within a *villa rustica* dated to the Late Roman and Byzantine periods (Fig. 1). The villa was of the atrium type with the oil press located south of the entrance and the wine press flanking the entire western side of the courtyard. The site shows a number of phases and the presses seem to be more clearly associated with the latter phases. The building and its installations were abandoned around the 6th cent. CE.

The Wine Press (Fig. 2)

The wine press is of the complex type with raised auxiliary floors flanking the central treading floor that measures 6.7×7.7 m (51.6 m^2). It had been hewn from the bed-rock. Four layers of superimposed mosaic floors had been laid showing a long period of use and the constant need to reseal the floor. The mosaic was of the white industrial type (25 tesserae to 10 cm^2) and was laid both parallel and at a diagonal to the walls of the press at different phases.

To the north and in the north-west corner were small rock-cut compartments that were higher than all, but the final mosaic. Their use will be discussed below. In the centre of the treading floor was an almost square depression (1.80×1.95 m and 0.5 m deep) in which had been installed a large square stone block (Fig. 3). This was the base for a single fixed screw press and thus a square mortice, widened to two sides, was cut in its centre.[3] Within the mortice was fixed the screw, held in position by wooden pegs.

Raised above and flanking the treading floor from the west were three auxiliary floors of which only small parts of the south-western one (2.4×3.3 m) survived and we can thus conclude that they all had been paved with plain white tesserae. Below each of these floors, on the edge of the main treading floor, were rock-cut bays in which semi-circular compartments had been constructed.[4] Based on comparison with other sites we can reconstruct them and suggest that they had vaulted superstructures. The bottom surface was again paved with a plain white mosaic. Ducts led from the floors above into the compartments and a gutter connected them to the main floor.[5]

A gap in the southern wall of the central treading floor led to an intermediate vat (1×1 m and a usable depth of 0.31 m). The pipe connecting it to the adjacent collecting vat was at the base of the tank, making its use as a settling vat impossible. The function of the tank could rather have been as a holding vat, to allow operation of the treading floor while the collecting vat was full and the mixing of the musts was not required. Another possibility is as a filtration tank where a plant such as Thorny Burnet (*Sacropoterium Spinosum*) was exploited as the filtering material to remove solids from the must. A ceramic pipe passed in turn into the collecting vat to the west.[6] This vat measures 1.8×2.2 m with a usable depth of 1.2 m and a volume of 4.8m^3. The floor was paved with a mosaic set square to the walls framing tesserae laid on the diagonal. On the southern wall of the vat, 0.6 m from the floor, was a niche. This was used as a step to enable exit from the vat when utilizing its bottom portion. In the north-eastern corner was a 0.4 m deep sump.

Operation of the Wine Press

After harvesting the grapes were left in the sun on the auxiliary floors for some three days. The static pressure of the bunches thus produced the "first must". The must poured through the ducts to collect in the vaulted compartments below. After fermentation this was drained separately directly into jars to produce a high quality *prototropum* or *mustum lixivium* sweet wine (Roll and Ayalon 1981: 119; Frankel and Ayalon 1988: 42; Frankel 1999: 42, 139–140).

The next stage was to shovel the grapes onto the main treading floor to be crushed under-foot. The must flowed into the intermediate vat to remove any solids and then into the collecting vat for initial fermentation of 3–4 days. Any solids still in the must sunk to be collected in the sump at the bottom of the vat. Once this finished the must was poured into jars for secondary fermentation, probably in the room to the south of the vat. This room was constructed simultaneously with the wine press and contained many broken jars.

The final stage was to collect the trodden rape and place them in a wooden frame constructed around the square stone base in the centre of the floor (Frankel 1984: 194; 1999: 140). As stated before a screw was fixed into the mortice. By turning a nut above the pressing board that was placed upon the grape remains, the lower quality wine (*mustum tortivum* or *circumsicium*) was produced from the must that dripped out to be collected in the vat (Frankel 1984: 194; 1999: 42). An additional very poor wine could be made by adding water to the pressed skins and then repressing the result.

In the small compartments were mixed and stored materials added to the wine during its processing, for example, marl used to help clear the must, neutralize its acidity and coagulate dirt (Frankel and Ayalon 1988: 20, 41; Frankel 1999: 42). Also utilized were flavoring additives such as water, persimmons, honey, scents, etc (Frankel 1984: 17–18; 1999: 42–43; Frankel and Ayalon 1988: 24).

Contemporary examples of complex wine presses of this type are not numerous but seem to centre in the Jerusalem and Shephelah area. They were found at e-Tur (Saller 1961), Bir el-Katt (Corbo 1955: 118–119), Qirya and the Valley of the Cross in Jerusalem (Rahmani 1991; Site 54) and two similar presses at Ras Abu-Ma'aruf (Seligman 1994; 1999: 144–148, 155–158, 162–164; Site 48). In addition three wine presses were found outside Jerusalem, at Kh. el-Mekhayyat (Saller 1941: 193–195), Emmaus (Hirschfeld 1983: 211–218) and Kh. Jalame (Weinberg 1988: 11–15; Site 20). The latter site is dated to around 350 CE, earlier then the other examples offered.

The Oil Press (Fig. 4)

Within a large room on the south-east corner of the building was constructed an oil press, cutting a mosaic dated to the original construction of the villa. This demanded structural changes. Thus the floor was cut in two places to allow the fixing of the various necessary installations and a massively built niche was positioned in the wall to anchor the beam. The lever and screw press type used here is not common in the southern part of Israel but this unit can be added to a growing group in the Jerusalem area.

The anchoring niche was built in the northern wall of the room, close to the corner. To withstand the huge pressures created by the screw on the pressing beam, the niche had to be massively constructed. Although little survives, the base of the niche is still extant and it consists of a heavy stone block in the eastern wall.

Cut into the bed-rock floor, 0.96 m below the niche, was a hole in which was ceramic basin in the shape of an inverted cone. At its base was a small sump. This was the collecting vat of the installation and it had a capacity of c. 110 litres. Above it was the press-bed consisting of a square stone slab with a hole in its centre and cut with one concentric and four radial channels that allowed the expressed liquid to flow into the vat below (Fig. 5). Around the edge of the central hole was an inset in which could be fitted a perforated board or slats.

Some 3.5 m to the south was hewn a round pit with a flat floor. The screw weight was located here (Fig. 6). This was no longer *in situ* but was found, broken into two and thrown outside the press. It had a diameter of 1.35 m and a height of 1 m. On its top surface is a round socket connected to two internal dovetail mortices. Dovetail tenons were wedged in these and then to a board that held the screw in position while allowing it to turn within the socket (Frankel, Patrich and Tsafrir 1990: 294).[7]

Operation of the Oil Press[8]

An essential feature used in the initial processing of the olive is missing. The crushing basin was not found and we do not know its former location. It was probably removed to be used elsewhere.

After crushing the olive pulp was placed in frails laid on the press-bed. A beam was anchored in the niche and placed above the pressing board laid upon the frails. Leverage was provided by turning the screw within a nut attached to the beam. The screw was held in position by the stone weight that lifted off the ground at a certain point. The oil and watery lees poured through a perforated board or slats to be collected in the vat below. After processing the unit was dismantled and the now separated floating oil was removed with a dipper juglet.

Regional Features[9]

In general the lever and screw press is a type common to Samaria and the north of the country (Frankel, Patrich and Tsafrir 1990: 292). Nevertheless, a number of similar installations have been found in Judea and Jerusalem, all dated from the Byzantine period. A specific feature worth noting is the central collecting vat that has a long regional history in the southern part of the country (Frankel 1996; 1999: 164–165; Frankel, Patrich and Tsafrir 1990: 292). Comparative presses were excavated at Kh. Siyar el-Ganam (Corbo 1955: 36–37), H. Beit Loya (Frankel, Patrich and Tsafrir 1990; Site 67), Ras Abu Ma'aruf (Seligman 1999: 149–154, 164–165; Site 48), Kh. el-Quseir (Frankel 1992: 49–59; Site 8) and probably French Hill, a site very close to Shu'afat (Frankel 1984: 346).

The screw weight with the round socket and two attached internal dovetail mortices seems to be a foreign introduction with a small local regional distribution. In Israel it is found only in southern Samaria and around Jerusalem, while many foreign examples are found in Asia Minor, southern France, northern Syria and at Sarepta (Frankel 1984: 56 and map 33; 1999: 114–115 and map 21; Frankel, Patrich and Tsafrir 1990: 294). Weights of this type were excavated at Qedumim (Magen 1993, Area F; Site 34) and French Hill (Oren 1970; 1971; Frankel 1984: 346).

Chronology

Both installations are clearly dated both by finds and comparative sites to the Byzantine period. The wine press, with its auxiliary floors and mosaic surfaces, is of a type that appears and disappears together with the Byzantines.[10] The other sites noted of this type are dated by the pottery to the 4th-7th cent. and this unit is no different.

While the oil press has a longer life span, it too characterizes the Byzantine period. Though installations like this one are found later then that time, none, in Israel, predate it. The other two excavated sites noted above with the same screw weights are both similarly dated.

Notes

1. I thank the excavator, Alexander Onn, for allowing me to publish these installations.
2. See Gibson and Edelstein 1985. Sites in the immediate area in which such installations were discovered include: Qalandiya (Baramki 1933; Site 46), 'Atara (Dalman 1928–1942, IV: 361), French Hill (Oren 1970; 1971), Ras et-Tawil (Gibson 1985–6; Site 47) and Pisgat Ze'ev – Deir Ghazali (Avner, Azmon and Savariego 1992; Avner 2000; Site 50) and Ras Abu Ma'aruf (Seligman 1994; 1999; Site 48).
3. Frankel 1984: 73, 187–188; 1999: 139, type 8.1.2.2 and T803, free square stone base for a single fixed screw press, square anchoring mortice widening in two directions. Five parallels of this type are given.
4. Frankel 1984: 68, 183; 1999: 144–145, type 8.0.1.2 and T831, auxiliary treading floors with collecting vats forming compartments of the main floor.
5. It has been noted that these compartments were placed on the west of the treading floor with their openings facing east, to protect the contents from prevailing westerly winds during fermentation (Rahmani 1990: 105).
6. This pattern, with the vat and settling tank flanking the treading floor in parallel, is known as the "four rectangle plan". It is a regional characteristic of the centre and south of Israel (Frankel 1984: 85, 196–197, type 9.1.1; 1999: 149, T91).
7. Frankel 1984: 56, 158–159; 1999: 114, T624, lever and screw press, screw base with a round socket connected to two internal dovetail mortices, seven parallels given.
8. See discussion in: Frankel 1984: 150–167; 1996, 1999: 107 ff.
9. Regional characteristics are comprehensively discussed by Frankel (1984: 204–217; 1996; 1999; Frankel, Patrich and Tsafrir 1990: 292–297).
10. Hirschfeld (1983: 216–218) dates all units with a single fixed screw base to the Byzantine period. It is clear that the reduced demand for wine following the Arab conquest caused the quick but gradual collapse of the wine industry and the closing of especially the larger presses (Ibid.: 218; Ayalon 1997).

Bibliography

Avner, R. 2000. "Deir Ghazali: A Byzantine Monastery Northeast of Jerusalem". *'Atiqot* 40: 25*-52* (Hebrew).

Avner, R., Azmon, S. and Savariego, A. 1992. "Jerusalem, Pisgat Ze'ev (Deir Ghazala) ". *ESI* 10: 128–130.

Ayalon, E. 1997. "The End of the Ancient Wine Production in the Central Coastal Plain". In: Y. Friedman, Z. Safrai and J. Schwartz (eds.). *Hikrei Eretz, Studies in the History of the Land of Israel Dedicated to Prof. Yehuda Feliks.* Ramat Gan: 149–166 (Hebrew).

Baramki, D.C. 1933. "A Byzantine Bath at Qalandia". *QDAP* 3: 105–109.

Corbo, V. 1955. *Gli Scavi di Kh. Siyar el-Ghanam.* Jerusalem.

Dalman, G. 1928–1942. *Arbeit und Sitte in Palästina.* 7 Vols. Güutersloh (repr. Hildesheim 1964).

Frankel, R. 1984. *The History of the Processing of Oil and Wine in Galilee in the Period of the Bible, the Mishna and the Talmud.* Unpublished Ph.D. Thesis, Tel Aviv University (Hebrew).

Frankel, R. 1992. "Some Oil Presses from Western Galilee". *BASOR* 286: 39–71.

Frankel, R. 1996. "Oil Presses in Western Galilee and Judaea - A Comparison". In: D. Eitam and M. Heltzer (eds). *Olive Oil in Antiquity, Israel and Neighbouring Countries from the Neolithic to the Early Arab Period.* Padova: 197–218.

Frankel, R. 1999. *Wine and Oil Production in Antiquity in Israel and Other Mediterranean Countries.* Sheffield.

Frankel, R. and Ayalon, E. 1988. *Vines, Winepresses and Wine in Antiquity.* Tel Aviv (Hebrew).

Frankel, R., Patrich, J. and Tsafrir, Y. 1990. "The Oil Press at Horvath Beit Loya". In: G.C. Bottini, L. Di Segni and E. Alliata (eds). *Christian Archaeology in the Holy Land. New Discoveries.* Jerusalem: 287–300.

Gibson, S. 1985–6. "Ras et-Tawil: Byzantine Monastery North of Jerusalem". *BAIAS* 5: 69–73.

Gibson, S. and Edelstein, G. 1985. "Investigating Jerusalem's Rural Enviroment". *Levant* 17: 139–155.

Hirschfeld, Y. 1983. "Ancient Wine Presses in the Park of Aijalon". *IEJ* 33: 207–218.

Magen, Y. 1993. "Qedumim – A Samaritan Site of the Roman-Byzantine Period". In: F. Manns and E. Alliata (eds.). *Early Christianity in Context, Monuments and Documents* (Studium Biblicum Franciscanum, Collectio Maior, 38). Jerusalem: 167–180.

Onn, A. and Rapuano, Y. 1995. "Jerusalem, Kh. er-Ras". *ESI* 13: 71.

Oren, E. 1970. "French Hill in Jerusalem". *HA* 34–35: 19 (Hebrew).

Oren, E. 1971. "French Hill in Jerusalem". *HA* 38: 16–19 (Hebrew).

Rahmani, L.Y. 1991. "Two Byzantine Winepresses in Jerusalem". *'Atiqot* 20: 95–110.

Roll, I. and Ayalon, E. 1981. "Two Large Wine Presses in the Red Soil Regions of Israel". *PEQ* 130: 111–125.

Saller, S. 1941. *The Memorial of Moses on Mount Nebo.* Jerusalem.

Saller, S. 1961. "The Archaeological Setting of the Shrine of Bethphage". *LA* 11: 194–208.

Seligman, J. 1994. "Jerusalem, Pisgat Ze'ev (East 'A)". *ESI* 12: 52–54.

Seligman, J. 1999. "Agricultural Complexes at Ras Abu Ma'aruf (Pisgat Ze'ev East 'A.)". *'Atiqot* 38: 137–170.

Weinberg, S.S. 1988. "The Buildings and Installations". In: G. Davidson Weinberg (ed.). *Excavations at Jalame, Site of a Glass Factory in Late Roman Palestine.* Columbia: 5–23.

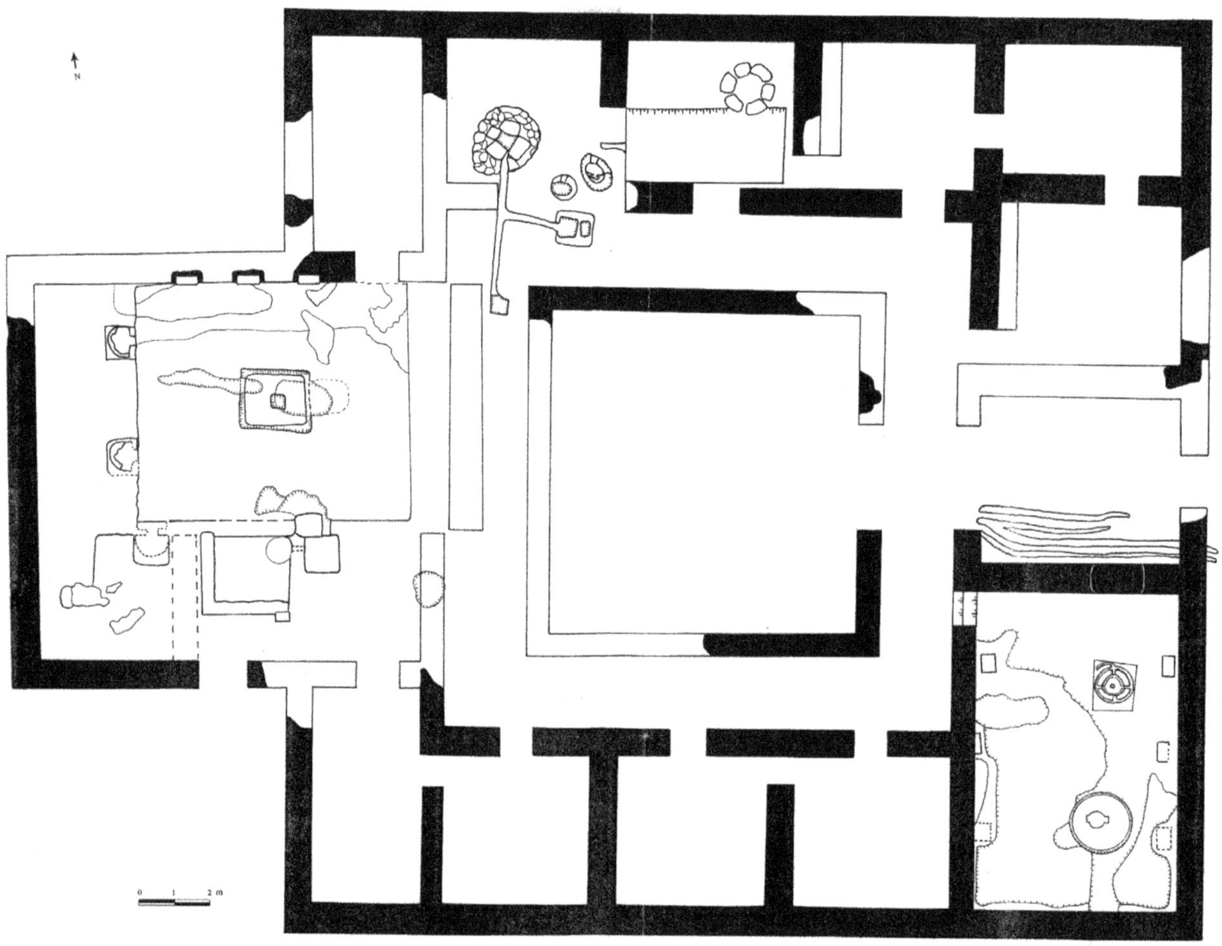

Fig. 51.1. Shu'afat, general plan of the building.

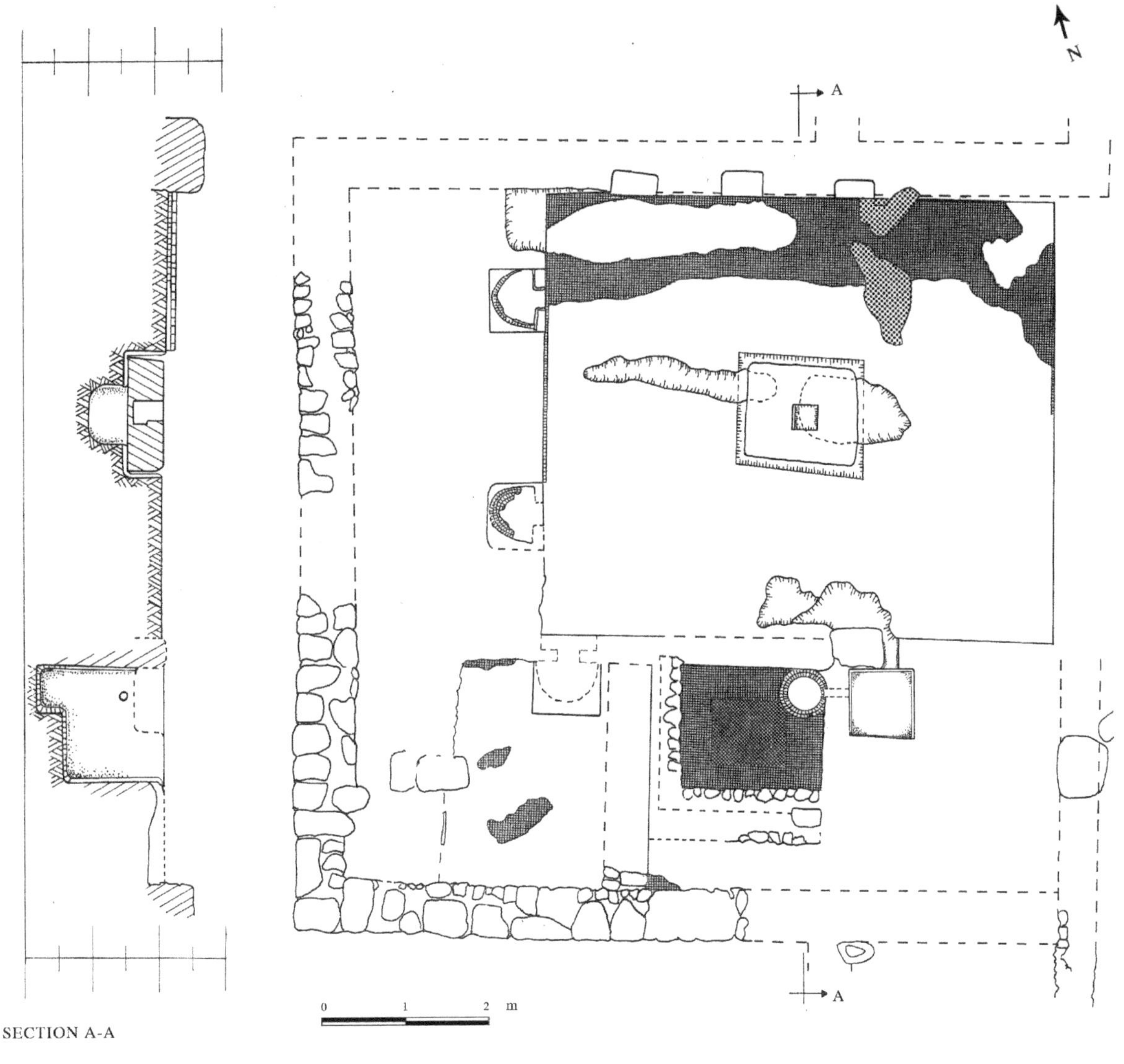

Fig. 51.2. Shu'afat, plan and section of the wine press.

Fig. 51.3. Shu'afat, screw base from the wine press.

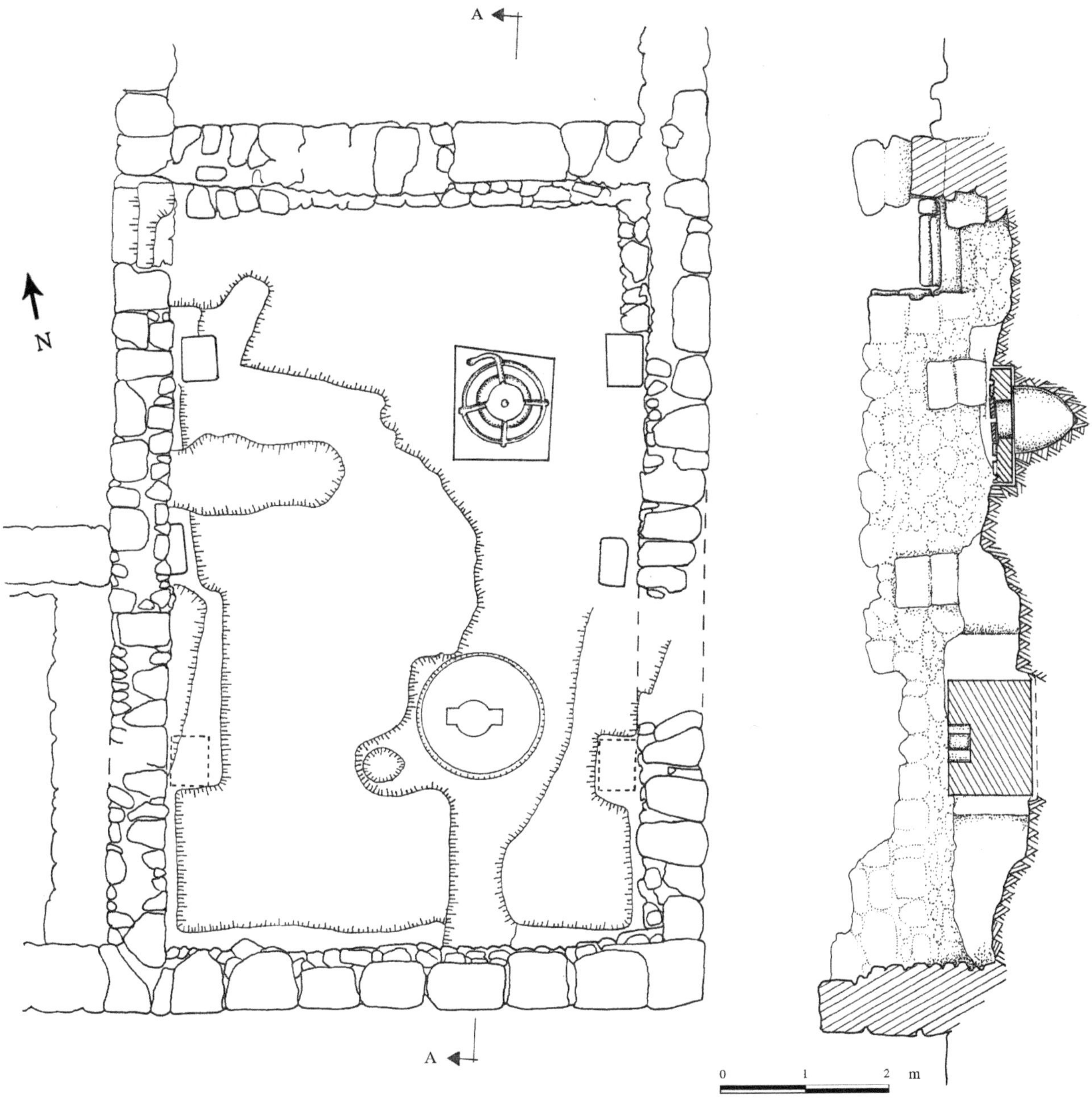

Fig. 51.4. Shu'afat, plan and section of the oil press.

Fig. 51.5. Shu'afat, press-bed of the oil press, looking west.

Fig. 51.6. Shu'afat, fragment of screw weight from the oil press.

52. The Byzantine Wine Press at Bethphage (e-Tur)

Jon Seligman and Taufik Deadle

The Site

The wine press at Bethphage was originally cleared in 1961 by Fr. S. Saller in the grounds of the monastery of that name in the e-Tur village on the eastern slope of the Mount of Olives (MR 1739/1317) (Saller 1961: 194–208). In 2004 the site was reexamined by Jon Seligman and Taufik Deadle following the building of housing for the overcrowded Christian population of the Old City.[1] The wine press is still visible between the newly built houses at the site.

It should be noted that the press was uncovered prior to the arrival of the archaeologists and thus the dating of the installation is based on its typology rather then any finds associated with it. Indeed a sewage system laid in 1950 to service the adjacent pigsty cut through the wine press and down through the cave below, further complicating understanding.

The site includes a wine press and a cave hewn under its surface which probably served as a storeroom in which jars were kept during secondary fermentation.

The Wine Press

The wine press (Fig. 1) is cut in the bed-rock, into an area that had previously served as a quarry. It belongs to the complex type and was built on two terraces formed by leveling the bed-rock, an upper western terrace and lower eastern. The press contains four treading floors – the main floor on the lower terrace, and three small floors on the upper terrace.

The central treading floor measures 6.07–6.23×6.23 m (38.3 m^2). The now fragmentary floor was originally paved with crude white tesserae averaging 2.2 cm^2. The edge of the mosaic has three to four rows of tesserae parallel to the walls forming a frame around the diagonally set mosaic in the centre. There is a possibility that a mortice for a single fixed screw had once existed here but a sewage trench had cut the floor at the point where a square hewn depression for the mortice may have originally been. On the western side of the floor were three auxiliary floors all measuring around 3×3 m (Frankel 1984: 183, type 8.0.1.2.; 1999: 139–140). These floors were cut on a rock terrace above the central treading floor. All were paved with the same white mosaic noted above, Saller (1961: 197) noting three different mosaic levels on the northernmost floor showing the press to have been in use over a long period of time. Under each of the floors was a vaulted compartment, a duct leading to it from the auxiliary floor above and a gutter leading from the compartment to the treading floor below. The lower part of the vaults was cut from the bed-rock while the upper part was built of small field stones set in a grey mortar. The walls of the compartments were coated with plaster. In the centre of the central auxiliary floor was a closed dovetail mortice for a single fixed screw press hewn in the bed-rock (Frankel 1984: 189, type 8.3; 1999: 140–141). From its base was an underground duct to the vaulted compartment below. The only other known example of a single fixed screw set into an auxiliary floor are the two from Ras Abu Ma'aruf (Seligman 1999; Site 48).

An additional compartment (diameter 0.54 m) was found on the northern edge of the main floor. No auxiliary floor seems to be connected with it and this is probably only a holding compartment. It was paved with white mosaic arranged in concentric circles to its base. In the base of the compartment is a duct connecting it to the treading floor.

On the eastern side of the main treading floor are two pipes through a bed-rock division connecting to two vats. One of the pipes was of ceramic and the other of lead. The vats are plastered with pink plaster, mixed with crushed ceramics and based on a grey bonding material set on flat stones. The floors of the vats are made of white mosaic on a pebble foundation set in grey mortar. The walls of the vats were partially hewn and partially built of plastered field stones.

At the south-east corner of the press was the settling vat (1.2×1.5×1.9 m with a volume of 3.4m^3). In the south-west corner of the floor of the vat was a conic sump with a mosaic floor (0.4 m diameter and 0.15 m deep). To its south was a collecting vat (1.8×2.2×1.9 m with a volume of 7.5m^3). On the eastern wall of the vat are stone slabs in the wall used as steps to allow access to it.

In the upper part of the division between the two vats was a blocked opening of 0.18 m diameter that connected them, allowing the filtered must to flow from the settling vat into the collecting vat for primary fermentation. On the bed-rock surface between the vats was a round depression of 0.42 m diameter with a sloped gutter back into the vat. This was used to place a wine storage jar during processing.

The Cave

To the south of the upper terrace was a deep cut in the bed-rock leading to a square entrance (1.5×0.52 m) cut in the southern façade of a cave. In the west jamb are three square bolt sockets. The threshold is built of four ashlar courses set in hewn pit in the bed-rock. At the base of the pit is a shallow division in the rock, dividing the pit into two round spaces. The western part is sealed with field stones while in the northern part is the built threshold. In the pit were ceramics of the Early Roman period (1st cent. BCE) and above, close to the entrance at the level of the threshold were ceramics of the Byzantine period (5th-6th cent. CE). The entrance led to a square cave (4.68×3.1 m, height 2.2 m at its apex) with a flat ceiling. In the corner at the eastern side of the cave was a hewn square opening (1.8×0.62 m) to a wide rounded *koch* (1.2 m diameter, 1.7 m high). In the northern jamb are two bolt holes. In the *koch* were human bone remains.

Operation of the Wine Press

The operation of the press and the discussion is identical to that at Ras Abu Ma'aruf (Site 48) and a full explanation with architectural parallels is brought in that article. This is a complex wine press of the same type noted and it should be dated architecturally to the Byzantine period.

Notes

1. In March 2004 an excavation was conducted by Taufik Deadle and Jon Seligman to re-document and complete the excavation conducted by Saller in 1961 (excavation permit 3930). Assistance was given by V. Esman and V. Pirsky (surveying), T. Sagiv (photography), A. Pickovsky (ceramic drawing), and Y. Nagar (Anthropology).

Bibliography

Frankel, R. 1984. *The History of the Processing of Oil and Wine in Galilee in the Period of the Bible, the Mishna and the Talmud.* Ph. D. thesis. Tel Aviv University (Hebrew).

Frankel, R. 1999. *Wine and Oil Production in Antiquity in Israel and Other Mediterranean Countries*. Sheffield.

Saller, S. 1961. "The Archaeological Setting of the Shrine of Bethphage". *LA* 11: 194–208.

Seligman, J. 1999. "Agricultural Complexes at Ras Abu Ma'aruf (Pisgat Ze'ev East 'A)". *'Atiqot* 38: 137–170.

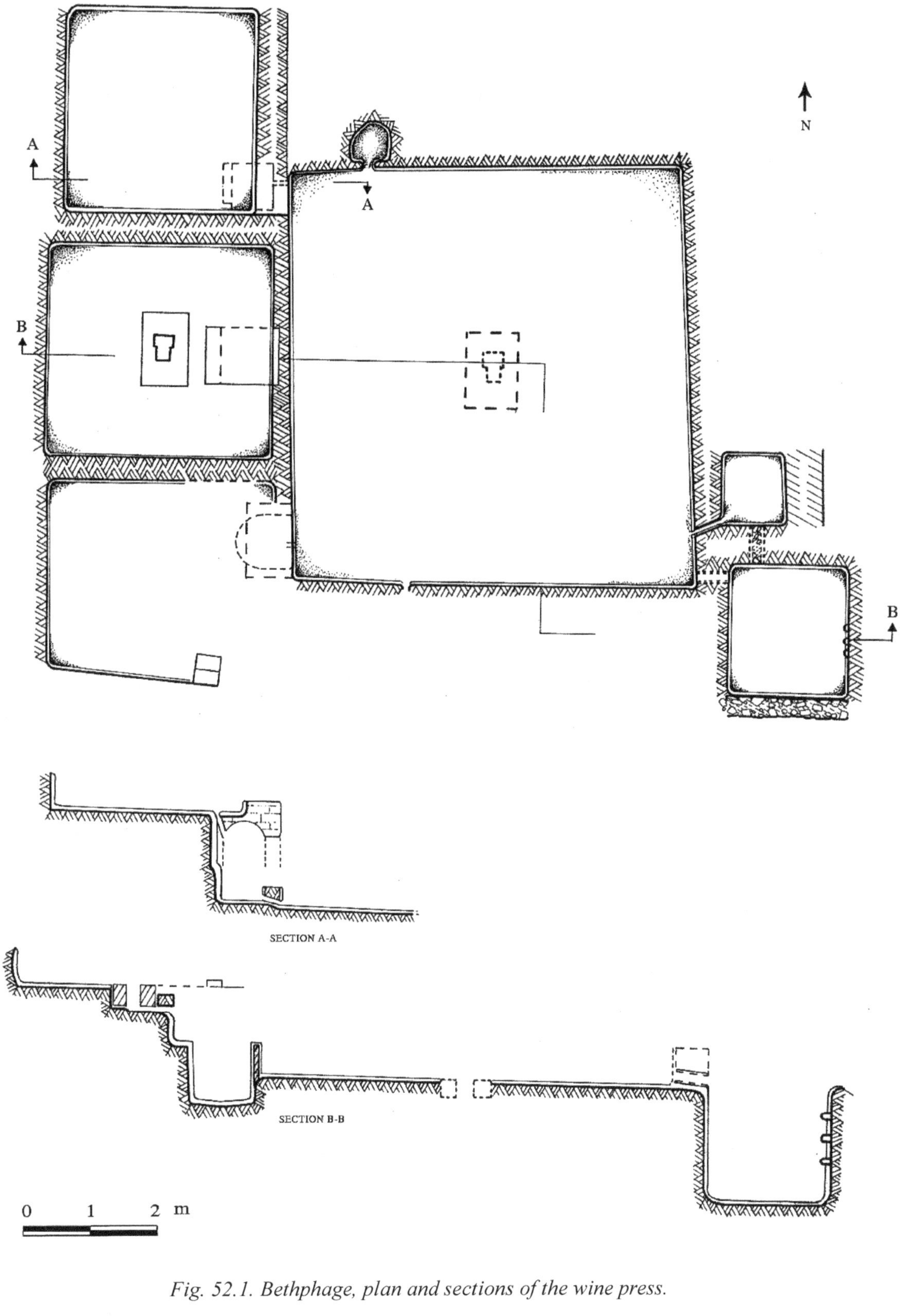

Fig. 52.1. Bethphage, plan and sections of the wine press.

53. Oil Presses at the Ophel, Jerusalem

Etan Ayalon, Rafael Frankel and Amos Kloner

In 1926 J. Garrow Duncan described a series of rock-cut chambers situated on the western slope of the eastern ridge south of the Old City of Jerusalem (MR 172/131) that today is usually called the City of David (Fig. 1). In the walls of one of these chambers (b) were cut five cross presses or as Garrow Duncan called them "cross and barrel" presses, three in the northern wall and two in the eastern wall. Below the crosses were rock-cut barrels in the shape of "decapitated cones". In these the frails of crushed olives were placed and the expressed liquid flowed out through openings at the bottom of the barrels. Garrow Duncan suggests that they probably all flowed into one cistern but did not clear the cave to examine if this was actually so. The upper arms of the crosses measured 0.91, 1.01, 0.99, 1.01, 1.01 and 0.79 m and were 0.3 m wide and 0.46 m deep. The cross arms measured 1.28 m across and were 0.3 m wide and deep and the barrels were approximately 0.79 m deep (for a detailed depiction of a cross press see Fig. 41.3 in the Introduction).

Bibliography

Garrow Duncan, J. 1926. "New Rock Chambers and Galleries on Ophel". *PEF QSt*: 7–14.

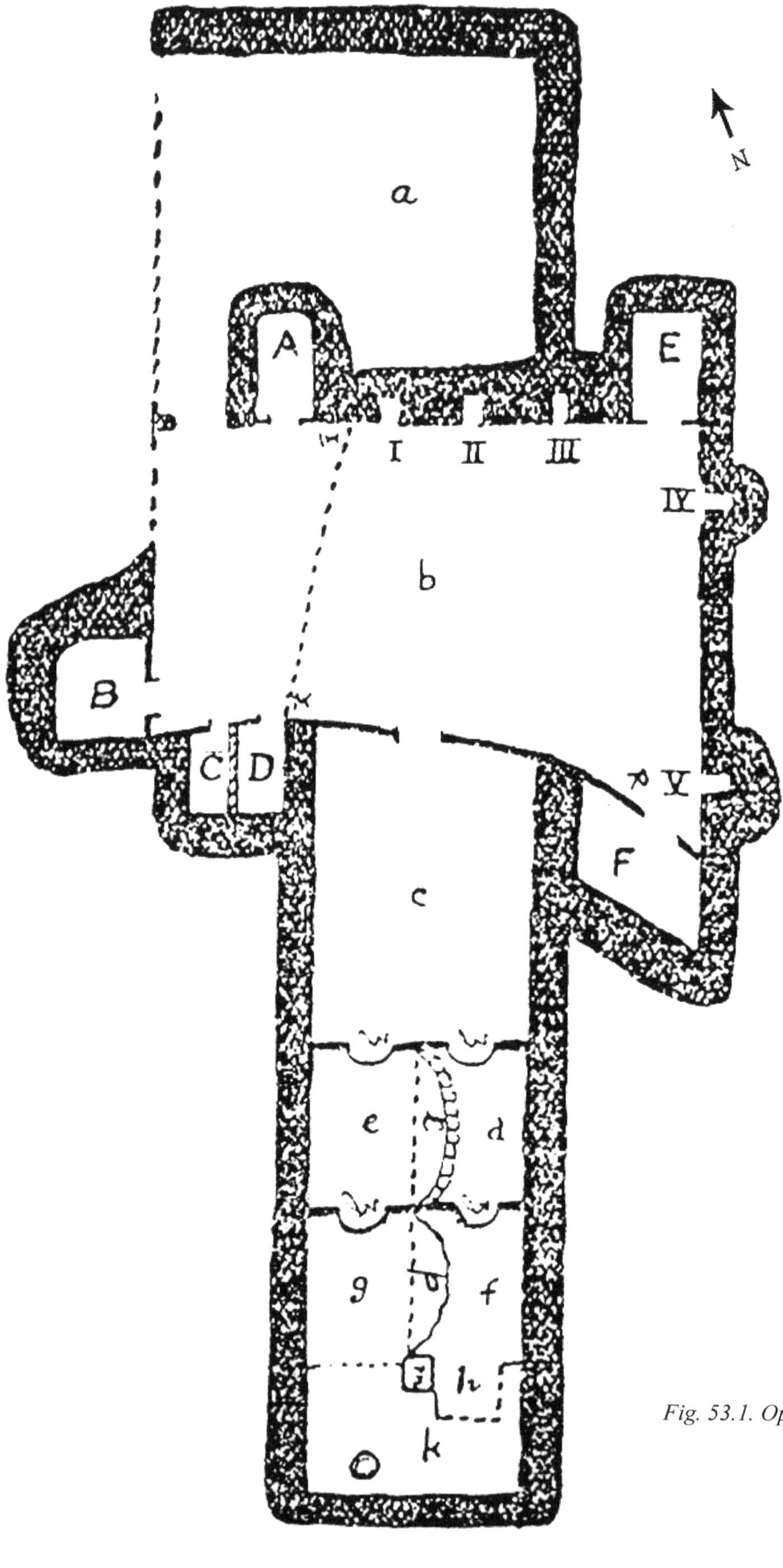

Fig. 53.1. Ophel, plan of the chambers with the cross presses.

54. The Wine Presses at the Qirya and in the Valley of the Cross – Jerusalem

Etan Ayalon, Rafael Frankel and Amos Kloner

These two wine presses are less than 500 m apart (MR 169/131) and are very similar having in common at least one feature that is apparently unique. The Qirya wine press was excavated by Ruth Amiran in 1952 and the Valley of the Cross installation by L.Y. Rahmani in 1955. They were published together by Rahmani (1991).

The Qirya Wine Press (Fig. 1)

The installation was discovered on the slope to the south of where the Knesset now stands during preparations made before erecting that building. It was badly damaged but the remains sufficed to reconstruct its original form.

It consisted of an approximately square treading floor (c. 5.5×6.0 m = 33m^2), an intermediary vat (c. 1.0×1.2 m, depth 0.5 m = 0.6 m^3) and a square collecting vat (c. 2×2 m, depth unknown). The latter two were connected by a shallow channel showing the intermediary vat to have served as a settling vat. A small channel leads off from the platform around the vats to the south. In the centre of the treading floor was a square depression doubtless to accommodate the base of a fixed screw press. To the north and west of the treading floor there were upper floors that drained into small collecting vats that were connected to the main floor by pipes 0.15 m above the floors of the vats and 0.5 m above the main floor. These small vats were roofed by conch shaped half domes facing the central floor. There were six such vats and although only four floors were discerned it is probable that there were originally also six floors. The upper floors varied in size. The only one that survived complete (No. 4) measured c. 2.3×4.3 m = c. 10 m^2. Each vat held about 250 litres.

The main treading floor and the upper floors and vats were all floored in crude white mosaics. The sherds found were dated to the 5th-6th cent. CE.

The Valley of the Cross Wine Press (Fig. 2)

The installation is located to the south of the Qirya wine press, above the slope to the west of the Monastery of the Cross and to the east of the perimeter fence of the Israel Museum.

It consists of a rectangular treading floor (c. 6.2×8.0 m = c. 50 m^2), a cubic intermediate settling vat (c. 0.9×0.9 m, depth 1.1 m = c. 0.9 m^3) and a collecting vat (first stage 1.5×1.9 m, depth 1.4 m = c. 4 m^3). In the centre of the treading floor was a square depression to accommodate the base of a fixed screw press. This depression was connected directly to the intermediary vat by a pipe partly rock-cut and partly made of clay. There were remains of seven niches around the treading floor, the three to the west clearly connected to upper floors, and thus similar in form to those of the Qirya wine press. It is not clear, however, if the others were also connected to upper floors.

The unusual component of the Valley of the Cross wine press is an additional vat (c. 0.7×0.8 m, depth 0.7 m = c. 0.4 m^3) located below a slope c. 6.5 m to the east of the collecting vat.

The excavator discerned three stages in the development of the press. In the first stage the additional vat was not yet functioning, in the second the collecting vat was connected to the additional vat by a lead pipe and in the third stage by a stone channel. Presumably the must fermented in the collecting vat and the additional vat was to fill the jars.

On the treading floor remains of three superimposed mosaic pavements were discerned. In the collecting vat in the first stage the floor was bed-rock, in the second plaster and in the third mosaic with a small sump in the corner. The intermediate vat was also paved in mosaic. The floor of the additional (filling?) vat consisted of one stone slab.

Of the pottery found only one item was dated earlier than the 5th cent. and none later than the early 7th cent. CE.

Conclusions

The plan of the two wine presses was very similar. The relationship between the three main elements, treading floor, settling vat and collecting vat in which the latter two were aligned along one side of the floor is typical of all the central regions of Israel ("the four rectangle plan"). The upper floors and vats are common in the Jerusalem area although not unknown in other areas.

The unusual element of the Valley of the Cross Winery is the additional (filling?) vat. Rahmani suggests that the channel to the south of the intermediate and collecting vats of the Qirya wine press led to a similar additional vat. Another point of interest, as Rahmani pointed out, is the fact that while the intermediate and collecting vats are to the south of the treading floor in the Qirya wine press they are to the east of the treading floor in the Valley of the Cross installation, but nevertheless in both cases the upper floors and hooded vats are to the north and west of the main floor. Rahmani assumed that the upper floors were originally surrounded by high walls and thus their contents were shielded from westerly winds as were those of the lower hooded vats. Both probably served in the production of *prototropum* or *mustum lixivium*, the first juice that seeped from the grapes by their own weight or of *passum*, raisin wine obtained from sun-dried fully ripe grapes.

Bibliography

Rahmani, L.Y. 1991. "Two Byzantine Winepresses in Jerusalem". *'Atiqot* 20: 95–110.

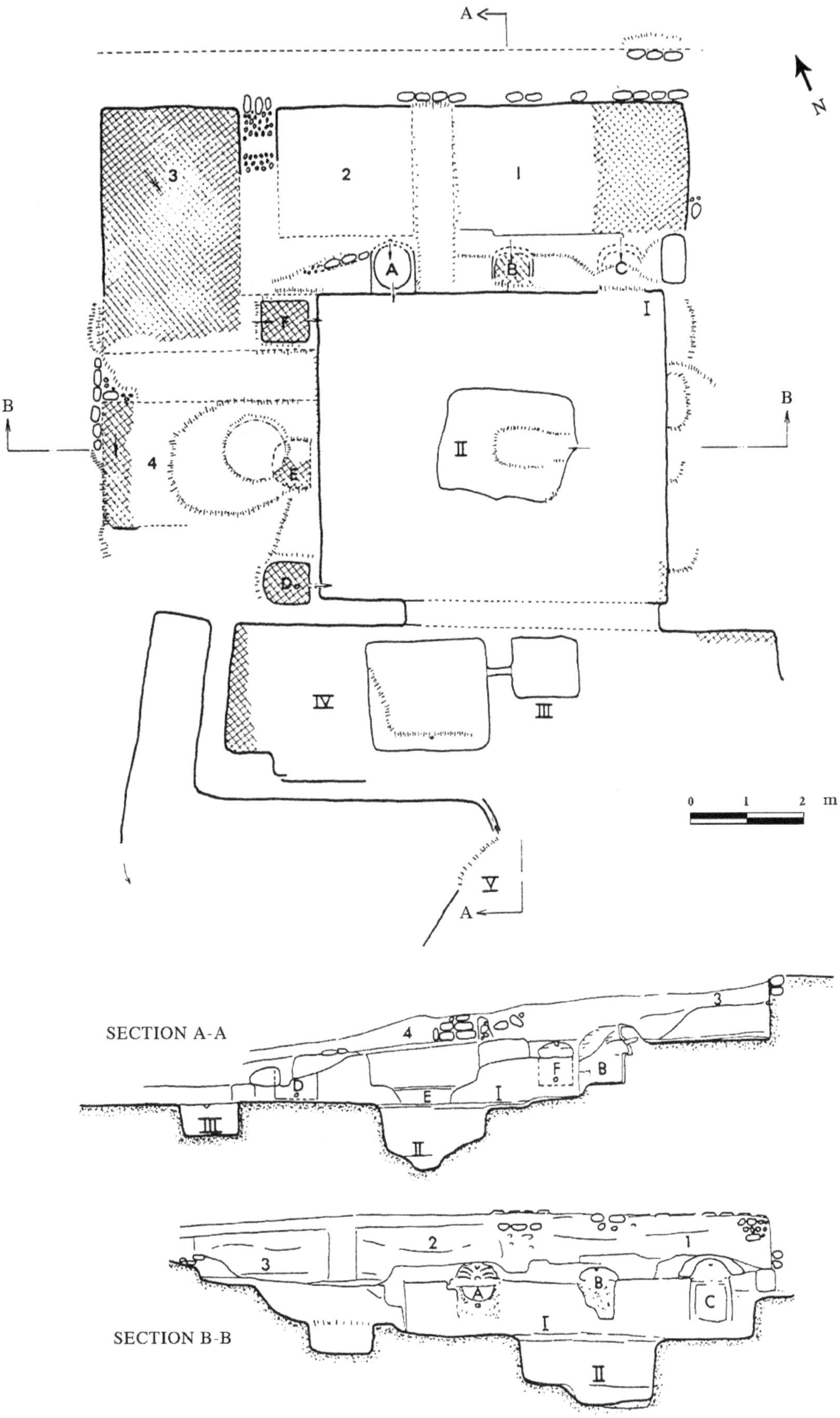

Fig. 54.1. Qirya, plan and sections of the wine press.

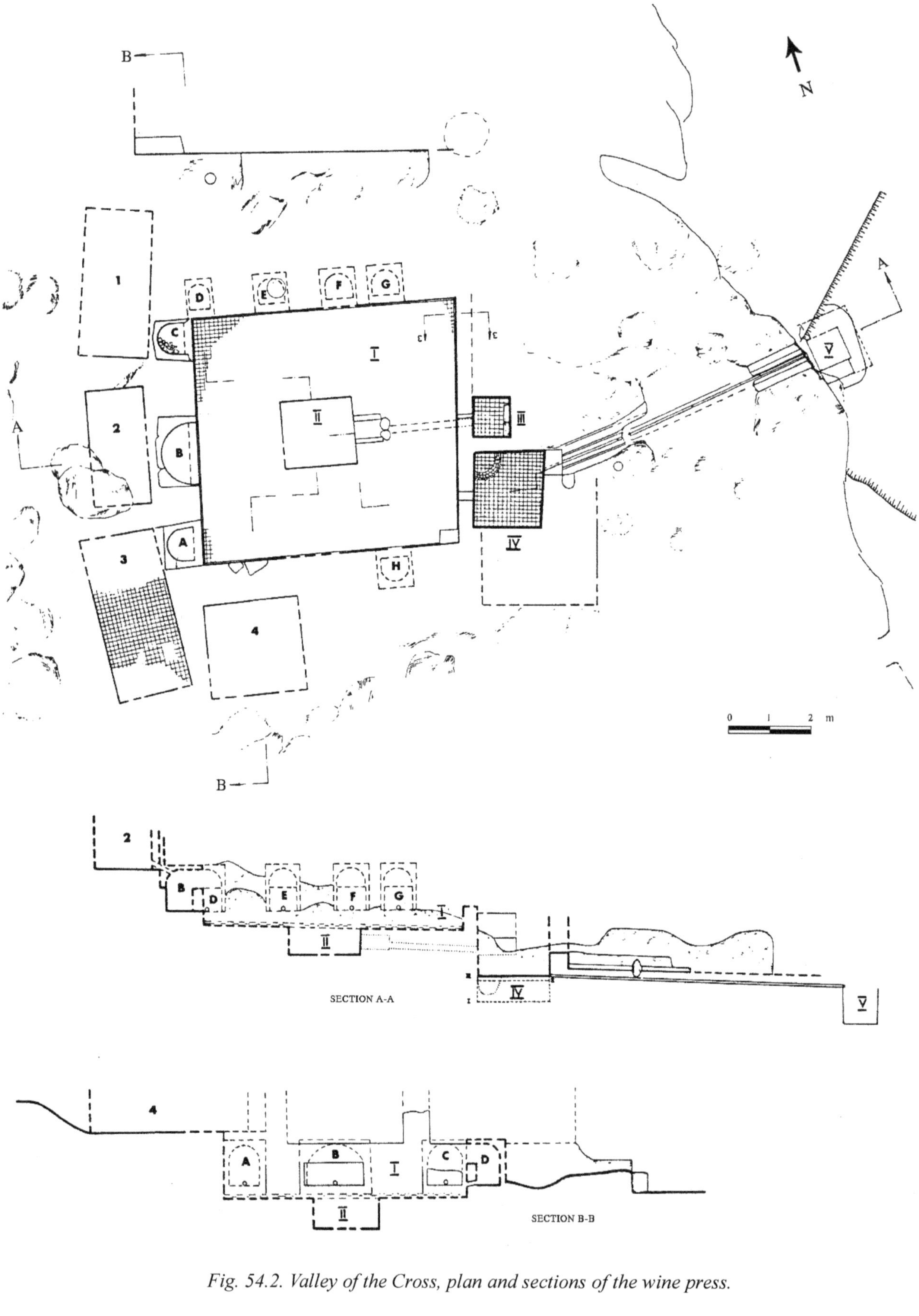

Fig. 54.2. Valley of the Cross, plan and sections of the wine press.

55. Wine and Oil Presses at 'Ain el-Jedide

Jon Seligman

The Site

The presses at 'Ain el-Jedide (MR 1634/1294) were discovered during a small excavation conducted by R.W. Hamilton (1935). 'Ain el-Jedide is a small spring on a hill 400 m to the south of Moshav 'Even Sapir, on the western outskirts of Jerusalem. The spring feeds an irrigation system of channels and pools that watered the agricultural terraces around it from at least the Byzantine period up until the present day. On these terraces were grown the vines and olive trees whose produce was processed in the presses described below.

The Presses

The presses were in a complex of buildings identified by the excavator as a small monastic establishment (Ibid.: 111) clearly dated by the ceramics and an inscription to the Byzantine period (Fig. 1). In the northern part of the building were two interconnected rooms that contained, according to Hamilton (Ibid.: 114) a wine press. If, as I believe, our general identification of such installations is correct, then we should associate the installations with both a wine press and olive press.

The wine press is of the simple type containing only a treading floor and a collecting vat. The treading floor measures 4.8×3.0 m (14.4 m^2) and it is paved with a simple white mosaic. A ceramic pipe passes through the western wall into a collecting vat measuring 2.4×2.1 m, depth 1.55 m with a volume of 7.8 m^3. At the bottom of the vat is a sump. Both the vat's floor and the sump are paved with mosaic. The only unusual aspect of the press is the fact that it was built into a closed building whereas most wine presses were open structures.

The vat occupies a corner of a room 0.6 m lower then the treading floor. This room (6.5×5.0 m) also contains the installations usually associated with an oil press, though, as noted above, they were attributed by Hamilton to be part of the wine press. In the centre of the room, probably *in situ* is a crushing basin with a diameter of 1.8 m. The plan and the limited number of photographs, both published and in the Mandatory archives, give little information as to the features of the basin. It clearly has a raised edge but if there was a raised or sunken pivot socket is unknown and it is probable that it had no socket at all. Basins of this type are infrequent and only known from five sites in Israel and an additional three in Syria (Frankel, Patrich and Tsafrir 1990: 290).

West of the crushing basin is a screw weight found lying on its side and out of position (Figs. 2–3). This is the only feature of the site clearly visible today. The weight is highly unusual (Frankel 1984: 55, 158, type 6.2.3; 1999: 113–114). It is square with diagonally cut corners so forming a sort of octagon measuring 0.75×1.05 m and 1 m high. A groove is cut along the top with a round socket at its centre and tapered mortices continuing from the top groove on two sides. On the top are the remains of four iron fittings – clamps that were probably fixed through the wooden block that surrounded the screw. The clamps prevented horizontal movement of the two blocks and thus prevented them from opening during operation. A proposed reconstruction is presented here (Fig. 4).

Missing from the original text are the press-bed, the collecting vat and the anchoring point of the press beam. These are not distinguishable from the surviving photographs.

Discussion

The wine press is easily understood and needs no further explanation. The major question still unanswered is if the screw weight and the crushing basin are associated with the production of oil or wine, as suggested by the excavator. Interestingly Frankel does not challenge this identification even though he notes that it is unknown elsewhere (Frankel, Patrich and Tsafrir 1990: 296). The burden of proof as to the use of the installation as a wine press falls, I believe, on those who support this theory. One could argue that the weight by itself could be used for the pressing of grapes, but together with a crushing basin it could only be used in the pressing of olives. They should thus be attributed to the latter use.

The screw weight at 'Ain el-Jedide is one of three related examples in Israel. Though lacking the groove along the top and having reversed T side mortices, the rectangular screw weight with cut corners in the courtyard of Emmaus is very similar. At Kabri (Frankel 1984: 55, 158; 1999: 113) is a cylindrical weight with a groove running between two T-shaped mortices. Frankel (Frankel, Patrich and Tsafrir 1990: 296–297; Frankel 1999: 113–114) discusses the 'Ain el-Jedide weight in some detail and suggests that the weight, with its rectangular shape with cut corners, mortices on two sides joined by an upper groove and metal clamp attachments, originates in Anatolia or the Aegean.

Callot (1984: 46, Pls. 41, 47) presents an example from

Sarfud (type 1b) in northern Syria with a groove across the top. He notes that this type is little utilized in this area though he does not differentiate it in his examples from type 1a, the most common type found in Israel with a socket and two trapezoid mortices. Another weight of this type was found in Behyo, also in northern Syria (Tchalenko 1953: 367, Pls. 119–120, 210.6). These weights were cylindrical and not rectangular. Anderson (1903: 15–16, Figs. 2–3) gives two examples from the Pontus, though he does not note their origin and a further example was found in Kalymnos in Asia Minor (Paton and Myres 1898: Fig. 1). Worth noting is the rectangular shape of these stone weights allowing them to be enclosed in an outer frame. It may thus be possible that our weight may also have had such an outer wooden frame. Further afield are a number of weights from Provence in southern France. Here a number of examples with dovetail mortices and a groove between them are noted by Brun (1986: Figs. 136, 155, 156, 168), though the small width of the groove demands a different type of reconstruction then that suggested by us (see above).

Frankel proposes that foreign influence is the reason for the unique features of the weight (Frankel, Patrich and Tsafrir 1990: 297; Frankel 1999: 114). The fact that the type is very uncommon even in these 'foreign' areas presents us with difficulties as to the place of origin for the weight. The only very similar example is that from Northern Syria (Callot 1984: 46). More information, especially from Asia Minor, is needed in assessing their origin.

The metal clamp, the remains of which were found on the top surface of the stone weight, is unknown elsewhere in Israel. Callot (1984: 47, Pls. 42, 49) provides an example from Ksegbe (type 2b) in northern Syria. Even here he notes that the clamp is exceptional (Ibid.: 48) and states that it is a sub-type of a commoner form (type 2c) as found in H. Beit Loya (Frankel, Patrich and Tsafrir 1990; Site 67). Clamp holes are also evident from two weight types in the Pontus (Anderson 1903: 15–16, Figs. 6–7) though with no side mortices or top groove showing the fixing of the weight to the screw to be of a completely different type. Brun (1986: Figs. 136, 155, 157, 168, 175, 182) provides a number of examples from Provence. Indeed three of these (Ibid.: Figs. 136, 155, 168) display the combination of clamp, dovetail mortices connected by a groove and a rectangular shape as found at 'Ain el-Jedide.

Bibliography

Anderson, J.G.C. 1903. "A Journey of Exploration in Pontus". *Studia Pontica* 1.

Brun, J.-P. 1986. *L'Oleiculture antique en Provence.* Paris.

Callot, O. 1984. *Huileries antiques de Syrie du Nord.* Paris.

Frankel, R. 1984. *The History of the Processing of Oil and Wine in Galilee in the Period of the Bible, the Mishna and the Talmud.* Unpublished Ph.D. Thesis. Tel Aviv University (Hebrew).

Frankel, R. 1999. *Wine and Oil Production in Antiquity in Israel and Other Mediterranean Countries.* Sheffield.

Frankel, R., Patrich, J. and Tsafrir, Y. 1990. "The Oil Press at Horvath Beit Loya". In: G.C. Bottini, L. Di Segni and E. Alliata (eds.). *Christian Archaeology in the Holy Land. New Discoveries.* Jerusalem: 287–300.

Hamilton, R.W. 1935. "Note on a Chapel and Winepress at 'Ain el Jedide". *QDAP* 4: 111–117.

Paton, W.R. and Myres, J.L. 1898. "On Some Karian and Hellenic Oil-Presses". *JHS* 18: 209–217.

Tchalenko, G. 1953. *Villages antiques de la Syrie du Nord.* Paris.

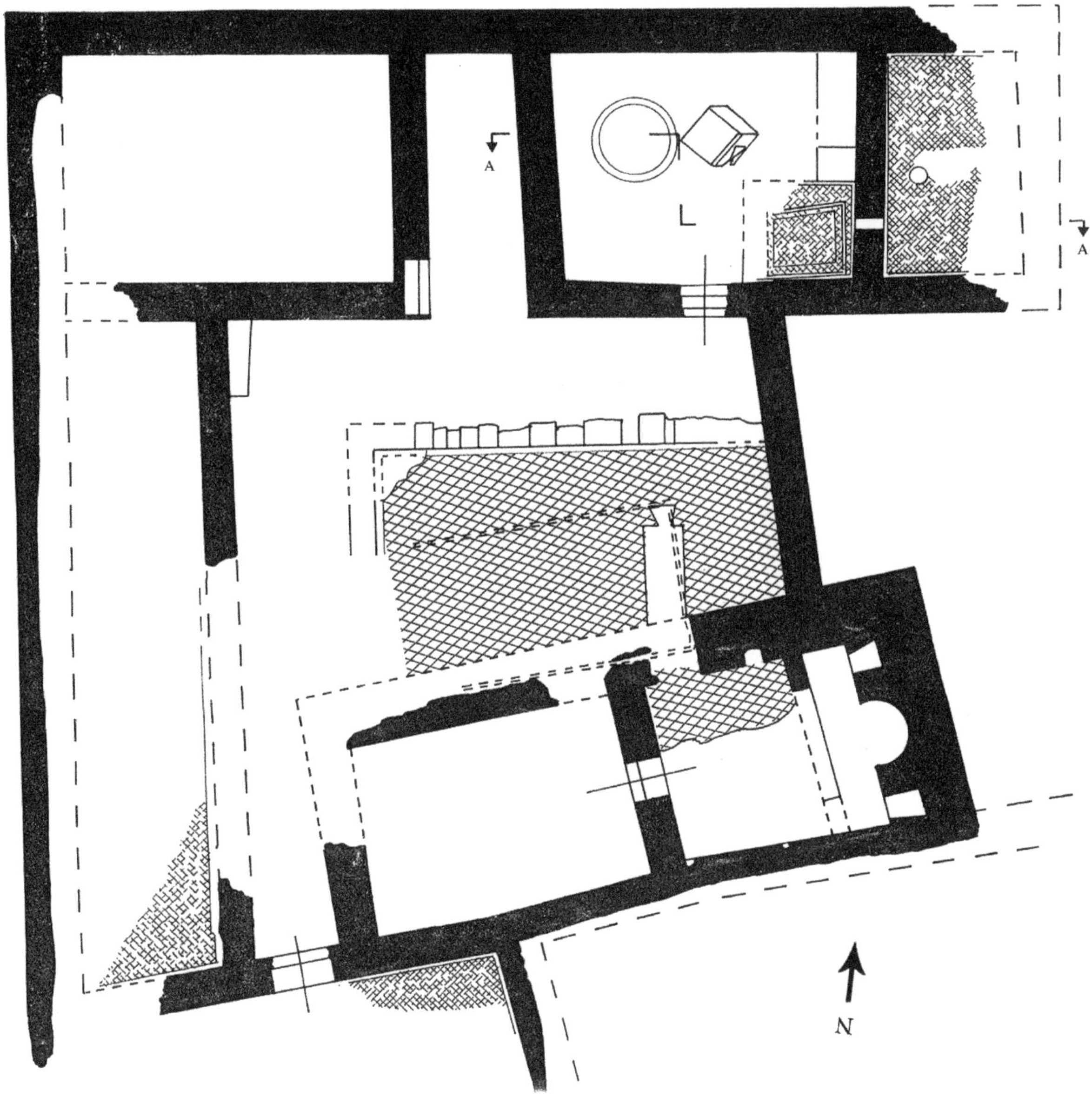

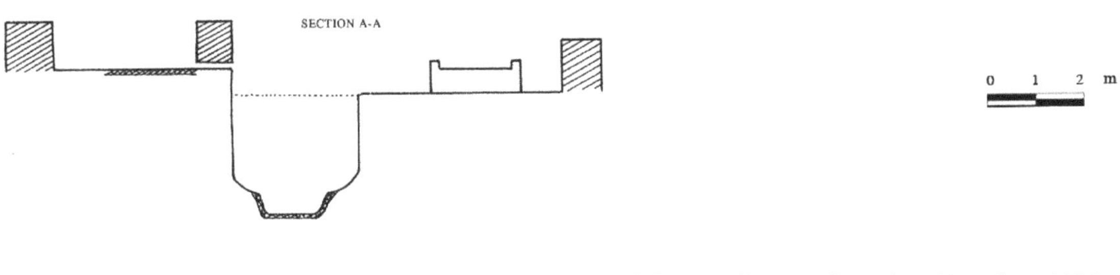

Fig. 55.1. 'Ain el-Jedide, plan of the building and section of the installations (based on Hamilton 1935: 113).

Fig. 55.2. 'Ain el-Jedide, screw weight, side.

Fig. 55.3. 'Ain el-Jedide, screw weight, top.

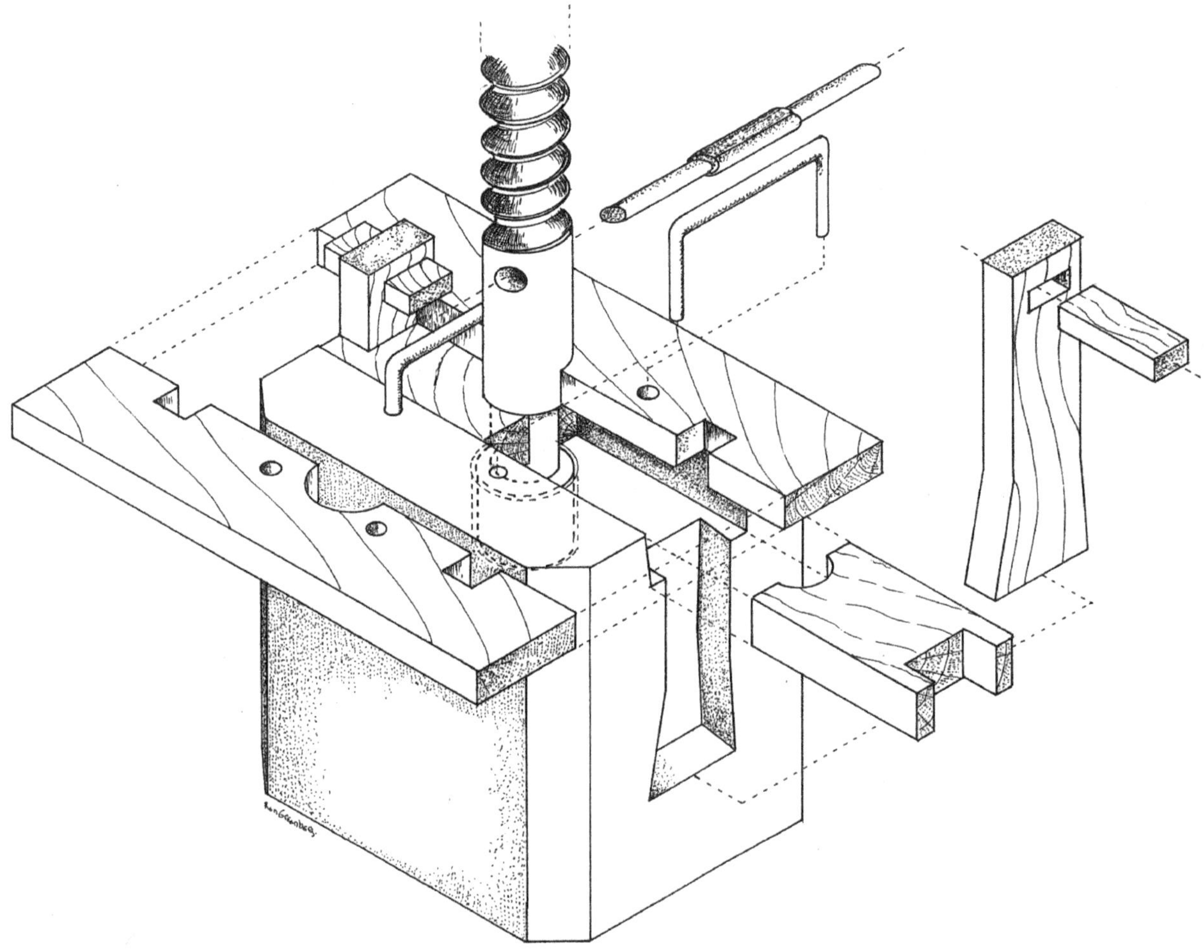

Fig. 55.4. 'Ain el-Jedide, reconstruction of screw weight.

56. Soreq Ridge – Installations From the Second Temple Period in the Jerusalem Hinterland

Daniel Weiss and Boaz Zissu

The Site

The Soreq Ridge site (MR 1597/1287) is situated in the Jerusalem hills, on an isolated hill surrounded by steep slopes (altitude 714 m). The site overlooks the confluence of the Soreq and Refa'im streams and commands a good overview of the southern Jerusalem hills.

The site consists of two farm building complexes. The upper complex, measuring approximately 57×88 m, is built on the hilltop. The lower (25×27 m) is located 200 m below it on the eastern slope.

Most of the pottery collected at the site dates to the Hellenistic period and therefore the farm can be dated to this period. In addition, however, in the upper structure some pottery from Iron Age II was found as well as from the Persian and Early Roman periods, while some of the pottery found in the lower structure dates to the Roman-Byzantine period.

The survey revealed seven wine presses hewn in bed-rock on the slopes of the hill, as well as the remains of an oil press and a round watchtower. During the excavation, several of these installations were cleared and prepared for visits by the public.[1]

Installation No. 1

This installation consists of two adjacent wine presses both hewn into the bed-rock (Figs. 1–2). The northern wine press has been completely excavated. It consists of two treading floors ("Two treading floors to one vat" Tosefta Ter. 3.7) separated by a partition hewn from the bed-rock, 0.4 m wide, a filtration vat and a collecting vat.

The eastern "treading floor" is rectangular and measures approximately 3×4 m (12 m^2 in area) and about 0.3 m in depth. In its northern wall are two niches about 0.4 m deep for anchoring the wooden beams used for secondary pressing. The niches open upwards and in both sides of each holes have been hewn that served to hold a horizontal rod to which the pressing beam was secured. The niches appear to have been damaged during use, as attested by repairs to the plaster. Remnants of a gray-to-light brown plaster were found in other places in the surface. The floor is not level and appears not to have been used for treading, but only for extraction of the must by beam presses. Flowing of the liquid from the floor to the filtration vat was assisted by extremely shallow radial grooves and a channel, also shallow.

The western floor is almost square in shape, measuring 3.2×3.5 m – 11.2 m^2 in area and 0.4 m deep, with a slight declivity toward the south. It was apparently intended for treading since all irregularities (depressions, cracks, etc.) had been removed by filling them with pebbles and coating them with white plaster that included gravel. A round niche for a pressing beam was carved into the eastern wall of the treading floor; it measures about 0.3×0.4 m and was 0.25 m deep. The treading floor drains through a channel into the rectangular filtration vat measuring approximately 0.6×0.9 m in area and 0.5 m in depth. From the filtration vat, the must flowed into the cylindrical collecting vat, 1.3 m in diameter and 1.9 m deep – i.e. volume of 2.5 m^3. A single step enabled descent into the vat. To the west of the collecting vat, a small cup-shaped depression was chiseled out of the rock; it may perhaps have contained marl that was added to the must to clarify it.

The two treading/pressing floors operated in parallel, but the use of different types of plaster attests to their having been hewn during two different periods of time.

The manner of fixing the press beams in the eastern treading floor is unusual, and is apparently unique to the Jerusalem area, where parallels have been found from the Iron Age and from the Hellenistic and Early Roman periods. Because of the proximity of the installation to the Hellenistic farmhouse, it, too, can be dated to the same period.

In the southern wine press, only the collecting vat was excavated. It is roughly rectangular, approximately 1.0×1.2 m in area and 1.3 m deep. Its total volume was thus 1.56 m^3. The treading floor of this installation appears to have been damaged in the course of later rock-cutting operations.

Installation No. 2

This was a complex wine press that went out of use when an oil press was carved out of the bed-rock within and alongside it (Figs. 3–5). The wine press includes a main treading floor, another treading floor beside it, a filtration vat and a collecting vat.

Because the rectangular main treading floor (c. 4.7×5.4 m, area 25.4 m^2) was cut into a rocky slope, the height of its walls varies, the western wall reaching a height of approximately 2 m. The secondary treading floor, south of the main floor and parallel to it, measures approximately 1.4×2.2 m, area c. 3 m^2; it drains into the main treading surface. There is a concave niche about 0.5 m high, 0.5 m wide and 0.3 m deep hewn in its western wall. This niche was apparently used to anchor the beam of the press that operated in the smaller floor, extracting more must from the remains of the grapes that had been previously trodden on the main floor.

There is also a rectangular niche carved into the western wall of the main treading floor, measuring approximately 0.6 m high, 0.35 m wide and 0.25 m deep. This niche served to anchor the beam of an additional lever and weights press that was probably also used for secondary pressing of rape. It is, however, possible that the niche is a later addition as part of the oil press described below.

A channel 0.4 m long, 0.35 m wide and 0.45 m deep directed the must into the rectangular filtration vat that was found to the east of the treading floor. It measures 0.9×1.0 m in area and 0.8 m deep. In the floor of the vat is a round sedimentation basin. A cylindrical conduit about 10 cm in diameter, hewn through the rock about 0.3 m above the floor of the filtration vat, conducted the liquid from it to the nearly cubical collecting vat. The latter measures about 1.7×1.7 m in area and 1.6 m deep, volume of 4.6 m^3. Around the vat on the south and east sides there is a hewn shelf at the height of the rim.

At the bottom of the collecting vat, sherds of two storage jars from the Early Roman period were found.

It appears that many cracks – the result of natural wear and tear – in the treading floor and other parts of the installation made operation of the wine press difficult, and it was probably for this reason it was converted into an outdoor oil extraction facility.

The Oil Press

From the crushing mill, only the crushing basin (diameter about 1.8 m) has survived. It was carved into a natural rock outcropping west of the wine press treading floor. The crushing basin is round with an elevated perimeter, and carved into its centre is a square socket 0.45×0.45 m and 0.25 m deep.

On the treading floor itself, a lever and weights press was installed, that included:

- A beam anchored in the niche in the western wall that was probably formerly used in the wine press;

- Three or four stone weights that were attached to the beam near the end (not found);

- A rectangular pit measuring 2.8×1.0 m, 0.4 m deep, cut into the bed-rock in the eastern part of the treading floor, for accommodating the beam weights;

- An almost cubical central collecting vat, measuring 0.5×0.5 m and 0.6 m deep (volume 150 litres), hewn into the floor between the niche and the pit, with grayish-white plaster applied to its walls;

- Two grooves, each about 2 cm deep, carved into the floor on both sides of the collecting vat. They were designed to accommodate the cover on which the frails were placed during the pressing operation.

In the floor, just north of the niche in which the beam was anchored, were found two flat-bottomed cylindrical vats approximately 0.6 m in diameter and 0.5 m deep, the purpose of which is not clear. Also indeterminate is the use made of two rectangular vats in the north-east section of the complex, hewn into the rock on a slightly lower level than the rest of the installation.

The wine press and the oil press can be dated to the Hellenistic and Early Roman periods respectively, because of the technology of their components and their proximity to the farmhouse that has been dated to those periods.

Installation No. 3

This is a simple wine press (Figs. 6–7) with a rectangular treading floor measuring approximately 4.8×3.4 m – 16.3 m^2 and a maximum depth of 1.2 m. Its round collecting vat measures 1.1×0.9 m and 1.1 m deep – a volume of 0.85 m^3. In the northern wall of the treading floor there is a rectangular niche hewn into the bed-rock for anchoring the wooden beam used in secondary expression of the must. The western wall of the wine press is constructed of large fieldstones; remains of fieldstone construction are also recognizable above its northern wall.

Summary

In the excavation of all three wine presses there was virtually no dateable pottery discovered, but their immediate proximity to the farmhouse of the Hellenistic and Early Roman periods as well as the technology of their components indicate that they belong to these two periods.

The site provides additional evidence for the existence of a series of agricultural settlements on the periphery of Jerusalem during the Second Temple period (Hasmonean-Herodian periods), and to the production of oil and wine to meet the needs of the city and its Temple.

Notes

1. The site was discovered by Harley Stark and was surveyed within the framework of the Archaeological Survey of Israel, Map 104 (Nes Harim). In April 1996, a number of agricultural installations were excavated at the

site within the framework of a Jewish National Fund (KKL) project to develop the Soreq Ridge area for recreation and tourism. The excavations were directed by the authors and carried out on behalf of the Israel Antiquities Authority. Also participating were Hanoch Tsoref (project manager for KKL) and Vadim Esman (surveyor). Our thanks to David Amit, Jon Seligman and Gideon Solimany for their constructive advice.

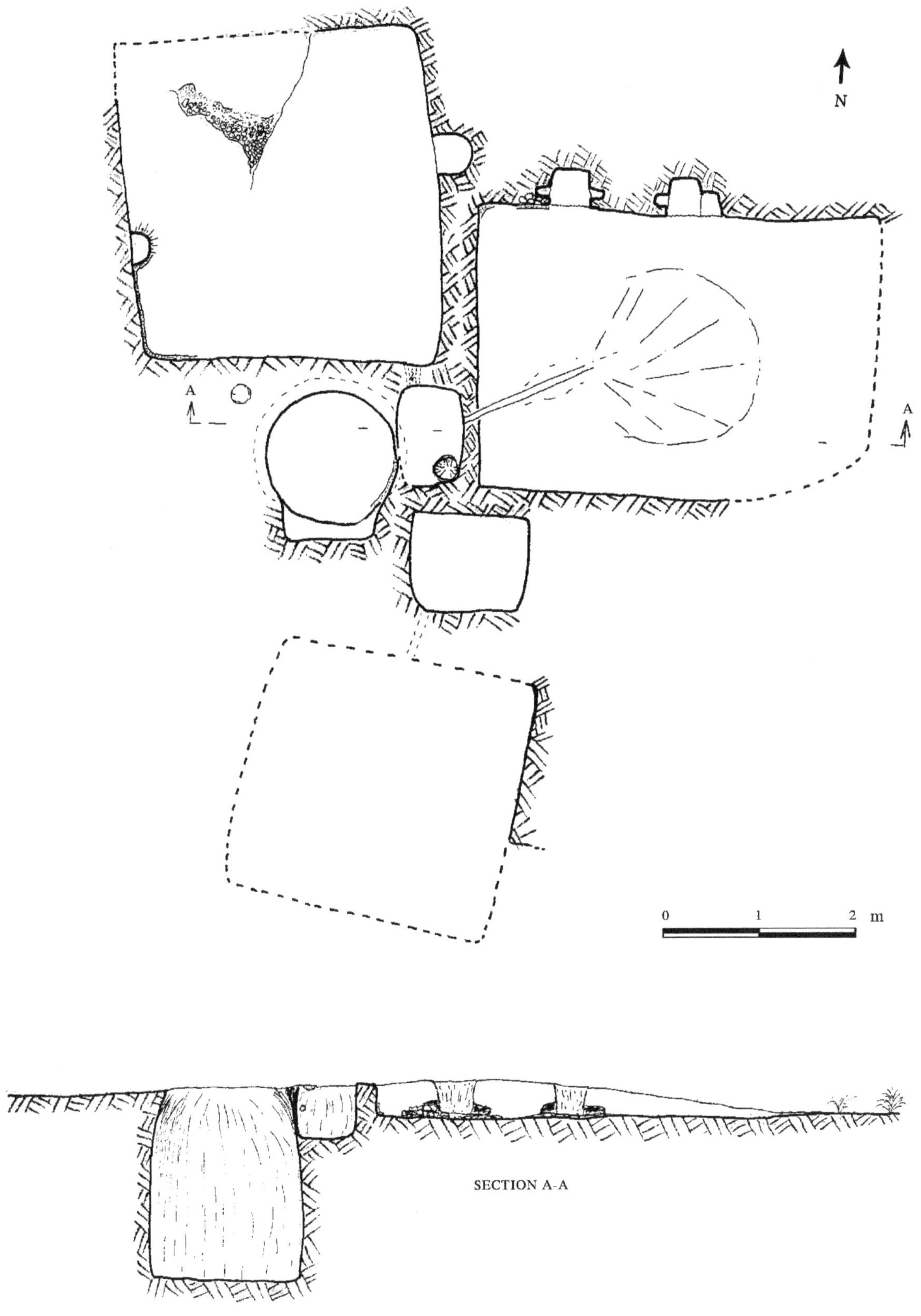

Fig. 56.1. Soreq Ridge, installation No. 1, plan and cross-section.

Fig. 56.2. Soreq Ridge, installation No. 1, view to the east.

Fig. 56.3. Soreq Ridge, installation No. 2, plan and cross-sections.

Fig. 56.4. Soreq Ridge, installation No. 2, view to the east.

Fig. 56.5. Soreq Ridge, installation No. 2, view to the west; on the left, the filtration vat and collecting vat, in the centre, the pit for the beam weights.

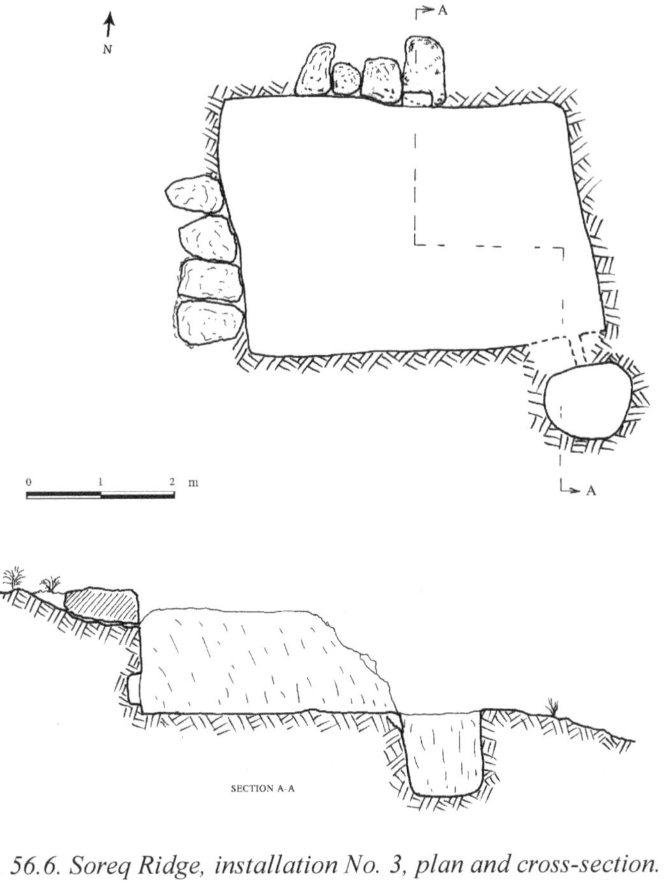

Fig. 56.6. Soreq Ridge, installation No. 3, plan and cross-section.

Fig. 56.7. Soreq Ridge, installation No. 3, view to the north.

57. An Oil Press and Ritual Bath at Kh. el-Najar in Beit Jala

Hananya Hizmi and Shabtai Zion

The Site

Kh. el-Najar is situated on a hill in north-west Beit Jala (MR 1680/1249) some 4 km from Jerusalem. In 1988, while the area was being leveled and graded as a plant nursery, an oil press and ritual bath were unearthed.[1]

The Oil Press

The Building

The oil press building measured 10×8 m. Three of its walls were built and preserved to a height of four courses (about 1.2 m); the fourth was hewn from bed-rock (Figs. 1–3). There the rock face was high, and the builders may have utilized a natural rock shelter. The entrance to the building no longer exists; it was probably on the south side, in the section where the wall is missing. The walls are made of roughly dressed stones and are about 0.8 m thick. Attached to the centre of both the eastern and the western walls and opposite each other there are piers 0.6 m wide, probably bases for an arch that supported the roof. In the eastern wall are two niches each 0.8 m wide. The sill of the northern niche is about 1 m above floor level; it was apparently a window opening. In the northern part of the western wall there is a niche similar in width.

The Crushing Installation

The olive crushing mill, built on an irregular surface, was found in the northern part of the building (Fig. 1:5). It was not preserved in its entirety and the crushing stone is missing. The basin is not monolithic, as is customary, but was constructed of stones. Similar basins have been found at several places in the vicinity of Jerusalem, e.g. Beitaniya (Saller 1957: 104, I. 22, Pls. 68–69) and En Ya'el. One section of its perimeter, a soft limestone *voussoir*-shaped stone, is preserved *in situ*. Its rim projects upward so that the olives would not fall out. The installation is 0.7 m high, the estimated diameter of the crushing surface is 1.8 m and the width of the rim is 0.12 m.

The Pressing Installations

Two different pressing installations were found in the building; The expressed liquid from both flowed into the same collecting vat 0.6 m in diameter and 0.71 m deep, volume c. 0.2 m^2 (Fig. 1:4). The walls of the vat were coated with light gray plaster. A stone slab was placed over its mouth; it had a square opening surrounded by a recess that apparently served to accommodate a lid. The eastern press is a direct pressure screw press with a press-bed 0.9 m in diameter carved into a slab of hard limestone set into the floor (Fig. 1:1). The groove for collecting the expressed liquid is about 5 cm both in width and depth. Projections that remained on the inside faces of the groove show that it was cut using a drill. The groove is connected to the collecting vat by a channel that passed under the cover plate of the vat. On both sides of the press-bed are mortices to hold the uprights of the press frame. Two parallelepiped-shaped stone piers were found in the vicinity of the press-bed, but not *in situ*. Each pier has a groove hewn the length of one of its narrow faces. It is possible that they originally stood on either side of the press-bed above the mortices with the grooves facing inward and served to support the press frame.

The second pressing installation is a lever and screw press. About 3.5 m to the north-west of the first press-bed, an additional press-bed was found, a little larger than the first (Fig. 1:3). It is made of hard limestone, about 1 m in diameter, with a circular groove. The press-bed rests on an earlier pit, poorly hewn from bed-rock and showing no remains of plaster; it is possible that it was never part of the oil press. The groove for collecting the expressed liquid seems to have been carved in the same way as was the first, terminating in a gutter from which a narrow channel approximately 2 m long leads to the collecting vat. The channel is coated with light gray plaster and its gradient is very gradual. About 2 m south of the press-bed, a cylindrical screw weight 1.1 m in diameter and 1.2 m tall was found *in situ* (Fig. 1:2), showing that the second press was of the lever and screw type. The weight – situated, as is usual, in a pit – has a round socket in the top in which the bottom end of the screw rotated and two trapezoidal mortices on the outside for holding the vertical tenons that secured the horizontal boards that attached the screw to the weight. The weight pit was enclosed by a rounded stone wall, of which only the south side was preserved. Finds in the oil press consisted mainly of shards of bowls, cooking pots, jars and roof tiles from the Byzantine period.

The Ritual Bath and Cave

The ritual bath (*mikveh*), located to the south of the oil press, was hewn out of the soft limestone bed-rock. It measured 2.0×1.3 m and 1.2 m deep and was coated with gray plaster typical of the Second Temple period. Four steps descend into the pool. Adjacent to the first step a plastered area was preserved and on it the remains of a channel built of medium-sized stones. Because of the extent of the destruction we could not be certain of the origin of the channel nor of the source of the water used to fill the ritual bath. The northern section of the bath was damaged when the oil press was built during the Byzantine period.

East of the ritual bath a cave was cut from the bed-rock. It has an antechamber and a rectangular entrance. In the cave lamps were found dating from the 1st cent. CE as well as Byzantine pottery.

Summary

The ritual bath and the cave preceded the construction of the oil press; the former attests to the presence of Jews here during the Second Temple period. There is at present no evidence for a settlement in the immediate area, but it is possible that there was an agricultural installation on the site, of which the ritual bath was part (compare to Qedumim [Magen 1993; Site 34]). It was destroyed when the Byzantine oil press was built several centuries later.

Two pressing installations were found in the oil press. Both used the screw principal, but in different ways. One press possibly utilized two stone piers, one fixed on either side of the press-bed; Wooden uprights stood in the grooves on the insides of the piers and above them heavy wooden board with a threaded hole in its centre was attached to accommodate the screw. Turning the screw applied pressure to the pressing board that lay on the olive mash frails. In the second installation, pressure was applied by a lever and screw arrangement. The rear end of the beam was anchored in the northern wall of the building. The frails were near the centre of the beam. Across the forked end of the beam was a plank with a threaded hole to accommodate the screw, the bottom end of which was attached to the stone weight. When the screw was turned it brought down the beam that exerted pressure on the olive mash frails. The beam was more than 7 m long.

The use of the two installations described above was already known during earlier periods (Frankel 1994; 1999). Why were there two different pressing installations in one complex? In our opinion, there are three possible answers: a) Because of the seasonal crop increase there arose a need for higher production capacity during the olive season, therefore a second press was installed; b) One of the presses, apparently the direct pressure screw press, went out of service and another press was installed alongside it; and c) Parts of an oil press that had survived from an earlier period, i.e. from the days of the Second Temple, were reused. It is clear that the lever and screw press was the later of the two.

Notes

1. The excavation was carried out on behalf of the Staff Officer for Archaeology in Judea and Samaria. The authors headed the excavation, and among the participants were Omar Saleh, Tanya Gornstein-Slotzka (illustrations and reconstruction), Paul Gratowsky (surveying) and Shlomi Amami (photographs, along with H. Hizmi).

Bibliography

Frankel, R. 1994. "Ancient Oil Mills and Presses in the Land of Israel". In: R. Frankel, S. Avitsur and E. Ayalon. *History and Technology of Olive Oil in the Holy Land.* Arlington and Tel Aviv: 19–89.

Frankel, R. 1999. *Wine and Oil Production in Antiquity in Israel and Other Mediterranean Countries.* Sheffield.

Magen, Y. 1993. "Qedumim – A Samaritan Site of the Roman-Byzantine Period". In: F. Manns and E. Alliata (eds.). *Early Christianity in Context, Monuments and Documents* (Studium Biblicum Franciscanum, Collectio Maior, 38). Jerusalem: 167–180.

Saller, S.J. 1957. *Excavation at Bethany (1949–1953).* Jerusalem.

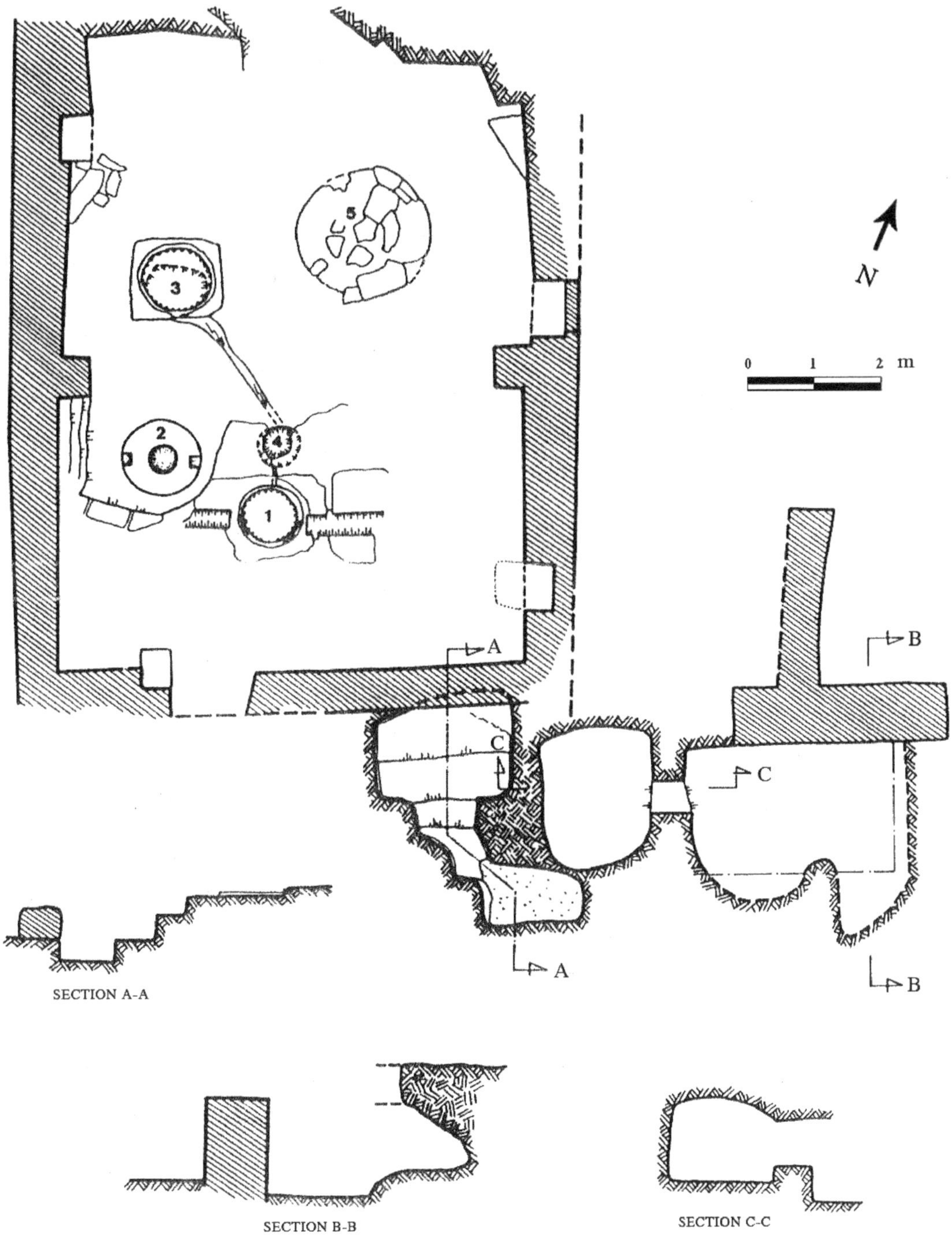

Fig. 57.1. el-Najar, plan and sections of the oil press and the ritual bath.

Fig. 57.2. el-Najar, overview of the oil press, looking north.

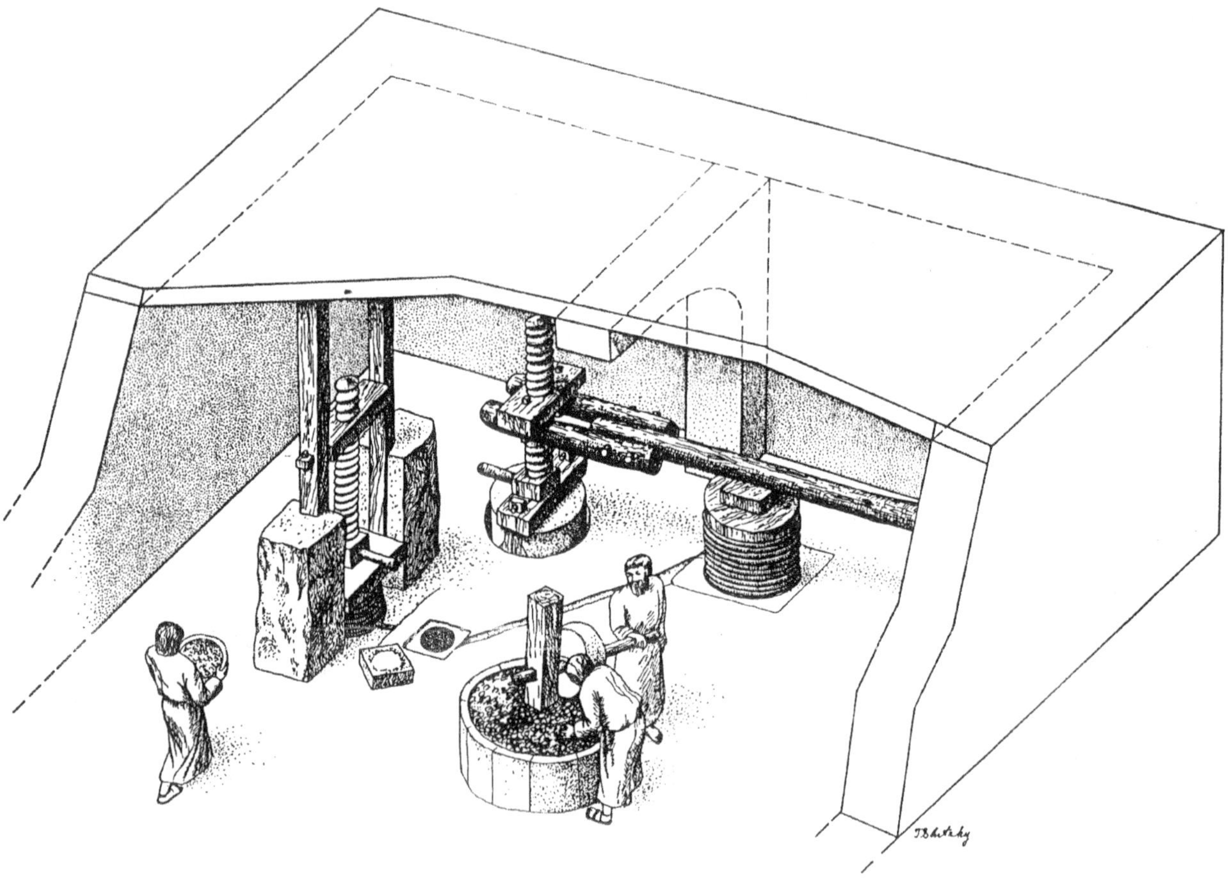

Fig. 57.3. el-Najar, suggested reconstruction of the oil press.

58. Oil Presses and a Wine Press at Kh. Marah el-Jum'a (Nebi Daniel)

David Amit

The Site

The site is situated about 5 km south-west of Bethlehem (MR 1642/1217), 980 m above sea level, to the east of the holy place Nebi Daniel and on the national water-shed. The settlement was built on terraces at the top of the slope that descends steeply eastwards. Recent agricultural activities have destroyed almost all the ancient site but man-made caves, rock-cut installations and shards from the site make it possible to reconstruct the history of the settlement.

In the southern part of the site several burial caves and some wine presses from the end of the Iron Age were recorded (Amit and Yezerski 2001) while in its northern part caves and various installations from later periods were surveyed. The latter can be dated from the Hellenistic to the Byzantine period and include the two oil presses (see below), two *mikva'ot* (Jewish ritual baths), columbaria caves, water cisterns and burial caves.

The ceramic evidence suggests that the later settlement was founded in the Hasmonean period. The spring found to the north of the site is called 'Ein Qusis and according to the SWP the name of the site itself was also Qusis (Conder and Kitchener 1881–1883, III: 118). As a result Rosenfeld (1991: 215–217) has identified the site with Bet Meqoshesh, the home of the priestly family of Bet Caiapha.

The Oil Presses

The two oil presses were found in large man-made caves situated near one another. In both cases the installations found are from two stages: 1) the earlier a lever and weights press, and 2) the later a direct pressure screw press that stood in a cross-shaped recess cut into the wall of the cave.

Oil Press A (Fig. 1)

The installation is located in a man-made cave, rectangular in shape (11×4 m, 3 m high). The entrance of the cave is through a passage (length 3.1 m, width 1.4 m) and there are six rounded compartments cut into the walls of the cave, three on either side.

The crushing mill is close to the entrance, a part that has been specially widened to leave room both for the movement of the animal or worker who operated the olive crusher and for those engaged in other activities. The round crushing basin is 1.5 m from the entrance and directly opposite it. It is 1.6 m in diameter and has a flat crushing surface and a square sunken socket typical of the region. Two fragments of the crushing stone were found nearby. It was originally 0.8 m in diameter and the bore was square as were the two shallow depressions that surrounded the bore on both sides. A complete crushing stone was found in secondary use as a capstone of a cistern. It could have come from this oil press or perhaps from Press B, where no crushing equipment was found.

The pressing equipment is in the inner, western part of the cave close to the southern wall (Fig. 2). In the western wall of the cave about 1 m above the floor is the niche for anchoring the press beam (0.9×0.7 m, depth 0.5 m). Most of the weights pit is covered in debris but the western end is visible and is about 4 m from the beam niche. One beam weight was found close to the pit. It has a bore in the shape of a reversed T consisting of a horizontal bore 0.2×0.2 m in the upper part of the stone joined to the top of the stone by a vertical bore. A rod was inserted into the horizontal bore and a rope that was attached to it went through the vertical bore and was tied to a drum hang from the beam end. The pressing point was between the beam niche and the weights pit 2.5 m from the former. This consisted of a cylindrical rock-cut central vat (diameter 0.7 m, depth 0.8 m, volume 0.3 m^3). Four small depressions were located at equal distances around the edge of the vat and doubtless served to support the frame, slats or perforated board on which the frails of olive mash were placed. On either side of the vat there were two stone piers (height 1.7 m, width 0.65 m, thickness 0.55 m) one of which was found *in situ* and the other was found lying between the vat and the entrance of the cave. They served to support the stack of olive pulp frails and also to support the raised beam when it was not in use. On the inner faces of these piers that were both concave, there were also grooves in the shape of a reversed L and vertical grooves connected to horizontal ones along the top edge of the pier (Fig. 3). These grooves are unusual and possibly served to hold a wooden framework to support the beam. Halfway between the beam niche and the central collecting vat there was an additional press-bed; circular groove (diameter 0.85 m, maximum width 6 cm, depth 4 cm) connected to the vat by a groove 0.75 m long. Frails of olive mash were probably placed on this auxiliary press-bed when awaiting pressing.

In the Byzantine period a screw press was installed in a cross-shaped recess in the western end of the northern wall of the cave (Fig. 4). The threaded board was placed in the horizontal arms of the cross and the screw in the vertical section. The frails were placed in the lower part of the vertical section that was wider than the upper. The oil collection in cross presses was usually central with a vat being at the lower end of the cross-shaped recess. In this case, however, the oil collection is lateral. A press-bed consisting of a circular groove (diameter 0.79 m) is situated at the bottom of the cross and is connected to the collecting vat of the lever and weights press by a narrow channel 2.5 m long, that joins that of the auxiliary press-bed.

It is not clear whether the screw press replaced the lever and weights press or whether the two installations operated together.

Oil Press B (Fig. 5)

This was situated in a cave two terraces below that of Press A. The cave was almost square in shape (7.0×7.5 m). A rock-cut rectangular pier was left un-cut in the centre to support the ceiling. The outer eastern side of the cave was originally open to the outside but it was later closed with a well built wall leaving an entrance 0.8 m wide. In the northern wall of the cave were three rounded compartments similar to those in the cave of Press A. In the south-eastern corner of the cave was a small but typical *mikveh* (Jewish ritual bath) (Amit 1996: 22–23). This installation shows that the oil press belonged to Jews and that it was from the period of the Second Temple. It allowed for the production of oil under conditions of a high degree of ritual purity so that it could be sold in the markets of Jerusalem.

The pressing installations were found close to the southern wall of the cave (Fig. 6). Presumably the crushing mill was in the northern part of the cave but it was not found. It is possible that it is covered with debris but it is more probable that it was taken to be used somewhere else at a later period._

The pressing equipment of both stages is very similar to that of Press A. However, only those components cut in the rock remain while the removable parts such as the weights and piers were not found.

From the first stage remain the beam niche, the central collecting vat and the weights pit. The collecting vat was also cylindrical but instead of the four small depressions found in Press A there was one square depression around the rim. From the later stage remain the cross-shaped recess in which the screw press was emplaced and on the floor below it is a rock-cut press-bed from which a channel led to the collecting vat of the lever and weights press.

To the north of the beam and weights press near the beam niche and press-bed there is a rectangular working area in the corner of which there is a small sump. These clearly served the same purpose as the auxiliary press-bed of Press A.

The Oil Presses – Conclusions

The oil presses found at this site, both the lever and weights presses and the screw presses are very similar to other installations found in Judea. As regards the former the anchoring of the beam in a niche, the central collection vat with piers on either side and the weight with a bore in the form of a reversed T are all found in large numbers in the Judean Shephelah. Their presence at this site shows, as is to be expected, that as regards this aspect of technical culture the Judean hills are not different from the Shephelah. The cross-shaped screw press is also found in the Shephelah and particularly in and around Jerusalem. It must be pointed out that while in the Roman period the installations in the Judean hills are very similar to those in the Shephelah, in the Byzantine period there is a change. In the northern Judean hills from Hebron northwards the Cross press appears while to the south of Hebron the Grooved Pier press (the "Judean press") is predominant. The unusual elements in these presses that are worthy of note are the L-shaped grooves in the piers in Press A, the auxiliary press-bed of the same press and the lateral collection in the Cross presses.

The Wine Press

On the lowest of the four terraces on which the caves and installations were found there are the remains of a wine press (Figs. 7–8). The southern wall of the treading floor is missing but it was probably square (4.5×4.5 m, area 20.5 m^2). It was partly cut into bed-rock and as a result the northern and western walls that are rock-cut reach a height of 1.1 m. On the floor there are remnants of a mosaic floor made up of large white and roughly made tesserae (25 per square decimetre). The collecting vat measured 1.1 m and widened to 1.5 m at its floor (depth 1.6 m, volume 2.7 m^3). Below the rim, at a depth of 0.45 m a triangular step was cut into the wall to facilitate entering the vat.

Of particular interest is the fact that the walls of the vat were plastered five times and the types of plaster can be dated, allowing for the dating of the stages when the wine press was used. In the two bottom layers the plaster is white with white stone particles and can be dated to the Hasmonean period. The third layer consists of grey plaster with stones and is from the Herodian period. The plaster of the fourth layer consists of pottery shards and is Byzantine. The top layer is of modern concrete suggesting that in recent years the vat served as a water cistern.

Bibliography

Amit, D. 1996. *Ritual Baths from the Second Temple Period in the Hebron Hills*. Unpublished Thesis, The Hebrew University of Jerusalem (Hebrew).

Amit, D. and Yezerski, I. 2001. "An Iron Age II Cemetery and Wine Presses at an-Nebi Danyal". *IEJ* 51: 171–193.

Conder, C.R. and Kitchener, H.H. 1881–1883. *The Survey of Western Palestine*. 3 Vols. London.

Rosenfeld, B.-Z. 1991. "The History of the Settlement of Two Families from the High Priesthood during the Second Temple Period". In: Y. Katz *et al.* (eds.). *Studies in the Historical-Settlement Geography of the Land of Israel*, Vol. 2. Jerusalem: 206–218 (Hebrew).

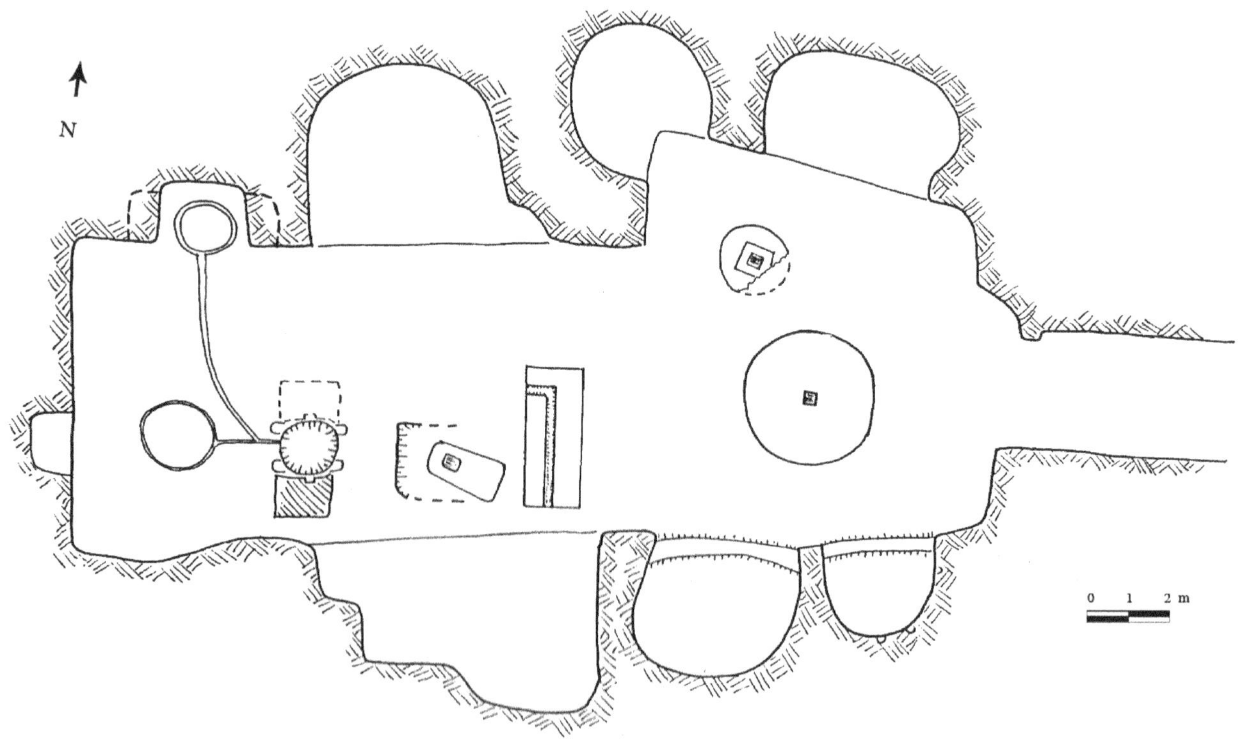

Fig. 58.1. Nebi Daniel, plan of oil press A.

Fig. 58.2. Nebi Daniel, oil press A, view to the west.

Fig. 58.3. Nebi Daniel, oil press A, grooved pier.

Fig. 58.4. Nebi Daniel, oil press A, cross press cut in the rock.

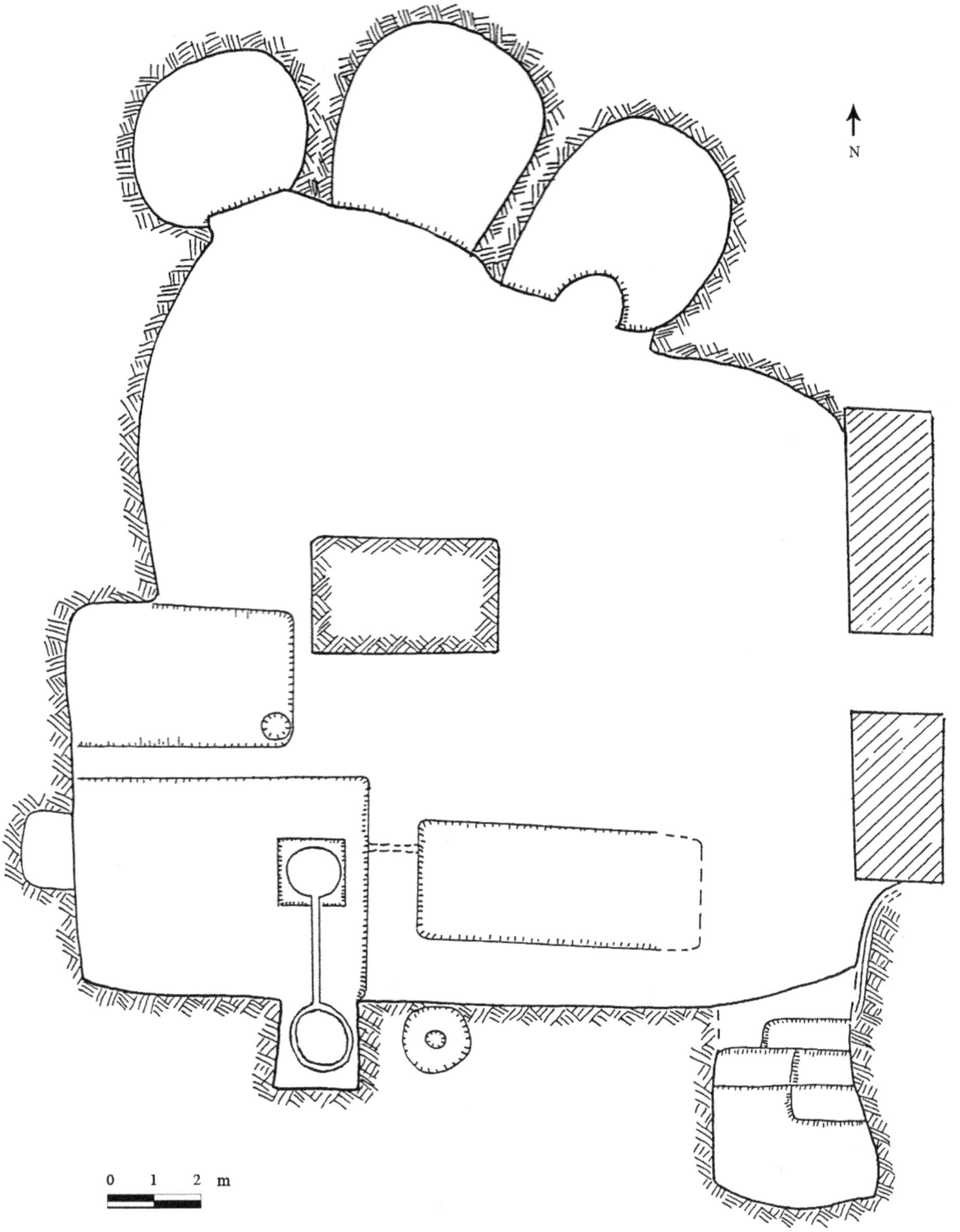

Fig. 58.5. Nebi Daniel, plan of oil press B.

Fig. 58.6. Nebi Daniel, oil press b, view to the south.

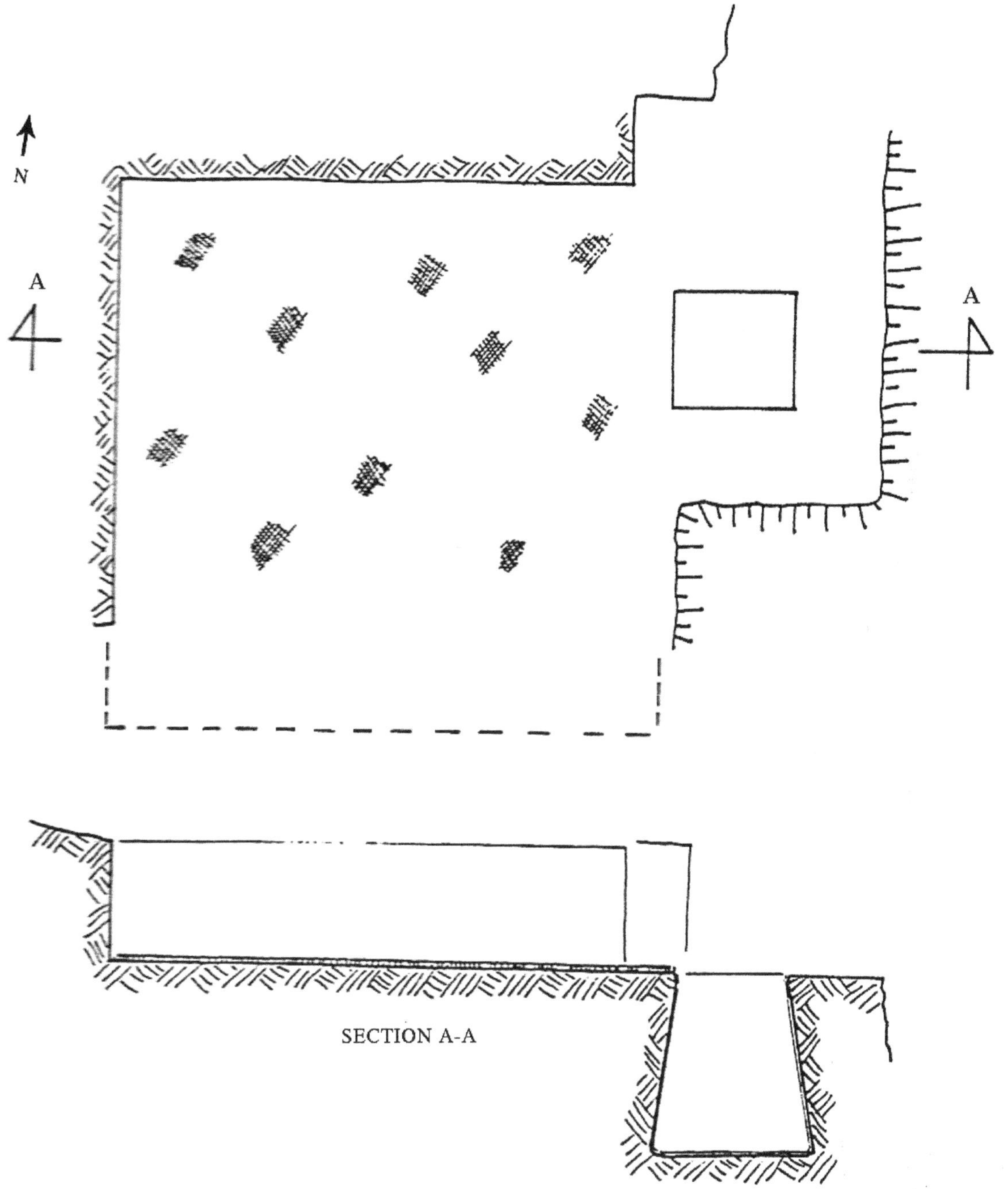

Fig. 58.7. Nebi Daniel, plan and section of the wine press.

Fig. 58.8. Nebi Daniel, the wine press, view to the north-east.

59. Byzantine Wine Presses at Ashqelon

Yigael Israel

The Site

The wine presses were unearthed in 1991 east of the Barne'a Quarter in Ashqelon (MR 1108/1216) in a rescue excavation headed by the author on behalf of the Israel Antiquities Authority (Israel 1993). They represent part of a large complex that included also an oil press (Site 60), four storage buildings, pottery kilns, a bathhouse and more (Fig. 1). Two large and sophisticated wine presses were unearthed in Areas C and E of the main complex. Another large wine press was discovered about 150 m south-east of Area E.

The Wine Press in Area C

The wine press was unearthed west of the storehouse complex and adjacent to it and north of the large pottery kiln (Figs. 2–3). Its overall area is 220 m^2, and its components were constructed in pits dug into the ground. Its walls were built of stones and hard mortar and well-coated with liquid proof plaster, and both the treading area and the vats were floored with slabs of limestone and marble. The wine press includes the following components:

- Compartments for "self-pressing" of the finest "first must" produced from grapes left standing there. This must, obtained without the grapes being trodden, may have been allowed to ferment in these compartments. Two of them, C206 and C207, were preserved south of the treading floor, and apparently there were two others to its north. They are rectangular, 3.5×1.4 m, and 1 m deep. Each compartment, therefore, could contain just less than 5 m^3. Each was covered by a vaulted roof constructed of stones and mortar, and apparently had two openings: one through which the grapes were put in, the other for removing them to the treading floor. A depression was built into the floor of each compartment for collecting dirt or remnants of must. The lead pipe connecting each compartment with the treading floor was probably used for draining the must or the water with which they were periodically rinsed. Compartments of this type are common in the wine presses of the Negev, the Shephelah and other places (Frankel 1999: 139).

- A plastered surface (C212) situated south-east of the treading floor. It is higher than the treading area, at the level of the top of its walls, and a step facilitated passing from one to the other.

- A treading floor (C202) measuring 6.5×6.0 m – 39 m^2. A pipe, 8 cm in diameter, in its western wall allowed the must to flow into the filtration vat. In the centre of the treading floor there was a large pit, in which originally stood a stone base for a wooden fixed screw press. At the bottom of the pit there was a channel running under the treading floor to the filtration vat, through which must remnants or rainwater could be drained off. The screw base itself was not preserved.

- A central filtration vat (C204), elliptical in shape, measuring 3.0×1.4 m and 0.5 m deep. A lead pipe was installed through the bottom of the wall on each of its long sides, through which the must flowed to the two collecting vats. The pipes opened into recesses built into the walls of the vats that served both as steps to enter them and to protect the floors from erosion by the steadily streaming must.

- The two large collecting vats (C203, C205) were in the shape of regular octagons – a rare characteristic in wine presses of this type. There are, however, similar collecting vats in the wine press at Negba (Site 61), about 8 km distant suggesting that this was a regional type. Each of the vats measures 3×3 m – a volume of approximately 10 m^3. In the centre of the floor of each vat is a double arrangement for settling out impurities (Fig. 4): a basin 1.4 m in diameter and 0.6 m deep with a marble floor in the centre of which was a small marble bowl 0.2 m deep.

- A working area (C201) extending around both vats. South of the collecting vats there were two large semi-circular recesses in the wall of compartment C206. There were probably similar recesses around the other sides of the area. The purpose of these recesses is not clear; they possibly served to hold large storage vessels.

The thickness of the walls suggests that the vat area may have been covered by a roof, perhaps a vault that has not survived. This would have aided in protecting the must from heat and sunlight and in controlling the fermentation process. It is the author's opinion that a lightweight roof of organic materials – wood, wool or other cloth, etc. – was also installed above the treading floor during the winemaking season.

The Wine Press in Area E

The wine press was discovered east of the oil press and the storehouse adjacent to it (Fig. 1, front). Its overall area is approximately 160 m^2. Originally it was larger but apparently when the storehouse was built the western section was destroyed. It is very similar to the wine press of area C both in form and size and consists of the following components:

- "Self-pressing" compartments. There were originally at least six, arranged on the west, south and east of the treading floor. Four of them were well preserved (E420, 422–424, 436, 463). Those on the western side, however, including vat E464 were destroyed when the storehouse was built to the west of the wine press.

- A large treading floor (E421) measuring 6.0×5.5 m – 33 m^2, with a pit in its centre from which the stone base of a wooden fixed screw was removed in the past.

- A rectangular sedimentation vat (E428) measuring 2.5×1.2 m, 0.8 m deep.

- Two square collecting vats (E427, 429) each 2.5×2.5 m and up to 2 m deep, each with a maximum volume of 12.5 m^3.

- Work surfaces.

Discussion

Aside from their handsome form and the caliber of their construction, the wine presses are outstanding for their ability to produce must on three different levels of quality: the choicest must, expressed by "self-pressing" in the small compartments; ordinary must, produced by "foot-power" on the treading floor; and the inferior type of must produced by squeezing the remnants of trodden grapes in a screw press. The multitude of compartments and vats enabled separate storage of each type of must. Another point that should be mentioned is the similarity of the wine presses in shape to those found in the Negev. Various scholars have already suggested that some of the "wines of Ashqelon and Gaza" (Mayerson 1985, 1993) that were so famous during the Byzantine period were actually produced in the wine presses of the Negev (Site 70), that represented the "hinterland" of these port cities. The discovery of wine presses at Ashqelon itself, similar in shape and construction to those of the Negev (as well as a kiln where the typical "Ashqelon Jars" [Mayerson 1992] were fired in which the wine was marketed) reinforces this connection. It should be pointed out, however, that in the Negev wine presses it was not possible to separate the various types of wine suggesting that they produced one type of wine only probably for export while the wine presses from Ashqelon near the urban centres catered for other markets also.

Bibliography

Frankel, R. 1999. *Wine and Oil Production in Antiquity in Israel and Other Mediterranean Countries*. Sheffield.

Israel, Y. 1993. "Ashqelon". *ESI* 13: 100–105.

Mayerson, P. 1985. "The Wines and Vineyards of Gaza in the Byzantine Period". *BASOR* 257: 75–80.

Mayerson, P. 1992. "The Gaza Wine Jar (Gazition) and the Lost Ashqelon Jar (Askalonium)". *IEJ* 42: 76–80.

Mayerson, P. 1993. "The Use of Ascalon Wine in the Medical Writers of the Fourth to the Seventh Centuries". *IEJ* 43: 169–173.

Fig. 59.1. Ashqelon, general view of the site, looking north-west. The wine press in Area E is in the front.

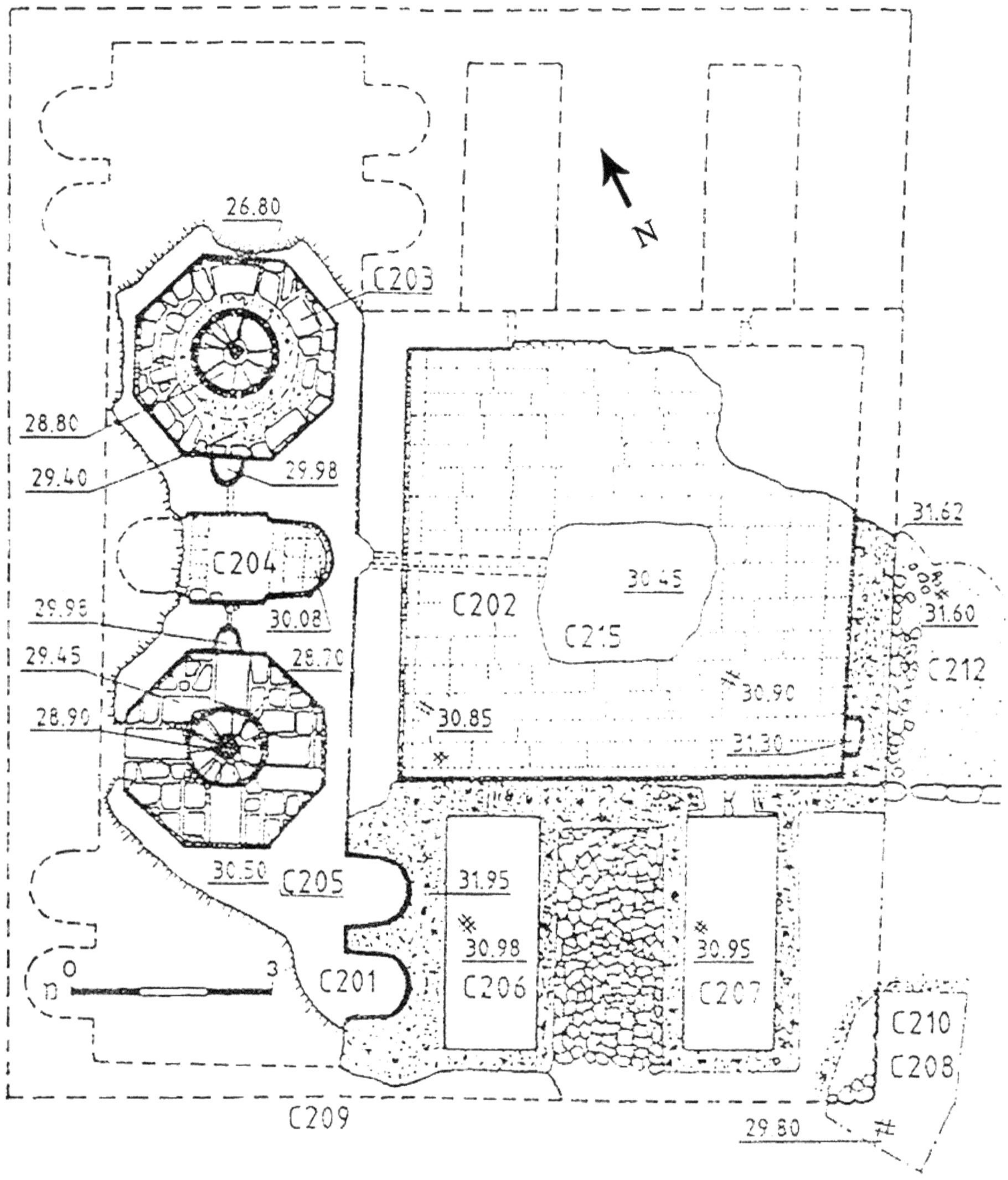

Fig. 59.2. Ashqelon, plan of the wine press in Area C.

Fig. 59.3. Ashqelon, the wine press in Area C, looking south-east.

Fig. 59.4. Ashqelon, wine press in Area C, paved bottom of collecting vat with settling basin.

60. A Byzantine Oil Press at Ashqelon

Yigael Israel

The oil press was unearthed in 1991 east of the Barne'a Quarter in Ashqelon (MR 1108/1216) in a rescue excavation headed by the author on behalf of the Israel Antiquities Authority (Israel 1993). It represents part of a large complex that also includes wine presses (Site 59), four storage buildings, pottery kilns, a bathhouse and more.

The oil press was discovered in the centre of the complex between the two large wine presses. Its total area is approximately 290 m^2 or, including the second storey, 330 m^2 and it consists of a large hall measuring 14.5×10 m plus additional rooms. To the east of the oil press was a wine press and between them stood one of the storage buildings. The walls of the oil press were built of dressed sandstone (*kurkar*) blocks and coated with white plaster. The following components were found in the main hall (Fig. 1):

- A crushing mill, the parts of which – crushing basin and crushing stone – were not found *in situ* but in a tank at the northern end of the hall (Fig. 2). It appears that originally the mill was situated in the centre of the hall south of the tank. The two items were well designed and finely cut. The basin, 2.2 m in diameter, had a raised rim and projecting socket. The diameter of the crushing stone is 1 m and it is 0.5 m thick.

- Two presses of the lever and weights[1] type, located at the two sides of the hall. The back end of each beam was apparently anchored in a niche located in the southern wall of the hall. The press-bed, hewn from hard limestone, is square, measuring 1.2 m on a side, with a circular groove terminating in a protruding channel (i.e. lateral collection). The expressed liquid flowed through the channel into a collecting vessel – a large ceramic jar set into a hole dug in the ground under the mouth of the channel and supported by stones and mortar. It is 0.8 m deep, with a maximum diameter of 0.6 m.

In each of the pressing installations there were three very unusual weights. They are of hard limestone, their average dimensions are 1.20×0.55×0.55 m, and therefore their average weight would be approximately 970 kg. Two dovetail mortices were cut into the narrow ends, terminating about one third of the way up and connected by a channel cut across the top of the weight. Both the channel and the mortices had rectangular cross-sections, ~0.10×0.15 m. A hole some 0.15 m in diameter was bored through the stone horizontally, connecting the mortices at their lower end. These weights are very different from the usual beam weights found in Israel. The mortices and upper channel are very similar to those of the Semana Weight found in the western Mediterranean (Frankel 1999: 102–106). However, the hole bored through the stone does not appear in the Semana weight but is found in nearly all weights in Israel. Nevertheless, these weights almost certainly functioned like the Semana weights in which the drum that lowered the beam and raised the weight was fixed to the weight and not to the beam as it was in other lever and weights presses in the Levant. The base plank of a drum fit into the top channel and its side planks were fixed in the mortices by a wooden rod passing through them and the hole in the stone weight. The drum was attached to the beam by a rope. Gradual operation of the drums of the three weights pulled the beam down, thus accomplishing the pressing. The location of the weights shows that the beam was about 10 m long. There was no evidence for piers as found in many presses that served to support and guide the beam.

- North of the western press were found the two components of a basalt "hourglass mill" of the Pompeiian type (Fig. 2). The conical lower stone was found further along the line of the three beam weights. A horizontal hole had been bored through it near the top, apparently to enable its use as an additional beam weight. The upper stone was found still further away, on top of a few stones. It may be that when it was needed, the installation was set up in the centre of the hall to be used for crushing small amounts of olives or for milling grain.

- A constructed, plastered tank was discovered at the centre of the northern part of the hall, measuring 3.0×2.5 m; its depth is unknown because the parts of the crushing mill thrown onto it precluded its excavation. Because of its proximity to the original location of the crushing mill it appears to have been used for holding the frails of olive mash before pressing. Thus it was possible to collect in the tank the fine "first oil" that seeped from the olive mash under its own weight.

North of the hall three rooms were found measuring 3.0–4.0×2.0–2.5 m. In the central chamber a round stone base was preserved, about 1 m in diameter that may have been intended to support a large vessel. Two other rooms were discovered south of the hall, measuring 3.5–4.0×4.0 m. The eastern one, in which a ceramic jar and strainer were unearthed, was found covered by debris from the collapse of the second storey that had a colored mosaic floor. Parts of this mosaic were also found in the south-east section of the hall.

The oil press was dated as was the complex as a whole to the Byzantine period. The Persian conquest of Palestine in 614 CE may have brought about its abandonment.

The oil press was well built and clearly "industrial" in character. This is attested by the presence of: the pair of presses that enabled continuous pressing of olive mash, each installation in turn; the tank; the adjacent rooms and the second storey, in which the oil was processed and refined and in which management activities may have been carried out; the adjacent storehouses; the pottery kilns discovered nearby, where the jars may have been made in which the oil was packaged and marketed. It should be noted that although the presses are from the Byzantine period they were lever and weights presses and not screw presses as was common at that period. The special shape of the weights should also be noted; they are not known from northern or central Israel. To date, similar weights have been discovered only in the Ashqelon area, near Kibbutz Erez and Sderot, and probably represent a local subtype. This type probably originated in the western Mediterranean and was of a type in which the drum was attached to the weight and not to the beam.

Notes

1. The author prefers to use the term "anchor" to describe the stone to which the winch was attached. However, for purposes of uniformity, the editors use the more common term "weight".

Bibliography

Frankel, R. 1999. *Wine and Oil Production in Antiquity in Israel and Other Mediterranean Countries*. Sheffield.

Israel, Y. 1993. "Ashqelon". *ESI* 13: 100–105.

Fig. 60.1. Ashqelon, the oil press, view to the north-east.

Fig. 60.2. Ashqelon, the crushing mill (in background) and the Pompeiian mill (in front).

61. The Byzantine Wine Press on Hill 113 (Negba)

Dov Meiron

On the fringes of Hill 113 (MR 1185/1188), adjacent to Kibbutz Negba and approximately 12 km east of Ashqelon, remains have been found of a settlement or farmstead from the Byzantine period. In 1970 a tomb from that period was unearthed in the area, but otherwise the site was not investigated. In 1983, during the course of agricultural work, remains were discovered of an installation that was subsequently excavated by the author in a salvage operation on behalf of the Israel Antiquities Authority. Although the tops of its walls have been destroyed, most of the installation was well preserved and is still visible today.

The installation (Fig. 1) is a wine press of the large "public" type; its total dimensions exceed 15×10 m. The walls are constructed of a solid mixture of mortar with small stones and a great many small and broken marine shells. They are coated with several layers of plaster reinforced with ribbed pottery shards and many bivalve shells. The wine press consists of the following components:

• Five rectangular compartments arranged west and north of the treading floor. Each measures 3–4×1.5 m and their walls have been preserved to a height of 0.35–0.45 m. They are floored with coarse white mosaic. Integrated into the floor of each compartment are two stone slabs, roughly rectangular and measuring approximately 0.3 m on a side. Where they are inserted into the floor, they are surrounded by two or three rows of tesserae parallel to the joints, while over the rest of the floor area the tesserae are set at random or in diagonal rows. The slabs may have served as bases for roof supports. The south wall of the southernmost compartment begins to curve upward (the upper part is missing), and in light of parallels found in similar wine presses of the same period it is possible that the compartments had vaulted roofs (Frankel 1999: 139). The compartments were probably used for storing grapes before treading or perhaps for initial fermentation of the must that leaked out of the grapes naturally. In each of the three western compartments a lead pipe 6 cm in diameter was preserved, through which the must flowed from the compartment to the treading floor. The walls between the other two compartments and the treading floor have not survived but it is probable that these compartments were also connected to the treading floor by pipes. The state of preservation of the remains in the area and the symmetry of the vat arrangement and the compartments suggest that there were no additional compartments to the south of the treading floor. In addition, it is possible that the north-west corner compartment was added at a later stage, since it differs from the others in its dimensions, its alignment, and in that its walls are coated with only a single layer of plaster.

• A square treading floor, 6×6 m – area 36 m², paved with well-fitted rectangular stone slabs (preserved only in the north-east corner) set into a solid infrastructure of small stones and mortar. It appears that its walls were also covered with stone slabs as the lower parts of several have been preserved on the western wall. A large pit measuring 2.0×1.5 m was discovered in the centre of the floor, from which a large stone had apparently been removed. Such a stone may have been the base for a fixed wooden screw used for secondary pressing of the rape that remained after the grapes had been trodden (Frankel 1999: 140–141). A channel that leads under the floor from the pit to the filtering vat probably served to drain the must produced in the treading floor. The channel, about 0.2 m high and wide, was lined and covered with stone slabs. In addition to that channel, there may originally have also been an upper channel that also connected the treading floor to the filtration vat – a phenomenon known from wine presses in the Negev – but no signs of it remain.

• An elliptical vat for filtering the must, measuring 3.0×1.7 m and aproximately 0.6 m deep, located between the two collecting vats. In the centre of its floor, that was tiled with white tesserae, was set a round vessel (diameter and depth 0.3 m) intended for collecting solid material that settled out of the must. Two pipes, 6 cm in diameter, emerged from the filtration vat at floor level and led to the two collecting vats situated north and south of it. Permanent filters could have been fitted to these pipes, or bundles of thorny plants could have been used.

• Two octagonal collecting vats, each measuring 3×3 m. The depth of the northern one is 1.2 m and of the southern 1.15 m; their volumes are therefore 8.9 m³ and 8.5 m³ respectively. In the centre of the floor of each vat is a two-level sedimentation basin with a maximum diameter of 1.25 m and maximum depth of 0.8 m. The floor of the southern basin is finished with coarse white mosaic tiles.

In the south wall of the northern collecting vat there is a semi-circular niche 0.55 m in diameter. Its floor, consisting of a stone slab, is at the level of the pipe coming from the filtration vat. The niche may have been built to enable the use of a pipe shorter than the thickness of the separating wall (0.65 m) or to aid in further filtration of the must. A similar arrangement apparently existed at one time in the southern vat but it has been destroyed and today there remains only a cavity.

This large wine press, because of its form and components, can be definitely dated to the Byzantine period. It belongs to a type of installation known from the Negev highlands and the southern Coastal Plain. The best parallels to it are the two wine presses from Ashqelon (Site 59) that were found only 8 km away to the north-west. Typical of these wine presses are the channel under the treading floor, the octagonal shape of the collecting vats, and the use of stone slabs flooring (that exists here only on the treading floor) instead of mosaic.

Bibliography

Frankel, R. 1999. *Wine and Oil Production in Antiquity in Israel and Other Mediterranean Countries*. Sheffield.

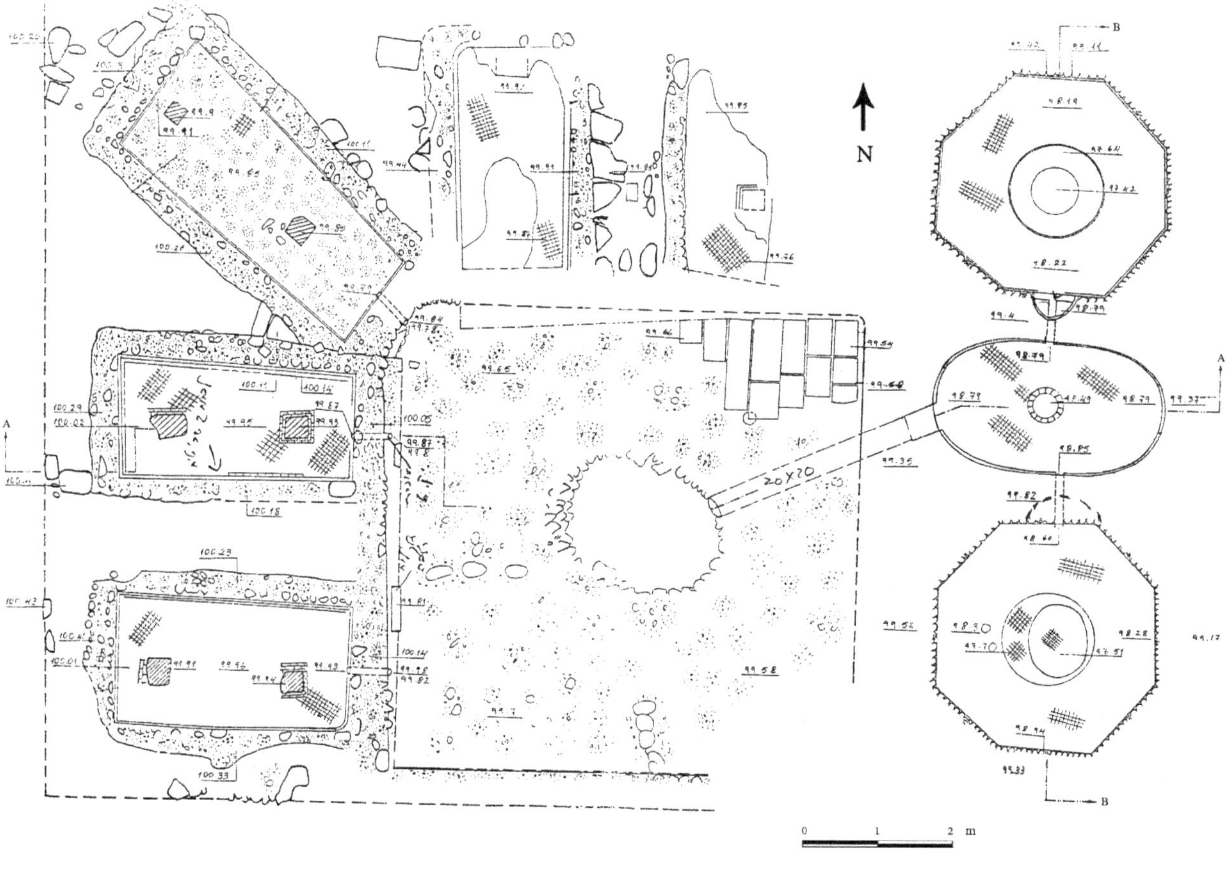

Fig. 61.1. Negba, plan of the wine press.

62. A Hellenistic Oil Press From 'Aderet

Jon Seligman

The Site

'Aderet is a farmhouse complex located on a hill (MR 1495/1185) to the south of the Elah Valley, 2 km south-east of Tel Socho and 11 km north-east of the important Hellenistic site at Maresha (Beth Guvrin).[1] This was an agricultural complex comprising two basic parts: a farmhouse and a courtyard with farm buildings, at one side of which was located an oil press. The important ceramic assemblage has yet to be processed and the house was thus dated by the numismatic evidence and group of three *morgantania* (medicine bottles) to the 3rd cent. BCE.

The Farmhouse

The farmhouse is a large stone rectangular building measuring 28×11 m (Fig. 1). It is sub-divided into two units of small rooms serviced by a single closed courtyard. The external wall continues on its east and west sides as the enclosure walls of a farmyard of unclear size. Entry to the farmyard was through a gate situated at the south-east corner of the farmhouse. An ancient road connects to the entrance to the complex.

On the western edge of the courtyard are a number of rooms that contain the installations of the oil press.

The Oil Press (Fig. 2)

The oil press complex consists of a round crushing basin and an improved lever press with a central collecting basin connected to a lateral vat.

On the bed-rock floor of a room at the southern end of the complex (3×4 m) is a crushing basin standing 0.75 m high, with a diameter of 1.35 m (Fig. 3). It is cut from hard limestone. The edge is raised above the crushing surface around 5–10 cm and there is a slightly raised socket in the centre that held the pivot for the power shaft that held the crushing stone. The walls, though close to the basin, are at a suitable distance to allow circular movement of the animal operating the mill. The floor was made of beaten earth and leveled rock. One entrance was found to the room. It led northwards over a stone threshold to an adjacent room where the press was situated. It is possible that a further opening existed to the south to an irregular, open courtyard that probably served to store the olives before crushing.

In the room to the north was a lever press used to extract the oil from the crushed olives brought from the basin to the south. The length of the room is unclear as the northern wall is unknown, but the width is 3 m. The room has one and possibly two entrances, both to the east. The first is opposite the pressing installation from the farmyard and the second, that is less certain, opens to the north of the press-bed. The press-bed was shallow, rectangular (1.55×1.10 m) and plastered. In its centre was a round shallow vat (diameter 0.55 m, depth only 0.3 m). On either side were plain stone piers, one of which was found fallen over the press-bed (Fig. 4). The bases of the piers were plastered similarly to the rest of the installation. A small gutter led from the central vat to a lateral collecting vat to the south. This vat was round with a diameter of 1.5 m and was 0.8 m deep (volume 1.4 m^3). It was covered with a hard plaster layer that continued over its edge to cover the floor around the pressing installation. No stone weights used in the pressing process were found in this area, though three were found in secondary use in the farmhouse, two with a reversed T shaped bore (Fig. 5) and the other doughnut shaped with a vertical bore. The pressing beam was probably anchored in the wall behind the collecting vat, though essential parts of this wall were missing.

West of the oil press were two rooms that contained many fragments of jars and large quantities of ash. The finds point to the utilization of these rooms as roofed storage areas for the oil produced here.

To the east of the oil press is an enclosed and leveled area that was not fully excavated. Cut into the bed-rock were three interconnected depressions to allow liquids to spill from the higher depression through to the lowest. It may well have been used to separate liquids of different specific gravities, such as oil from watery lees.

Regional and Chronological Features

Oil presses clearly dated from the early part of the Hellenistic period are not common. Nevertheless a number of installations, some in the immediate vicinity have the same chronological range, and improved lever presses similar to that discovered at 'Aderet became standard in the Hellenistic and Roman periods (Frankel 1984: 127; 1999: 76). A central feature of the new technology was the construction of stone piers on either side of the press-bed, usually monoliths as in 'Aderet, to support the pile of frails and to hold the beam in place while working beneath it before and after the pressing.

While this press type is quite common, and the earliest

examples are to be dated to the Hellenistic period, comparable presses of a similar date are few and far between. An early lever and weights press was found at Tirat Yehuda (Yeivin and Edelstein 1970). Though the whole installation is enclosed within a single hall it was made up of a crushing basin, a press-bed between two plain stone piers and a lateral collecting vat, slightly to one side. The weights have reversed T shaped bore, similar to two at 'Aderet. The site is dated to the Hellenistic period (330–180 BCE).

A number of oil presses found at the nearby site of Maresha are of similar date. These are located within caves, using the bed-rock in place of built parts of the installation (Kloner and Sagiv 1993; 2003; Sagiv and Kloner 1996; Site 65). All have typical southern central collecting vats (Frankel 1984: 204–217; 1999: 84–85; Frankel, Patrich and Tsafrir 1990: 292–297). The press at 'Aderet, located as it is in the southern Shephelah district, certainly falls within this regional grouping. An unusual feature at 'Aderet, though, is an additional larger lateral collecting vat connected to the central vat by a small gutter. This would have allowed continuous operation of the press without dismantling the frails from the press-bed over the central collecting vat.

The 'Aderet oil press falls into a large group of improved lever and weights presses with plain piers, central collection of the expressed liquid and weights with a reversed T bore. Frankel gives this group the type number T401110002 and attributes 19 other presses to it (Frankel 1999: 68 of attached CD-ROM). The importance of 'Aderet is clear being that it is one of the earliest presses of this group.

Notes

1. A salvage excavation was conducted at the site by Ora Yogev of the Department of Antiquities and Museums in 1981 following the expansion of a chicken house at the nearby village of 'Aderet (Yogev 1982).

Bibliography

Frankel, R. 1984. *The History of the Processing of Oil and Wine in Galilee in the Period of the Bible, the Mishna and the Talmud.* Unpublished Ph.D. Thesis. Tel Aviv University (Hebrew).

Frankel, R. 1999. *Wine and Oil Production in Antiquity in Israel and other Mediterranean Countries*. Sheffield.

Frankel, R., Patrich, J. and Tsafrir, Y. 1990. "The Oil Press at Horvath Beit Loya". In: G.C. Bottini, L. Di Segni and E. Alliata. (eds.). *Christian Archaeology in the Holy Land. New Discoveries*. Jerusalem: 287–300.

Kloner, A. and Sagiv, N. 1993. "The Olive Presses of Hellenistic Maresha". In: M.-C. Amouretti and J.-P. Brun (eds). *Oil and Wine Production in the Mediterranean Area* (BCH Supplément 26). Paris: 119–135.

Kloner, A. and Sagiv, N. 2003. "Subterranean Complexes 44 and 45". In: A. Kloner. *Maresha Excavations Final Report I. Subterranean Complexes 21, 44, 70* (IAA Reports No. 17). Jerusalem: 51–72.

Sagiv. N. and Kloner, A. 1996. "Maresha: Underground Olive Oil Production in the Hellenistic Period". In: D. Eitam and M. Heltzer (eds.). *Olive Oil in Antiquity. Israel and Neighbouring Countries from the Neolithic to the Early Arab Period.* Padova: 255–292.

Yeivin, Z. and Edelstein, G. 1970. "Excavations at Tirat Yehuda". *'Atiqot* (HS) 6: 56–67 (Hebrew).

Yogev, O. 1982. "Aderet". *ESI* 1: 1.

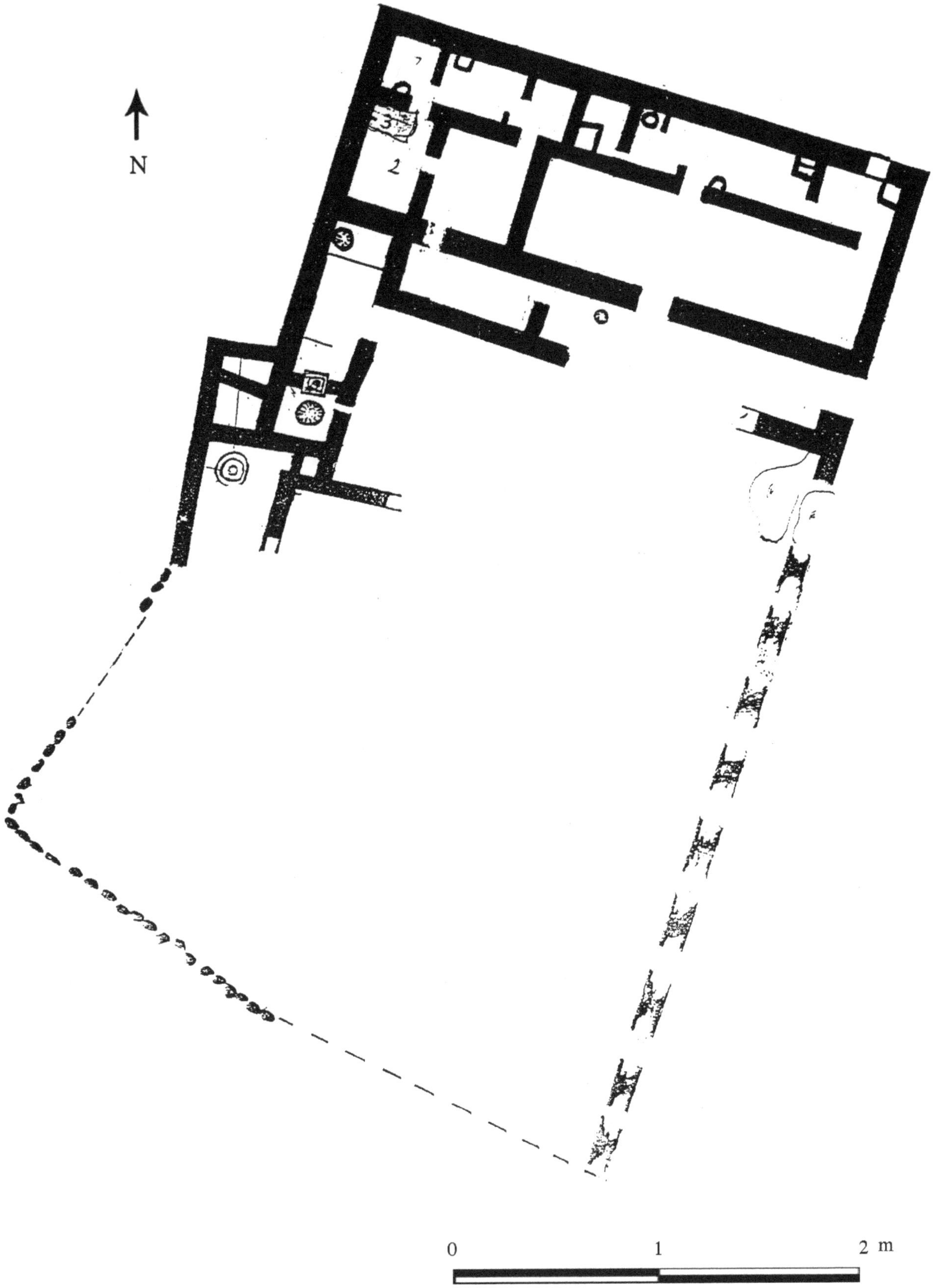

Fig. 62.1. 'Aderet, plan of the farmhouse.

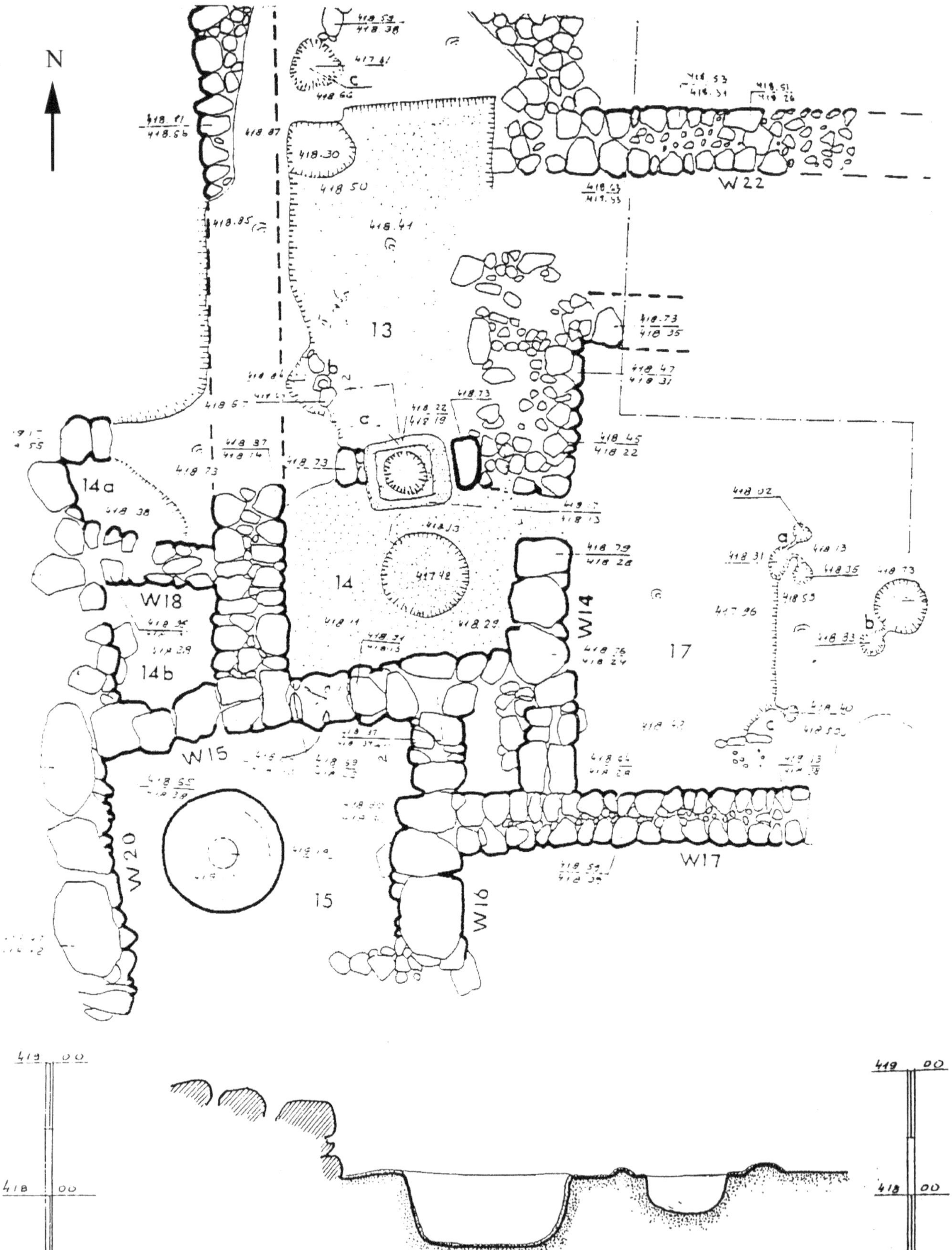

Fig. 62.2. 'Aderet, plan and section of the oil press.

Fig. 62.3. 'Aderet, the crushing basin, looking north.

Fig. 62.4. 'Aderet, the oil press; the east pier rests over the central collecting vat with lateral vat at front, view to the north.

Fig. 62.5. 'Aderet, press weight with reversed T bore in secondary use in late wall of the farmhouse.

63. Jedur*

Etan Ayalon, Rafael Frankel and Amos Kloner

A twin wine press (Fig. 1) dated to the Byzantine period was excavated in 1933 at Jedur (MR 1588/1156) and published by Baramki (1935). Both installations are partly cut in bed-rock and partly built and plastered. Each consists of a treading floor (A & B), an intermediate vat (a & c) and a collecting vat (b & d) reached by steps. In both there is also a niche in the northern wall, in A built and in B cut into the rock face. These served to anchor beams of lever and weights presses. The connection between the treading floor (A) and intermediate vat (a) of the western press consists of a bore that leads to a channel cut into a protruding lip. The bore made it possible to control the flow of the must from the treading floor to the vat and the protruding lip made it possible to direct the must directly into a jar. The latter arrangement was probably to keep separate the inferior must from the rape left after treading that was pressed using the lever and weights press. The intermediate vat of the western wine press (a) is clearly a settling tank as the outlet is at the rim of the vat and not at the bottom as it is in those intermediate vats that were used to clean the must by sieving. No section is given for the intermediate vat of the eastern wine press so that it is not certain that it functioned in the same manner.

Biblography

Baramki, D.C. 1935. "Recent Discoveries of Byzantine Remains in Palestine – 3. A Winepress at Khirbet Jedur, near Beit Ummar". *QDAP* 4: 121.

Dimensions of Wine Presses

	Treading Floor			Intermediate Vat				Collecting Vat			
	Length	Width	Area (m^2)	Length	Width	Depth	Volume (m^3)	Length	Width	Depth	Volume (m^3)
West	A			a				b round part			
	4.35	4.15	18.05	1.2	1.35	0.5	0.81	2	2	2.3	7.22
								b steps			
								1.1	1.35	1	1.48
								b total			8.7
East	B			c				d			
	3.1	3.8	11.78	0.9	1.2	?	?	1.9	0.9	1.2	2.05

* Dimensions taken from plans (Baramki 1935)

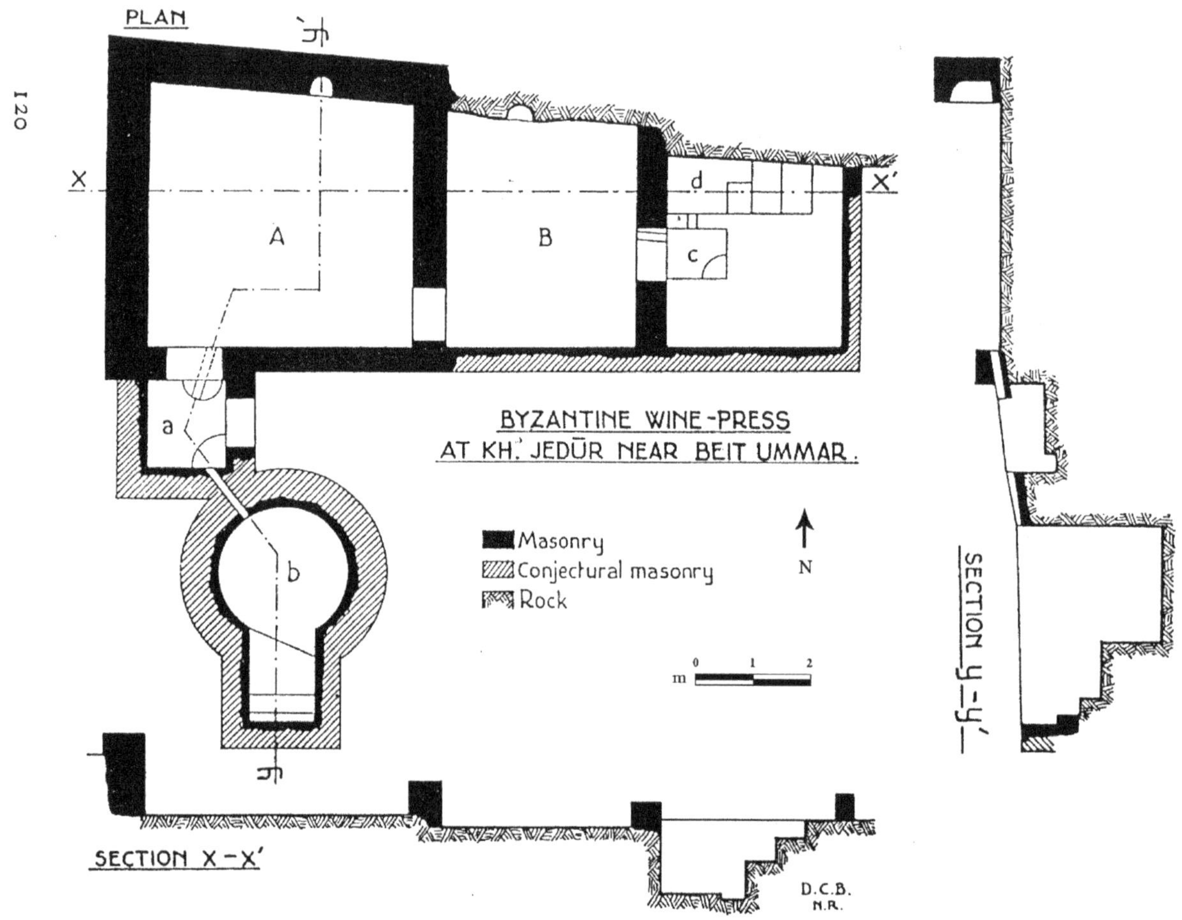

Fig. 63.1. Jedur, plan and sections of the wine press.

64. Wine Press near 'Iyye Nahash

Boaz Zissu

While engaged in routine inspection A. Klein of the Israel Antiquities Authority discovered burial caves and rock-cut installations in the vicinity of 'Iyye Nahash, 2.5 km east of Beit Guvrin (MR 1427/1131). The wine press was surveyed by the author in May 1997. It is situated on the upper part of the northern slope of a hill south of the site. A section of the eastern aqueduct of Beth Guvrin / Eleutheropolis passes close by to the south of the press.

The installation (Figs. 1–2) is cut in bed-rock and consists of a central treading floor, two shallow auxiliary floors, a settling tank and a collecting vat. The treading floor is square (3.5×3.5 m – area of 12.25 m^2, depth 0.3 m) and paved in white mosaic (tesserae 2×3 cm in size) of which only remnants survived at the edges. The mosaic was laid on a bed of white cement that included many small pebbles. There was no evidence for equipment for secondary pressing – neither a beam niche nor a screw base were found.

The auxiliary treading floor south of the main floor measures 2.1×1.75 m – area 3.7 m^2 and that to the east 1.4×1.6 m – area of 2.25 m^2. In neither case does a partition separate the floor from the main treading floor.

North of the main treading floor there is a rectangular cavity that is divided into three parts by an L-shaped partition:

1. A settling tank (0.7×0.7 m, depth 0.3 m). The partition between the treading floor and the vat has not survived but the bottom of the bore that connected them remained showing that it entered the tank at the south-eastern corner. The settling tank was paved in mosaic and its walls were plastered.

2. A collecting vat the original dimensions of which were 1.5×1.4 m, depth 1.5 m – volume 3.2 m^3. The floor of the vat was also paved with mosaics and the walls plastered but little remains of either.

3. A flight of steps located north of the settling tank descended to the collecting vat but only one step survived (width 0.5 m, height 0.25 m).

At a later stage the collecting vat was deepened to become the entrance shaft from which a burial cave was cut consisting of a central area and three *arcosolia*. At yet a later stage the whole complex served as a water cistern.

Conclusions

The wine press is probably to be dated to the Late Roman or Byzantine period. The burial cave is typical of the same period and the installation was probably used as a water cistern till modern times.

The wine press represents an interesting development of the "four rectangle plan" (Frankel 1999: 149–150) in that the settling vat, collecting vat and flight of steps all fit into yet another rectangle. It was also well built paved with mosaic and plastered. Therefore two questions arise that are difficult to answer: Why did the wine press lack equipment for secondary pressing? And why was it abandoned and turned into a burial cave at a period when wine production in the country as a whole was at a peak?

Bibliography

Frankel, R. 1999. *Wine and Oil Production in Antiquity in Israel and Other Mediterranean Countries*. Sheffield.

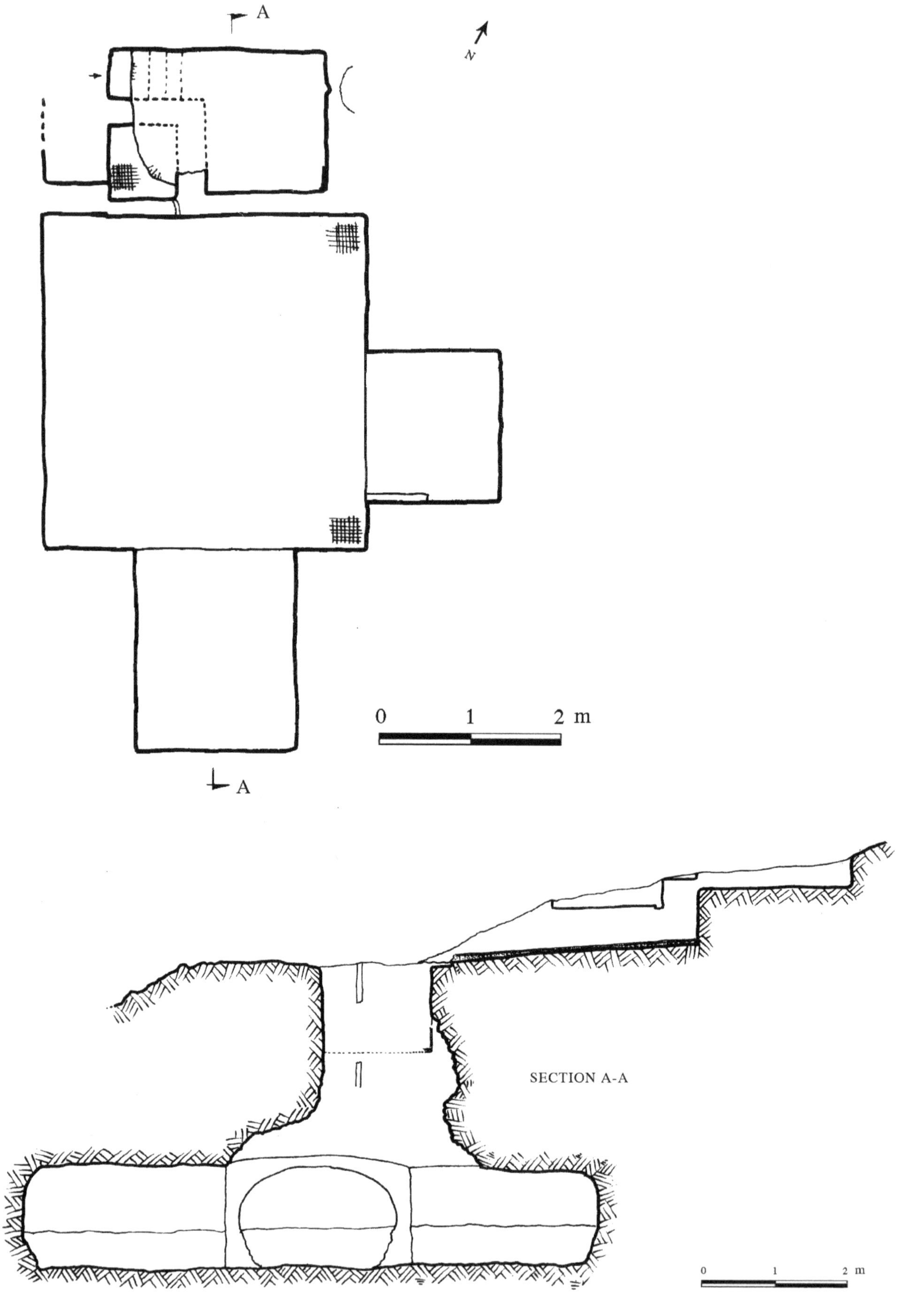

Fig. 64.1. 'Iyye NaHash, plan and section of the wine press.

Fig. 64.2. 'Iyye NaHash, the wine press and cavity, view to the south-east.

65. The Oil Presses of Maresha

Amos Kloner

The Site

The ancient city of Maresha is identified with Tell Sandahannah that is situated in the Shephelah region 39 km east of Ashqelon (MR 140/111). It was a Biblical city. After the destruction of the first temple Maresha with all southern Judea became Edomite territory and in the Hellenistic period it became the main city of Idumea (Edom). There is evidence for the presence of a Sidonian community in the city during this period. The city was captured by the Hasmonean king John Hyrcanus I in 125 BCE and destroyed by the Parthians in 40 BCE never to be rebuilt.

The site consists of an upper city, a lower city and a large number of man-made subterranean complexes consisting of caves, columbaria, water cisterns, store rooms, oil presses, etc. (Fig. 1). The upper city was excavated in 1900 by Bliss and Macalister who identified three strata, two Hellenistic and one from the Iron Age. The city in its last two phases was fortified, almost square in plan and measured about 24 dunams. The lower city that measures some 320 dunams has been partly excavated during recent years.

Macalister identified 63 subterranean complexes and 25 more have been identified since. In these, 25 oil presses were found that will be described here briefly (Figs. 2–10).

The Maresha Oil Press

The oil presses of Maresha are all very similar in character. As they are in man-made caves for archaeologists and for students of ancient technologies they have the great advantage that except for the perishable components that were of wood, leather and similar materials they have survived complete up to the ceiling. There is also no doubt as to the date of the presses, all being from the Hellenistic period. The caves and presses were also beautifully cut.

The crushing basins had a concave crushing surface and the crushing stones were convex in order to fit them (Figs. 4, 9). There was it seems only one crushing stone to each crushing basin and usually one crushing basin for two presses. The crushing basins are sometimes placed in the main press room but in some cases they are in specially cut round adjuncts often with large semi-circular niches in the walls doubtless to place olives both before and after crushing.

The presses are lever-and-weights presses (Fig. 5, 10). The beam that varied in length approximately from 4.5–7.5 m was anchored in a niche cut in the wall. The niches were usually elongated vertically but their final shape was almost certainly a result of wear and apparently originally they were circular. A rock-cut wall about 1.5 m high divided the area at the back where the beam niche was found from the main room of the press. The collecting vat was at the bottom of a cylindrical aperture that was cut into this wall and that was open both to the back and the front. The frails of olive pulp were piled up over the opening of the collecting vat and held in place in the cylindrical aperture, the sections of wall on either side acting as press piers. The beam was raised and lowered in the opening in the cylindrical aperture pressing down on the frails and the expressed liquid seeped straight down into the vat. In front of the collecting vat there was often an elongated shallow rock-cut basin in which the jars for the oil and watery lees probably stood. Each press was usually equipped with three weights of the type with a bore in the form of a reversed T. The weights were in a rectangular pit that allowed the beam to descend to the maximum. There was usually an entrance in the dividing wall to enable passage between the back room and the main press room. The presses often came in pairs and then the approach to both back rooms was through one entrance in the wall that was usually between the two presses. One unusual feature in the Maresha press rooms was a small rock-cut altar located in a niche in the wall often above the entrance in the dividing wall (Fig. 5).

In several presses it was possible to distinguish signs of repair. As already mentioned above the niches in which the beams were anchored were often enlarged and elongated as a result of wear and in some cases indentations were cut on either side of the niche in which a cross piece was inserted in order to make the niche smaller. This would also have allowed the beam to be operated at two different heights, below and above the cross piece. Similarly in several cases the central protrusion of the crushing basin was replaced by a new one specially cut for the purpose.

One other element found in one of the subterranean complexes at Maresha (press 14) and also connected to the production of olive oil is a rock-cut store room with round depressions in the floor to stand oil jars.

Presses, crushing basins and storage rooms similar to those at Maresha are found at other sites in the surrounding region, the Shephelah.

Bibliography

Kloner, A. and Sagiv, N. 1991. "The Technology of Oil Production in the Hellenistic Period at Maresha, Israel". In: M.-C. Amouretti, J.-P. Brun and D. Eitam (eds.). *Oil and Wine Production in the Mediterranean Area from the Bronze Age to the End of the XVIth Century. Pré-Actes.* Aix-en-Provence-Toulon: 61–68.

Kloner, A. and Sagiv, N. 1993. "The Olive Presses of Hellenistic Maresha". In: M.-C. Amouretti and J.-P. Brun (eds.). *Oil and Wine Production in the Mediterranean Area* (BCH Suppl. 26). Paris: 119–136.

Kloner, A. and Sagiv, N. 2003. "Subterranean Complexes 44 and 45". In: A. Kloner. *Maresha Excavations Final Report I. Subterranean Complexes 21, 44, 70* (IAA Reports No. 17). Jerusalem: 51–72.

Sagiv. N. and Kloner, A. 1996. "Maresha: Underground Olive Oil Production in the Hellenistic Period". In: D. Eitam and M. Heltzer (eds.). *Olive Oil in Antiquity. Israel and Neighbouring Countries from the Neolithic to the Early Arab Period.* Padova: 255–292.

The Oil Presses

Press	Subter. Complex	Presses	Basins	Weights	Beam Length	Altar	Crushing Basin	Crusher
1	104	2					1	
2	21	3+1		2	4.5 m		1	
3	29	2+1	3					
4	44	2				1	1	
5	84					2?		
6	139	2						
7	89	1+1					1	
8	61	2	2	3	5.5 m	1		
9	61	2+1	2	3+3+3	7.5 m		2	1
10	99	2+1						
11	1	2						
12	97	1						
13	75						1	
14	Kh. Sandahana	2+1					1	
15	74	2	2	3+3	4.5 m	1	1	
16	74	1+1					1	1
17	128							
18	84							
19	161							
20	90							
21	107							
22–26	154, 62, 127, 131, 139							

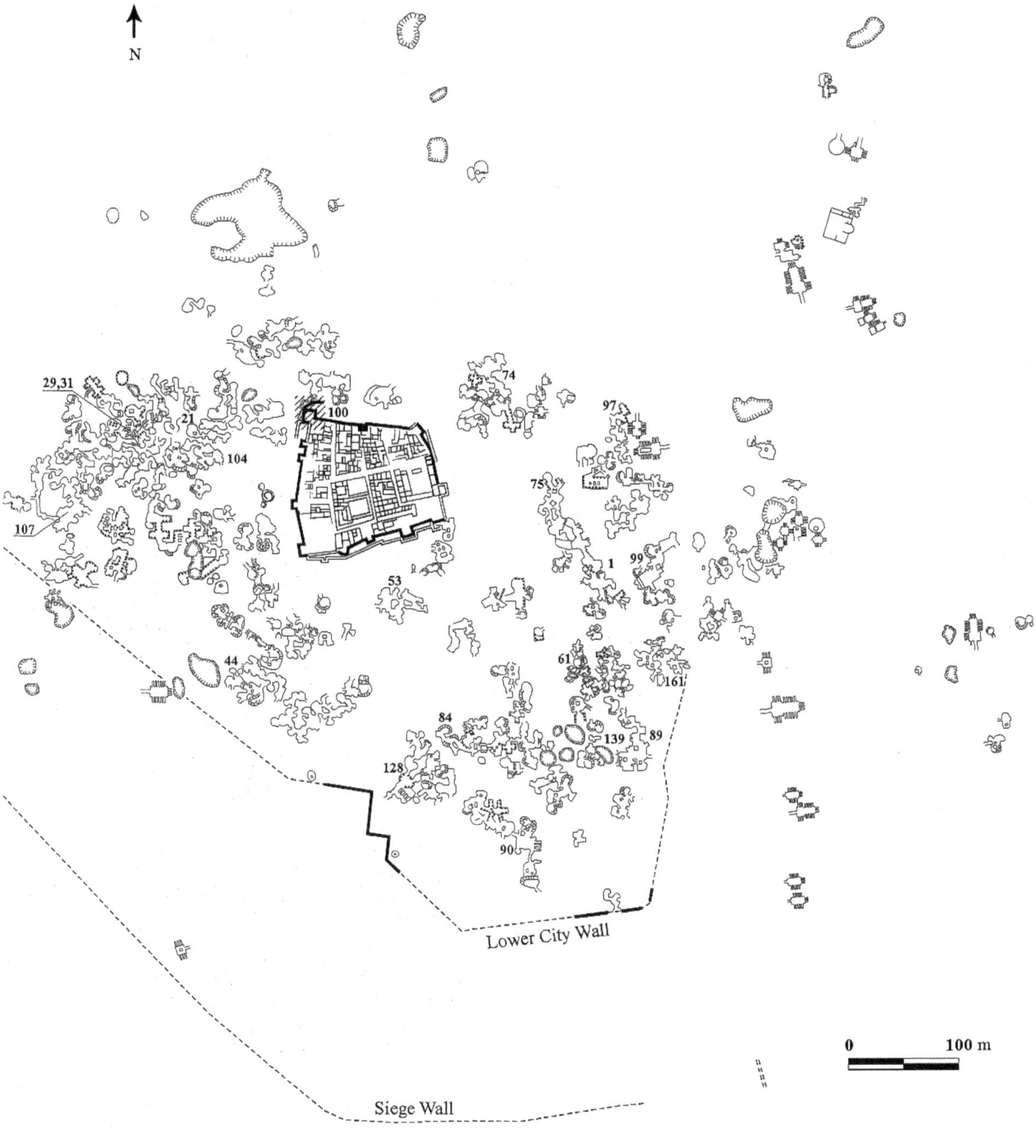

Fig. 65.1. Maresha, general plan of the city and the subterranean complexes.

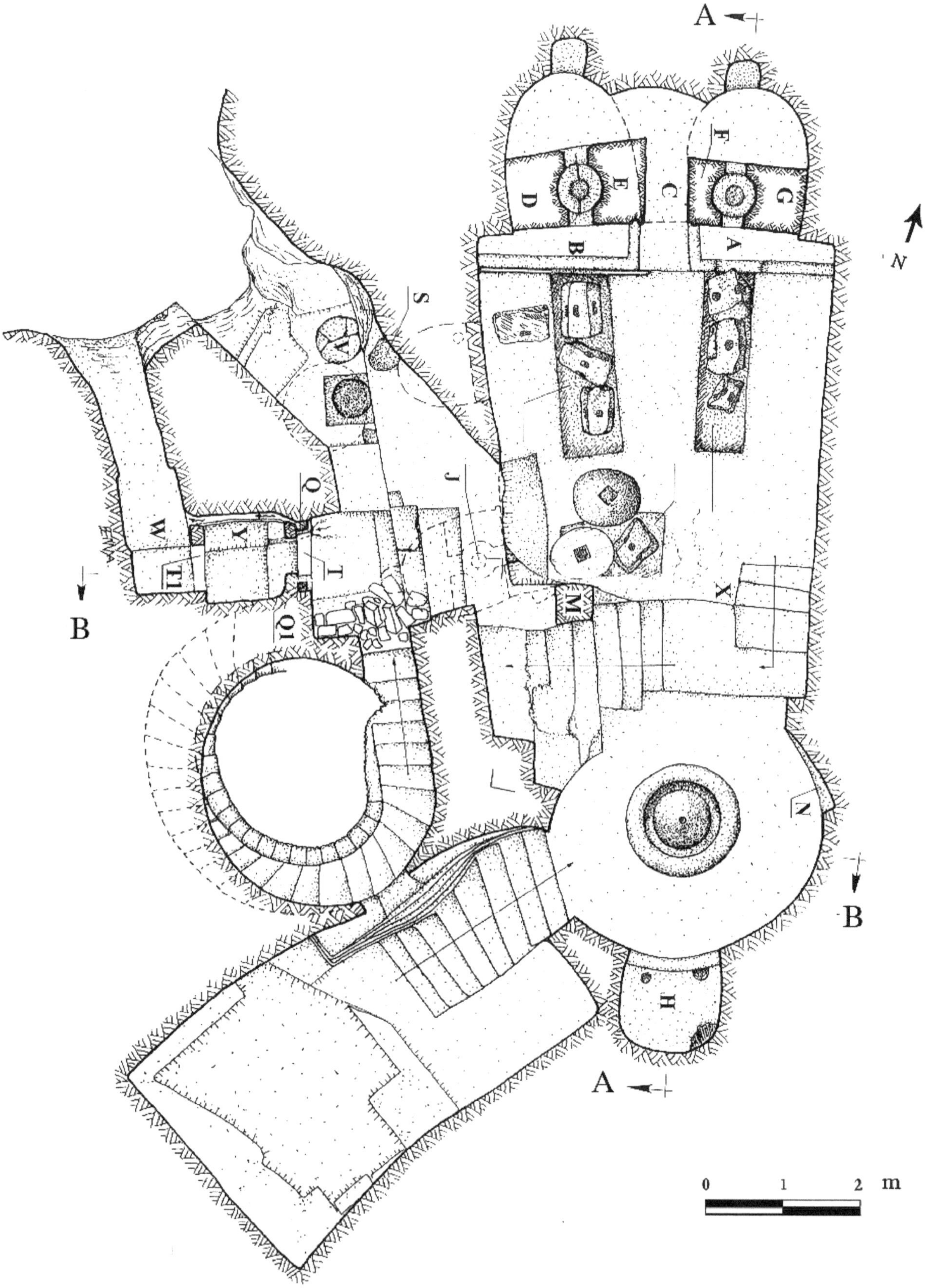

Fig. 65.2. Maresha, plan of Complex 44 oil press after excavation.

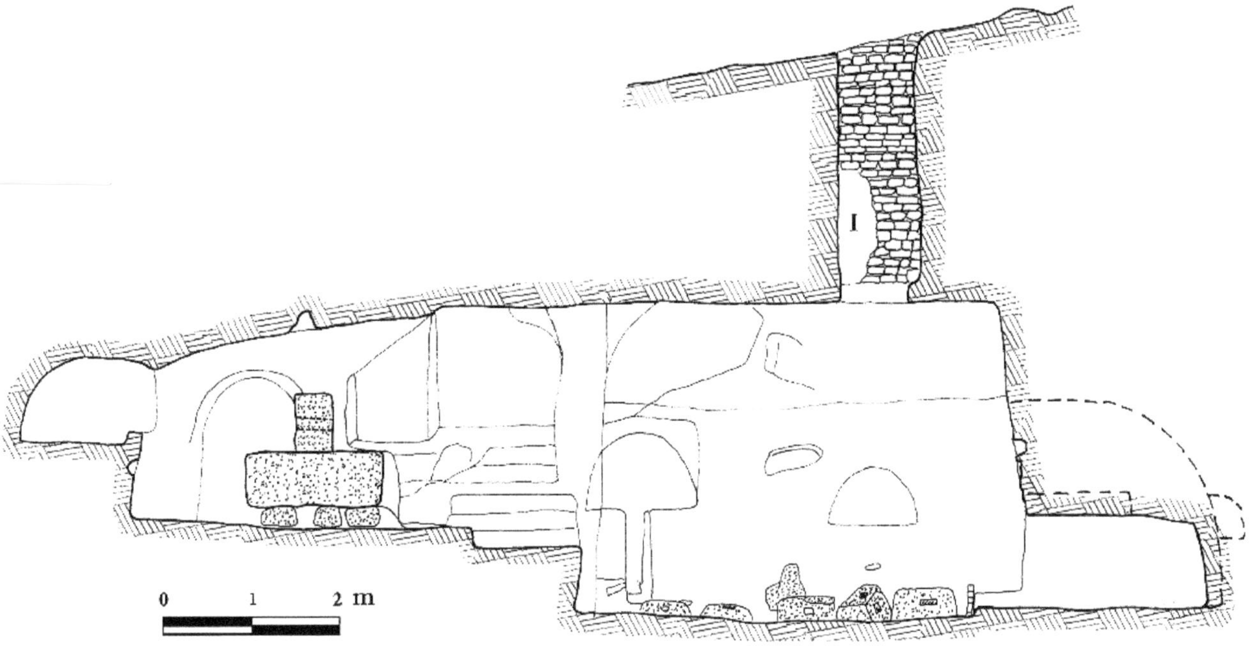

SECTION A-A

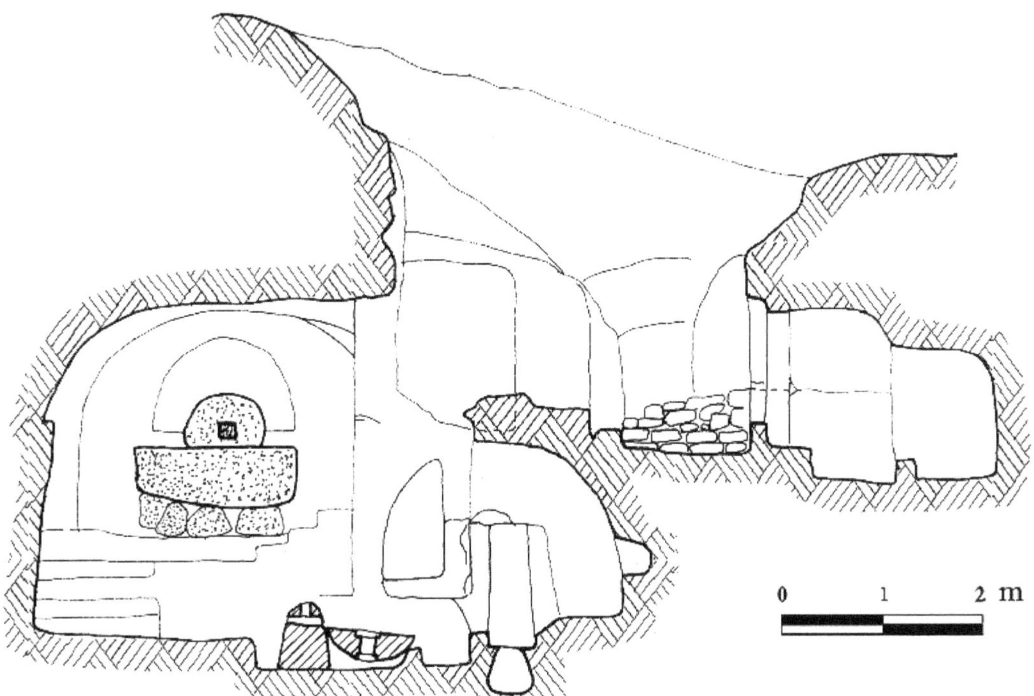

SECTION B-B

Fig. 65.3. Maresha, sections of Complex 44 oil press after excavation.

Fig. 65.4. Maresha, Complex 44, the crushing mill, looking south.

Fig. 65.5. Maresha, Complex 44, the three pressing installations, looking west.

Fig. 65.6. Maresha, Complex 44, the third pressing installation, looking south-west.

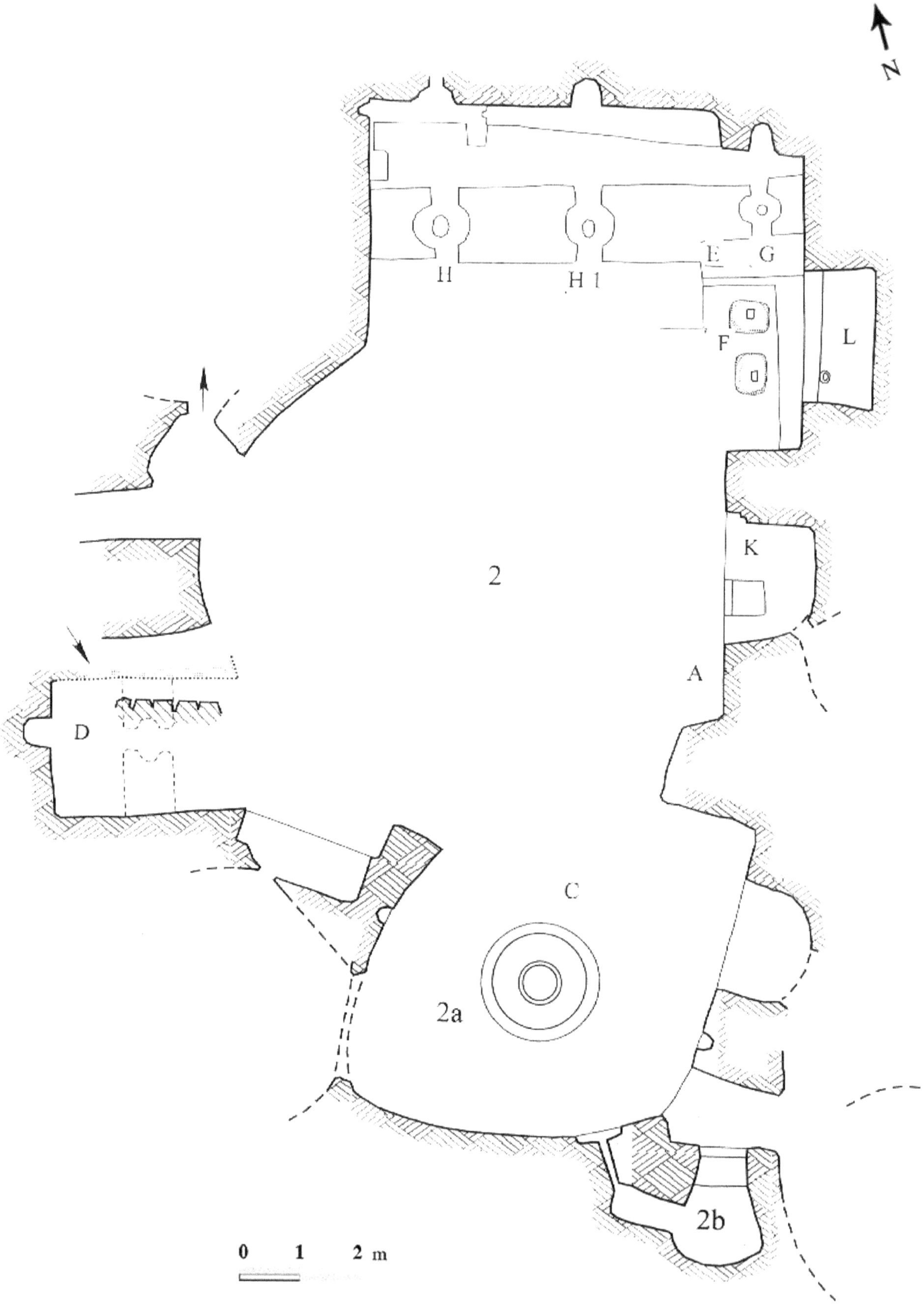

Fig. 65.7. Maresha, plan of the oil press (System A) in Complex 21.

Fig. 65.8. Maresha, installations H, H1 and G of the oil press (System A, Room 2), looking north.

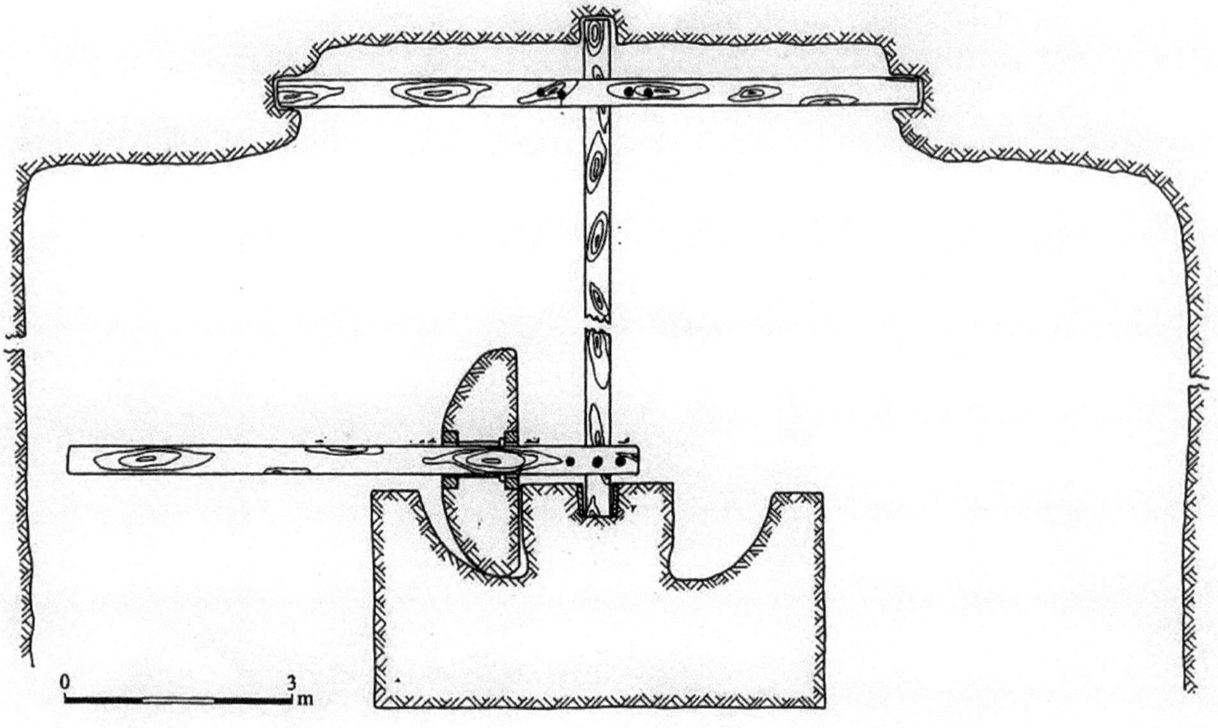

Fig. 65.9. Reconstruction of the Maresha type crushing mill.

Fig. 65.10. Reconstruction of the Maresha oil press (drawing: Anna Yamim).

66. The Oil Press at Kh. Dukas

Etan Ayalon, Rafael Frankel and Amos Kloner

The oil press at Kh. Dukas (MR 1516/1097) was excavated by Benny Frankel (1981a). It consisted of two typical standard grooved pier presses and a large crushing basin (diameter 4 m). The presses were in use during the Byzantine and Early Arab periods. Between the piers there were round central collecting vats within square depressions and in one case a crushing stone (diameter 0.52 m) was found on the central vat. The excavator suggests (Frankel 1981b) that it was placed there in order that the frails of olive pulp could be placed upon it. Four small square depressions were found in the floor opposite the four corners of the collecting vats measuring 0.12×0.18 m. These were placed exactly below the anterior and posterior grooves of the piers. The purpose of these grooves had in the past led to considerable controversy. Dalman (1928–1942, IV: 226–228), for instance, suggested that a guiding device connected to the pressing board (the platen) ran up and down in these grooves while Peleg (1981) showed that this was not possible. On the basis of his excavation at Kh. Dukas Benny Frankel suggested that the anterior and posterior grooves were to hold rods the bottoms of which were placed in the depressions and the tops in the threaded board (Figs. 1–2). He suggested also that the purpose of these rods was to stabilize the press piers and especially to withstand lateral pressures during pressing (see Kloner and Dray, Chapters 71 and 72 in this book).

Bibliography

Dalman, G. 1928–1942. *Arbeit und Sitte in Palästina.* 7 Vols. Gütersloh (repr. Hildesheim 1964).

Dray, Y. 1993. "The Judean Grooved Pier Press: A New Suggestion for the Reconstruction of its Operation". *NZ* 19: 91–95 (Hebrew).

Frankel, B. 1981a. "Kh. Dukas – Excavation Report". IAA Archives.

Frankel, B. 1981b. "Notes and Remarks on Olive Oil Production in Judea." *Teva Va'Aretz* 23: 183 (Hebrew).

Kloner, A. 1989. "The Upright Pillar Press in Caves in Judaea". *NZ* 15: 66–74 (Hebrew).

Peleg, Y. 1981. "How Ancient Olive Presses Worked". *Israel, Land and Nature* 6: 98–103.

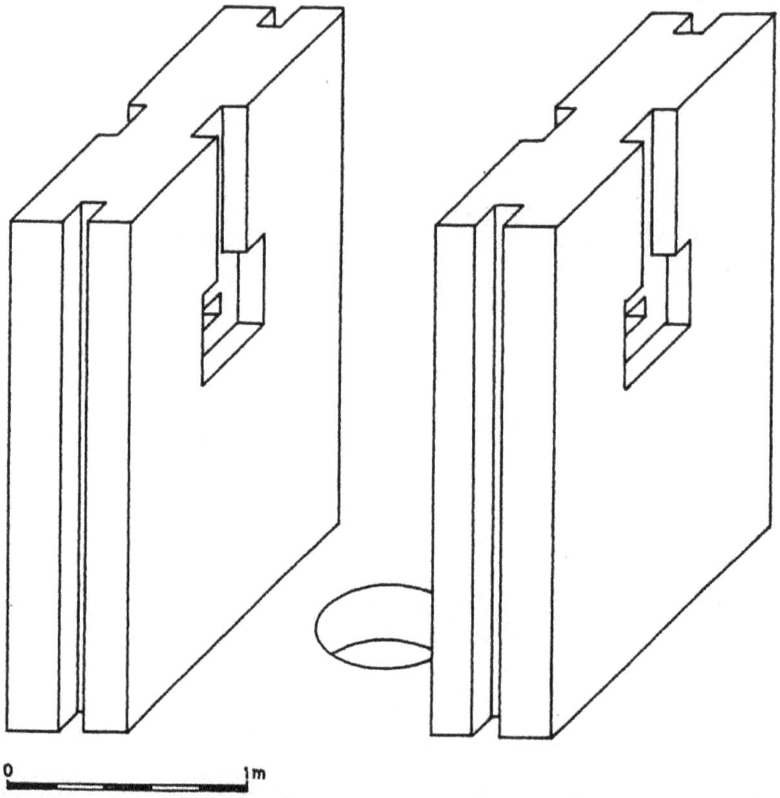

Fig. 66.1. Grooved piers of standard Judean type.

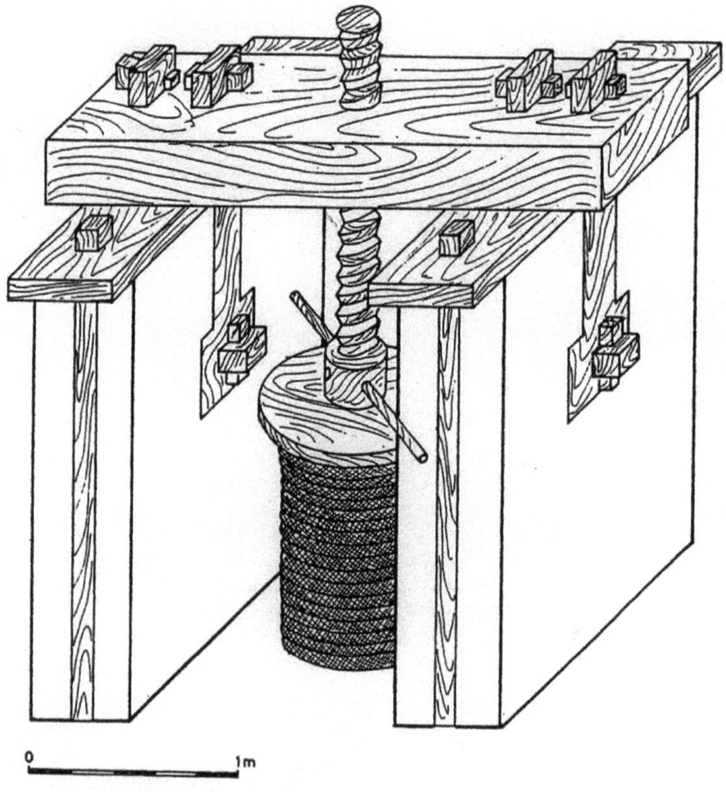

Fig. 66.2. Suggested reconstruction of a standard Judean grooved pier press.

67. The Oil and Wine Presses at H. Beit Loya

Rafael Frankel

The Site

H. Beit Loya, situated in the southern Shephelah (MR 1433/1080), was excavated in 1983 by J. Patrich and Y. Tsafrir (1985: 106–111). The ancient village was occupied from the Hellenistic till the Mameluke periods. On its western edge a complex from the Byzantine period was uncovered which was almost certainly a monastery. It consisted of a church, a courtyard and ancillary buildings including an oil press surrounded by an enclosure wall. A wine press was uncovered 20 m east of the church.

The Oil Press

The oil press was in the area between the southern wall of the church and the enclosure wall (Frankel, Patrich and Tsafrir 1990: 287–300). The fact that external plaster layers that covered the southern wall of the church were cut when building the oil press shows that it was a later addition.

The crushing basin was found at the west end of the atrium of the church, apparently not in its original position. It was 1.9 m in diameter and of an unusual form lacking not only the central socket but also the central protrusion. Two crushing stones probably rotated in the basin although it has been suggested that in crushing basins of this type rollers were used (Callot 1984: 29, Pls. 24–26). A groove in the rim of the basin allowed the liquid extracted during crushing to be collected and a small horizontal cut in the wall of the basin made it possible to place a vessel below the groove to avoid wastage. It is difficult to explain the presence of this device in the crushing basin at H. Beit Loya when it is almost unknown in other installations. It should be pointed out, however, that a similar arrangement has been reported from North Syria also in a crushing basin that lacks a socket (Ibid.: Pl. 24) suggesting that in this type of olive mill the crushing procedure was in some way different from that usually practiced.

The press was a lever and screw press (Figs. 1–2). The niche in which the beam (lever) was anchored (the fulcrum) is in the eastern enclosure wall. It is 0.8 m high, 0.9 m wide, 0.92 m deep and 1.2 m above the floor. At a distance of 2 m from the niche is a cylindrical central collecting vat made of hard limestone that was inserted into the soft bed-rock so that its rim was 0.2 m above the floor. The central vat is 0.92 m in diameter and 0.67 m deep (volume 445 litres). Its rim is 0.20–0.25 m thick and on it are four grooves. On the bottom of the vat at the north-west side is a sump 0.23 m in diameter and 0.13 m deep. The vat stands between two plain piers 1.65 m high and 0.7–0.9 m wide.

Two screw weights were found. Weight A was on the line of the beam 8.3 m from the inner wall of the niche showing that the beam was at least 9 m long. The second weight (B) was found upside down 0.45 m from the vat suggesting that at the final stage of the use of the press it served perhaps as a work table. Presses with two screw weights are known, however, and it is possible that the two weights were at one time used in unison. Between the vat and weight A (3.4 m from the centre of the vat and 1.4 m from the weight) are two square sockets on either side of where the beam would have been. The sockets are 0.28 m wide, 0.32 m deep and 0.35 m apart and clearly served to secure the bottom of two wooden uprights that stood on either side of the beam and kept it in line.

The screw weights (Fig. 3) are of a very unusual type having as well as a central socket, a rectangular upper depression, two internal mortices, two external mortices and a bore connecting all four.

It is of great significance that installations very similar to the two unusual elements found at Beit Loya – the crushing basin without socket or protrusion and the sophisticated screw weights – have been published from North Syria (Ibid.: 29, Pls. 24–26, 48–49, Pl. 43), suggesting that these elements were brought to the region by monks from Syria. This could also be the reason for the use of a lever and screw press also not usual in the Shephelah at this period.

The Wine Press (Fig. 4)

The treading floor of the wine press is paved in mosaics and measures 7.0×5.5 m (area 38.5 m^2). In the centre of the floor is a round stone base with a square mortice to secure a fixed screw press. The mortice is connected directly to the intermediary settling vat by a clay pipe. The treading floor is also connected to the settling vat by an open channel that terminates in a protruding spout. The intermediary vat is square (0.95×0.90 m, depth 0.75 m – volume 0.64 m^3), and the collecting vat round (diameter 2.25 m, depth 1.55 m, volume 6.1 m^3). The wall of the collecting vat is plastered and the floor of mosaic. In the centre of the floor of the collecting vat there is a sump 0.65 m in diameter and 0.45 m deep and around its rim are small depressions that clearly served to stand jars in.

The wine press of H. Beit Loya is very similar to that of area G at H. Zikhrin (Fisher 1985: 119), the main difference being the lack of compartments around the treading floor at H. Beit Loya.

Bibliography

Callot, O. 1984. *Huileries antiques de Syrie du Nord.* Paris.

Fisher, M. 1985. "Excavations at Horvat Zikhrin". *Qadmoniot* 18: 112–121 (Hebrew).

Frankel, R., Patrich, J. and Tsafrir, Y. 1990. "The Oil Press at Horvath Beit Loya". In: G.C. Bottini, L. Di Segni and E. Alliata (eds.). *Christian Archaeology in the Holy Land. New Discoveries* (Studium Biblicum Franciscanum Collectio Maior 36). Jerusalem: 287–300.

Patrich, J. and Tsafrir, Y. 1985. " A Byzantine Church and Agricultural Installations at Khirbet Beit Loya". *Qadmoniot* 18: 106–112 (Hebrew).

Fig. 67.1. Beit Loya, oil press during excavation, looking east.

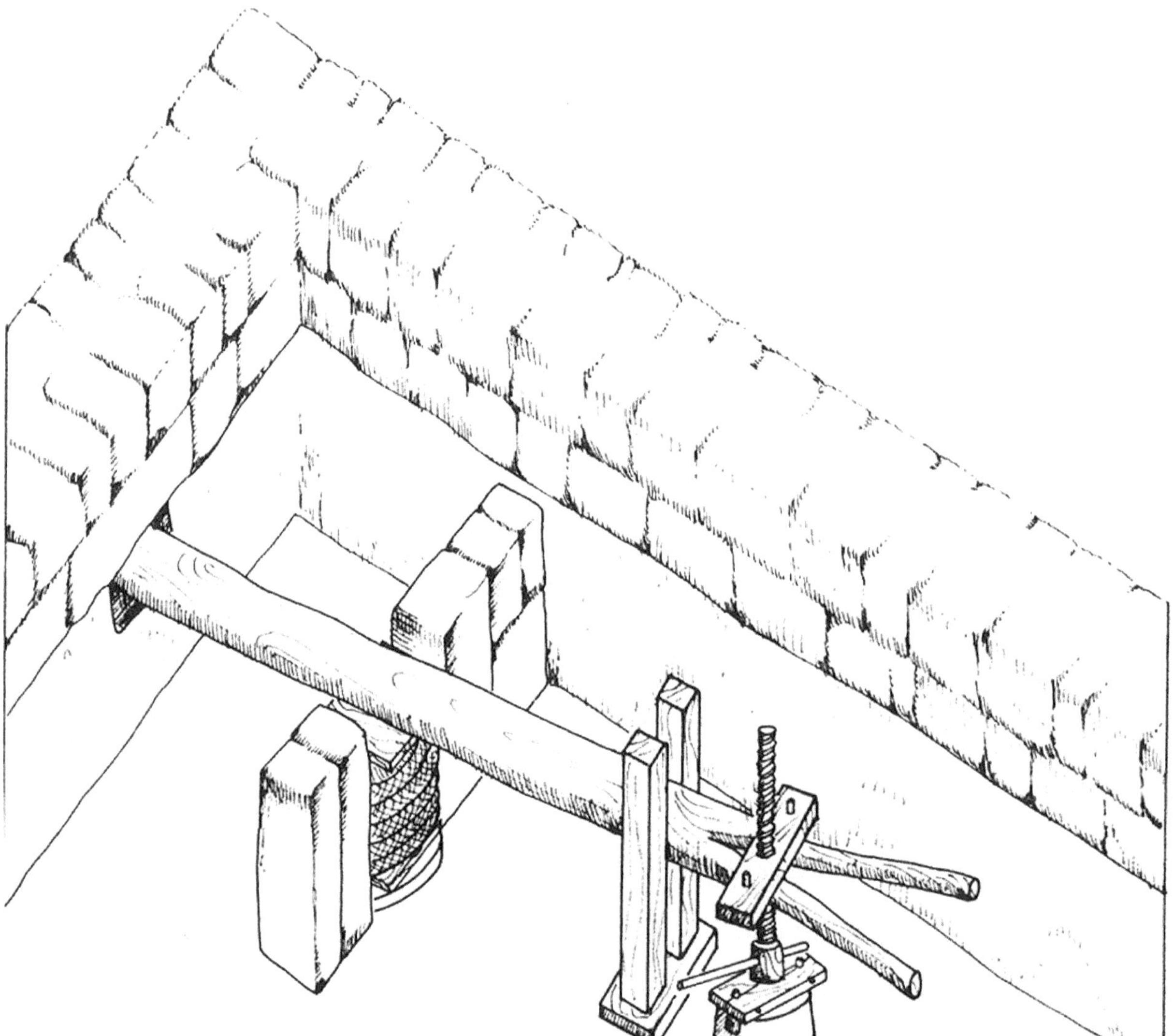

Fig. 67.2. Beit Loya, oil press, reconstruction of the beam and screw press.

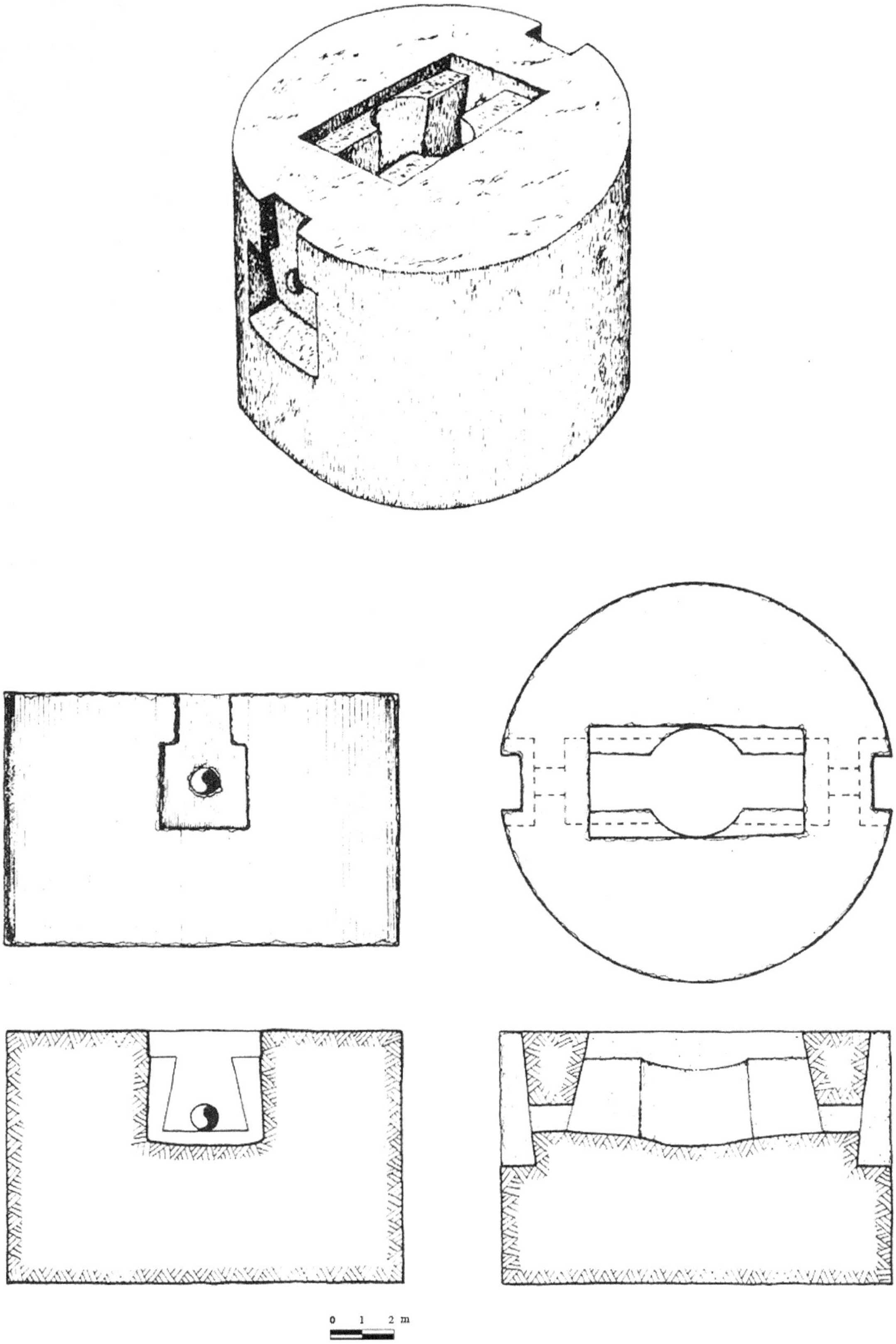

Fig. 67.3. Beit Loya, oil press, the screw weight.

Fig. 67.4. Beit Loya, wine press, looking south.

68. Underground Oil Press at 'Ahuzat Hazzan

Gideon Avni and Shlomo Gudovitz

The Site

'Ahuzat Hazzan is situated in the southern Shephelah, approximately 2 km south of the village of Amazia (MR 1409/1028). The oil press is part of a settlement or farm that was occupied during the 1st-8th cent. CE. Beneath the settlement was a complex underground system that was excavated and a plan of which was drawn up in the years 1980–1984 (Gudovitz *et al.* 1986; Avni *et al.* 1987; Avni and Gudovitz 1996). The oil press that was operated using a beam and weights press was in a large underground room hewn in soft limestone rock and connected to the surface by a stepped passage that enabled easy access and operation.

In the immediate vicinity there were several storage rooms also cut out of bed-rock in which jars of olive oil were stored. These storage rooms were connected to the press room via a series of tunnels and to the surface by a number of shafts.

The press and storage rooms were in use during the 1st cent. CE and ceased functioning at the beginning of the 2nd cent. CE. It was at this time that all the rooms were joined to form a larger system that was apparently used as a hideout. The connecting of earlier underground systems to later ones was common in sites in the Judean Shephelah and is almost certainly to be associated with the preparations for the Bar Kokhba revolt (Kloner 1983; Kloner and Tepper 1987). In the 4th cent. CE the press was renewed and continued in use until the end of the Byzantine period, when the roof in the press room collapsed.

The Oil Press

The press room was located in the centre of the underground system, and measured 10.5×6.5 m and 3.5 m in height (Figs. 1–2). In its north-west part stood a limestone crushing basin 1.7 m in diameter and 0.7 m high (Fig. 3). Alongside it was found a round crushing stone with flat crushing surface, 1.1 m in diameter and 0.35 m wide. The crushing apparatus stood on a base of small stones that filled a shallow depression in the floor. Between the crushing basin and the walls of the room there was a gap 0.8 m wide leaving room for the passage of the beast or man who turned the crushing stone.

In the south-east corner of the room was a square niche that served to anchor the beam of the press. The beam was placed between two square stones of hard limestone through one of which a hole was drilled. It is possible that these stones belonged to a later stage of use and that originally the beam was anchored directly in the niche in the bed-rock.

The centre of the central collecting vat was 2.5 m from the beam niche and it consisted of a plastered bell-shaped cistern, 1 m deep and 0.9 m in diameter (volume 0.63 m^3). The method of oil collection employed here was central (Frankel 1999: 63). The frails of olive mash were placed directly on the opening of the collecting vat and the expressed liquid was obtained by the pressure of the beam on them. Since no way of draining the cistern was found, the oil and watery lees were apparently removed directly from the vat using cups or ladles.

Unlike Hellenistic oil presses found at other sites in the Judean Shephelah in which there were plain piers on either side of the central vat, no piers were found in the oil press at 'Ahuzat Hazan. However, changes in quarry marks on the walls near the central vat show that the earlier press did include two such piers that were later removed employing rough wide chisels.

Adjacent to the bell-shaped central vat to the west there was an additional small plastered basin measuring 1.0×1.8 m with a small depression in the centre, diameter 0.3 m. This basin probably served to produce "first oil". The frails of olive mash would have been placed on the basin for a short time before they were pressed. The oil that seeped out at this stage was considered of the highest quality and collected separately.

Three stone weights found scattered in the room were larger than usual (0.8×0.8×0.5 m). Both vertical and horizontal holes were drilled through the top of each weight forming a reversed T. Probably originally the weights were placed in a weights pit in the floor. However, later quarrying activity destroyed all signs of the pit.

The presence of one pressing installation *vis-a-vis* one set of crushing apparatus is unusual for the Judean Shephelah where the prevailing relationship between pressing and crushing equipment is 2:1 or 3:1 (Kloner and Sagiv 1993).

Next to the northern wall of the room there was a round bell shaped cistern, depth 2 m, diameter at base 2.8 m. On its floor there were 39 round depressions each measuring 0.25–0.30 m in diameter and 0.2–0.3 m in depth connected to each other by channels. These clearly served to stand storage jars of oil and indeed several jars dating to the 1st-2nd cent. were found in the cistern.

Similar depressions also with channels were found in four unique storerooms that were connected to the press room by a network of tunnels (Fig. 4). The storerooms were 6–8 m long, 3–4.5 m wide and 2 m high and in each there were 145–150 such depressions. The floors of the rooms sloped, the side farthest from the entrance being 0.2–0.3 m higher than that near the entrance. The connecting channels all led to small cisterns 0.5 m deep that were in the corner of the rooms closest to the entrances.

Access to the press was through a narrow stepped passageway, the upper part of which built with stones and the lower hewn in the rock. Before entering the press room one passed through a doorway with two stone door jambs, placed 2.5 m before the entrance to the press room itself. On the north side of the passageway between the doorway and the entrance to the press room there was a rock hewn *mikveh* (Jewish ritual bath) measuring 4×3.5 m. Seven wide steps descended from the entrance to a rectangular pool. In the south-west corner of the *mikveh* there was a niche containing a plastered basin, 0.7×1.1 m. At the bottom of the basin there was a hole that allowed water to flow from the basin to the pool. The bath was filled from the surface by a channel that runs along the north side of the passageway. The proximity of the *mikveh* to the press room suggests that during the first stage of the use of the press in the 1st-2nd cent. the ritual cleanness that immersion in the bath imparted was connected to work in the oil press (Reich 1990). In the secondary stage of use during the Byzantine period the *mikveh* room was blocked with fill.

Bibliography

Avni, G. and Gudovitz, S. 1996. "The Oil Press and the Subterranean Storage Complexes at "Ahuzat Hazzan"". In: D. Eitam and M. Heltzer (eds.). *Olive Oil in Antiquity. Israel and Neighbouring Countries from the Neolithic to the Early Arab Period.* Padova: 137–147.

Avni, G., Gudovitz, S., Mintzker, Y. and Kloner, A. 1987. "The Hiding Complex at Ahuzat Hazzan". In: A. Kloner and Y. Tepper. *The Hiding Complexes in the Judean Shephela.* Tel Aviv: 115–128 (Hebrew).

Gudovitz, S., Mintzker, Y. and Avni, G. 1986. "Horvat Hazzan". *ESI* 4: 46–48.

Frankel, R. 1999. *Wine and Oil Production in Antiquity in Israel and Other Mediterranean Countries.* Sheffield.

Kloner, A. 1983. "The Subterranean Hideways of the Judean Foothills and the Bar-Kokhba Revolt". *The Jerusalem Cathedra* 3: 114–135.

Kloner, A. and Sagiv, N. 1993. "The Olive Presses of Hellenistic Maresha". In: M.-C. Amouretti and J.-P. Brun (eds.). *Oil and Wine Production in the Mediterranean Area* (BCH Suppl. 26). Paris: 119–136.

Kloner, A. and Tepper, Y. 1987. *The Hiding Complexes in the Judean Shephela.* Tel Aviv (Hebrew).

Reich, R. 1990. *Miqwa'ot (Jewish Ritual Immersion Baths) in Eretz-Israel in the Second Temple and the Mishnah and Talmud Periods.* Unpublished Ph.D. Thesis, Hebrew University, Jerusalem.

Tepper, Y. 1987. "The Oil Presses at the Maresha Region". In: M. Heltzer and D. Eitam (eds.). *Olive Oil in Antiquity – Israel and Neighbouring Countries from the Neolithic to the Early Arab Period.* Haifa: 25*-42* (Hebrew).

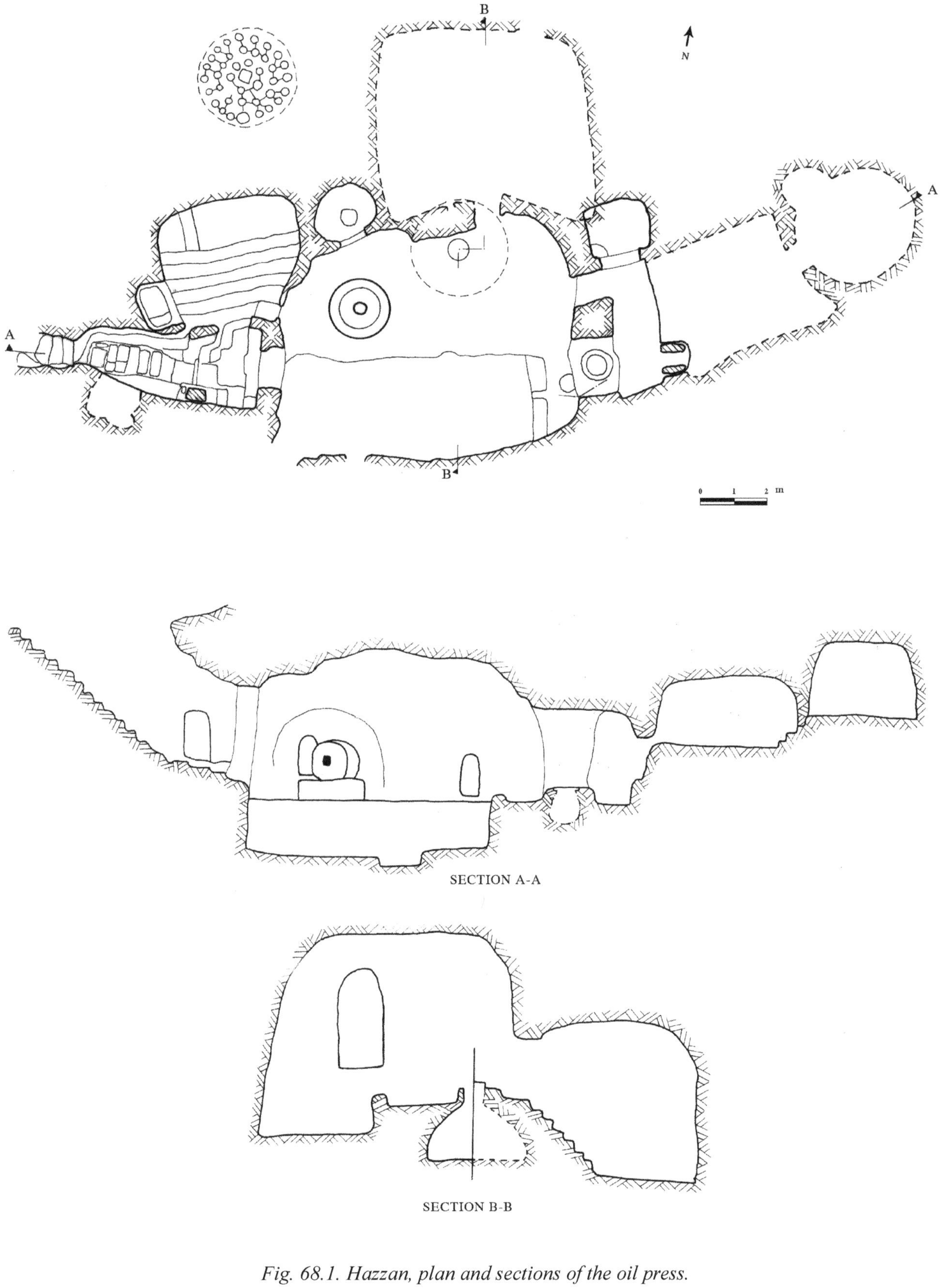

Fig. 68.1. Hazzan, plan and sections of the oil press.

Fig. 68.2. Hazzan, the oil press, looking west.

Fig. 68.3. Hazzan, the crushing mill.

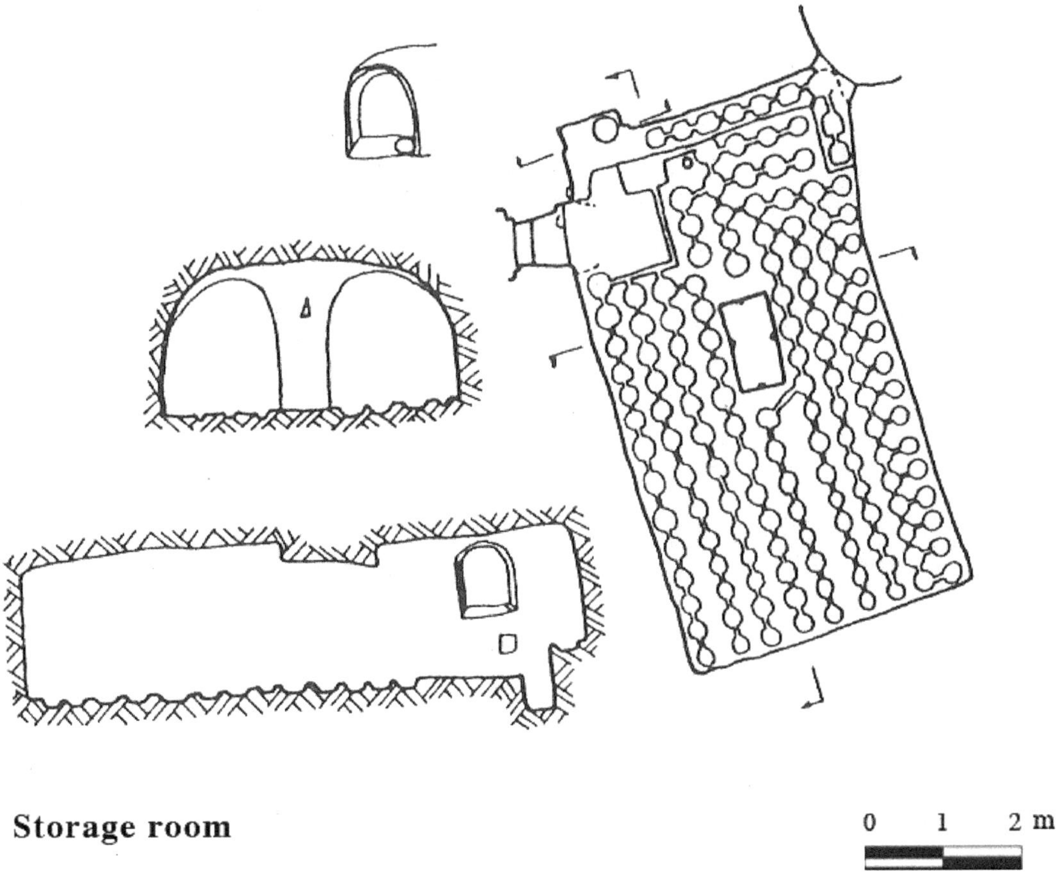

Storage room

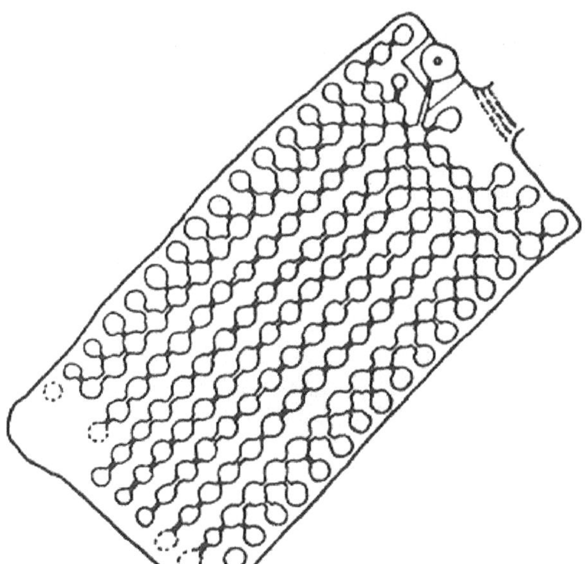

A second storage room

Fig. 68.4. Hazzan, storage rooms equipped with depressions in the floor for standing jars.

69. A Wine Press at Kh. Umm Halasa

David Amit

The Site

In the border region between the Hebron mountains to the north-west and the Wilderness of Judea to the east there were in the Byzantine period three large flourishing villages: Aristobulias (Kh. Istabul), Ziph and Caphar Baricha (Bani Na'im). Further east several monasteries were surveyed in the 1970's by Hirschfeld (1990; 1992) and by the author: El-Qasrein (identified with the Monastery of Severianus), Kh. Umm Rukba and Kh. el-Quneitira. In 1996 a survey was carried out by the author at Kh. Umm Halasa (MR 1659/0995) that is situated 1.5 km north of Kh. Umm Rukba and to the east of the settlement Peney Hever. This ancient architectural complex proved also to be a Byzantine monastery. The complex is almost rectangular (45×38 m, area almost 2,000 m^2) and surrounded by a strong stone wall. It consists of buildings, caves, water cisterns and other installations (Amit 1997).

The Wine Press (Figs. 1–2)

Approximately 50 m north of the built enclosure a wine press was discovered. It is similar to other complex wine presses of the Byzantine period but it is surprising to find it at this site on the edge of the wilderness. It was partly covered by later buildings but the components of the installation could be discerned and recorded. The treading floor was square and cut in the rock (4.2×4.2 m, area 17.6 m^2). As the ground is sloping the height of the walls vary reaching a maximum of 1.3 m in the south-east corner. The floor was paved in mosaic. The tesserae were exceptional – not cubic as is usual but in the form of a prism (average size 1.5×1.5×3.7 cm). In the centre of the treading floor there was a stone base for a fixed screw press for the secondary pressing of the rape after treading. The stone is approximately rectangular (1.1×0.8 m). The mortice is square (0.4×0.4 m, depth 0.55 m) and it widens at the bottom on two adjacent sides, the northern and western. A narrow channel connected the mortice to the collecting vat. Two rock-cut compartments were adjacent to the northern wall of the treading floor and one adjacent to the western wall. They were apsidal in shape, 0.50–0.55 m in width, 0.55–0.60 m in length and 0.45–0.50 m in depth. Their walls were plastered and they were paved in similar mosaic to that of the treading floor. The compartments were connected to the treading floor by bores that were at the bottom of the compartments but 0.5 m above the treading floor. The collecting vat was parallel to the northern half of the eastern wall of the treading floor. It was square (1.85×1.85 m) but as it was not excavated the depth is at present unknown. Similarly the channel that connected the floor to the vat is hidden by later building but the outlet of the channel from the screw mortice is visible in the wall of the vat. Next to the northern wall of the vat there is a small rectangular depression (0.3×0.6 m, depth 0.25 m) that probably contained marl to be spread on the must. Close to the south-east corner of the vat a rock-cut corner is discernable, probably of another element connected to the wine press.

Another wine press was found 50 m north of the one already described and much smaller than the latter. It consisted of a square treading floor (2.1×2.1 m – area 4.4 m^2; depth again varying according to the slope of the ground 0.1–0.4 m) from the north-east corner of which a short narrow channel (length 0.4 m, width 5 cm) leads to a round collecting vat (diameter 1 m).

Bibliography

Amit, D. 1997. "Kh. Umm Halasa: An Additional Monastery in the Wilderness of Ziph". In: Y. Eshel (ed.). *Judea and Samaria Research Studies – Proceedings of the 6th Annual Meeting 1996*. Kedumim–Ariel: 259–270 (Hebrew).

Hirschfeld, Y. 1990. "List of the Byzantine Monasteries in the Judean Desert". In: G.C. Bottini, L. Di Segni and E. Alliata (eds.). *Christian Archaeology in the Holy Land, New Discoveries*. Jerusalem: 1–90.

Hirschfeld, Y. 1992. *The Judean Desert Monasteries in the Byzantine Period*. New Haven and London.

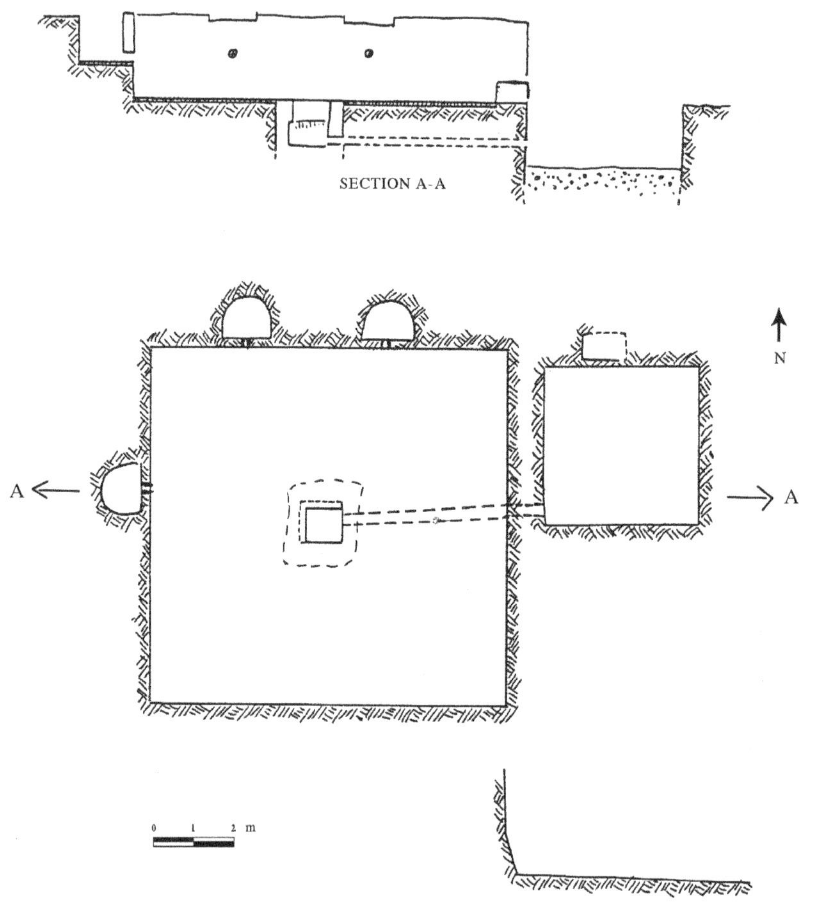

Fig. 69.1. Umm Halasa, plan and section of the wine press.

Fig. 69.2. Umm Halasa, the wine press, looking south.

70. Byzantine Wine Presses in the Negev

Gaby Mazor

Around the cities of Halusa, Shivta and 'Avdat in the northern and central Negev, Byzantine wine presses have been discovered that are distinct in their forms, their dimensions and their ashlar construction. These installations were investigated in 1977 by the author assisted by 'Eran Erez. Because of the differences between these wine presses and those in the center and the north of Israel, they are treated here as a separate group. First the installations will be numbered and located one by one, then their components and characteristics will be discussed together. For their measurements see Table 1.

The Wine Presses

Wine press No. 1 (Fig. 1): Located in the agricultural area east of Halusa (MR 1183/0563). It was unearthed by the author in 1973–1974.

Wine press No. 2 (Fig. 2): Located near the south-west corner of Shivta (MR 1143/0322). In 1938–1939 it was excavated by the Colt expedition, but the results of the excavation were never published. The expedition saw it as a bath, whereas Wiegand, who surveyed Shivta, thought it was a tomb.

Wine press No. 3 (Fig. 3): Located in the agricultural area of Shivta, 500 m north-east of the southern church (MR 1157/0331). It, too, was excavated by the Colt expedition.

Wine press No. 4 (Fig. 4): Located within a complex of buildings and courtyards (a monastery?) south of Shivta's northern church (MR 1154/0329). It, like Nos. 2 and 3, was unearthed by the Colt expedition.

Wine press No. 5 (Fig. 5): This installation is part of a complex of buildings and courtyards about 50 m south-east of the southern gate of the acropolis at 'Avdat (MR 1282/0226). It was excavated in 1958–1960 by M. Avi-Yonah and A. Negev (see also Tahal 1995: 131).

Wine press No. 6 (Fig. 6): Located within the confines of the agricultural area situated north-west of 'Avdat (MR 1281/0230). This wine press, too, was excavated by Avi-Yonah and Negev.

Wine press No. 7 (Fig. 7): Located in the agricultural area south of 'Avdat (MR 1277/0222). It, too, was excavated by Avi-Yonah and Negev.

Wine press No. 8 (Fig. 8): This installation was constructed in the eastern wing of rooms in a large structure found north of the Nabatean potter's workshop at 'Avdat (MR 1287/0227). It was excavated in 1977 by A. Negev and R. Cohen.

Installation No. 9 (Fig. 9): About 1 km west of 'Avdat (MR 1272/0229) there is a round industrial installation that was unearthed in 1959–1960 by A. Negev.

Characteristics

Most of the wine presses have three sections, constructed on three different levels: the upper level comprised compartments for storing grapes; the middle level was the treading floor and the system for draining the must; the lower level held the collecting vats (Fig. 10).

The treading floor is paved with limestone slabs that were fitted with great precision, so that the must would not penetrate the joints. The slabs are 0.3–1.0 m long, 0.3–0.4 m wide and 7–10 cm thick. The floor slopes slightly toward the collecting vats; the height differential is 0.13–0.26 m. The treading floor is surrounded by a low wall (bench?) 0.5–0.6 m high, faced with stone slabs. Centered in the front wall of the treading floor is a trough through which the must flows to the collecting vat. One or two additional drainage openings (see editors' note below) are installed in or near the centre of the treading floor, through which the must flows down a shaft 0.6–0.8 m deep and along a sub-floor channel measuring 0.2×0.2 m. Both drainage systems lead to a small intermediate basin measuring close to 0.6×0.8 m, from which the must ultimately flows through one or two troughs into the collecting vat(s). In the bottoms of the vats, that are about 1 m deep, are sedimentation basins. All the vats are surfaced with impermeable plaster and some of them are floored with stone slabs.

The collecting vats are surrounded by a walled courtyard paved with stone slabs. Sometimes additional wine presses, storerooms, etc. were built in the courtyard. Most of the wine presses had small built-in compartments for storing grapes. Cisterns, constructed or hewn, were found near some of the wine presses, and sometimes a roofed storage area to hold the jars of wine was built in the courtyard.

Presumably, secondary pressing of the rape was also carried out in the wine presses of the Negev, although it is not clear what method was used. A niche was found in the western wall of wine press No. 1, that could have been used for anchoring the beam of a lever and weights press.

Until now, no stone bases with sockets for fixing wooden screws (such as were used for secondary pressing of the rape in the wine presses of central and northern Israel) have been discovered in the Negev installations (see editors' note below). It may be that at the time of pressing the screw was mounted in the drainage opening of the treading floor and the upper trough used for draining the must. However, in this opening there is actually no evident means of anchoring the screw so that it would not pull out while it was operating.

Except for No. 9, the wine presses can be divided into three types, apparently reflecting stages in their technological and functional development.

A. Wine Presses Without Compartments

Nos. 1 and 4. In light of the absence of compartments it is reasonable to assume that each of these wine presses belonged to a private viticulturist or a group of vineyard owners (a monastery?). In wine press No. 1 improvised compartments were installed at a later stage, carelessly built in contrast to the ashlar construction of the installation. What appears to be a farmhouse that has not been excavated is located south of the wine press. West of it is a constructed cistern and east of the treading floor is a side entrance hall. No. 1 is exceptionally well constructed. Its walls are built of field stones and mortar, faced on both sides by well-cut stone slabs. On the east, south and west the walls are about 3 m high, apparently to prevent the infiltration of sand. Two niches were built in the southern wall and one in the western. The niches are approximately 0.5 m in height and 0.3 m in width and depth.

Two arch support piers, one on the northern wall and one on the southern, attest to the fact that the wine press was roofed. The northern pier is double: its southern half extends toward the treading floor, its northern half toward the collecting vats. Apparently the roofing was made of wooden beams and thatch.

On the inside face of the front wall, above the trough, is carved a rosette – the one and only decoration found to date in all the wine presses of the Negev.

There is a correlation between the number of drainage openings in the treading floor and the number of collecting vats. In wine press No. 1 there are two openings and two vats, while in No. 4 there is one opening and one vat.

Wine press No. 4 was installed in an existing structure, the walls of which were not built at 90° angles. In order to correct this situation, wedge-shaped interior walls were added. The collecting vat is floored with stone slabs fitted with extreme precision.

B. Wine Presses With Compartments

Nos. 2, 5, 6 and 8 were built with 6–10 compartments on three sides of the treading floor. Their wall heights ranged from 0.4 to 0.6 m and their floors were made of stone slabs. They ranged in area from 4 to 7 m^2 and were connected to the treading floor by openings about 0.5 m wide. In these wine presses, too, there is a direct relationship between the number of openings in the treading floor and the number of collecting vats.

Sometimes storerooms were built alongside the wine presses, indubitably used for storing jars of wine. In wine press No. 6 all that remains of the storeroom is a few foundation layers, but in No. 8 the storeroom walls are preserved to a substantial height. Six piers attached to the walls show that the storeroom was roofed with stone slabs. A few flooring slabs were also preserved. The door of the storeroom opens onto the collecting vat courtyard.

Wine press No. 8 was built within an existing edifice and it was necessary to make various changes in the rooms of the building to make it suitable for its new function. The treading floor was installed in the central room that was then especially well-plastered for the purpose. The collecting vat was hewn out in the courtyard of the building and alongside it a square intermediate basin was constructed. The south-east room was apparently used as a storeroom. Other rooms were converted for use as grape storage compartments.

C. Expanded Wine Presses

Nos. 3 and 7 were, it seems, originally wine presses with compartments, to each of which an additional treading floor was later added, either for treading different varieties of grapes or because of expanded production.

In the western wing of No. 7 a second small wine press was built at a later stage; it contained a treading floor, intermediate basin and collecting vat. An isolated compartment was built south of the treading floor. It may be that the definite separation between the two wine presses indicates a segregation of grape varieties, and not the expansion of production. To accelerate production processes (such as sedimentation of impurities) a third collecting vat was built, to which no drainage arrangement led. The must was apparently brought to it in containers.

A second wine press (only partially excavated) was also built in the courtyard of wine press No. 3. In the north-east corner of the courtyard a treading floor was unearthed, in the centre of which is a pit or depression, but neither an opening nor a drainage channel was found, nor were an intermediate basin and collecting vat discovered. The eastern collecting vat of the larger wine press may also have served to collect must from the smaller one. South and east of this treading floor, five compartments had been built. It is entirely possible that this is either an installation for drying fruit or a wine press whose construction was never completed. In the main installation of No. 3, piers were found in the collecting vats that indicate, apparently, that they had been roofed. A storehouse was built in the courtyard and to the north of the wine press a constructed cistern was discovered.

D. Installation No. 9

This is a circular structure with low walls, diameter ~9 m. Partitions divide it into seven compartments that are shaped like the segments of a circle and that presumably were originally floored with stone. Their access openings are situated in the outer (perimeter) wall. The inner circle comprises a round pit, ~1.5 m in diameter, hewn into the bed-rock on three step-like levels. In the lowest level an opening leads to a cistern found under the installation. The connection between the cistern and the installation is unclear but it is possible that the cistern was hewn out at a later stage. In the opinion of Amos Kloner, the installation was used for drying fruits, but the reason for its extraordinary construction is indeterminate; it is also difficult to explain its distance from the city and from the other wine presses. Its connection with the wine industry, if any, cannot be established with certainty.

Summary

The Byzantine wine presses of the Negev are marked by their almost uniform layout that perhaps hints at centralized organization of the wine industry, despite the fact that no written sources have yet been found attesting to this. It should be noted that a similar uniformity is found in other architectural elements in the cities of the Negev, e.g. houses, public buildings and cisterns. The differences between the various wine presses are connected to the solutions found for technical problems encountered in the production process, e.g. the compartments alongside the treading floors that were apparently intended to separate the crops of various growers or to isolate different varieties of grapes.

The development of the wine presses is indicated by the diverse changes and additions found in some of them: storehouses, supplementary wine presses in secondary construction and cisterns for storing the water with which the installation was washed. The wine press of the Negev is a large complicated structure with a lengthy tradition that required many resources for its construction. It attests to the extent and importance of the winemaking industry in the regional economy. Apparently the enterprise was centralized, grapes bought from vineyard owners thus allowing for the production of wine in large quantities. Such enterprises may have been owned by a single affluent private individual, jointly owned by a group or clan, publicly owned (as by a municipal authority) or owned by an organization like a monastery. In the area of Ashqelon (Site 67), Kissufim and Gaza, large pottery kilns were found in which the typical jars used for exporting wine were fired – one more facet of the complex economy of Negev wine production during the Byzantine period.

During the Byzantine period the Negev enjoyed a flourishing economy and a large population deriving, among other factors, from the development of defense systems along the southern border of the empire and the installation of a wide road network. Viticulture and winemaking were clearly a major branch of commerce in the region, possibly even through governmental planning and encouragement. This is hinted at by the installations described above and by written sources that mentioned the wine of Halusa as well as the wines of Gaza and Ashqelon that were probably produced in part in the Negev (Mayerson 1985, 1992, 1993).

Judging from the evidence from the few wine presses excavated, wine production in the Negev ceased at the end of the Byzantine period.

Editors' Note

1. Note, however, that in the introduction to this book it is suggested that the wooden bases of single fixed screw presses were installed in these shafts.

Bibliography

Mayerson, P. 1985. "The Wine and Vineyards of Gaza in the Byzantine Period". *BASOR* 257: 75–80.

Mayerson, P. 1992. "The Gaza Wine Jar (Gazition) and the Lost Ashkelon Jar (Askalonium)". *IEJ* 42: 76–80.

Mayerson, P. 1993. "The Use of Ascalon Wine in the Medical Writers of the Fourth to the Seventh Centuries". *IEJ* 43: 169–173.

Mazor, G. 1981. "The Wine-Presses of the Negev". *Qadmoniot* 14: 51–60 (Hebrew).

Tahal, G. 1995. "'Avedat". *ESI* 14: 130–133.

Table 1. Measurements of Wine Presses in the Negev

No	Thickness of Wall (M)	Treading Floor Area (M²)	Average Compart. Area (M²)	Overall Compart. Area (M²)	Volume of Coll. Vat A Coll. Vat B (M³)	Treading Floor Incline (M)	Ave. Dimensions Treading Floor Openings A & B (M)
1	0.5	28.5	?	?	(A) ? (B) ?	0.13	(A) 0.32×0.45 (B) 0.2×0.2
2	1	33	5.75	46	(A) 6.7 (B) 6.7	?	(A) 0.4×0.4 (B) 0.3×0.3
3	0.6	35	4.75	38	(A) 5 (B) 3.4	0.26	(A) 0.5×0.5 (B) 0.4×0.4
3a	0.35–0.8	16	4	16	(B) 3.4 (?)	?	—
4	0.6–0.9	25.5	—	—	(A) 3.6	?	(A) 0.32×0.32
5	0.6	33	7.4	66	(A) 8.8	0.26	(A) 0.35×0.35
6	0.85–1	33	6.5	65	(A) 6.5 (B) 4	? ?	(A) 0.4×0.4 (B) 0.3×0.3
7	0.8	33	5.5	22	(A) 2.6 (B) 5.6	?	(A) 0.5×0.5
7a	0.6–0.8	13	5	5	3.1	?	0.4×0.4
8	0.6–0.7	16.3	4.25	22	4.2	?	0.4×0.4

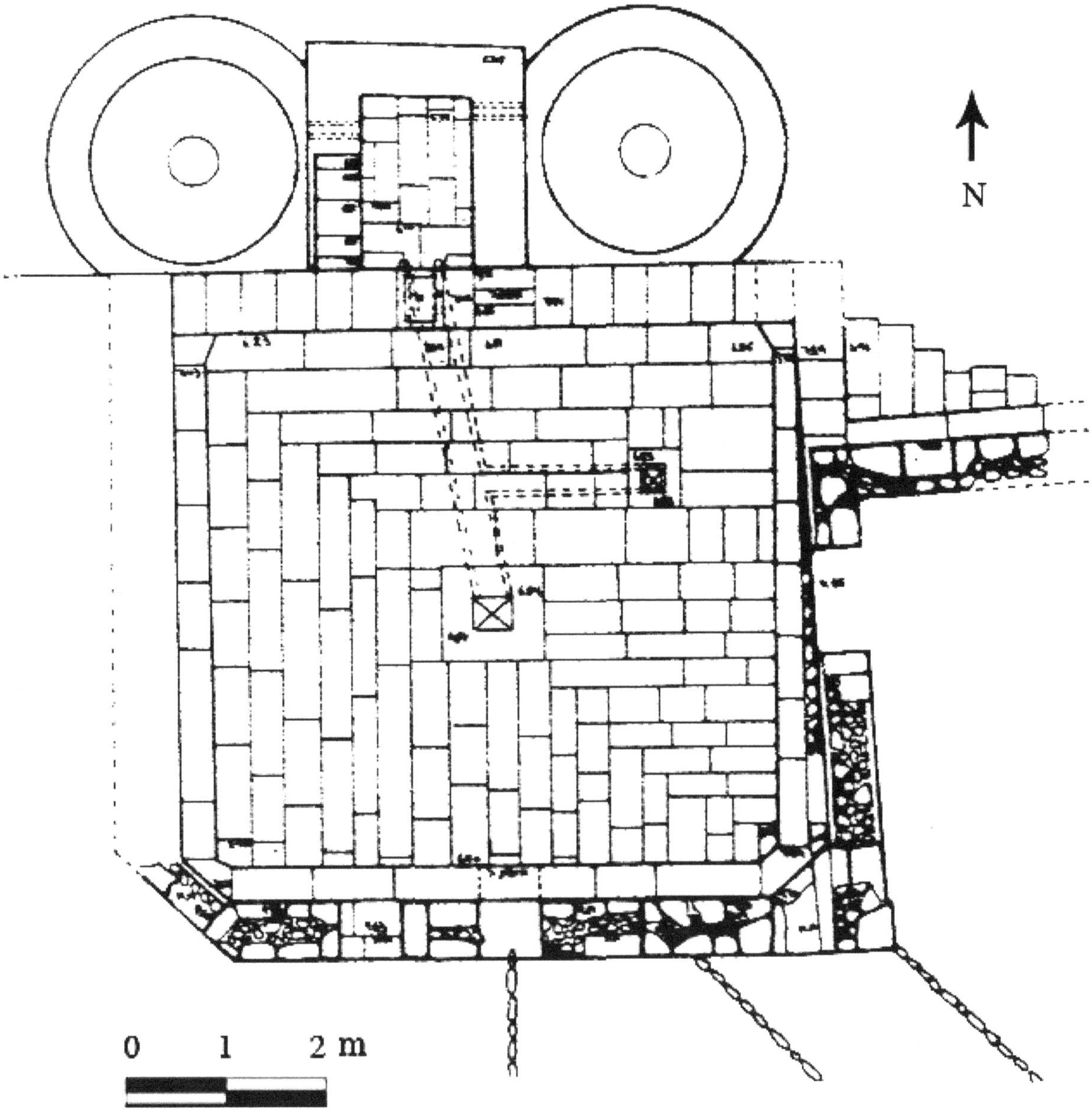

Fig. 70.1. Negev, wine press No. 1.

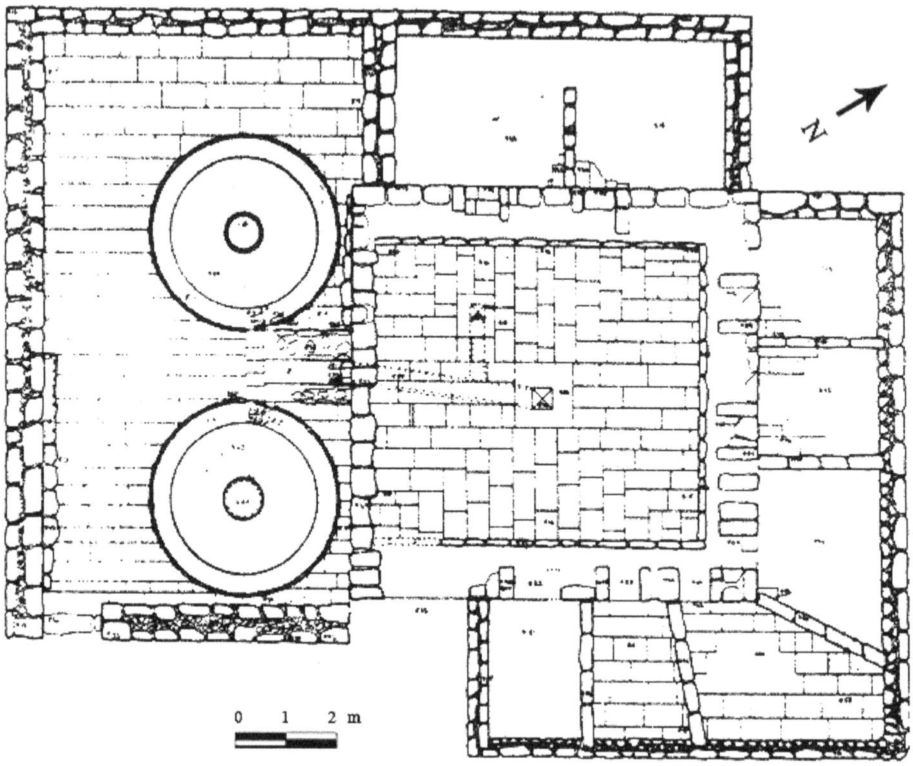

Fig. 70.2. Negev, wine press No. 2.

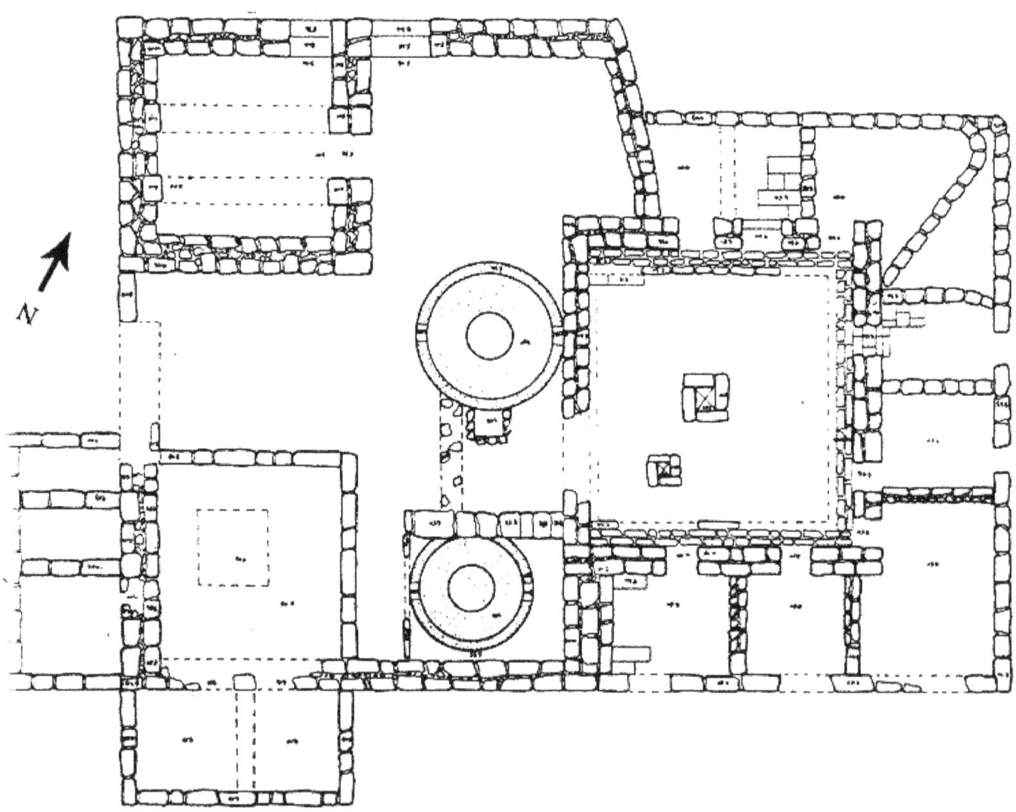

Fig. 70.3. Negev, wine press No. 3.

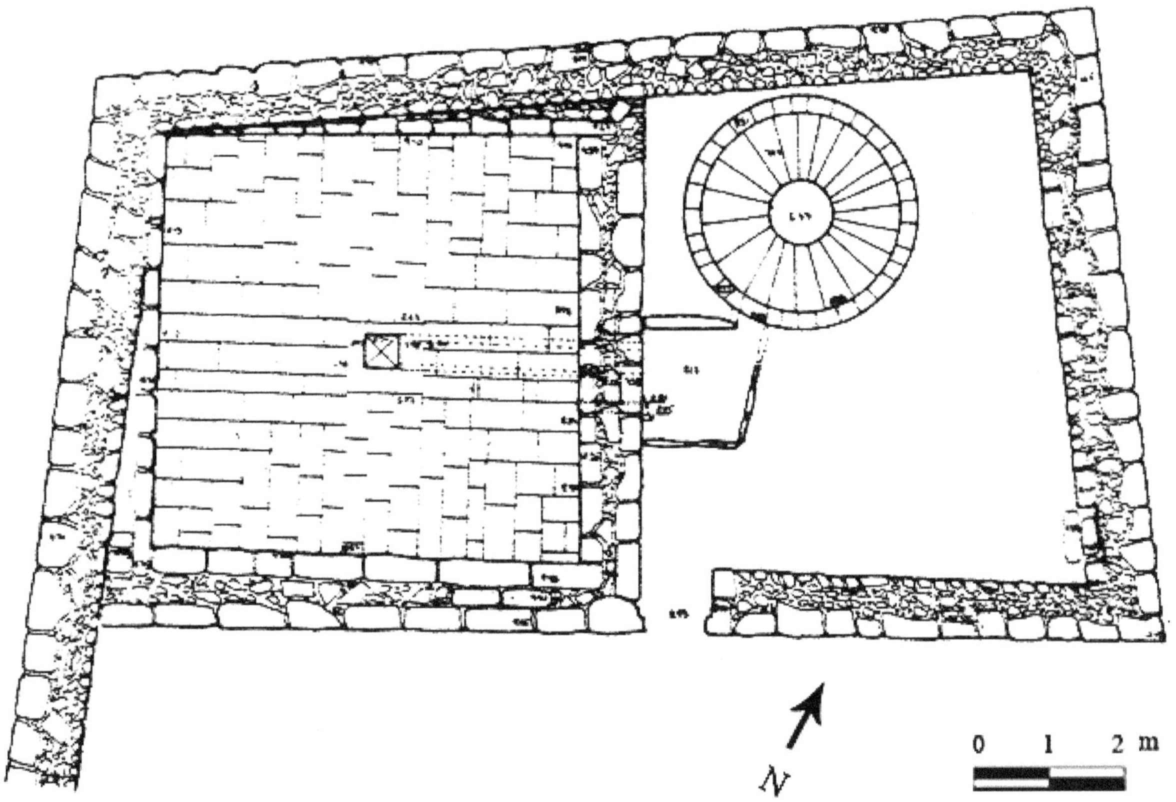

Fig. 70.4. Negev, wine press No. 4.

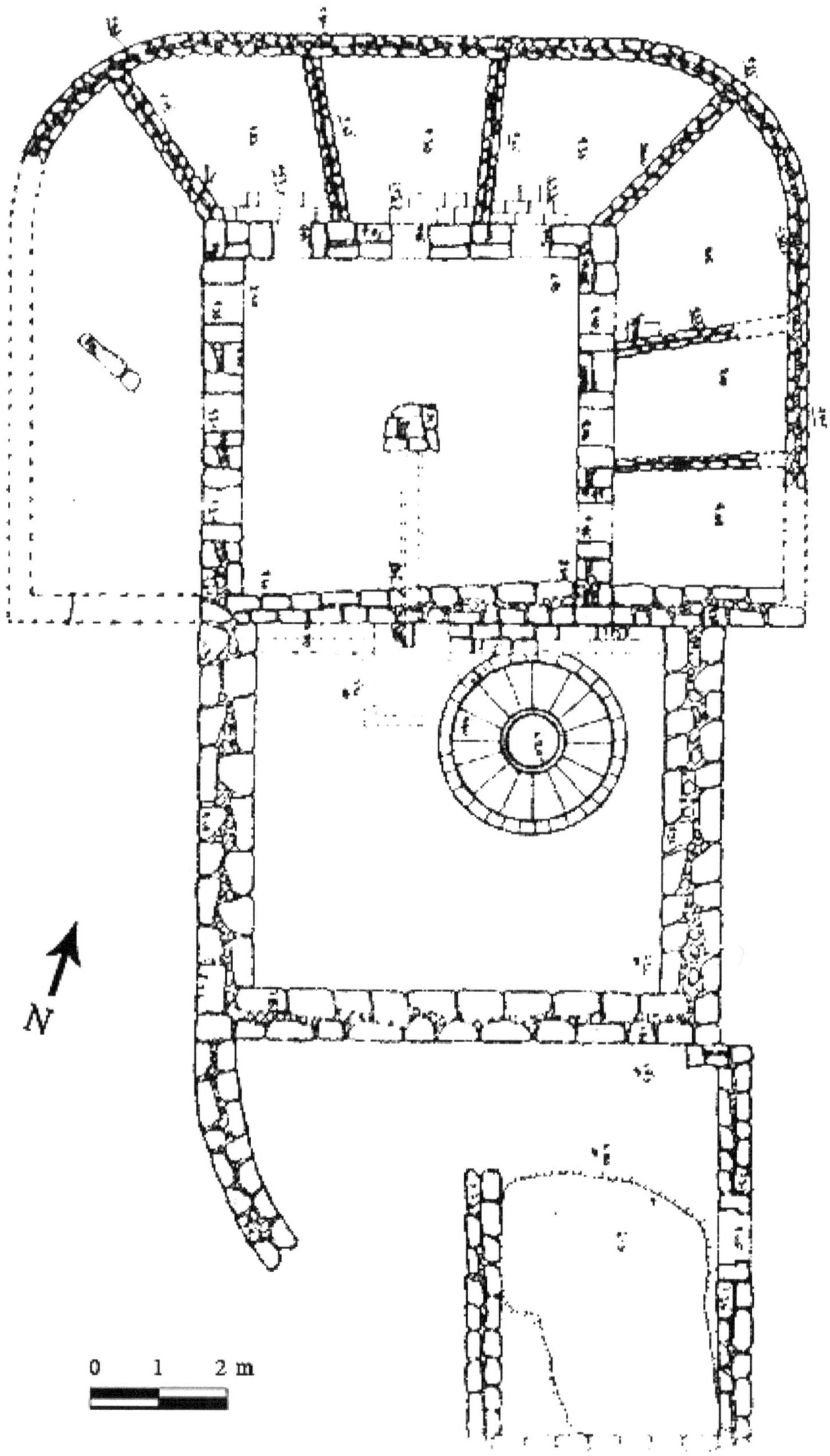

Fig. 70.5. Negev, wine press No. 5.

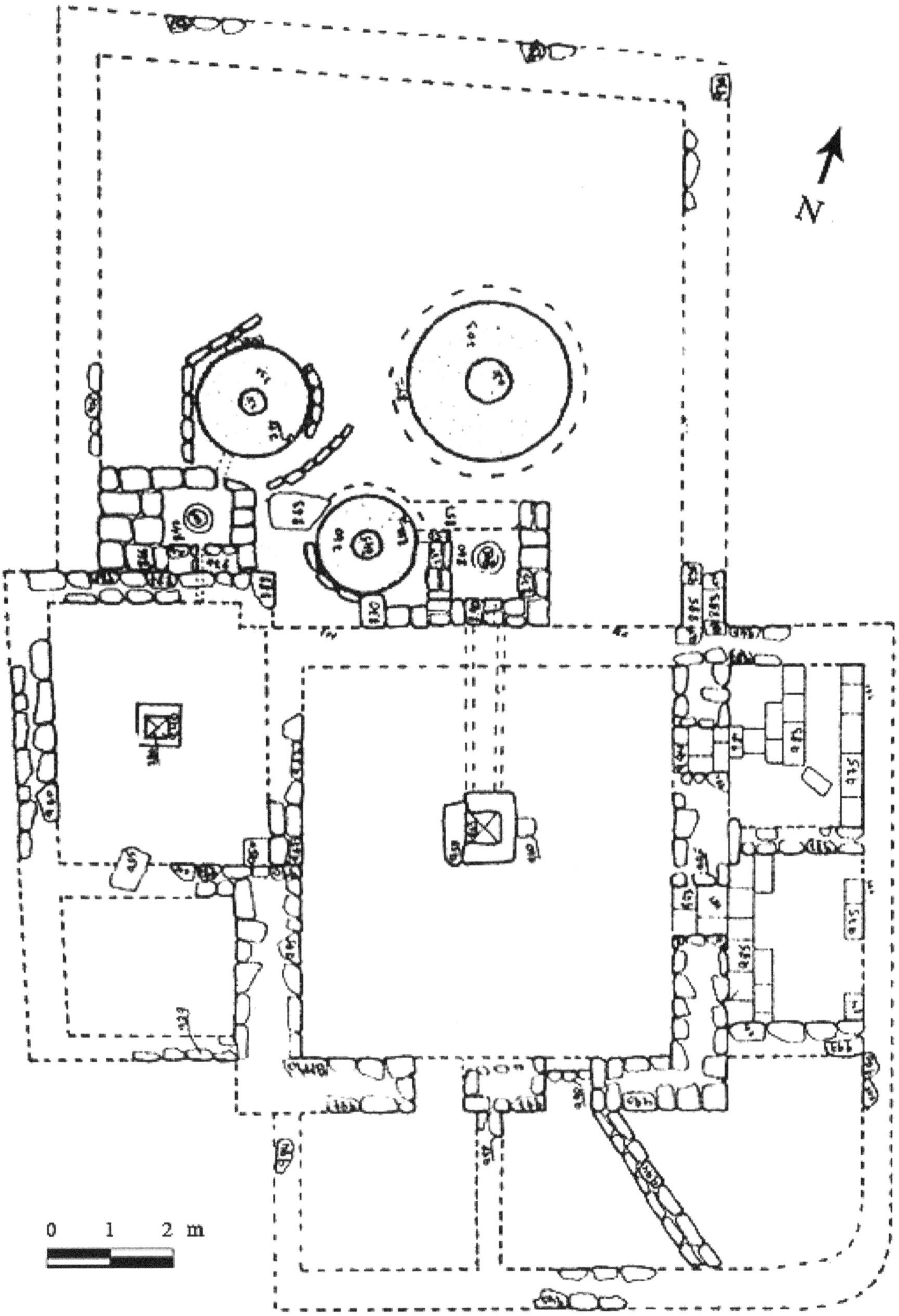

Fig. 70.6. Negev, wine press No. 6.

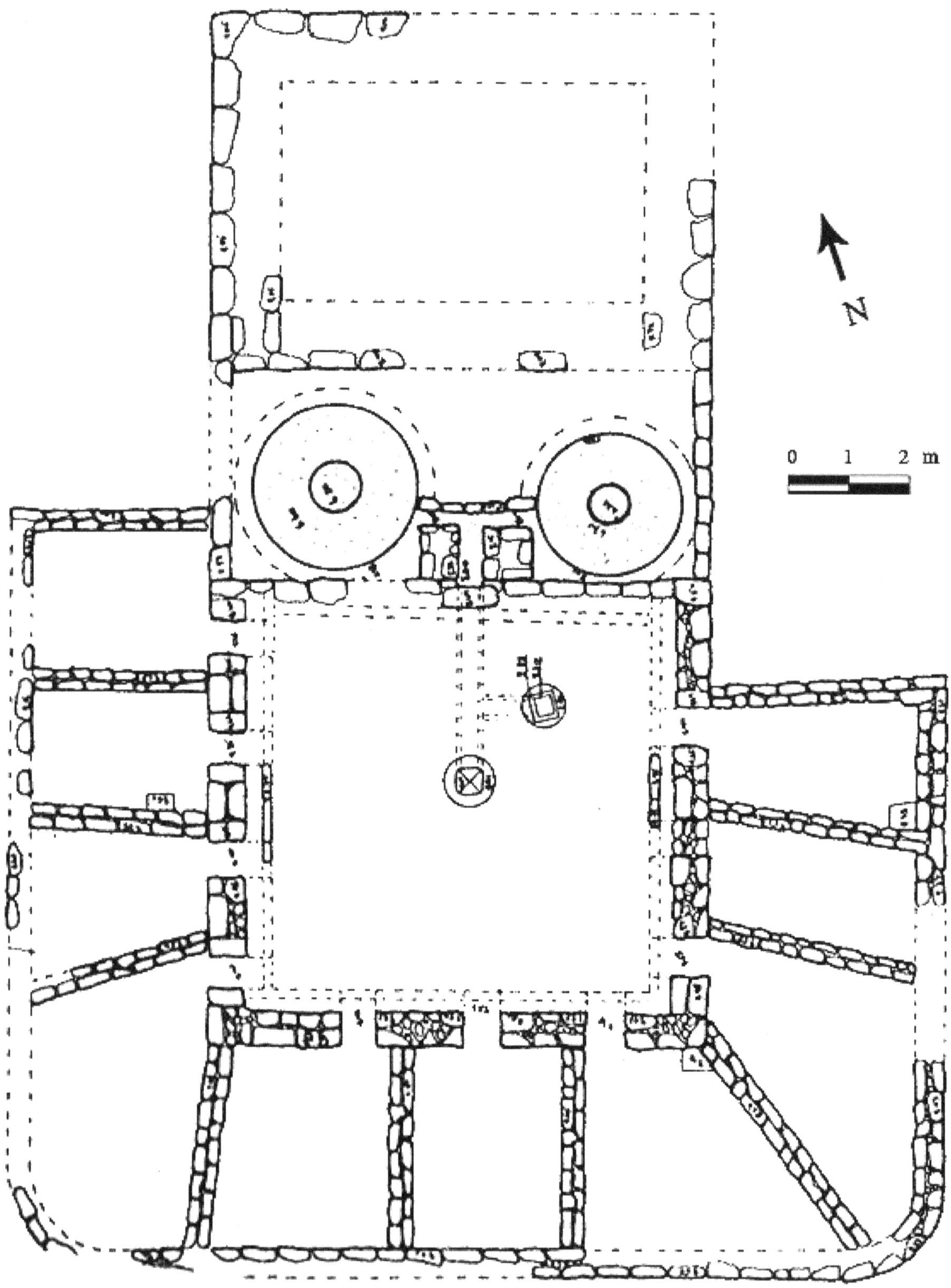

Fig. 70.7. Negev, wine press No. 7.

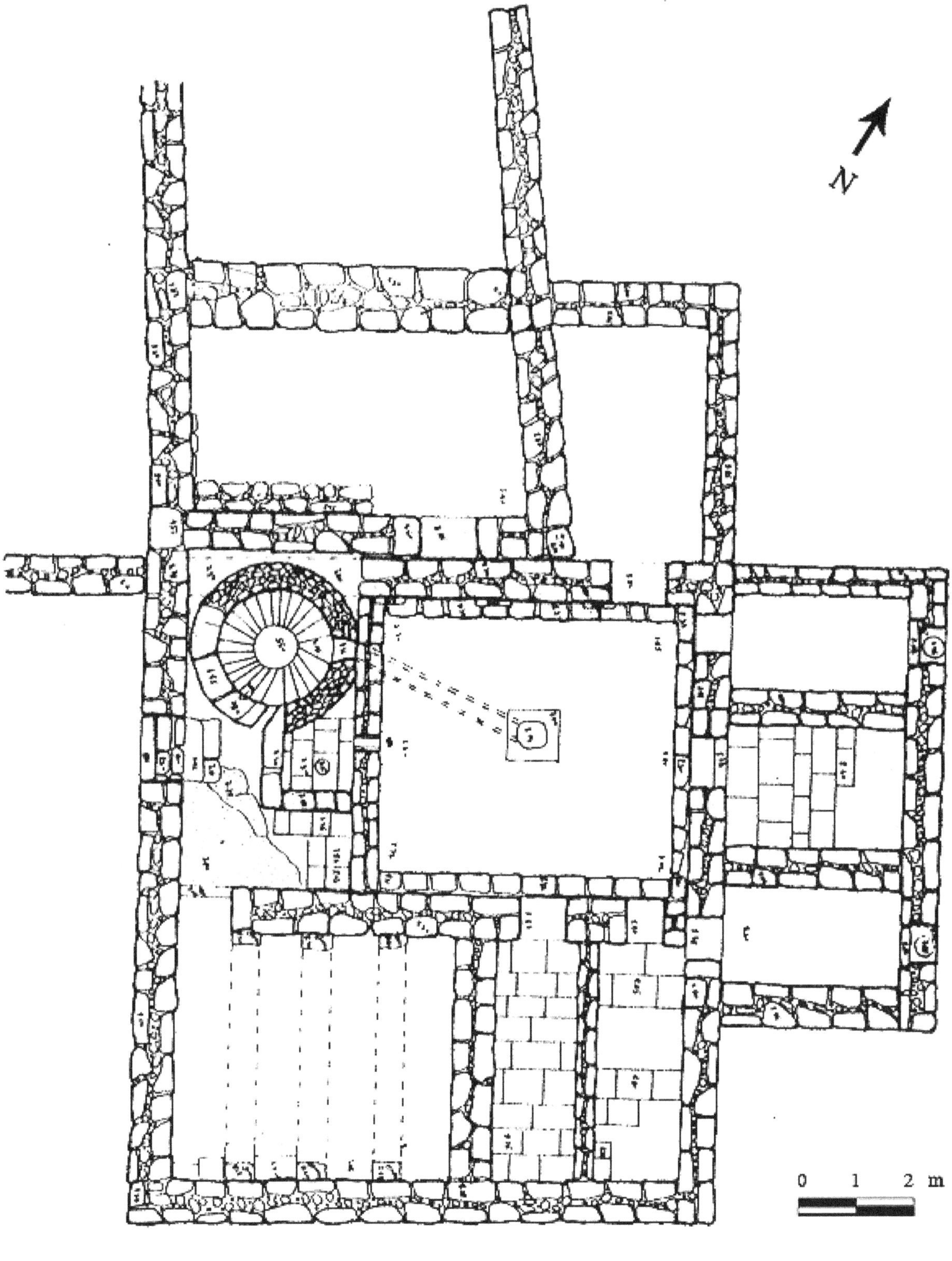

Fig. 70.8. Negev, wine press No. 8.

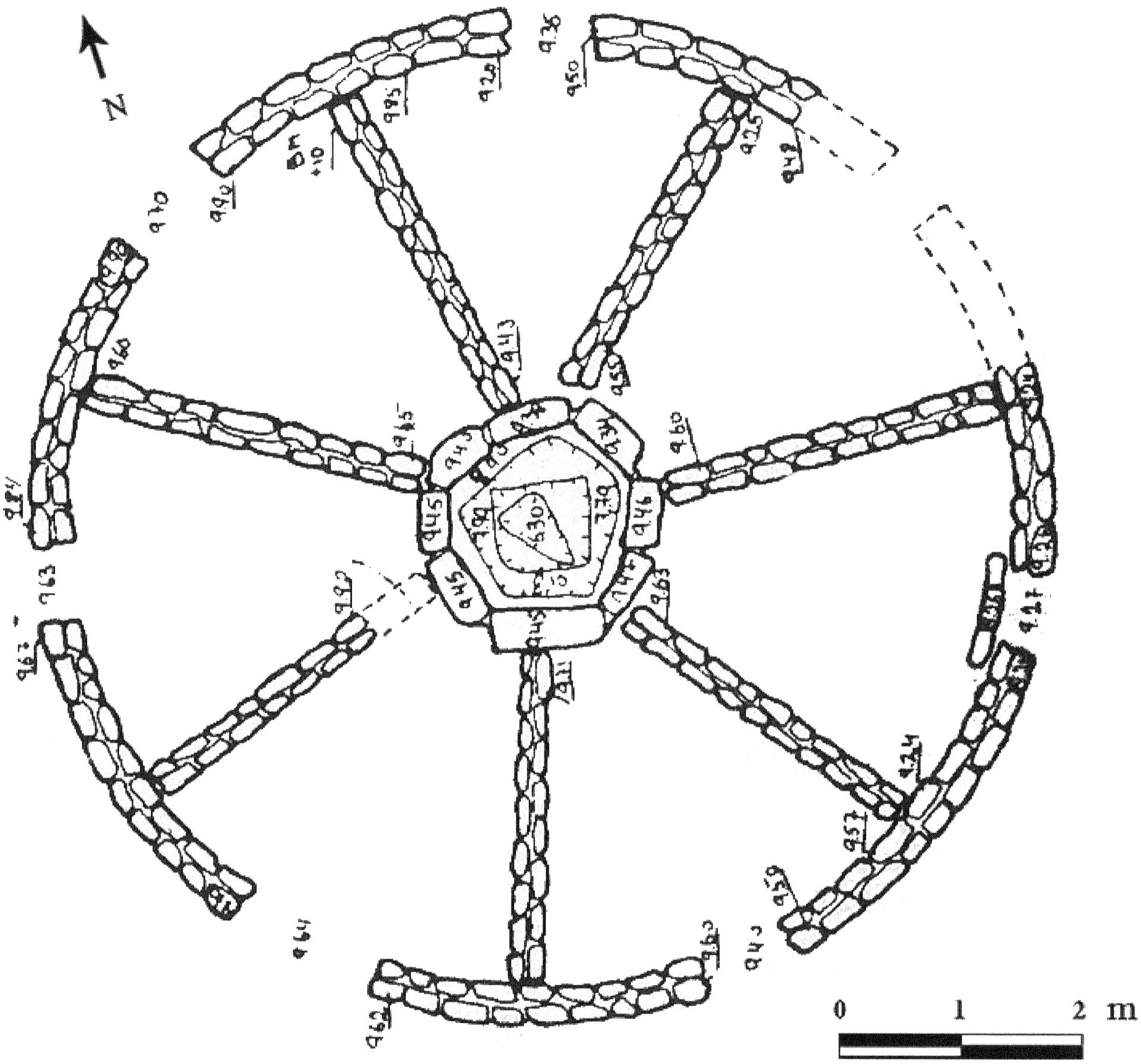

Fig. 70.9. Negev, installation No. 9.

Fig. 70.10. Negev, wine press in Shivta.

71. The "Judean Grooved-Pier Press" – A New Proposal for its Method of Operation

Yehoshua Dray

Olive oil presses from the Roman and Byzantine periods in which the oil was expressed by a "grooved-pier press" are found at many sites, mainly in the Judean Hills and the Shephelah (see Kloner, "Judean Press" in this book). They have been widely discussed in recent years (Frankel 1988–89; 1994: 68–70; 1999: 126–130), mainly because of the difficulty in reconstructing the exact method in which they worked. Today the accepted view is that these installations can be numbered among the direct pressure frame presses in which a screw does the pressing by moving up and down through a threaded plank held in place by two fixed stone piers. This article intends to present a different proposal for reconstructing the operation of this press. According to it, the presses belong to the group in which pressing is implemented by the gravitational power of hanging weights.

Description of the Installation

Many installations of this type are known today. Most of them were discovered in surveys and not during archaeological excavations, a circumstance that makes it difficult both to date them and to reconstruct the way in which they worked. They are usually found in the ruins of settlements, in many cases on the surface, while others have been discovered in caves. The most common type of press consists of two large rectangular stone blocks (that functioned, according to the reconstruction presented below, as weights and not just as "piers") located on two sides of a central collecting vat. Each stone block had square mortices cut into both broad lateral sides. The mortices were connected to the top of the block by narrow grooves and to each other by horizontal holes, and there were also vertical grooves for the whole length of the narrow posterior and anterior sides. In a few presses found in caves like those of Susiya, H. Horan and 'Anim, small sockets were discovered cut into the rock floors and the ceilings, under and above the vertical grooves carved into the narrow sides of the "piers".

Conventional Operational Reconstruction

The reconstruction presently accepted by most scholars was first suggested by B. Frankel (1981: 183). On top of two stone "piers" was laid a thick plank, in the centre of which was a threaded hole. In the hole was mounted a wooden screw that was lowered by turning it in one direction and raised by turning it in the other. When it was lowered by the operator it exerted pressure on the frails of olive mash that had been placed underneath it, and the oil (mixed with watery lees) was expressed and ran into the collecting vat. The upper plank is fixed in place on the "piers" by wooden tenons fitted to the mortices on their broad sides. The tenons are held in place by wooden rods inserted into the horizontal holes. The tenons on the sides are fastened to the upper plank by means of dowels, fixing it to the "piers". Thus the turning of the screw and the consequent pressing operation were enabled without the plank moving in any direction.

In the grooves carved into the narrow faces of the "piers" were fixed wooden rods, whose bases were inserted into the sockets in the floor and whose upper ends were also doweled to the upper plank. In caves, the upper ends of the rods were inserted into sockets in the roof. The purpose of these rods, according to Frankel, Kloner and others, was to help stabilize the "piers" during pressing (that entailed the exertion of tremendous lateral forces) and to prevent them from being moved out of place.

New Reconstruction Proposal

The main flaw in the conventional reconstruction that sees the "piers" as immobile pillars is the effort invested in the precision with which the "piers", with their mortices and channels, were cut. This, at a time when oil presses could have operated with common direct pressure frame presses (Frankel 1994: 63–68) that were far easier to install. And indeed there is evidence that direct pressure screw presses with wooden poles worked contemporaneously with the presses under discussion here, sometimes as described above and at other times in conjunction with a press beam (Tepper 1982: 82). In addition, there is reasonable doubt as to the capability of the slender rods that, according to the usual reconstruction, fitted into the channels on the narrow sides of the "piers" to withstand the tremendous pressures generated during operation of the installation. Also, the sockets designated to receive the ends of those slender rods are not deep enough to support them under the tons of pressure involved.

It is, therefore, necessary to reconstruct the action of the installation in a different way, in which these special devices will be exploited far more efficiently, a way that will justify the effort invested in their preparation.

In the author's opinion, this is a screw and weights press in which continuous pressing was carried out by capitalizing on the weight of these two massive blocks of stone that averaged 2.5 tons each. Pressing the olive mash began with successive turning of the screw with the aid of a wooden handle that was inserted in holes bored in its thickened lower end (Fig. 1). The screw descended and started exerting pressure on the frails of olive mash. When this pressure – or actually the resistance of the full frails to the pressure – equaled the weight of the heavy stone blocks, further turning of the screw resulted in the weights being lifted into the air. At this point, the installation became a normal press in which expression of the liquid was effected by the gravitational power of heavy stone weights and not by the direct pressure of the screw itself. In this way was attained the maximum continuous load capacity that the installation was capable of exerting – that is to say, maximum utilization of the stone weights.

The gradual descent of the suspended weights continued as long as the expressed liquid from the olive mash was still being collected, enabling its efficient extraction without investing additional effort (that was required according to the accepted reconstruction, in which the screw had to be turned more or less continuously in order to maintain the required extraction pressure). The rods fixed in the channels carved into the narrow sides of the weights and in the sockets in the floor and ceiling (or in a horizontal plank above the installation) were used, according to the reconstruction suggested here, as rails upon which the weights moved up and down. In this way, the rotational movement of the entire installation is eliminated and the vertical movement of the weights enabled. This type of installation is characterized by its stability, largely because of the fact that the weights are suspended from points near their top ends while their centres of gravity are found close to their bottom ends, thus preventing the danger of overturning.

Concentrating the weight of the heavy stone "piers" over the area of the pressing frails results in a maximum pressure of ~one kg/cm^2 when the weights are suspended in the air. Comparison with other types of press reveals that this is about the same pressure exerted by a complex lever and screw press with a large stone weight hanging from the bottom of the screw, of the type quite common during the Byzantine period – except that the "pier press" took up a lot less space.

This reconstruction proposal was first examined in a practical way by constructing a 1:10 model of the press and measuring the pressures generated in various directions during its operation. The model was built of original materials: the wooden parts were made from Calliprinos oak and the weights from hard limestone. It was operated – as far as possible – under "real conditions": the olive mash was wrapped in a sheet of coarse cloth instead of being placed in woven frails – a method known both from earlier times and from use by contemporary traditional oil presses. When the screw was turned, the weights rose up into the air and the expressed liquid began to flow from the olive mash.

In the next stage the author physically reconstructed the original press, full-size, found in a cave at Susiya on Mt. Hebron (Fig. 2). The weights were made of cast concrete, the wooden parts of oak. On January 31, 1996 the installation was used for the first time under real-life conditions. Some 40 kg of fresh olives were mashed in a crushing basin and loaded into the press, and with the rotation of the screw the weights rose and remained suspended in the air. They remained in this condition for several hours, with the liquid from the olive mash constantly draining into the collecting vat. The quantity of liquid expressed proved the efficiency of the press.

The installation under discussion is more sophisticated and efficient than a normal direct pressure frame screw press since it does not demand continuous attention or repeated retightening of the screw. From the aspect of the annals of technology, it may be appropriately classified as a late development of the lever and screw press that appeared in the Land of Israel during the Late Roman period and became the principal type of press in use during the Byzantine period (Frankel 1994: 50 ff). In these, also, turning the screw caused the weight to be lifted into the air, thus exerting continuous pressure on the frails (in contrast to the opinion of those who feel that the huge stone was simply a fixed base for anchoring the wooden screw).

We feel that much can be learnt from this example and are of the opinion that research on ancient installations, and the theoretical reconstruction of their operation, must be accompanied by practical testing. The evaluation should be carried out using genuine equipment, carefully and completely reconstructed – preferably as full-sized devices, or if that is impractical, at least as reduced-size models.

In recent years it has become more common to reconstruct ancient installations by adding the parts made of organic materials such as wood, leather, rope, etc., that have disappeared over the centuries, a subject in which the author has acquired much expertise. This adds an important dimension to the work of researchers, enabling them to examine their reconstruction proposals in a practical way and in "real time," as was done in this case.

Bibliography

Frankel, B. 1981. "Notes and Remarks on Olive Oil Production in Judea". *Teva Va'Aretz* 23: 183 (Hebrew).

Frankel, R. 1988–1989. "An Oil Press at Tel Safsafot". *Tel Aviv* 15–16: 77–91.

Frankel, R. 1994. "Ancient Oil Mills and Presses in the Land of Israel". In: R. Frankel, S. Avitsur and E. Ayalon. *History and Technology of Olive Oil in the Holy Land.* Arlington and Tel Aviv: 19–89.

Frankel, R. 1999. *Wine and Oil Production in Antiquity in Israel and Other Mediterranean Countries*. Sheffield.

Tepper, Y. 1982. "The Oil Plant at Give'on". *NZ* 18: 66–86 (Hebrew).

Fig. 71.1. "Judean grooved-pier press" – proposed reconstruction of its operation.

Fig. 71.2. "Judean grooved-pier press" reconstructed in Susiya.

72. Olive Oil Presses with "Judean Piers"

Amos Kloner

Presses of the grooved "Judean Piers" type have been found in oil presses in the Shephelah, the Judean Hills and the Judean Desert, installed in roofed or open-air structures as well as in caves[1]. Approximately forty sites (thirty-one were discovered prior to 1983) are now known to have had presses of this type (Frankel 1984: 171–172, maps 36–37; 1994: 68–70; 1999: 126–130). In each of these complexes, the extraction device was of the type known as a "direct pressure screw press." The name "Judean Piers" Press derives from the fact that it is typical only of Judea and that its major components were two piers – rectangular vertical blocks of stone upon which the screw and other wooden parts of the press were mounted.

During his researches in the region the author has recorded installations of this type at several sites in the Judean Shephelah. They are very similar both in form and dimensions to six pairs of stones published in the past (Clermont-Ganneau 1896: 452–454; Dalman 1928–1942, IV: 226–228; Gichon 1979–1980: 209–220).

The piers, hewn from local limestone, are installed opposite each other with their broad faces parallel and about 1 m apart. Occasionally they are discovered slightly out of line, the movement having occurred after they were no longer in use. In every pair of stones, the shape and dimensions of both are identical (See Site 66, Fig. 1). Each stone is a rectangular parallelepiped set vertically on one of its two smallest faces and is approximately 2 m high, 1 m wide and 0.5 m thick. A groove about 10 cm wide and deep is cut into each of the narrow faces along its entire height. In a few oil presses installed in caves, depressions were found carved into the ceiling and floor precisely above and below these grooves; such a find was reported by D. Amit from a cave at Susiya (see Yeivin 1993: 28–31). A mortice in the shape of a reversed T is cut from the top to about the center on each of the broad faces of the pier. The horizontal component of the T is square or rectangular in shape, each side measuring 0.3–0.4 m. The mortices on the two lateral faces are connected by an opening 0.15×0.15 m that penetrates the pier through the center of the head of the T. A central collecting vat is hewn between the two piers.

The pair of stone piers served to support the wooden parts of the press. A wooden platform of one or more heavy planks was placed across the top of the piers and connected to them by means of rods fixed in the grooves on their narrow sides and tenons in the mortices on their broad sides. These rods and tenons were locked in place by pegs where they passed through the platform and the two tenons on each pier were secured by an additional rod – also pegged in place – that connected them through the opening that penetrated the pier through the head of the T. In the centre of the platform between the two piers there was a threaded bore ("nut") in which the wooden screw turned up or down. Approximately a dozen pressing frails containing olives previously mashed in the crushing mill were stacked between the piers above the collecting vat. The operator turned the screw with a handle inserted into one of the holes at its bottom end; the screw moved downward, exerting pressure on a pressing board laid on top of the frails. The expressed liquid – oil and watery lees – flowed through the frails into the central collecting vat. The oil, floating on the surface of the watery lees, was either removed with a utensil or flowed into a separate vat. The oil press workers carefully separated the oil from the watery lees that were discarded.

There have been differences of opinion among scholars and investigators in the past as to the exact manner in which the vertical grooves on the narrow faces of the piers functioned. Dalman suggested that extensions at the corners of the pressing board moved up and down in these grooves. In his opinion the grooves served as guides for these extensions, stabilizing the pressing board while it was being raised or lowered (Dalman 1928–1942, IV: 226–228, Pl.69). With the excellent technology in carpentry in the Roman and Byzantine periods such extensions could have been built of wood but their strength would have been limited. Iron bands could have been attached to the board with their ends bent to fit into the grooves in which they moved and some scholars have reconstructed the press in this way (Gichon 1980: Fig. 9).

Dalman's reconstruction was rejected by Y. Peleg in an article (Peleg 1981: 98–103) and in a letter quoted by A. Stahl (1980: 81). Because of the relatively large distance of the grooves from the point where the screw exerted pressure, any deflection of the pressing board would have moved the extensions out of line, thus, in Peleg's opinion, impeding the movement of the board. In addition, while making a wooden pressing plate with extensions as Dalman proposed was possible, it would not have been easy. Peleg reasoned that if guide grooves were needed, it would have been more convenient to locate them on the broad inner face of the stone pier. Peleg did not, however, suggest an alternative function for the vertical grooves.

Stahl (1980: 81) also believed, following B. Frankel, that the absence of abrasions along the length of the vertical grooves indicates that they did not serve as guides for moving parts. However, his reconstruction of the press with an additional upper board that stabilized the screw and rested on four supports placed in the vertical grooves lacks both archaeological proof and technological merit.

B. Frankel excavated a press of this type at Kh. Dukas (Site 66) north of Tarqumiye and found four small depressions cut in the bed-rock immediately below the four vertical grooves. He suggested that the press be reconstructed with wooden rods in the narrow vertical grooves, their lower ends fixed in the depressions in the rock and their upper ends pegged into the wooden platform resting on top of the stone piers. According to this reconstruction, the screw had two points of stabilization: the upper one the platform resting across the piers, the lower one the moving pressing board (Frankel 1981: 183). The rods placed in the vertical grooves served to withstand lateral pressures exerted during the pressing operation. This reconstruction is accepted by most scholars today.

Y. Dray recently suggested that this installation should not be included in the category of "direct pressure screw presses" in which the screw descends between two stationary piers, but in a category of presses in which pressing is carried out with the aid of hanging weights (Dray 1993: 91–95; see Dray in this book). Dray suggests that turning the screw and the resultant pressure on the olive frails caused the "piers" to rise into the air like hanging weights, and in this way the maximum use was made of their weight for purposes of pressing. The wooden rods installed in the grooves on the narrow faces of the pillars served as tracks to guide the movement of the weights up and down. Dray accompanied his proposal with a small model, and later reconstructed the grooved-pier press at Susiya in this fashion. The author of this article still has some doubts concerning Dray's proposal because of technical problems such as the tremendous pressure exerted on the wooden screw and the incompatibility of the shape of the piers with this *modus operandi*, and because of the lack of archaeological evidence such as the absence of wear and tear in the grooves on the piers.

Sites have been surveyed in the Judean Desert in recent years in which oil presses with grooved pier presses have been preserved on the surface: Kh. el-Qasr (Hirschfeld and Kloner 1988–9: 11–15; Fig. 1) and Kh. Quneitra (Hirschfeld 1985). Several oil presses of this type dated to the Byzantine period (4th cent. CE and later) have been discovered in the Judean Shephelah but have not yet been published.

A pair of piers was found on the surface (Fig. 2) at Kh. Judraya (MR 1498/1218); another installation of this type (Fig. 3) was found there in a cave (other caves served as water cisterns, storehouses and hiding complexes). The cave is large, measuring about 6–7×12 m (Fig. 4–5). A crushing basin 1.9 m in diameter and a pair of piers *in situ* were found in line with the entrance and arranged in a single row. The distance from the basin to the closest pier is 1.5 m; the pillars were placed 1.2 m apart. Setting up the three elements of the installation in a straight line may have been common practice; they were also discovered positioned in this manner at other sites. The discovery of two installations at the same site should also be noted; it is not clear whether both were in operation at the same time but this is a reasonable assumption.

A crushing basin and two pairs of stone piers were discovered (Fig. 6) at Giv'at Sharret in Bet Shemesh (MR 1478/1274). The building in which they were found (destroyed in development operations) was dated on the basis of potsherds and roof tiles to the 5th-6th cent. CE (Hirschfeld and Kloner 1988–9: 15). The piers were removed to the Har Giloh Field School by the Israel Antiquities Authority and the crushing stone to the Shephelah Museum at Kibbutz Kefar Menachem.

A pair of grooved piers (Figs. 7–8) was found *in situ* in an oil press at Moshav Roglit (Kh. el-Jurfa). Finds at the site are from the Roman and Byzantine periods, including a Byzantine church (MR 1504/1202).

An oil press with one or two pairs of piers (Figs. 9–10) was discovered at H. Benaya (MR 1426/0982). Wine presses and hiding complexes were also found at the site (Kloner 1986: 12–14).

An oil press was discovered at Kh. Umm Ba'ala (MR 1407/0934) in which a crushing basin, two crushing stones and the pier of a pressing installation were found. Various installations such as wine presses and hewn and plastered reservoirs, as well as potsherds from the Roman and Byzantine periods and the Middle Ages were also found.

An oil press (Figs. 11–12) was found in a large cave measuring 15×17 m at H. Za'aq (MR 1375/0914). Hiding complexes had been hewn out of the rock (Kloner and Tepper 1987: 260) before its installation. The cave was investigated by A. Navon in 1978 and in 1982 by Y. Goren and P. Fabian on behalf of the Joe Alon Center and the Israel Antiquities Authority. The crushing basin was discovered *in situ* and nearby a single press pier was found lying on its side. Adjacent to it and along the same line, as at Kh. Judraya and Kh. el-Qasr, a second pier can be reconstructed.

Two piers were found *in situ* (Fig. 13) in a cave at Kh. Hauran (MR 1448/1171). Small depressions had been carved at the height of the top of the piers in the rock wall of the cave on their left and in a large hewn boulder on their right. Crossbars were placed in these depressions to help stabilize the wooden threaded board attached to the top of the piers. Opposite the space between the piers a niche was hewn in the wall of the cavern for anchoring a large wooden beam of the kind used in the lever and weights presses typical of the Shephelah (Kloner and Sagiv 1991: 61–68). Y. Tepper has suggested that this press combines elements of the two types of presses found in the region – the lever and weights press and the grooved-pier screw press – to produce an unusual type of lever and screw press in which the beam anchored in the niche was operated by a screw descending between the two piers. (It should be noted that lever and screw presses were usually operated by weights in Israel).

Another pier installation in a cave at Kh. el-Bass (MR 1444/1070) was found that may also have combined a long beam anchored in a niche in the opposite wall with a screw descending between two piers (Fig. 14). There are no data available for a precise dating of the installation except that the oil press is later than the hiding complex conjoining the cave (Kloner and Tepper 1987: 86).

The last two examples from Kh. Hauran and Kh. el-Bass indicate the possibility that in these presses a long beam was utilized to press upon the frails in which the pressing point was close to the anchor point, and not by direct screw pressure in which the frails are placed between the two piers. This method is similar to the latter except that the pressing power of the installation is increased through the use of a beam working on the lever principle. In either case, pressing by means of a screw between two "Judean Piers" was the most advanced of its type in ancient times and its pressing capability the highest.

It is worth noting that the upper end of the screw was never mounted or anchored in the ceiling of the cave. The ceilings of caves in the Judean Shephelah are composed of a thin layer of soft limestone covered by a crust of hard Pleistocene limestone. Those who constructed the presses were undoubtedly familiar with the relative weakness of the rock and aware of the danger inherent in the great pressure the screw could exert upon it. On the other hand, they did carve depressions in the ceiling above the grooves in the sides of the piers, and these are evidence that the pier supports in those grooves (unlike the screws, they exerted little if any vertical force) were anchored in the ceiling and in similar depressions in the cave floor.

Finds from structures adjacent to the oil presses in which "Judean Pier" presses had been installed and finds from the installations themselves show that they were in most cases constructed during the 4th cent. CE and continued in operation throughout the Byzantine period and – in some cases – into the Early Arab period. The large numbers found are evidence for the economic importance of olive growing and processing in the region during the Roman and Byzantine periods. The settlements of the Judean Shephelah produced more olive oil than was consumed by the population of the region; large quantities were marketed and exported during the Iron Age and the Hellenistic, Roman and Byzantine periods. In parallel with the pier presses, other pressing installations were employed in Judaea during the Byzantine period (for example see Frankel, Patrich and Tsafrir 1990: 287–300). However, most Judean presses of that period were of the type reported here, showing the high technological level attained in Judaea during the Late Roman period and particularly during the Byzantine period.

Notes

1. Since there is nothing unusual in the crushing process employed in these installations, this article deals only with the special presses used in them.

Bibliography

Clermont-Ganneau, Ch. 1896. *Archaeological Researches in Palestine* II. London.

Dalman, G. 1928–1942. *Arbeit und Sitte in Palästina.* 7 Vols. Gütersloh (repr. Hildesheim 1964).

Dray, Y. 1993. "The Judean Grooved Pier Press: A New Suggestion for the Reconstruction of its Operation". *NZ* 19: 91–95 (Hebrew).

Frankel, B. 1981. "Notes and Remarks on Olive Oil Production in Judea". *Teva Va'Aretz* 23: 183 (Hebrew).

Frankel, R. 1984. *The History of the Processing of Wine and Oil in Galilee in the Period of the Bible, the Mishna and the Talmud.* Unpublished Ph.D. Thesis, Tel Aviv University (Hebrew).

Frankel, R. 1994. "Ancient Oil Mills and Presses in the Land of Israel". In: R. Frankel, S. Avitsur and E. Ayalon. *History and Technology of Olive Oil in the Holy Land.* Arlington and Tel Aviv: 19–89.

Frankel, R. 1999. *Wine and Oil Production in Antiquity in Israel and Other Mediterranean Countries.* Sheffield.

Frankel, R., Patrich, J. and Tsafrir, Y. 1990. "The Oil Press at Horvath Beit Loya". In: G.C. Bottini, L. Di Segni and E. Alliata (eds.). *Christian Archaeology in the Holy Land, New Discoveries.* Jerusalem: 287–300.

Gichon, M. 1979–1980. "The Upright Screw-Operated Pillar Press in Israel". *SCI* 5: 206–243.

Hirschfeld, Y. 1985. "Khirbet el-Quneitra – A Byzantine Monastery in the Wilderness of Ziph". *EI* 18: 243–255 (Hebrew).

Hirschfeld, Y. and Kloner, A. 1988–9. "Khirbet el-Qasr: A Byzantine Fort in the Judean Desert". *BAIAS* 8: 5–20.

Kloner, A. 1983. "The Subterranean Hideaways of the Judaean Foothills and the Bar Kokhba Revolt". *The Jerusalem Cathedra* 3: 135–144.

Kloner, A. 1986. "Horvat Benaya". *ESI* 5: 12–14.

Kloner, A. and Sagiv, N. 1991. "The Technology of Oil Production in the Hellenistic Period at Maresha, Israel".

In M.-C. Amouretti, J.-P. Brun and D. Eitam (eds.). *Oil and Wine Production in the Mediterranean Area from the Bronze Age to the End of the XVIth Century. Pré-Actes. Aix-en-Provence-Toulon*: 61–68.

Kloner, A. and Tepper, Y. 1987. *The Hiding Complexes in the Judean Shephelah*. Tel Aviv (Hebrew).

Peleg, Y. 1981. "How Ancient Olive Presses Worked". *Israel, Land and Nature* 6: 98–103.

Stahl, A. 1980. "Olive Oil Production in Judea". *Teva Va'Aretz* 23: 80–81 (Hebrew).

Yeivin, Z. 1993. *Horvat Susiya, a Town of the Talmudic Period (Guide)*. Jerusalem (Hebrew).

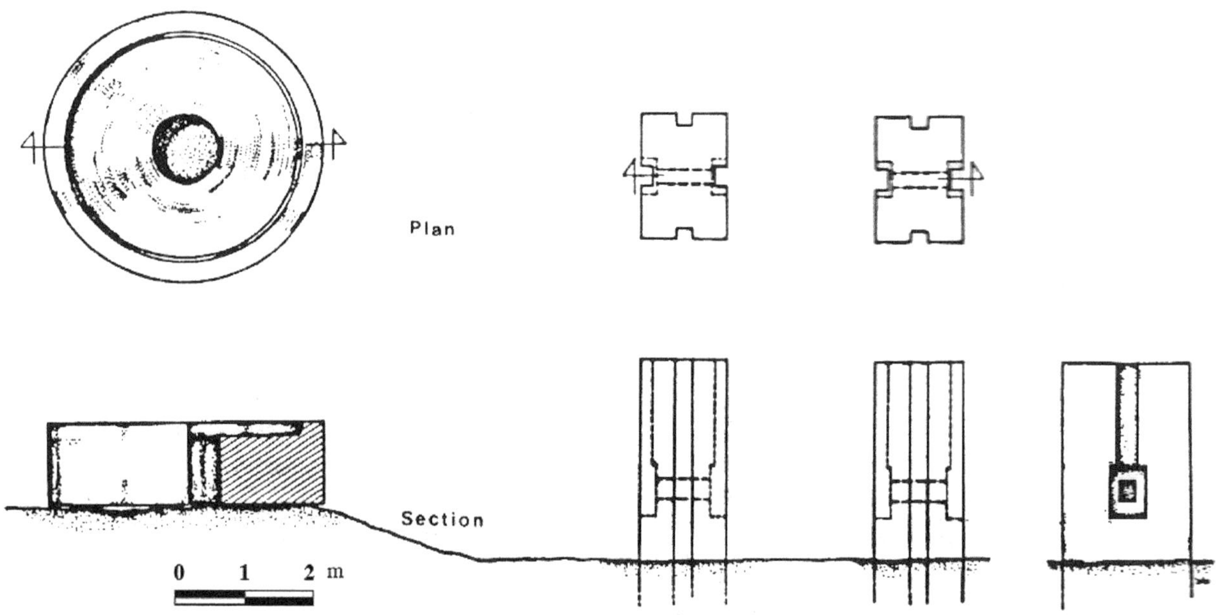

Fig. 72.1. "Judean piers press," Kh. el-Qasr, plan and section of the oil press.

Fig. 72.2. "Judean piers press," Kh. Judraya, stone pier in situ.

Fig. 72.3. "Judean piers press," Kh. Judraya, two piers in situ *in cave.*

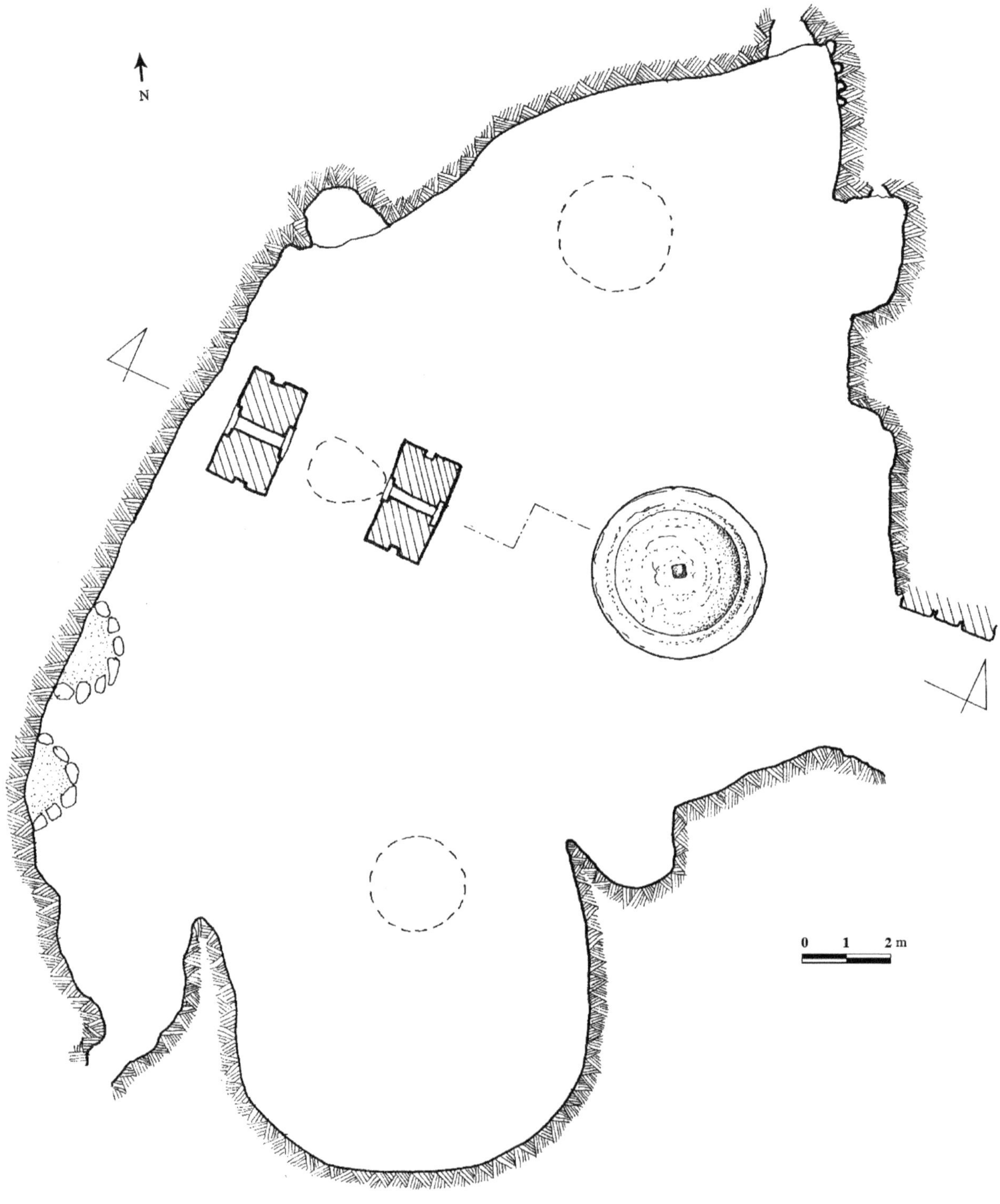

Fig. 72.4. "Judean piers press," Kh. Judraya, plan of the oil press cave.

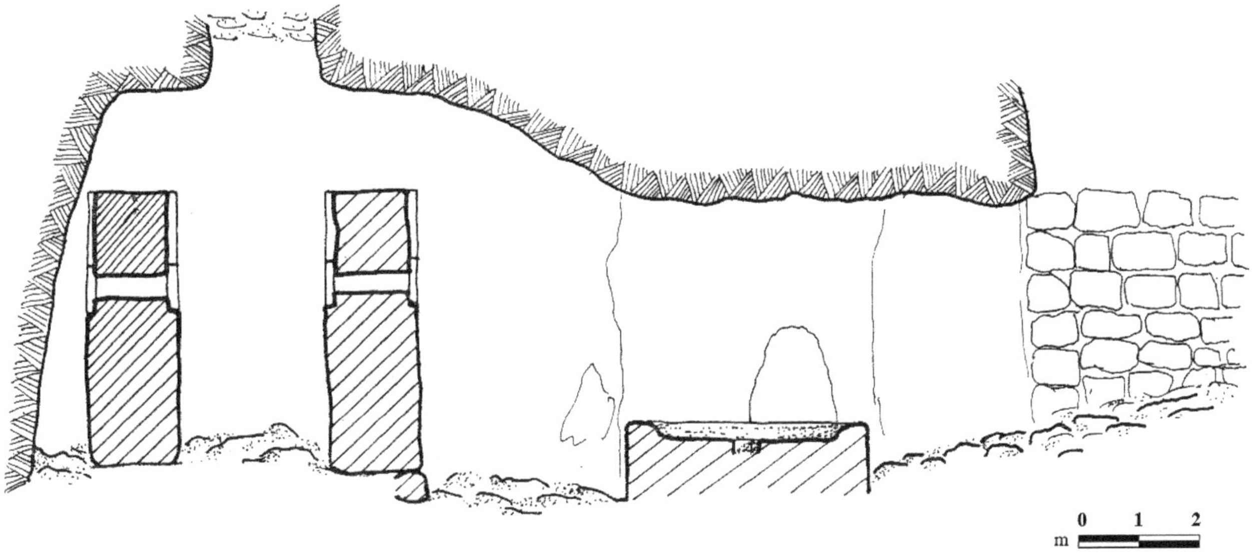

Fig. 72.5. "Judean piers press," Kh. Judraya, section of the oil press cave.

Fig. 72.6. "Judean piers press," Giv'at Sharet, three piers (of four) removed from original installation.

Fig. 72.7. "Judean piers press," Kh. El-Jurfa (Roglit), two piers in situ.

Fig. 72.8. "Judean piers press," Kh. El-Jurfa (Roglit), two piers in situ.

Fig. 72.9. "Judean piers press," H. Benaya, overview of site with one pier in situ.

Fig. 72.10. "Judean piers press," H. Benaya, overview of site with one pier in situ.

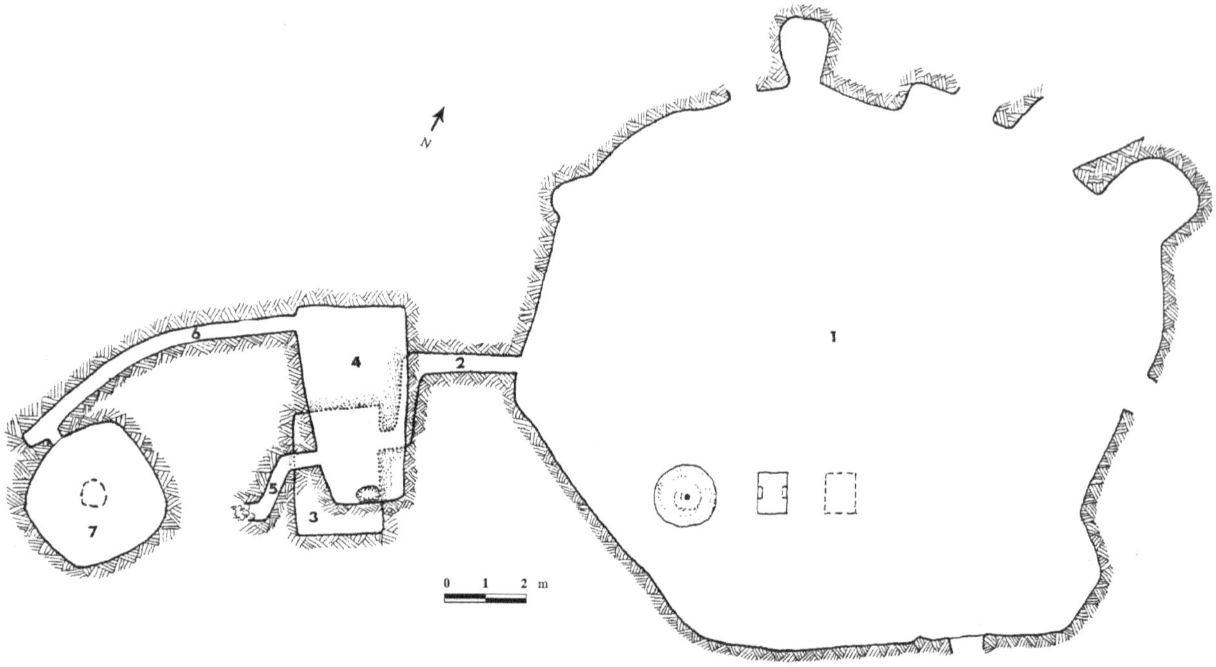

Fig. 72.11. "Judean piers press," H. Za'aq, plan of the oil press cave.

Fig. 72.12. "Judean piers press," Kh. Za'aq, cave with crushing basin (in foreground) and pier.

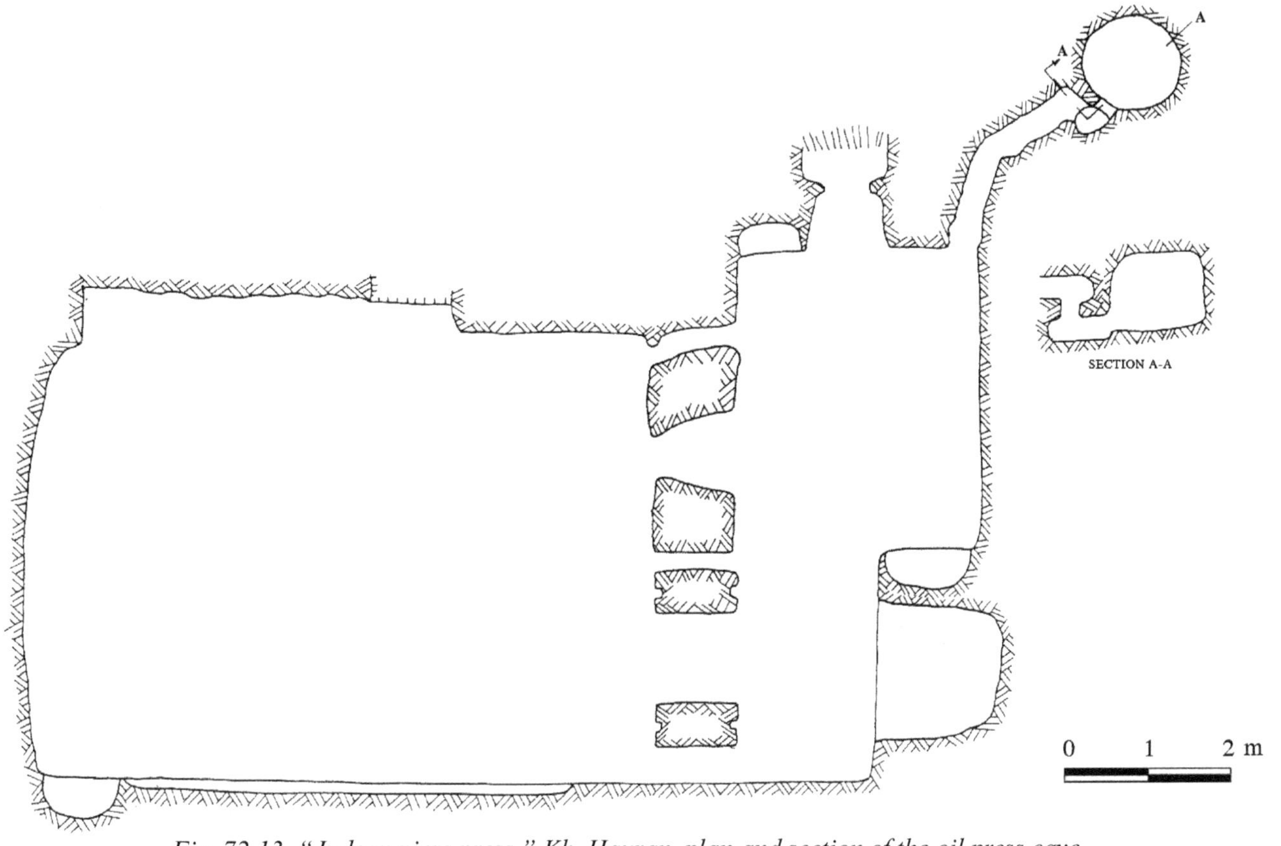

Fig. 72.13. "Judean piers press," Kh. Hauran, plan and section of the oil press cave.

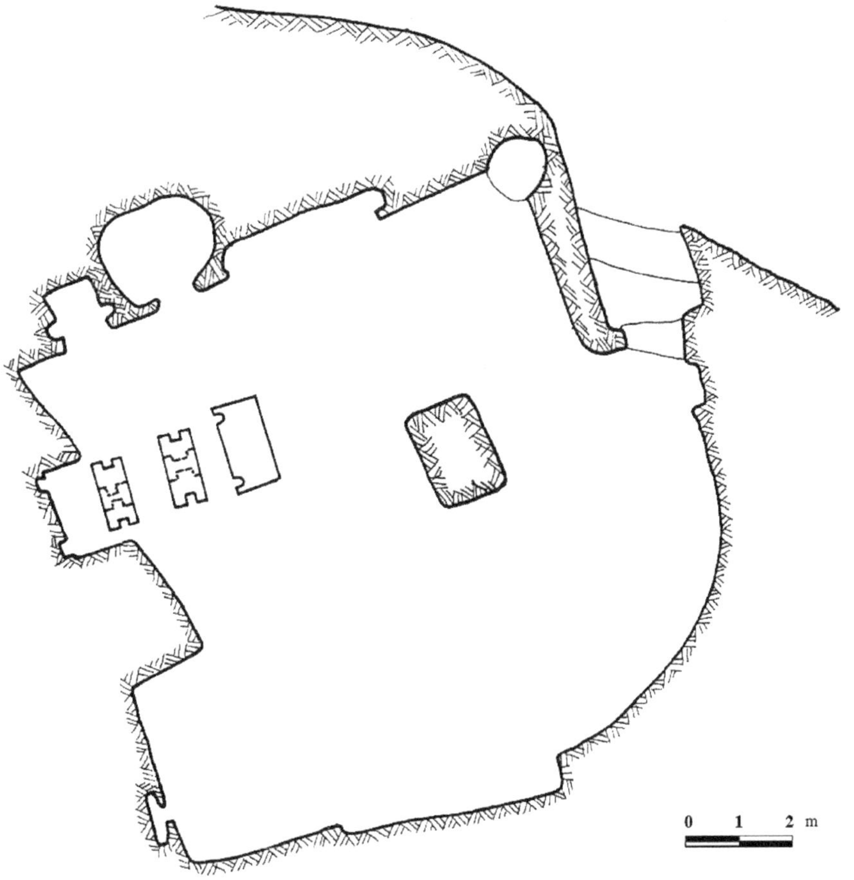

Fig. 72.14. "Judean piers press," Kh. el-Bass, plan of the oil press cave.

73. Wine Presses with Stone Rollers –An Ancient Phenomenon Seen in a New Light

David Amit and Yuval Baruch

Introduction

Until recently finding stone rollers next to ancient wine presses was considered a rather rare phenomenon. Even in the few instances when scholars mentioned such rollers, they deemed them to be an element unrelated to the wine press and the wine-making process and suggested they have a function that is not associated with the work of a wine press.

It seems the first one to deal with these stone rollers, indeed the kind that is specific to the Carmel, was L. Oliphant in 1883. He described the remains at H. Sumaqa on Mt. Carmel and wrote:

"I often found near these mills huge limestone rollers about three feet in diameter and seven feet long. On the sides of these were four vertical lines of sunk grooves, four or five grooves in each line. Taking 2.7 as the specific gravity of the stone, they must have weighed about two tons each. What their functions were, or whether they had anything to do with the olive-crushing process, I am at a loss to conjecture" (Oliphant 1887: 95).

Approximately 120 years later E. Netzer published installations used to produce date wine that he exposed in Jericho. He described a roller that was found in one of the installations as evidence that corroborates his conclusion that this was not a press for manufacturing grape wine because "a stone that is meant for crushing or pressing has no place in wine presses used to produce grape wine" (Netzer 2002: 73).

These two quotations, separated by 120 years, are sufficient to illustrate to what extent stone rollers were considered unusual and foreign to the landscape of ancient agricultural installations in the Land of Israel. It was against this background that we were surprised to discover, during study trips we conducted in southern Mount Hebron, a series of wine presses from the Roman period in the region of Susiya, in each of which there were a number of large stone rollers (Amit and Baruch 2007; Figs. 1–6). It was clear that these rollers are part of the wine press installations and are connected to the process of wine production. They were intended for crushing the grape skins and extracting the rest of the must from the fruit after treading on it by foot.

In our search for parallels of the phenomenon, we were surprised to discover that the distribution of the wine presses with rollers is much more extensive than it seemed until now. It turns out that wine presses of this type are located in relatively dense concentrations and in defined regions that cover most parts of the settled country.

The rollers were made of hard limestone and resemble a column; however, the stone-dressing is quite crude. In many cases only broken pieces of the original rollers have remained, but the existence of complete rollers in the different regions where the phenomenon has been identified shows that the rollers usually have a uniform cigar-like shape, in other words, they are thicker in the middle and narrower at the ends. In addition to the unique group of grooved or slotted rollers on the Carmel, which are conspicuous due to their size, the dimensions of the other rollers elsewhere in the country are quite similar: 1.5–1.7 m long, 0.55–0.65 m in diameter in the middle, 0.45–0.55 m in diameter at the ends, that is to say, c. 10 cm less than the centre.

History of Research

The Grooved Columns on Mt. Carmel

From a standpoint of the history of the research we should begin the discussion with the grooved stone rollers of the Carmel, which are a phenomenon unto themselves. It is possible that this phenomenon and its scope in that limited region is what diverted attention from the general picture of rollers in the other regions of the country and have delayed their study until now. The 19th cent. explorers of the Land of Israel have already expressed their opinion regarding the phenomenon on the Carmel. More than anyone else L. Oliphant, whose quotation is presented above, dealt with these rollers.

The largest concentration of such grooved columns (c. 30 items) is located at H. Sumaqa on Mount Carmel. The excavator of the site, S. Dar, discusses the grooved columns in the context of the production installations at the site and its vicinity, as well as at H. Raqit (Dar 1999: 77–94, 145–174; 2004: 85–86). Dar surveyed a dozen installations which he refers to as "workshops" and excavated five of them. He dated them to the Late Roman period. The workshops where the grooved columns were used are different than the wine and oil presses that were found at the site and in its vicinity, although some of the components that are common to each of the agricultural production installations are somewhat similar. Because of this similarity Dar (1999: 77) first rejected the proposal put forth by R. Frankel and E. Ayalon (1988: 43) that these are wine presses. He also rejected his original suggestion that these are tanning installations because of the absence of any resemblance to known installations that were used for working skins. By way of elimination Dar proposed that these workshops were used to produce dye from the vegetation that grows on Mt. Carmel; however, he does admit the topic is still open to further discussion and even mentions the possibility that they were used for grape pressing (Dar 2004: 86).

Similar Installations in Northern Syria

Dar sought parallels outside of the Land of Israel and he mentioned installations located in northern Syria where both grooved columns and smooth columns were found that are similar to the rollers in the Land of Israel. The installations were discovered at six sites discussed by O. Callot in his book summarizing the oil industry in northern Syria (Callot 1984). Callot described a number of installations whose plan is very similar to the traditional plan of the wine press: a rock-hewn installation composed of a treading floor and a number of collecting vats. The installations in northern Syria are exceptional in that their massive and well-built construction surrounds the hewn part of the wine press and in each one of them a number of cigar-like columns were discovered.

Callot interpreted the installations where the rollers were discovered as oil presses (Callot 1984, 20–23, Pls. 8–12). He believes the olive crop was piled onto the treading floor and then crushed by means of the heavy rollers that were rolled back and forth on top of it using rods that were inserted into the grooves. After the fruit was crushed the rollers served as a press which was placed on top of the olive paste in order to extract the maximum amount of liquid in a process that involved secondary pressing. However, Frankel (Brun 1993: 519–520) has already shown that some of the installations are wine presses.

Wine Presses with Rollers in Other Parts of Israel

Until recently we had very little knowledge regarding wine presses elsewhere in the country where stone rollers were found, apart from the unique group of installations with slotted rollers on Mt. Carmel. Their random publication and the assumption that these are anomalous components that are unsuitable for a wine press resulted in a situation whereby scholars who were studying wine presses paid almost no attention to the phenomenon.

Clear evidence of the roller phenomenon having been ignored by researchers is demonstrated by the fact that Frankel does not mention it in his doctoral dissertation, which constitutes a basis for all study of wine and oil presses in the Land of Israel (Frankel 1984). It was only five years later, in a joint publication by him and Ayalon, that there is a short paragraph dedicated to the rollers (Frankel and Ayalon 1988: 42–43).

Another decade passed and in his comprehensive book, which is based on his dissertation and updated, Frankel dedicated a special chapter to rollers in wine presses (Frankel 1999: 146–147). In this chapter he differentiated between two kinds of rollers that were documented in two adjacent, but separate regions: 1) simple rollers (T861) that were found at six sites, five of which are in the central Lower Galilee; 2) slotted rollers (T862) from sixteen sites in the western Lower Galilee and on Mt. Carmel.

With the discovery of the group of wine presses with rollers in southern Mt. Hebron and the beginning of our research of this topic, we were astonished at the large area that ostensibly remained between the Lower Galilee and Mount Carmel, where Frankel wrote about the rollers, and the southern part of Mount Hebron where we discovered them. It would seem that from this region – which is actually all of the central mountain range, its inner valleys and its lowlands – there is almost no information about wine presses with rollers. We therefore began a methodic search of the preliminary and final excavation reports and enquired among our colleagues with the aim of discovering the lost "rollers". Our search has slowly uncovered finds from dozens of wine presses where rollers were discovered and the southern mountain range has changed from a "blank page" to a distribution map that is checkered with finds. Although the study is still ongoing, we have already discerned that the find spots are concentrated in groups. On the distribution map (Fig. 6) and in Table 1, in which we have included both the wine presses that were previously known and the new finds, these groups are apparent in the following regions of the country (from north to south): the western Galilee, Lower Galilee, Mt. Carmel, Modi'in region, Judean Shephelah, central Mt. Hebron, southern Mt. Hebron and Yattir region. Isolated sites were found in Nahal 'Iron, in the Samaria Hills and western Samaria, in the Jerusalem hills and on the coastal plain. A discrete group consists of the presses that were used in the production of wine or honey from dates that were discovered in Jericho and the northern Dead Sea.

Conclusions

Since we are dealing with installations that have distinct and quite unique typological characteristics, they could ostensibly be considered installations that were built in the same period. Despite this, the excavation and survey results indicate the prolonged use of the installations, which first began in the Early Roman period and continued until the Byzantine period.

Rollers were also found in four complex wine presses where there is also a screw-press: at Ha-Solelim (No. 8 in Table 1), at Geva (No. 28 in Table 1), at Zur Natan (No. 33 in Table 1; Site 31 in this book), and at Bet Dagan (No.

41 in Table 1).

The rollers that were discovered in various excavations seemed to several scholars to be foreign elements that have nothing to do with wine production and therefore the installations where the rollers were found were often interpreted as other kinds of production installations and not as wine presses. The current state of research has reached a point where one can no longer doubt that these installations are wine presses. The main arguments for this are:

1. The similarity between them and rock-hewn wine presses, and the existence of the basic wine press components in them: treading floor and collecting vat.

2. Similar installations were discovered in different regions of the country and it is therefore clear they were used for the production of a widespread and known product. The possibility that they were intended for the manufacture of oil, which was also a very common crop, does not stand up to the test of criticism.

3. The existence of rollers at what are unmistakably wine presses, such as those at Geva, Zur Natan and Bet Dagan, corroborates the proposed connection between the rollers and the manufacture of wine.

Based on the aforementioned reasons the suggestion that these installations were used to make a product that was exclusive to a particular region such as date wine (Netzer 2002) or persimmon products (Hirschfeld 2004; Bèlis 2006) should be rejected.

Bibliography

Amit, D. and Baruch, Y. 2007. "Wine Presses with Integrated Rolling Stones from South Mount-Hebron". In Y. Eshel (ed.), *Judea and Samaria Researches Studies*, Vol. 16. Ariel. (Forthcoming).

Assaf, E. 1998. "Ard el-Muàllaq (B)". *ESI* 20: 75–76.

Baruch, Y. 1997. "A Site of the Early Roman and Byzantine Periods at ed-Deir, Hebron". *'Atiqot* XXXII: 109–117 (Hebrew).

Bèlis, M. 2006. "The Production of Indigo Dye in the Installations of 'Ain Feshka". In K. Galor, J.B. Humbert and J. Zangenberg (eds.), *Qumran, The Site of The Dead Sea Scrolls; Archaeological Interpretations and Debates*. Leiden and Boston: 253–262.

Brun, J.-P. 1993. "Les vestiges archéologiques". In M.-C. Amouretti and J.-P. Brun (eds.), *Oil and Wine Production in the Mediterranean Area* (BCH Suppl.). Paris: 518–521.

Callot, O. 1984. *Huileries antiques de Syrie du Nord*. Paris.

Dagan, Y. and Avganim, A. 1998. "Nahal Zanoah (Site 04/12)". *ESI* 17: 127–129.

Dar, S. 1999. *Sumaqa, A Roman and Byzantine Jewish Village on Mount Carmel, Israel* (BAR IS 815). Oxford.

Dar, S. 2004. *Raqit, Marinus' Estate on the Carmel, Israel* (BAR IS 1300). Oxford.

Dar, S. and Ziegelmann, A. 1997. "H. el-Kerak in the Carmel". In S. Dar and Z. Safrai (eds.), *The Village in Ancient Israel*. Tel Aviv: 161–191 (Hebrew).

Eisenberg, E. 2000. "Nahal Yarmut". *HA-ESI* 112: 91*-93*.

Eitam, D. 1996. "A Survey of Rural Installations – The Eastern Valleys and the Fringes of the Desert". In A. Zertal. *The Manasseh Hill Country Survey, The Eastern Valleys and The Fringes of The Desert*. Tel Aviv: 681–738 (Hebrew).

Frankel, R. 1984. *The History of the Processing of Oil and Wine in Galilee in the Period of the Bible, the Mishna and the Talmud*. Unpublished Ph.D. Thesis. Tel Aviv University (Hebrew).

Frankel, R. 1999. *Wine and Oil Production in Antiquity in Israel and Other Mediterranean Countries*. Sheffield.

Frankel, R. and Ayalon, E. 1988. *Vines, Winepresses and Wine in Antiquity*. Tel Aviv (Hebrew).

Gershuny, L. 2006. "Excavations at Khirbat Marmita". *'Atiqot* 53: 139–178.

Golani, A. 2005. "Salvage Excavations in the Modi'in Landscape". *'Atiqot* 50: 73–97.

Hirschfeld, Y. 2004. "Excavations at 'Ein Feshkha, 2001: Final Report". *IEJ* 54: 37–74.

Kochavi, M. 1972. "The Land of Judah". In: M. Kochavi (ed.). *Judaea, Samaria and the Golan, Archaeological Survey 1967–1968*. Jerusalem: 19–91 (Hebrew).

Kogan-Zehavi, E. 2000. "Road 2 (North)". *ESI* 20: 88.

Magen, Y., Tzionit, Y. and Sirkis, O. 2004. "Khirbet Badd 'Isa – Qiryat Sefer". In: Y. Magen, *et al. The Land of Benjamin*. Jerusalem: 179–241.

Netzer, E. 2002. "Date "Winepresses" in the Royal Estate at Jericho". In: Y. Eshel (ed.). *Judea and Samaria Researches Studies*, Vol. 11. Ariel: 69–80 (Hebrew).

Oliphant, L. 1887. *Haifa or Life in Modern Palestine*. Edinburgh and London.

Peilstöcker, M. and Kapitaikin, A. 2000. "Bet Dagan". *ESI* 20: 59*-60*.

Peleg, Y., Feller, Y. and Srukh, I. 2004. "Karmé Zur". *HA-ESI* 116: 61*-62*.

Safrai, Z. and Linn, M. 1988. "Excavations and Surveys in the Mishmar Ha-'Emeq Area". In B. Mazar (ed.) *Geva, Archaeological Discoveries at Tell Abu-Shusha, Mishmar Ha-'Emeq*. Jerusalem: 167–214 (Hebrew).

Sion, O. 2001. *Settlement History in Central Samarian Region in the Byzantine Period.* Ph.D. Thesis, The Hebrew University. Jerusalem (Hebrew).

Sion, O. 2002. "Excavations at Rogem Ganim, Kiryat Menahem, Jerusalem". In: Y. Eshel (ed.). *Judea and Samaria Researches Studies*, Vol. 11. Ariel: 115–122 (Hebrew).

Table 1: Wine Presses and Stone Rollers in the Land of Israel*

Region	No.	Map Reference	Site	Description	Date/ Period	Reference
Western Galilee	1	16820/25970	H. Waziya, near Julis	A grooved column		Aviam in Dar 1999: 145
	2	16860/24920	North of 'Iblin	A column near a concentration of wine presses		Ibid.
	3	16720/24790	North of 'Iblin	A grooved column (not *in situ*)	Late Roman	Ibid.
	4	17085/25366	(Triang. Point 109.4) Gush Segev	A grooved column		Frankel *et al.*, pers. comm.
	4	17085/25366	(Triang. Point 109.4) Gush Segev	A grooved column		Ibid.
	5	17022/25404	Gush Segev	A grooved column		Ibid.
	6	17110/25374	Gush Segev	A grooved column		Ibid.
Lower Galilee	7	17231/23905	Ha-Solelim area	Wine Press No. 4		Caspi, pers. comm.
	8	17263/23899	Ha-Solelim area	In a complex wine press (No. 5) with a screw press	Late Roman	Ibid.
	9	17221/24074	Ha-Solelim area	Ritual bath(?) next to Wine Press No. 6	Roman	Ibid.
	10	17405/24074	Ha-Solelim area	Wine Press No. 23, c. 5 rollers		Ibid.
	11	17490/24144	Ha-Solelim area	Wine Press No. 28, 4 rollers		Ibid.
	12	17471/24031	Ha-Solelim area	Wine Press No. 30		Ibid.
Mt. Carmel	13	15280/24000	Khureibe			Olami and Lahav in Dar 1999: 145
	14	15040/23840	H. Luvim			Ibid.
	15	15170/23700	Mahzevot Kedumim	A grooved column (near wine press?)		Ibid.

*The dating of the installations is in accordance with the researchers' reports

Region	No.	Map Reference	Site	Description	Date/ Period	Reference
	16	15160/23550	H. Raqit	Grooved columns (not *in situ*)	Late Roman?	Dar 2004: 85–86
	17	15590/23560	Ancient quarry between 'Isfiya and Nahal Nesher			Ibid.
	18	15883/23310	H. 'Ala ed-Din			Ibid.
	19	15380/23240	H. Devela, Daliyat el-Karmil			Ibid.
	20	15690/23140	H. Kerak	Grooved columns (not *in situ*)	Not later than Late Roman	Dar and Ziegelmann 1997: 169–174
	21	15840/23150	H. Dubba			Olami and Lahav in Dar 1999: 145
	22	15850/23020	H. Mansura			Ibid.
	23	15340/22920	H. Dereg			Ibid.
	24	15530/22950	Givat Ha-Kochavim			Ibid.
	25	14970/23240	H. Heglon			Ibid.
	26	15700/22930	H. 'Irak el-Natif			Ibid.
	27	14790/22340	H. Talimon			Ibid.
	28	16330/22450	Geva	A roller in a complex wine press with a screw-press	Byzantine	Safrai and Linn 1988: 205–207
	29	15390/23065	H. Sumaqa	6 workshops with 30 grooved columns	Late Roman	Dar 1999: 77–94
	30	15449/23073	H. Sumaqa	Wine Press 471: 4 columns (3 grooved); Wine Press 472: an ungrooved column	Late Roman	Dar 1999: 188–194
Nahal 'Iron	31	15630/21020	Rujum el-Ahmar	A grooved column (not *in situ*)		Olami and Lahav in Dar 1999: 145
Samaria Hills	32	19030/19860	'Irak el-Hamam	3 wine presses with 4 cylinders and a screw press(?)		Eitam 1996: 703–706, Sites 36/2, 36/4, 36/7a; Sion 2001: 198
Western Samaria	33	15050/18280	Zur Natan	Complex wine press with screw press	Byzantine	Dar 1999: 190; Ayalon, this book (Site 31)
Modi'in	34	15051/14521	Modi'in	Wine Press (f2)	Byzantine	Kogan-Zehavi 2000
	35	15040–52/ 14605–28	'Ard el-Mu'àllaq	Wine Press A		Assaf 1998
	36	14807/14787	Modi'in	Wine Press 162		Zelinger and Avner unpublished

Region	No.	Map Reference	Site	Description	Date/ Period	Reference
	37	14889/14805	Modi'in	Wine Press 168		Ibid.
	38	15050/14396	Modi'in	F53B		Golani 2005: 76
	39	15000/14365	Kh. Umm el-Omdan		Early Roman	Onn and Weksler-Bdolah, unpublished
	40	15410/11480	Kh. Badd 'Isa, Qiryat Sefer	Wine Press No. 3	Late Roman	Magen *et al.* 2004: 36, Fig. 8
Coastal Plain	41	13340/15675	Bet Dagan	A complex wine press with a screw press	Byzantine	Peilstöcker and Kapitaikin 2000: 59–60
Judean Shephelah	42	15454/13310	Kefar Uriyah	Roller fragment in secondary use in a wine press re-used to produce carob honey	Early Roman	Zilberbod, unpublished
	43	14800/12390	Nahal Yarmut		Early Roman	Eisenberg 2000
	44	15005/12430	Nahal Yarmut		Late Roman - Byzantine	Dagan and Avganim 1998: 127–129
	45	14950/12450	H. Kikh			Kogan-Zehavi, pers. comm..
	47	15135/13045	Kh. Marmita, No. 81			Gershuny 2006: 153–157
	48	16570/12950	Rogem Ganim, Kiryat Menahem	A wine press with a roller *in situ*	Early Roman	Sion 2002: 118
Jerusalem region	48	17067/12647	Mâr Eliâs Monastery	A cigar-shaped roller (not *in situ*) near rock-hewn wine presses		Baruch, unpublished
	49	16435/12625	Kh. Abū Shawān		Late Roman	Baruch, unpublished
Mount Hebron	50	16300/11770	Gush Etzyon	A cigar-shaped roller beside a hewn vat	Late Roman	Amit, unpublished
	51	11290/15970	Karmé Zur	A wine press with a roller *in situ*	Late Roman	Peleg *et al.* 2004: 61–62
	52	16175/10570	ed-Deir	A roller (not *in situ*) near a farmhouse and agricultural installations	Late Roman	Baruch 1997
	53	11330/16920	Nahal Si'īr	Near a farmhouse (?): two cylinders near a hewn wine press (?)		Amit, unpublished
Northern Dead Sea and Jericho	54	19130/13980	Royal estate at Jericho	Wine press in an area of workshops	No date	Netzer 2002: 72–74
	55	19300/12490	'Ein Feshkha	A farmhouse	Early Roman	Hirschfeld 2004: 37–74
Southern Mt. Hebron	56	10000/16380	Kh. Birrin	A roller (not *in situ*) near a farmhouse(?)		Amit, unpublished
	57	16380/0999	West from Kh. Muneizil	A roller at the foot of a tower		Hirschfeld, pers. comm.

Region	No.	Map Reference	Site	Description	Date/ Period	Reference
	58	16008/09004	H. Susiya	A wine press with a roller	Late Roman - Byzantine	Baruch, unpublished
	59	16010/08920	Rujum el-Hamiri	3 rollers near rock-hewn installations	Byzantine	Baruch, unpublished
	60	16035/08893	Susiya	A wine press with a roller *in situ*		Baruch, unpublished
	61	16150/08835	Ras e-Tawamin	A wine press with 3 rollers	Early-Late Roman	Baruch and Amit, unpublished
	62	16195/08720	Ras e-Tawamin	A wine press with 3 rollers	Early Roman	Baruch and Amit, unpublished
	63	08865/16170	Ras e-Tawamin	A wine press with 2 rollers		Baruch and Amit, unpublished
	64	16250/08765	Ras e-Tawamin	A wine press with a roller *in situ*		Baruch and Amit, unpublished
	65	15950/08730	Elevation Point 809	A wine press with rollers *in situ* near a stronghold	Early-Late Roman	Kochavi 1972: 80.
	66	16280/08958	Next to road to Kh. Umm el-Amad	Two roller fragments		Amit, unpublished

Fig. 73.1. Ras e-Tawamin, wine press No. 61, looking south.

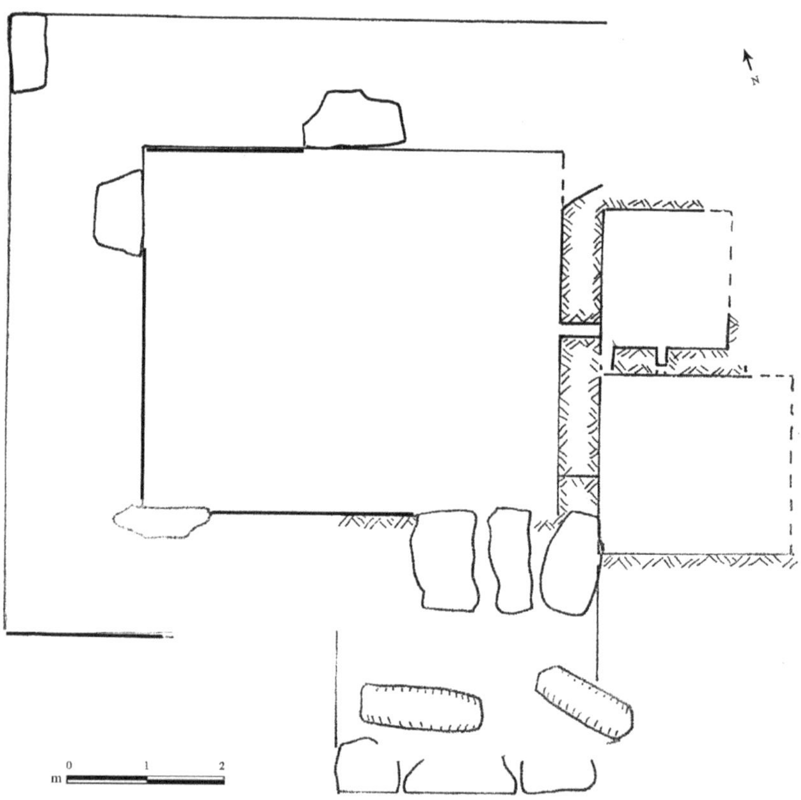

Fig. 73.2. Ras e-Tawamin, plan of wine press No. 62.

Fig. 73.3. Ras e-Tawamin, wine press No. 62, looking south.

Fig. 73.4. Ras e-Tawamin, wine press No. 62, looking east.

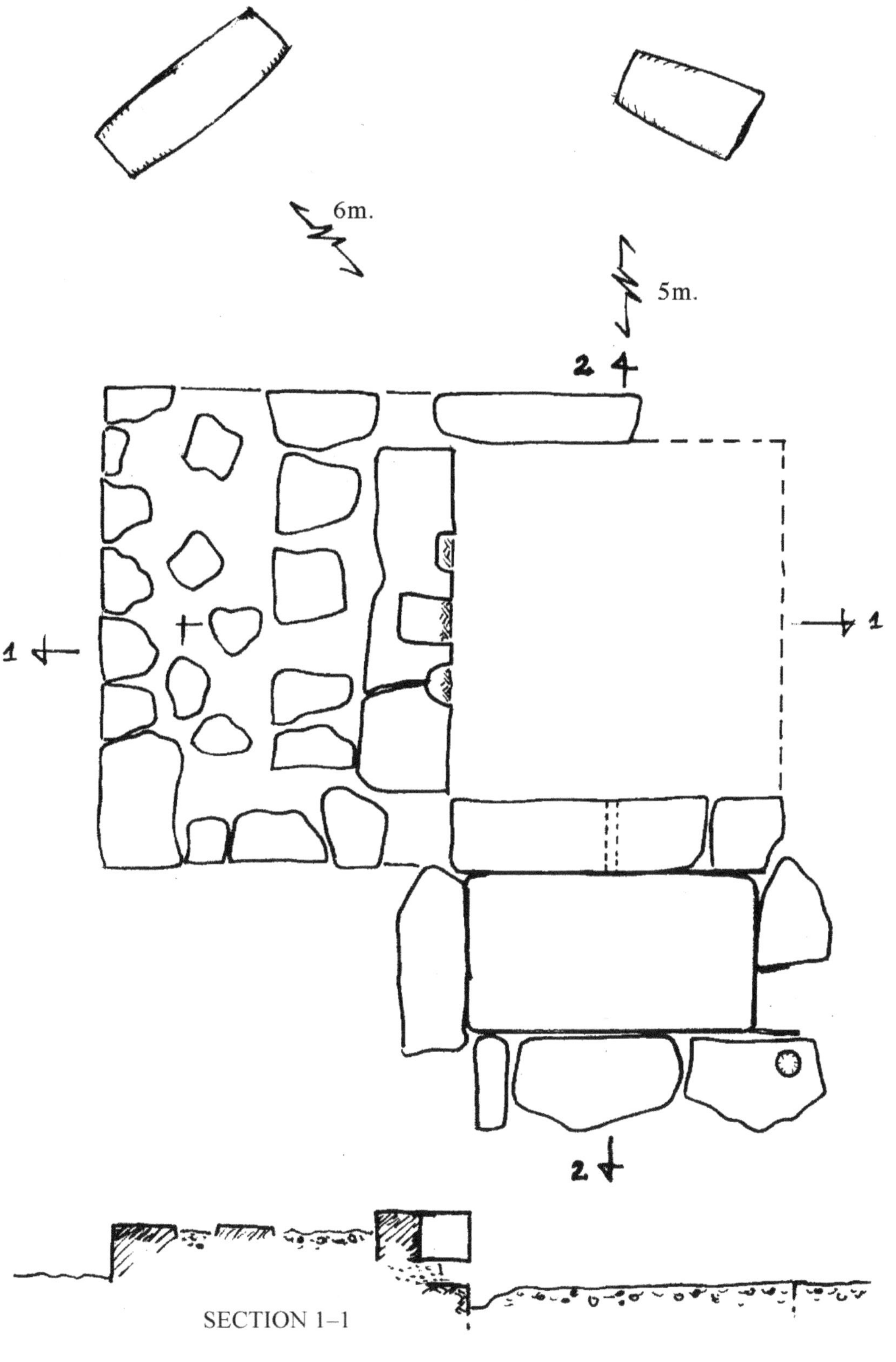

Fig. 73.5. Elevation Point 809, plan and section of wine press No. 65.

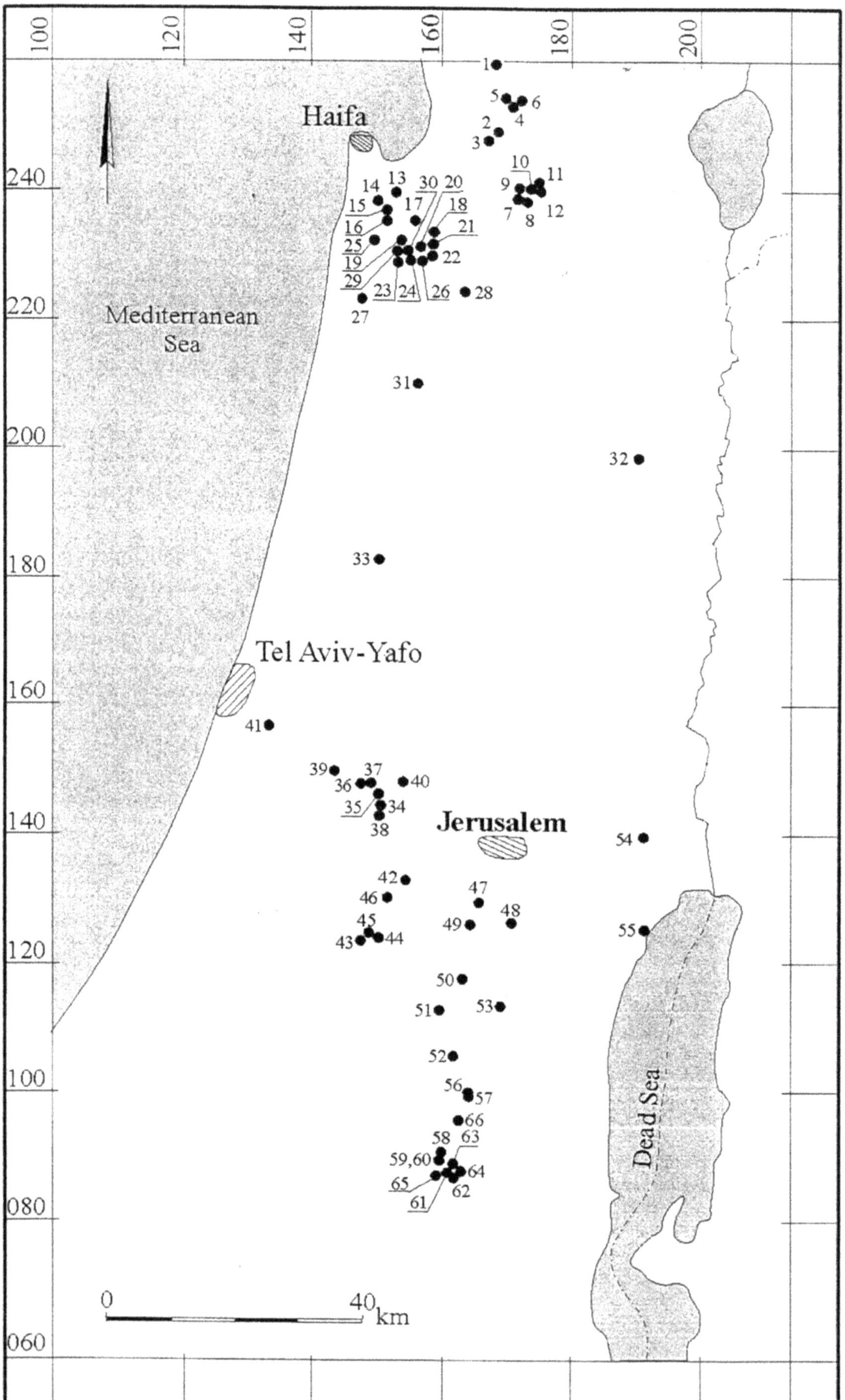

Fig. 73.6. Distribution map of wine presses with rollers.

74. The Olive (*Olea europaea*) in Eretz Israel During the Hellenistic-Early Arab Periods

Nili Liphschitz

Introduction

Olea europaea (Olive) occurs today in Israel throughout the Mediterranean region and there constitutes a common component of the maquis and garigue formations. Most populations are of the domesticated olive or of escapes from cultivation and it is difficult to define stands of wild olive trees. Wild olive was one of the components of the dominant climax association of *Quercus calliprinos* (Kermes oak) - *Pistacia palaestina* (Terebinth) of the Mediterranean region during antiquity (Liphschitz and Biger 1990).

Remnants of the olive appear in the archaeological profile of Israel starting from about 45,000 years ago (Liphschitz and Waisel 1977) but those from early contexts are clearly of the wild olive.

The olive is one of the earliest fruit trees to be domesticated and was used both for its oil and its edible fruits. In order to determine the date of domestication, botanical examination of pollen grains, stones and wood gathered from living trees and from archaeological contexts is needed. The size of the stones was found to be the only reliable parameter that makes it possible to distinguish between wild and cultivated varieties. The unusually small dimensions of the Chalcolithic stones collected at Shoham (Liphschitz *et al.* 1996), Nahal Zehora II (Liphschitz and Bonani 2000) and Nevallat (van den Brink *et al.* 2001) are characteristic of the wild variety of olive. The stones of cultivated varieties are much larger. Moreover, the marked increase in the quantities of olive wood found in archaeological contexts since the Early Bronze Age and the appearance of pottery vessels for oil separation, oil lamps and olive oil installations in this period indicate that the earliest widespread cultivation of the olive in Israel was at that time (Liphschitz, *et al.* 1991).

Wood remains found in archaeological excavations supply much more information than that obtained from olive stones concerning the quantity and location of the growth habitat of trees. Olive stones only indicate the diet of the inhabitants of a site at the relevant period but not necessarily the existence of olive groves in the vicinity, as the fruit could well have been brought from afar. However, timber in everyday use such as for construction of simple structures and for fire wood clearly derived from the vicinity of the sites and therefore represents the local vegetation (Liphschitz 1986). The timber could have been taken from wild trees as well as from cultivated olives that had ceased to give fruit, especially in areas that are poor in timber and where every possible source was exploited. Research carried out on the use of timber in the construction of houses in the last few centuries shows intensive exploitation of olive wood in Arab villages, mainly for roofing and framing (Biger and Liphschitz 1992). Trunks of olive trees were also used as press beams in pre-industrial oil presses (Liphschitz, Biger and Ayalon 1994). This phenomenon is particularly marked in Galilee that is characterized by a high number of olive groves.

The relative proportion of olive wood in relation to wood that derived from other trees can indicate the relative quantity of olive trees in the landscape during the different periods. A rise in the proportion of olive wood used in construction points to an increase in the number of cultivars and to the establishment of orchard agriculture in the vicinity of a site.

From the Early Bronze Age onwards many sites located in the Mediterranean region of Israel show a high proportion of olive wood as compared to wood of other local trees. Surveys made recently by the author in various closed reservations in these regions showed an almost complete absence of the olive and a predominance of a maquis of Kermes oak (*Quercus calliprinos*). The percentages of olive wood in the wood assemblages dating from before the Early Bronze Age are similar to that found today in these closed reservations. In later periods the percentages increase to 40–50%, clearly pointing to extensive organized olive groves (Liphschitz 1996).

Palynological research carried out in the Lake of Galilee (Baruch 1986) showed a marked increase in the percentages of olive pollen grains during the period 350 BCE–550 CE, i.e. Hellenistic to Byzantine periods, clearly indicating a considerable expansion of olive cultivation in the region during this period. Historical records also point to Galilee as the major centre of olive cultivation at that period (Josephus, *Wars* II 21 2; *Gen. R.* 20 6).

It is, therefore, of interest to examine the occurrence of olive wood in archaeological contexts from these periods at sites in the Mediterranean region of Israel.

Olive Wood Remains in the Hellenistic-Early Arab Periods

A comprehensive survey of the occurrence of *Olea europaea* remains in archaeological excavations in Israel was published several years ago (Liphschitz 1996). However, most of the sites included in the survey were from early periods, i.e. Bronze and Iron Ages. Since that publication numerous sites have been excavated and consequently a great amount of additional information has amassed. The present state of knowledge enables us, therefore, to focus on the later historic periods: Hellenistic, Roman, Byzantine and Early Arab.

Archaeobotanical data concerning the Hellenistic to Early Arab periods are available from twenty sites located in the Mediterranean region of the country, the natural habitat of *Olea europaea*, and from four sites located in the northern Negev – a semi-arid Irano-Turanian region where olives could thrive especially under cultivation (Liphschitz 1986; 1996).

The proportion of olive wood in relation to the wood of other trees was calculated for each period at each site. Comparison was made on one hand between sites located in the Mediterranean region and those in the Irano-Turanian region (Northern Negev) and on the other hand between

Table 1. Location and number of samples & percentage of olive wood (*Olea europaea*) in Israel during Hellenistic, Roman, Byzantine and Early Arab periods.

Mediterranean Region

Location / Period	Hel	Rom	Rom-Byz	Byz	EA
Northern Israel					
Bir Ansuba (Mt. Hermon)			1/2		
H. Kanaf (Golan)					1/7
Kazerin (Golan)				1/5	14/21
Gamla (Golan)	25/65				
Snir (Golan)	2/5				
Zippori (Galilee)		2/12			
Meiron (Galilee)	8/12				
Mizpe Yamim (Galilee)	1/15				
H. Sumaqa (Carmel)				3/13	
Central Israel					
Zur Natan (Sharon)				1/11	
Tel Aphek (Sharon)	6/16	12/37			3/7
Kfar Saba (Sharon)					1/5
Apollonia (Coastal Plain)				6/18	4/34
Tel Michal (Coastal Plain)	1/8				
Jaffa (Coastal Plain)	2/20	7/31			
H. Zichrin (Shephelah)			25/35		
H. Mezad (Shephelah)			11/22		
H. Hermeshit (Shephelah)			34/159		
H. Boraq (Judea)					5/7
Herodion (Judea)	5/61				
Number	50/202	19/80	71/218	11/47	28/81
Percentage	24.7%	23.7%	32.5%	23.4%	34.5%
Irano-Turanian Region					
Northern Negev					
'Aro'er	1/5				
H. 'Uza	1/4	1/2			
Tel Masos				3/41	
Beer Sheba	7/106	2/28			
Number	9/115	3/30		3/41	
Percentage	7.8%	10.0%		7.3%	

early periods (Bronze Age to Iron Age) and late periods (Hellenistic to Early Arab) in multi-period sites (Aphek-Antipatris, Tel Michal, Jaffa, Beer Sheba, H. 'Uza and Tel Masos).

During the Hellenistic to Early Arab periods (Tables 1–2) olive wood constituted 27.78% of the wood assemblages in the Mediterranean sites. In the Northern Negev olive wood constituted only 8.36%. Comparison of olive percentages between the four periods shows no significant differences.

Comparison between the Early Bronze Age – Iron Age to Hellenistic – Early Arab periods in the multi-period sites (Tables 3–4) shows a small reduction in the percentages of olive wood from the earlier to later periods: from 30.3% to 21.4 % in the Mediterranean region and from 13.7% to 5.2% in the Northern Negev.

It should be noted that there is no preference in the percentage of olive wood in sites in Galilee as compared to other sites in the Mediterranean region. Abandoned olive groves are spread today all over the hilly regions of the country testifying to the economic importance of the olive in the past.

Table 2. Number of samples and percentage of olive wood (*Olea europaea*) in the Hellenistic to Early Arab periods in the Mediterranean and Irano-Turanian regions.

Region	Olive Wood	
	Number	**Percentage**
Mediterranean region	184/530	29.20%
Northern Negev	15/186	8.06%

Table 3. Number of samples and percentage of olive wood along the archaeological profile at six sites.

Mediterranean Region

Period / Location	EB	MB	LB	Ir	Per	H	R	EA
Tel Aphek	8/16	27/70	14/61	7/33	-	6/16	12/37	3/7
Tel Michal	-	1/1	3/9	0/8	2/29	1/9	0/2	0/2
Jaffa	-	-	-	-	2/10	2/20	7/31	-
Number	8/16	28/71	`17/70	7/41	4/39	9/45	19/70	3/9
Percentage	50.0%	39.4%	24.3%	17.1%	10.2%	20.0%	27.1%	33.3%

Irano-Turanian Region

Period / Location	EB	MB	LB	Ir	Per	H	R	EA
Beer Sheba	-	-	-	94/856	0/4	7/106	2/28	0/7
H. 'Uza	-	-	-	9/46	-	1/4	1/2	
Tel Masos	-	-	-	31/72	-	-	3/41	-
Number				134/976	0/4	8/110	6/78	0/7
Percentage				13.7%	0.0%	7.3%	7.7%	0.0%

Table 4. Number of samples and percentages of olive wood in early and late archaeological periods.

	EB-Ir		Per-EA	
Region & Sites	**No.**	**%**	**No.**	**%**
Mediterranean				
Aphek-Michal-Jaffa	60/198	30.3%	35/163	21.4%
Irano-Turanian				
Beer Sheba-'Uza-Masos	134/976	13.7%	10/192	5.2%

Bibliography

Baruch, U. 1986. "The Late Holocene Vegetational History of Lake Kinneret (Sea of Glilee) Israel". *Paleorient* 12: 37–48.

Biger, G. and Liphschitz, N. 1992. "The Rural Dwelling Houses in 19th Century Palestine – A Dendrohistoric View". *Area* 24: 45–55.

Liphschitz, N. 1986. "Overview of the Dendrochronology and Dendroarchaeology in Israel". *Dendrochronologia* 4: 37–58.

Liphschitz, N. 1996. "Olives in Ancient Israel in View of Dendroarchaeological Investigations". In: D. Eitam and M. Heltzer (eds.). *Olive Oil in Antiquity. Israel and Neighbouring Countries from the Neolithic to the Early Arab Period.* Padova: 7–13.

Liphschitz, N. and Biger, G. 1990. "Dominance of *Quercus calliprinos* (Kermes oak) – *Pistacia palaestina* (Terebinth) Association in the Mediterranean Territory of Eretz Israel during Antiquity". *JVS* 1: 67–70.

Liphschitz, N., Biger, G. and Ayalon, E. 1994. "The Use of Timber in the Oil Press Installations in Eretz Israel". *Israel – People and Land, Eretz Israel Museum Yearbook* 7–8 (25–26): 199–216 (Hebrew).

Liphschitz, N. and Bonani, G. 2000. "Dimensions of Olive (*Olea europaea*) Stones as a Reliable Parameter to Distinguish Between Wild and Cultivated Varieties: Further Evidence". *Tel Aviv* 27: 23–25.

Liphschitz, N., Gophna, R., Hartman, M. and Biger, G. 1991. "The Beginning of Olive (*Olea europaea*) Cultivation in the Old World: A Reassessment". *JAS* 18: 441–453.

Liphschitz, N., Gophna, R., Bonani, G. and Feldstein, A. 1996. "Wild Olive (*Olea europaea*) Stones From a Chalcolithic Cave at Shoham, Israel and Their Implications". *Tel Aviv* 23: 135–142.

Liphschitz, N. and Waisel, Y. 1977. "Dendroarchaeological Investigations in Israel: Central Negev: Nahal Zin". In: A. Marks (ed.). *Prehistory and Palaeoenvironments in the Central Negev* Vol II. Minnesota: 355–356.

van den Brink, E.C.M., Liphschitz, N., Lazar, D. and Bonani, G. 2001. "Chalcolithic Dwelling Remains, Cup Marks and Olive (*Olea europaea*) Stones at Nevallat". *IEJ* 51: 36–43.

Abbreviations

ABSA – Annual of the British School of Archaeology in Athens

BA – The Biblical Archaeologist

BAIAS – Bulletin of the Anglo-Israel Archaeological Society

BAR IS – British Archaeological Reports, International Series

BASOR – Bulletin of the American Schools of Oriental Research

BCH – Bulletin de correspondence hellenique

BIES – Bulletin of the Israel Exploration Society (Yedi'ot, Hebrew)

BJ – Bonner Jahrbücher

CRAIBL – Comptes rendus à l'Acadie des Inscription et des Belles Lettres

DOP – Dumbarton Oak Papers

EI – Eretz-Israel

ESI – Excavations and Surveys in Israel

HA – Hadashot Arkheologiyot

HANE/S – History of the Ancient Near East / Studies

IEJ – Israel Exploration Journal

JAS – Journal of Archaeological Science

JHS – Journal of Hellenic Studies

JRA – Journal of Roman Archaeology

JVS – Journal of Vegetation Science

LA – Liber Annuus

NEAEHL – New Encyclopedia of Archaeological Excavations in the Holy Land

NZ – Niqrot Zurim

PEF QSt – Palestine Exploration Fund, Quarterly Statement

PEQ – Palestine Exploration Quarterly

QDAP – Quarterly of the Department of Antiquities in Palestine

RB – Revue biblique

SCI – Scripta Classica Israelica

SWP – Survey of Western Palestine

General Bibliography

Abel, F.M. 1938. *Géographie de la Palestine.* Vol. 2. Paris.

Adan-Bayewitz, D. 1986. "The Pottery from the Late Byzantine Building (Stratum 4) and its Implications". In: L.I. Levine and E. Netzer (eds.). *Excavations at Caesarea Maritima 1975, 1976, 1979 – Final Report* (Qedem 21). Jerusalem: 90–121.

Adan-Bayewitz, D. 1993. *Common Pottery in Roman Galilee — A study in Local Trade*. Ramat Gan.

Adan-Bayewitz, D. and Aviam, M. 1997. "Iotapata, Josephus and the Siege of 67; Preliminary Report on the 1992–94 Seasons". *JRA* 10: 131–165.

Aharoni, Y. 1954. "Byzantine Monastery near Beit Hashita". *BIES* 18: 209–215 (Hebrew).

Ahmet, K. 2001. "A Middle Byzantine Olive Press Room at Aphrodisias". *Anatolian Studies* 51: 159–167.

Amit, D. 1996. *Ritual Baths from the Second Temple Period in the Hebron Hills*. Unpublished M.A. Thesis, The Hebrew University of Jerusalem (Hebrew).

Amit, D. 1997. "Kh. Omm Halasa: An Additional Monastery in the Wilderness of Ziph". In: Y. Eshel (ed.). *Judea and Samaria Research Studies – Proceedings of the 6th Annual Meeting 1996*. Kedumim–Ariel: 259–270 (Hebrew).

Amit, D. *et al.* 1998. "Mazor – 1993–1996". *ESI* 18: 57–66.

Amit, D. *et al.* 1999. "Mazor (El'ad) 1996–1997". *HA-ESI* 109: 55*-64*.

Amit, D. and Baruch, Y. 2007. "Wine Presses with Integrated Rolling Stones from South Mount-Hebron". In Y. Eshel (ed.), *Judea and Samaria Researches Studies*, Vol. 16. Ariel. (Forthcoming).

Amit, D. and Yezerski, I. 2001. "An Iron Age II Cemetery and Wine Presses at an-Nebi Danyal". *IEJ* 51: 171–193.

Amouretti, M.-C., Brun, J.-P. and Eitam, D. (eds.). 1991. *Oil and Wine Production in the Mediterranean Area from the Bronze Age to the End of the XVIth Century. Pré-Actes*. Aix-en-Provence-Toulon.

Amouretti, M.-C. and Brun, J.-P. (eds.). 1993. *Oil and Wine Production in the Mediterranean Area* (BCH Suppl. 26). Paris.

Anderson, J.G.C. 1903. "A Journey of Exploration in Pontus". *Studia Pontica* 1.

Anderson, F.G. 1985. *Shiloh II. The Remains from the Hellenistic to the Mamluk Periods*. Copenhagen.

Applebaum, S. 1988. "Selected Inscriptions from the Temples and Villages of Mt. Hermon". In: S. Dar (ed.). *The Hermon Settlements in Antiquity*. Tel Aviv: 33–53 (Hebrew).

Ariel, D.T. 1987. "Coins from the Synagogue at 'Ein Nashut". *IEJ* 37: 147–157.

Ariel, D.T. 1998. "Coins from the Survey and Excavations at Horvat Hermeshit (1988–1990)". *'Atiqot* 34: 1*-4*.

Ariel, D.T. In preparation. "Coins from the Survey and Excavations at Horvat Hermeshit".

Assaf, E. 1998. "Ard el-Muàllaq (B)". *ESI* 20: 75–76.

Avi-Yonah, M. 1960. "Hulda". In: *Louis M. Rabinowitz Fund Bulletin* 3: 57–60.

Avi-Yonah, M. 1966. "An Addendum to the Episcopal List of Tyre". *IEJ* 16: 209–210.

Avi-Yonah, M. 1970. "Synagogues: Hulda". In: B. Mazar (ed.). *The Encyclopedia of Archaeological Excavations in Eretz-Israel*: 106 (Hebrew).

Aviam, M. 1988. "Olive Growing and Viticulture in Upper Galilee in Ancient Times". *Israel – People and Land, Eretz Israel Museum Yearbook* 4 (22). Tel Aviv: 197–210 (Hebrew).

Aviam, M. 1990. "Tel Yodefat, Oil Press". *ESI* 9: 106.

Aviam, M. 2002. "Yodefat/Jotapata: The Archaeology of the First Battle". In: A.M. Berlin and J.A. Overman (eds.). *The First Jewish Revolt*. London and New York: 121–133.

Avitsur, S. 1976. *Man and his Work, Historical Atlas of Tools and Workshops in the Holy Land*. Jerusalem (Hebrew).

Avitsur, S. 1994. "Olive Oil Production in the Land of Israel, Traditional to Industrial". In: R. Frankel, S. Avitsur and E. Ayalon. *History and Technology of Olive Oil in the Holy Land*. Arlington and Tel Aviv: 91–157.

Avner, R. 2000. "Deir Ghazali: A Byzantine Monastery Northeast of Jerusalem". *'Atiqot* 40: 25*-52* (Hebrew).

Avner, R., Azmon, S. and Savariego, A. 1992. "Jerusalem, Pisgat Zeev (Deir Ghazala)". *ESI* 10: 128–130.

Avni, G. and Gudovitz, S. 1996. "The Oil Press and the Subterranean Storage Complexes at "Ahuzat Hazzan"". In: D. Eitam and M. Heltzer (eds.). *Olive Oil in Antiquity, Israel and Neighbouring Countries from the Neolithic to the Early Arab Period* (HANE/S VII). Padova: 137–147.

Avni, G. *et al*. 1987. "The Hiding Complex at Ahuzat Hazzan". In: A. Kloner and Y. Tepper. *The Hiding Complexes in the Judean Shephela*. Tel Aviv: 115–128 (Hebrew).

Ayalon, E. 1984. "Two Wine Presses from the Roman Period at the Haaretz Museum Center". *Tel Aviv* 11: 173–182.

Ayalon, E. 1988. "A Byzantine Road and Other Discoveries in the Vicinity of Tell Qasile". *Israel – People and Land, Eretz Israel Museum Yearbook* 4 (22). Tel Aviv: 9–34 (Hebrew).

Ayalon, E. 1996. "The "Samaritan" Olive Oil Press". In: D. Eitam and M. Heltzer (eds.). *Olive Oil in Antiquity, Israel and Neighbouring Countries from the Neolithic to the Early Arab Period* (HANE/S VII). Padova: 149–156.

Ayalon, E. 1997. "The End of the Ancient Wine Production in the Central Coastal Plain". In: Y. Friedman, Z. Safrai and J. Schwartz (eds.). *Hikrei Eretz, Studies in the History of the Land of Israel Dedicated to Prof. Yehuda Feliks*. Ramat Gan: 149–166 (Hebrew).

Ayalon, E. 1999. "The Open-Air Oil Plants of Western Samaria". In: Y. Eshel (ed.). *Judea and Samaria Research Studies, Proceedings of the Eighth Annual Meeting 1998*. Kedumim-Ariel: 115–126 (Hebrew).

Ayalon, E. 2008. "Migdal, Horvat". In: E. Stern (ed.). *NEAEHL* 5. Jerusalem: 1950–1952.

Ayalon, E., Charvit, Y. and Qidron, A. 1990. "Grouped Installations from the Roman-Byzantine Period in the fields of Zur Natan". *Israel – People and Land, Eretz Israel Museum Yearbook* 5–6 (23–24): 93–120 (Hebrew).

Ayalon, E., Neidinger, W. and Matthews, E. 1989/1990. "Horvat Migdal". *ESI* 9: 137–138.

Ayalon, E., Neidinger, W. and Matthews, E. 1993. "Zur Natan (Horvat Migdal)". *ESI* 13: 45–46.

Ayalon, E. and Rosen, B. 2007. "The Wines of the Land of Israel in the Roman-Byzantine Period – Sweet but Perhaps Poisonous". In: A. Baruch (ed.). *Jerusalem and Eretz Israel 4–5*. Ramat Gan: 135–156 (Hebrew).

Balut, S. 2005. "Two Unique Olive Oil Presses from the Border Region of Lycia-Pamphylia-Pisidia". *Adalya* 8: 191–219 (Turkish with English summary).

Baramki, D.C. 1933. "A Byzantine Bath at Qalandia". *QDAP* 2: 105–109.

Baramki, D.C. 1935. "Recent Discoveries of Byzantine Remains in Palestine – 3. A Winepress at Khirbet Jedur, near Beit Ummar". *QDAP* 4: 121.

Baruch, E. 1999. "Bone Tools in Talmudic Sources". In: E. Ayalon and C. Sorek. *Bare Bones, Ancient Artifacts from Animal Bones*. Tel Aviv: 73–76 (Hebrew).

Baruch, U. 1986. "The Late Holocene Vegetational History of Lake Kinneret (Sea of Galilee) Israel". *Paleorient* 12: 37–48.

Baruch, Y. 1997. "A Site of the Early Roman and Byzantine Periods at ed-Deir, Hebron". *'Atiqot* 32: 109–117 (Hebrew).

Bèlis, M. 2006. "The Production of Indigo Dye in the Installations of 'Ain Feshka". In K. Galor, J.B. Humbert and J. Zangenberg (eds.), *Qumran, The Site of The Dead Sea Scrolls; Archaeological Interpretations and Debates*. Leiden and Boston: 253–262.

Ben David, H. 1989. *Olive Cultivation and Oil Production in Roman and Byzantine Golan*. Unpublished MA thesis, Bar Ilan University, Ramat Gan (Hebrew).

Ben David, H. 1998. "Oil Presses and Oil Production in the Golan in the Mishnaic and Talmudic Periods". *'Atiqot* 34: 1–61 (Hebrew).

Ben David, H. 1999. *Settlement in the 'Lower Golan' during the Hellenistic, Roman and Byzantine Periods*. Unpublished Ph.D. Dissertation, Bar Ilan University. Ramat Gan (Hebrew).

Ben Yosef, Y. 1966. "Lohamei Hageta'ot". *HA* 17: 21 (Hebrew).

Biger, G. and Liphschitz, N. 1992. "The Rural Dwelling Houses in 19th Century Palestine – A Dendrohistoric View". *Area* 24: 45–55.

Biran, A. 1980. "Two Discoveries at Tel Dan". *IEJ* 30: 89–98.

Borowski, O. 1982. "A Note on the Iron Age Cult Installation at Tell Dan". *IEJ* 32: 58.

Brøndsted, J. 1928. "La basilique des cinq martyrs à Kapljuc". In: E. Dyggve and J. Brønsted. *Recherches à Salone I*. Copenhagen: 33–176.

Brun, J-P. 1986. *L'Oleiculture antique en Provence*. Paris.

Brun, J.-P. 1993. "Les vestiges archéologiques". In M.-C. Amouretti and J.-P. Brun (eds.), *Oil and Wine Production in the Mediterranean Area* (BCH Suppl. 26). Paris: 518–521.

Brun, J.-P. 2003a. *Archéologie du vin et de l'huile de la préhistoire à l'époque hellénistique*. Paris.

Brun, J.-P. 2003b. *Le vin et l'huile dans la Méditerranée antique, Viticulture, oléiculture et procédés de transformation*. Paris.

Brun, J.-P. 2004. *Archéologie du vin et de l'huile dans l'Empire romain*. Paris.

Brun, J.-P. 2005. *Archéologie du vin et de l'huile en Gaule romaine*. Paris.

Buchenino, A. 2007. "Building Remains and Industrial Installations from the Early Islamic Period at Kh. Deiran, Rehovot". *'Atiqot* 56: 119–144 (Hebrew); 84*-85* (English Summary).

Callot, O. 1984. *Huileries antiques de Syrie du Nord.* Paris.

Callot, O. 1987. "Les huileries du Bronze Récent à Ougarit: Premiers éléments pour une étude". In: M. Yon. *La centre de la ville Ras Shamra, Ougarit III, 38e-44e campagnes (1978–1984)*. Paris: 197–212.

Cato 1979. "Marcus Porcius Cato: De Agri Cultura". In: *Cato and Varro* (trans. W.D. Hooper). London: 2–157.

Chiat, M.J.S. 1982. *Handbook of Synagogue Architecture* (Brown Judaic Studies 29). Chico, California.

Chidiosan, N., Ayalon, E. and Yosef, A. 1990. "Two Byzantine Winepresses in the Eretz Israel Museum, Tel Aviv". *Israel – People and Land, Eretz Israel Museum Yearbook* 5–6 (23–24). Tel Aviv: 23–36 (Hebrew).

Clermont-Ganneau, Ch. 1896. *Archaeological Researches in Palestine II*. London.

Columella 1969–79. *Columella, Lucius Junius Moderatus: De Re Rustica* (trans. H.B. Ash, E.S. Forster and E. Heffner), 3 Vols. London.

Conder, C.R. and Kitchener, H.H. 1881–1883. *The Survey of Western Palestine.* 3 Vols. London.

Corbo, V. 1955. *Gli Scavi di Kh. Siyar el-Ghanam*. Jerusalem.

Crowfoot, J.W., Crowfoot, G.M. and Kenyon, K.M. 1957. *Samaria-Sebaste III. The Objects from Samaria*. London.

Crowfoot, J.W., Kenyon, K.M. and Sukenik, E.L. 1942. *Samaria-Sebaste I. The Buildings at Samaria*. London.

Dagan, Y. and Avganim, A. 1998. "Nahal Zanoah (Site 04/12)". *ESI* 17: 127–129.

Dahari, U. 1998. "Horbat Tinshemet, Church of St. Bacchus". *ESI* 18: 67–68.

Dalman, G. 1928–1942. *Arbeit und Sitte in Palästina*. 7 Vols. Gütersloh (repr. Hildesheim 1964).

Danby, H. 1933. *The Mishnah*. London.

Dar, S. 1986. *Landscape and Pattern. An Archaeological Survey of Samaria 800 B.C.E. - 636 C.E.* (BAR IS 308). Oxford.

Dar, S. 1993. *Settlements and Cult Sites on Mt. Hermon, Israel* (BAR IS 589). Oxford.

Dar, S. 1999. *Sumaqa, A Roman and Byzantine Jewish Village on Mount Carmel, Israel* (BAR IS 815). Oxford.

Dar, S. 2004. *Raqit, Marinus' Estate on the Carmel, Israel* (BAR IS 1300). Oxford.

Dar, S., Safrai, Z. and Tepper, Y. 1986. *Um Rihan: A Village of the Mishnah*. Tel Aviv (Hebrew).

Dar, S. and Ziegelmann, A. 1997. "H. el-Kerak in the Carmel". In S. Dar and Z. Safrai (eds.), *The Village in Ancient Israel*. Tel Aviv: 161–191 (Hebrew).

Davidson Weinberg, G. (ed.). 1988. *Excavations at Jalame, Site of a Glass Factory in Late Roman Palestine*. Columbia.

Dorpfeld, W. 1895. "Die Ausgrabungen am Westabhange der Akropolis II". *Mittelungen des kaiserlich deutschen Archäologischen Institut* 20: 161–206.

Drachman, A.G. 1932. *Ancient Oil Mills and Presses*. Copenhagen.

Dray, Y. 1993. "The Judean Grooved Pier Press: A New Suggestion for the Reconstruction of its Operation". *NZ* 19: 91–95 (Hebrew).

Dray, Y. 1994. "The Oil Presses of Zur Natan". In: W. Neidinger, E. Matthews and E. Ayalon. *Report on TFAHR Excavations at Zur Natan, Israel*. Houston: 14–15.

Dray, Y. 2003. "Wine Production in the Improved Byzantine Wine Press". In: Y. Eshel (ed.). *Judea and Samaria Research Studies* 12: 219–228 (Hebrew).

Eisenberg, E. 2000. "Nahal Yarmut". *HA-ESI* 112: 91*-93*.

Eisenberg, E. and Ovadiah, R. 1998. "A Byzantine Monastery at Mevo Modi'im". *'Atiqot* 36: 1*-19* (Hebrew).

Eitam, D. 1996. "A Survey of Rural Installations – The Eastern Valleys and the Fringes of the Desert". In A. Zertal. *The Manasseh Hill Country Survey, The Eastern Valleys and The Fringes of The Desert*. Tel Aviv: 681–738 (Hebrew).

Eitam, D. 1997. "Mass Production of Cereal Groats in Ancient Israel during the Roman and Byzantine Periods". In: Y. Eshel (ed.). *Judea and Samaria Research Studies, Proceedings of the Sixth Annual Meeting*. Kedumim-Ariel: 191–202 (Hebrew).

Eitam, D. 1999. "The Bread of the Bible – of Groats or of Dough". In: Y. Avishur and D. Deutsch (eds.). *Michael – Papers in Honour of Michael Heltzer*. Tel Aviv: 119*-125* (Hebrew).

Elizur, Y. 1986. "Oil Press Caves – "Oil Press Built into the Rock"". *NZ* 13: 179–190 (Hebrew).

ESI 2 (1983): 61: "Kefar Ya'bes (Kiludiya)".

Féderlin, L. 1906. "À propos d'Isaie X 29–31". *RB* 15: 266–273.

Feig, N. 1990. "Wine Presses of the Hellenistic Period from Tell es-Samariya". *Israel – People and Land, Eretz Israel Museum Yearbook* 5–6 (23–24). Tel Aviv: 73–91 (Hebrew).

Feig, N. 2002. "Salvage Excavations at Meron". *'Atiqot* 43: 87–114.

Finkielsztejn, G. 2005. "Les mosaïques de la komopolis

de Porphyreon du Sud (Kfar Samir ; Haïfa, Israël): un évêché (?) entre village et cité". In: H. Morlier (ed.). *La mosaïque gréco-romaine IX. Collection de l'École Française de Rome* 352 (2 Vols.). Paris: 435–452.

Fischer, M. 1985. "Excavations at Horvat Zikhrin". *Qadmoniot* 18: 112–121 (Hebrew).

Fitzgerald, G.M. 1939. *A Sixth Century Monastery at Beth-Shan*. Philadelphia.

Foerster, G. 1986. "A Painted Christian Burial Cave near Kibbutz Lohamei Hageta'ot". In: M. Yedaya (ed.) *The Western Galilee Antiquities*. Tel Aviv: 416–431 (Hebrew).

Frankel, B. 1981a. "Kh. Dukas – Excavation Report". *IAA Archives*.

Frankel, B. 1981b. "Notes and Remarks on Olive Oil Production in Judea." *Teva Va'Aretz* 23: 183 (Hebrew).

Frankel, R. 1984. *The History of the Processing of Oil and Wine in Galilee in the Period of the Bible, the Mishna and the Talmud.* Unpublished Ph.D. Thesis. Tel Aviv University (Hebrew).

Frankel, R. 1985a. "The Hellenistic Aqueduct of Acre-Ptolemais". *'Atiqot* 17: 134–138.

Frankel, R. 1985b. "Western Galilee, Oil Presses". *ESI* 4: 110–114.

Frankel, R. 1986a. "Horvat Din'ila". *ESI* 5: 21–23.

Frankel, R. 1986b. "Khirbet el-Quseir". *ESI* 5: 88–91.

Frankel, R. 1988–1989. "An Oil Press at Tel Safsafot". *Tel Aviv* 15–16: 77–91.

Frankel, R. 1992. "Some Oil Presses from Western Galilee". *BASOR* 286: 39–71.

Frankel, R. 1993a. "The *Trapetum* and the *Mola Olearia*". In: M.-C. Amouretti and J.-P. Brun (eds.). *Oil and Wine Production in the Mediterranean Area* (BCH Suppl. 26). Paris: 477–481.

Frankel, R. 1993b. "Screw Weights from Israel". In: M.-C. Amouretti and J.-P. Brun (eds.). *Oil and Wine Production in the Mediterranean Area* (BCH Suppl. 26). Paris: 107–118.

Frankel, R. 1994. "Ancient Oil Mills and Presses in the Land of Israel". In: R. Frankel, S. Avitsur and E. Ayalon. *History and Technology of Olive Oil in the Holy Land.* Arlington and Tel Aviv: 19–89.

Frankel, R. 1996. "Oil Presses in Western Galilee and Judaea – A Comparison". In: D. Eitam and M. Heltzer (eds). *Olive Oil in Antiquity, Israel and Neighbouring Countries from the Neolithic to the Early Arab Period* (HANE/S VII). Padova: 197–218.

Frankel, R. 1997. "Presses for Oil and Wine in the Southern Levant in the Byzantine Period". *DOP* 51: 73–84.

Frankel, R. 1999. *Wine and Oil Production in Antiquity in Israel and Other Mediterranean Countries*. Sheffield.

Frankel, R. 2001. "Cato's Press a Reappraisal". In: J.-P. Brun and P. Jockey (eds.). *Techniques et sociétés en Mediterranée, Hommage à Marie-Claire Amouretti.* Paris: 313–325.

Frankel, R. 2003. "Olives and Olive Oil in Israel and the Southern Levant in Antiquity". In: J.M. Blazquez Martinez and J. Remesal Rodriguez (eds.). *Estudio Estudios sobre el Monte Testaccio (Rama) III.* Barcelona: 637–662.

Frankel, R. 2006. "Two installations for the production of olive oil". In: I. Finkelstein, D. Ussishkin and B. Halpern (eds.). *Megiddo IV, The 1998–2002 Seasons.* Tel Aviv: 618–629.

Frankel, R, Avitsur, S. and Ayalon, E. 1994. *History and Technology of Olive Oil in the Holy Land.* Arlington and Tel Aviv.

Frankel, R. and Ayalon, E. 1988. *Vines, Winepresses and Wine in Antiquity*. Tel Aviv (Hebrew).

Frankel, R., Getzov, N. and Friedman, D. 1989. "The Aqueducts of Acre and the Western Galilee". In: D. Amit, Y. Hirschfeld and J. Patrich. *The Aqueducts of Ancient Palestine*. Jerusalem: 81–96 (Hebrew).

Frankel, R., Patrich, J. and Tsafrir, Y. 1990. "The Oil Press at Horvath Beit Loya". In: G.C. Bottini, L. Di Segni and E. Alliata (eds.). *Christian Archaeology in the Holy Land, New Discoveries*. Jerusalem: 287–300.

Gal, Z. 1986. "Vineyard Cultivation at 'Emek Harod and its Vicinity During the Roman-Byzantine Period". *Israel – People and Land, Haaretz Museum Yearbook* 2–3 (20–21). Tel Aviv: 129–138 (Hebrew).

Garrow Duncan, J. 1926. "New Rock Chambers and Galleries on Ophel". *PEF QSt*: 7–14.

Geoponica. *Geoponica sive cassiani bassi scholastici de resticos ecologae* (ed. H. Beckh). Leipzig 1895.

Gershuny, L. 2006. "Excavations at Khirbat Marmita". *'Atiqot* 53: 139–178.

Gibson, S. 1985–6. "Ras et-Tawil: Byzantine Monastery North of Jerusalem". *BAIAS* 5: 69–73.

Gibson, S. 1995. *Landscape Archaeology and Ancient Agricultural Field Systems in Palestine.* Unpublished Ph.D. Thesis, University College of London. London.

Gibson, S. and Edelstein, G. 1985. "Investigating Jerusalem's Rural Landscape". *Levant* 17: 139–155.

Gichon, M. 1979–1980. "The Upright Screw-Operated Pillar Press in Israel". *SCI* 5: 206–243.

Giveon, R. 1988. "Area 1 – An Oil Press and Domestic Buildings". In: B. Mazar (ed.). *Geva, Archaeological Discoveries at Tell Abu-Shusha, Mishmar Ha-'Emeq.* Tel Aviv: 215–218 (Hebrew).

Golani, A. 2005. "Salvage Excavations in the Modi'in Landscape". *'Atiqot* 50: 73–97.

Gophna, R. and Beit-Arieh, Y. 1997. *Archaeological Survey of Israel, Map of Lod (80)*. Jerusalem.

Goren, D. 1994. *The Urban Character of Gamla during the First Century BCE and the First Century CE.* Unpublished MA Thesis, Bar-Ilan University, Ramat Gan.

Greenhut, Z. 1998. "Horvat Hermeshit (1988–1990)". *'Atiqot* 34: 121–172 (Hebrew).

Gudovitz, S., Mintzker, Y. and Avni, G. 1986. "Horvat Hazzan". *ESI* 4: 46–48.

Guérin, V. 1880. *Description géographique, historique et archéologique de La Palestine.* 7 Vols. Paris (repr. Amsterdam 1969).

Gutman, S. 1994. *Gamla – A City in Rebellion.* Tel Aviv (Hebrew).

HA 11 (1964): 15–17: "American Expedition to Study Ancient Glass in Israel" (Hebrew).

HA 13 (1965): 10–11: "Mosaic Pavement Discovered at Modi'im" (Hebrew).

HA 16 (1965): 10–12: "The Expedition to Study Ancient Glass: Kh. Jalame ('Asafne)" (Hebrew).

HA 21 (1967): 13–14: "The Study of Ancient Glass" (Hebrew).

HA 25 (1968): 8–9: "The Expedition to Study Ancient Glass" (Hebrew).

HA 34–35 (1970): 19: "French Hill in Jerusalem" (Hebrew).

HA 38 (1971): 16–19: "French Hill" (Hebrew).

HA 65–66 (1978): 5–6: "H. Massah" (Hebrew).

Hadas, A. In press. "Methods of Wine Production in Ancient Wine Presses". In: S. Gudovitz, "The Excavations at Mulabbis". *'Atiqot*.

Hadjisavvas, S. 1992. *Olive Oil Production in Cyprus from the Bronze Age to the Byzantine Period*. Nicosia.

Hamilton, R.W. 1935. "Note on a Chapel and Winepress at 'Ain el Jedide". *QDAP* 4: 111–117.

Heltzer, M. and Eitam, D. (eds.). 1987. *Olive Oil in Antiquity, Israel and Neighbouring Countries from Neolith to Early Arab Period*. Haifa.

Hero of Alexandria – *Herons von Alexandria Mechanik und Katoptrik* (L. Nix and W. Schmidt, trans.). Leipzig 1900.

Herzog, Z., Rapp, G. Jr. and Negbi, O. (eds.). 1989. *Excavations at Tel Michal, Israel*. Minneapolis and Tel Aviv.

Hirschfeld, Y. 1983. "Ancient Wine Presses in the Park of Aijalon". *IEJ* 33: 207–218.

Hirschfeld, Y. 1985. "Khirbet el-Quneitra – A Byzantine Monastery in the Wilderness of Ziph". *EI* 18: 243–255 (Hebrew).

Hirschfeld, Y. 1990 "List of the Byzantine Monasteries in the Judean Desert". In: G.C. Bottini, L. Di Segni and E. Alliata (eds.). *Christian Archaeology in the Holy Land, New Discoveries*. Jerusalem: 1–90.

Hirschfeld, Y. 1992. *The Judean Desert Monasteries in the Byzantine Period*. New Haven and London.

Hirschfeld, Y. 2000. *Ramat Hanadiv Excavations, Final Report of the 1984–1998 Seasons*. Jerusalem.

Hirschfeld, Y. 2004. "Excavations at 'Ein Feshkha, 2001: Final Report". *IEJ* 54: 37–74.

Hirschfeld, Y. and Kloner, A. 1988–9. "Khirbet el-Qasr: A Byzantine Fort in the Judean Desert". *BAIAS* 8: 5–20.

Humbel, X. 1976. *Vieux pressoirs sans frontières*. Paris.

Huttenmeister, F. and Reeg, G. 1977. *Die Antiken Synagogen in Israel* I. Weisbaden.

Ilan, Z. 1991. *Ancient Synagogues in Israel*. Tel Aviv (Hebrew).

Iliffe, J.H. 1935. "Cemeteries and a 'Monastery' at the YMCA, Jerusalem". *QDAP* 4: 70–80.

Israel, Y. 1993. "Ashqelon". *ESI* 13: 100–105.

Josephus, *Jewish War* (Trans. J. Thackery). London 1927.

Josephus, *The Life of Josephus*. In: *The Works of Flavius Josephus* (Trans. W. Whiston). Philadelphia 1831: 128–161.

Jungst, E. and Thielscher, P. 1954 "Catos Keltern und Kollergange". *BJ* 154: 32–93.

Kaplan, J. 1993. "Tel Aviv". *NEAEHL* 4: 1451–1457.

Khalil, L.A. and al-Nammari, F.M. 2000. "Two Large Wine Presses at Khirbet Yajuz, Jordan". *BASOR* 318: 41–57.

Kloner, A. 1983. "The Subterranean Hideways of the Judean Foothills and the Bar-Kokhba Revolt". *The Jerusalem Cathedra* 3: 114–135.

Kloner, A. 1986a. "Horvat Benaya". *ESI* 5: 12–14.

Kloner, A. 1986b. "The Structure at Hulda and its Use as a Wine Press". In: A. Kasher, A. Oppenheimer and U. Rappaport (eds.). *Man and Land in Eretz-Israel in Antiquity*. Jerusalem: 197–208 (Hebrew).

Kloner, A. 1989. "The Upright Pillar Press in Caves in Judaea". *NZ* 15: 66–74 (Hebrew).

Kloner, A. 2006. "The Structure and Installations at Hulda and Their Function as a Jewish Winepress". In: E.M. Maeir and P. de Miroschedji (eds.). *"I Will Speak the Riddle of Ancient Times", Archaeological and Historical Studies in Honor of Amihai Mazar*. Winona Lake: 853–864.

Kloner, A. and Sagiv, N. 1991. "The Technology of Oil Production in the Hellenistic Period at Maresha, Israel". In: M.-C. Amouretti, J.-P. Brun and D. Eitam (eds.). *Oil and Wine Production in the Mediterranean Area from the Bronze Age to the End of the XVIth Century. Pré-Actes*. Aix-en-Provence-Toulon: 61–68.

Kloner, A. and Sagiv, N. 1993. "The Olive Presses of Hellenistic Maresha". In: M.-C. Amouretti and J.-P. Brun (eds.). *Oil and Wine Production in the Mediterranean Area* (BCH Suppl. 26). Paris: 119–136.

Kloner, A. and Sagiv, N. 2003. "Subterranean Complexes 44 and 45". In: A. Kloner. *Maresha Excavations Final Report I. Subterranean Complexes 21, 44, 70* (IAA Reports No. 17). Jerusalem: 51–72.

Kloner, A. and Tepper, Y. 1987. *The Hiding Complexes in the Judean Shephelah*. Tel Aviv (Hebrew).

Kochavi, M. 1972. "The Land of Judah". In: M. Kochavi (ed.). *Judaea, Samaria and the Golan, Archaeological Survey 1967–1968*. Jerusalem: 19–91 (Hebrew).

Kogan-Zehavi, E. 2000a. "Rehovot, Khirbet Deiran". *HA-ESI* 111: 59*.

Kogan-Zehavi, E. 2000b. "Road 2 (North)". *ESI* 20: 88.

Kohl, H. and Watzinger, C. 1916. *Antike Synagogen in Galilea*. Leipzig.

Lauffer, S. 1971. *Diokletians Preisedict*. Berlin.

Levine, L.I. and Netzer, E. 1986. *Excavations at Caesarea Maritima 1975, 1976, 1979 – Final Report* (Qedem 21). Jerusalem.

Liphschitz, N. 1986a. *Dendroarchaeological Research Report #143*. Tel-Aviv University (Hebrew).

Liphschitz, N. 1986b. "Overview of the Dendrochronology and Dendroarchaeology in Israel". *Dendrochronologia* 4: 37–58.

Liphschitz, N. 1996. "Olives in Ancient Israel in View of Dendroarchaeological Investigations". In: D. Eitam and M. Heltzer (eds.). *Olive Oil in Antiquity, Israel and Neighbouring Countries from the Neolithic to the Early*

Arab Period (HANE/S VII). Padova: 7–13.

Liphschitz, N. and Biger, G. 1990. "Dominance of *Quercus calliprinos* (Kermes oak) – *Pistacia palaestina* (Terebinth) Association in the Mediterranean Territory of Eretz Israel during Antiquity". *JVS* 1: 67–70.

Liphschitz, N., Biger, G. and Ayalon, E. 1994. "The Use of Timber in the Oil Press Installations in Eretz Israel". *Israel – People and Land, Eretz Israel Museum Yearbook* 7–8 (25–26): 199–216 (Hebrew).

Liphschitz, N. and Bonani, G. 2000. "Dimensions of Olive (*Olea europaea*) Stones as a Reliable Parameter to Distinguish Between Wild and Cultivated Varieties: Further Evidence". *Tel Aviv* 27: 23–25.

Liphschitz, N. *et al.* 1991. "The Beginning of Olive (*Olea europaea*) Cultivation in the Old World: a Reassessment". *JAS* 18: 441–453.

Liphschitz, N. *et al.* 1996. "Wild Olive (*Olea europaea*) Stones from a Chalcolithic Cave at Shoham, Israel and Their Implications". *Tel Aviv* 23: 135–142.

Liphschitz, N. and Waisel, Y. 1977. "Dendroarchaeological Investigations in Israel: Central Negev: Nahal Zin". In: A. Marks (ed.). *Prehistory and Palaeoenvironments in the Central Negev* Vol II. Minnesota: 355–356.

Magen, I. 1982. *The Archaeological Discoveries at Qedumim – Samaria, Qedem Museum*. Jerusalem.

Magen, Y. 1984. "Kalandia – A Vineyard Farm and Winery of Second Temple Times". *Qadmoniot* 17 (66–67): 61–71 (Hebrew).

Magen, Y. 1993a. "Mount Gerizim and the Samaritans". In: F. Manns and E. Alliata (eds.). *Early Christianity in Context, Monuments and Documents* (Studium Biblicum Franciscanum, Collectio Maior, 38). Jerusalem: 91–148.

Magen, Y. 1993b. "Qedumim – A Samaritan Site of the Roman-Byzantine Period". In: F. Manns and E. Alliata (eds.). *Early Christianity in Context, Monuments and Documents* (Studium Biblicum Franciscanum, Collectio Maior, 38). Jerusalem: 167–180.

Magen, Y. 2000. "Mt. Gerizim – A Temple City". *Qadmoniot* 33: 74–118 (Hebrew).

Magen, Y. 2004. "Kalandia – A Second Temple-period Viticulture and Wine-manufacturing Agricultural Settlement". In: Y. Magen *et al. The Land of Benjamin*. Jerusalem: 29–144.

Magen, Y. 2008. "Oil Production in the Land of Israel in the Early Islamic Period". In: *Judea and Samaria Researches and Discoveries* (JSP 6). Jerusalem: 217–257.

Magen, Y., Tzionit,Y. and Sirkis, O. 2004. "Khirbet Badd 'Isa – Qiryat Sefer". In: The Land of Benjamin. Jerusalem: 179–241.

Manns, F. 1984. *Some Weights of the Hellenistic, Roman and Byzantine Periods*. Jerusalem.

Ma'oz, Z.U. 1988. "Ancient Synagogues of the Golan". *BA* 52: 116–128.

Ma'oz, Z.U. 1993. "'Ein Nashut". *NEAEHL* 2: 412–414.

Mayerson, P. 1985. "The Wines and Vineyards of Gaza in the Byzantine Period". *BASOR* 257: 75–80.

Mayerson, P. 1992. "The Gaza Wine Jar (Gazition) and the Lost Ashkelon Jar (Askalonium)". *IEJ* 42: 76–80.

Mayerson, P. 1993. "The Use of Ascalon Wine in the Medical Writers of the Fourth to the Seventh Centuries". *IEJ* 43: 169–173.

Mazar, B. (ed.). 1988. *Geva, Archaeological Discoveries at Tell Abu-Shusha, Mishmar Ha-'Emeq*. Tel Aviv (Hebrew).

Mazar, E. 1994. "A Burial Ground of the Roman Period at Gesher Haziv". *'Atiqot* 25: 77–93.

Mazor, G. 1981. "The Wine-Presses of the Negev". *Qadmoniot* 14: 51–60 (Hebrew).

Messika, N. 1995. "Lohame Hageta'ot (Giv'at Yassaf)". *ESI* 14: 134–135.

Mettens, A. 2004. "Horbat Nashé". *HA-ESI* 116: 29*-33*.

Meyers, E.M., Strange, J.F. and Meyers, C.L. 1981. *Excavations at Ancient Meiron, Upper Galilee, Israel 1971–72, 1974–75, 1977*. Cambridge, Mass.

Mishnah – see Danby 1933.

Mommsen, T. and Kreuger, P. 1985. *The Digest of Justinian* (trans. A. Watson). Philadelphia, Pennsylvania.

Nadelman, Y. 2000. "Rehovot, Kh. Deiran (A)". *ESI* 20: 93*.

Neidinger, W., Matthews, E. and Ayalon, E. 1990. *Preliminary Report on the Excavations at Zur Natan, 1989 and 1990 Seasons*. Houston.

Neidinger, W., Matthews, E. and Ayalon, E. 1994. *Report on TFAHR Excavations at Zur Natan, Israel*. Houston.

Netzer, E. 2002. "Date "Winepresses" in the Royal Estate at Jericho". In: Y. Eshel (ed.). *Judea and Samaria Researches Studies*, Vol. 11. Ariel: 69–80 (Hebrew).

Oliphant, L. 1887. *Haifa or Life in Modern Palestine*. Edinburgh and London.

Onn, A. and Rapuano, Y. 1995. "Jerusalem, Kh. er-Ras". *ESI* 13: 71.

Oren, E. 1970. "French Hill in Jerusalem". *HA* 34–35: 19 (Hebrew).

Oren, E. 1971. "French Hill in Jerusalem". *HA* 38: 16–19 (Hebrew).

Paley, S.M. and Porath, Y. 1993. "Tel Hefer". *NEAEHL* II: 609–614.

Palladius – Owen, T. 1807 (trans.). *The Fourteen Books of Palladius Rutilius Taurus Aemilianus on Agriculture*. London.

Palladius. 1898. *Agricultura* (ed. J.C. Schmitt). Leipzig.

Paton, W.R. and Myres, J.L. 1898. "On Some Karian and Hellenic Oil-Presses". *JHS* 18: 209–217.

Patrich, J. and Tsafrir, Y. 1985. "A Byzantine Church and Agricultural Installations at Khirbet Beit Loya". *Qadmoniot* 18: 106–112 (Hebrew).

Peilstöcker, M. and Kapitaikin, A. 2000. "Bet Dagan". *ESI* 20: 59*-60*.

Peleg, M. 1984. "Lohamei Hageta'ot". *ESI* 3: 72.

Peleg, M. 1991. "Persian, Hellenistic and Roman Burials at Lohamei Hageta'ot". *'Atiqot* 20: 131–152.

Peleg, Y. 1981. "How Ancient Olive Presses Worked". *Israel, Land and Nature* 6: 98–103.

Peleg, Y., Feller, Y. and Srukh, I. 2004. "Karmé Zur". *HA-ESI* 116: 61*-62*.

Pifano, S. 1983. "Kefar Saba". *HA* 83: 27 (Hebrew).

Pliny the Elder. *Naturalis Historiae*. 10 Vols. (Trans. H. Rackham, W.H.S. Jones and D.E. Eichholz). London and Cambridge 1958–1963.

Prausnitz, M.W. 1993. "Achzib". *NEAEHL* I: 32–35.

Rahmani, L.Y. 1991. "Two Byzantine Winepresses in Jerusalem". *'Atiqot* 20: 95–110.

Reich, R. 1990. *Miqwa'ot (Jewish Ritual Immersion Baths) in Eretz-Israel in the Second Temple and the Mishnah and Talmud Periods*. Unpublished Ph.D. Thesis. Hebrew University, Jerusalem.

Robinson, D.M. 1938. *Excavations at Olynthus VII: The Hellenic House*. London.

Roll, I. 1999. "Introduction: History of the Site, Its Research and Excavations". In: I. Roll and O. Tal. *Apollonia-Arsuf, Final Report of the Excavations Vol. I. The Persian and Hellenistic Periods* (Tel Aviv University Institute of Archaeology Monograph Series No. 16). Tel Aviv: 1–62.

Roll, I. and Ayalon, E. 1981. "Two Large Wine Presses in the Red Soil Regions of Israel". *PEQ* 130: 111–125.

Rosenfeld, B.-Z. 1991. "The History of the Settlement of Two Families from the High Priesthood during the Second Temple Period". In: Y. Katz *et al.* (eds.). *Studies in the Historical Settlement Geography of the Land of Israel*, Vol. 2. Jerusalem: 206–218 (Hebrew).

Rosenson, I. 1996. "The Essence and Siting of Oil Producing Facilities in the Mishnaic and Talmudic Period Based on Authentic Sources". In: D. Eitam and M. Heltzer (eds.). *Olive Oil in Antiquity, Israel and Neighbouring Countries from the Neolithic to the Early Arab Period* (HANE/S VII). Padova: 113–118.

Rossiter, J.J. 1981. "Wine and Oil Processing at Roman Farms in Italy". *Phoenix* 35: 345–361.

Rossiter, J.J. and Haldenby, E. 1989. "A Wine-Making Plant in Pompeii Insula II, 5". *Échos du Monde Classique / Classical Views* 33/8: 229–239.

Rozenson, I. 1987. "Vines on the Hermon – Literary Allusions to Jewish Agricultural Activity". *Cathedra* 46: 191–193 (Hebrew).

Safrai, Z. and Linn, M. 1988. "Excavations and Surveys in the Mishmar Ha-'Emeq Area". In: B. Mazar (ed.). *Geva, Archaeological Discoveries at Tell Abu-Shusha, Mishmar Ha-'Emeq*. Jerusalem: 167–214 (Hebrew).

Sagiv, N. and Kloner, A. 1996. "Maresha: Underground Olive Oil Production in the Hellenistic Period". In: D. Eitam and M. Heltzer (eds.). *Olive Oil in Antiquity, Israel and Neighbouring Countries from the Neolithic to the Early Arab Period* (HANE/S VII). Padova: 255–292.

Saller, S.J. 1941. *The Memorial of Moses on Mount Nebo*. Jerusalem.

Saller, S.J. 1957. *Excavation at Bethany (1949–1953)*. Jerusalem.

Saller, S. 1961. "The Archaeological Setting of the Shrine of Bethphage". *LA* 11: 194–208.

Scheltema, H.J. and Holwerda, D. 1957. *Basilicorum Libri, LX* (Series B volumen III in Libr. XV-XX). Groningen and The Hague.

Schumacher, G. 1888. *The Jaulan*. London.

Seligman, J. 1994. "Jerusalem, Pisgat Ze'ev (East 'A)". *ESI* 12: 52–54.

Seligman, J. 1997. "Nahal Haggit". *ESI* 16: 61–63.

Seligman, J. 1999. "Agricultural Complexes at Ras Abu Ma'aruf (Pisgat Ze'ev East 'A)". *'Atiqot* 38: 137–170.

Seligman, J. Forthcoming. "Nahal Haggit – A Roman and Medieval Farming Community in the Southern Carmel". *IAA Reports*. Jerusalem.

Shavit, E. 1995. "Rishon LeZiyyon". *ESI* 13: 57.

Shenhav, E. 1984. "Ya'ar Lavi Survey". *HA* 84: 18 (Hebrew).

Shenhav, E. 1993. "Shuni". *NEAEHL* 4: 1382–1384.

Shukron, E. and Savariego, A. 1994. "Jerusalem, Pisgat Ze'ev (Villa Quarter)". *ESI* 12: 56–58.

Sidi, N., Amit, D. and 'Ad, U. 2003. "Two Winepresses from Kefar Sirkin and Mazor". *'Atiqot* 44: 253–266.

Siegelmann, A. 1998. "An Oil Press of the Byzantine Period in Qiryat Ata". *'Atiqot* 34: 101–106 (Hebrew).

Sion, O. 2001. *Settlement History in Central Samarian Region in the Byzantine Period*. Ph.D. Thesis, The Hebrew University. Jerusalem (Hebrew).

Sion, O. 2002. "Excavations at Rogem Ganim, Kiryat Menahem, Jerusalem". In: Y. Eshel (ed.). *Judea and Samaria Researches Studies*, Vol. 11. Ariel: 115–122 (Hebrew).

Sion, O. and 'Amar, Z. 1997. "Treading Installations at Khirbet Hablâtâ (Hallamish)". *'Atiqot* 32: 205–207 (Hebrew).

Stager, L.E. and Wolff, S.R. 1981. "Production and Commerce in Temple Courtyards: An Olive Press in the Sacred Precinct at Tel Dan". *BASOR* 243: 95–102.

Stahl, A. 1980. "Olive Oil Production in Judea". *Teva Va'Aretz* 23: 80–81 (Hebrew).

Sussman, V. 1969. "An Ancient Burial Cave at Rehovot". *'Atiqot* 5: 69–71 (Hebrew).

Syon, D. 1994. "The Coins of Gamla". In: S. Gutman. *Gamla – A City in Rebellion*. Tel Aviv: 151–172 (Hebrew).

Syon, D. 1998. "A Winepress at Akhziv". *'Atiqot* 34: 85–99 (Hebrew).

Tahal, G. 1995. "'Avedat". *ESI* 14: 130–133.

Tchalenko, G. 1953. *Villages antiques de la Syrie du Nord*. Paris.

Tchernia, A. and Brun, J.-P. 1999. *Le vin romain antique*. Grenoble.

Tepper, Y. 1982. "The Oil Plant at Giv'eon". *NZ* 18: 66–86 (Hebrew).

Tepper, Y. 1987. "The Oil Presses at the Maresha Region". In: M. Heltzer and D. Eitam (eds.). *Olive Oil in Antiquity, Israel and Neighbouring Countries from the Neolithic to Early Arab Period*. Haifa: 25*–42* (Hebrew).

Tosefta (trans. J. Neusner). Hoboken, N.J., 1977.

Tzaferis, V. 1984. *The Excavations at Kursi-Gergasa* (*'Atiqot* 16). Jerusalem.

Tzaferis, V. and Shai, T. 1976. "Excavations at Kafr er-Rameh". *Qadmoniot* 9: 83–85 (Hebrew).

Urman, D. 1974. "Presses for the Production of Grape Syrup in the Golan". *Teva Va'Aretz* 16: 173–176

(Hebrew).
Urman, D. 1979. *The Golan During the Roman and Byzantine Periods*. Unpublished Ph.D. Thesis, New York University. New York.
Urman, D. 1985. *The Golan: A Profile of a Region during the Roman and Byzantine Periods* (BAR IS 269). Oxford.
van den Brink, E.C.M. *et al.* 2001. "Chalcolithic Dwelling Remains, Cup Marks and Olive (*Olea europaea*) Stones at Nevallat". *IEJ* 51: 36–43.
Varro. "Marcus Terentius Varro: Rerum Rusticarum". In: *Cato and Varro* (trans. W.D. Hooper). London 1979: 161–528.
Villeneuve, F. 1988. "Fouilles à Khirbet Edh-Dharia (Jordanie) 1984–1987". *CRAIBL*. Paris: 458–479.
Vitruvius. *De Architectura* (trans. F. Granger), 2 Vols. Cambridge 1962.
Wace, A.J.B. 1921–1923. "Excavations at Mycenae". *ABSA* 25.
Wagner, D. 1996. "Oil Production at Gamla". In: D. Eitam and M. Heltzer (eds.). *Olive Oil in Antiquity, Israel and Neighbouring Countries from the Neolithic to the Early Arab Period* (HANE/S VII). Padova: 301–306.
Weinberg, S.S. 1988. "The Buildings and Installations". In: G. Davidson Weinberg (ed.). *Excavations at Jalame, Site of a Glass Factory in Late Roman Palestine*. Columbia: 5–23.
White, K.D. 1975. *Farm Equipment of the Roman World*. Cambridge.
Yadin, Y. 1965. *Masada: First Season of Excavations, 1963–64*. Jerusalem.
Yalqut HaPirsumim ("List of Monuments and Historical Sites") 1091, Jerusalem 1964 (Hebrew).
Yeivin, Z. 1966. "Two Ancient Oil Presses". *'Atiqot* (HS) 3: 52–62 (Hebrew).
Yeivin, Z. 1984. "Korazim 1983–1984". *ESI* 3: 66–71.
Yeivin, Z. 1993. *Horvat Susiya, A Town of the Talmudic Period (Guide)*. Jerusalem (Hebrew).
Yeivin, Z. and Edelstein, G. 1970. "Excavations at Tirat Yehuda". *'Atiqot* (HS) 6: 56–67 (Hebrew).
Yevin, Z. and Finkielsztejn, G. 1999a. Catalogue of the exhibition *Castra at the Foot of Mount Carmel, the City and its Secrets*. The National Maritime Museum. Haifa.
Yevin, Z. and Finkielsztejn, G. 1999b. "Horbat Castra – 1993–1997". *HA-ESI* 109 :23*-27*.
Yevin, Z. and Finkielsztejn, G. 2001. "Kfar Samir", "Castra Samaritanorum" and "Porphyreon of the South". In: A. Negev and S. Gibson (eds). *Archaeological Encyclopedia of the Holy Land*. New York-London: 116, 279–280, 395.
Yevin, Z. and Finkielsztejn, G. 2008. "Kafr Samir". In: E. Stern (ed.). *NEAEHL* 5. Jerusalem: 1895–1899.
Yogev, O. 1982. "Aderet". *ESI* 1: 1.
Yogev, O. and Rochman, A. 1985. "Tell es-Samariya". *ESI* 4: 103–104.
Yron-Lubin, M. 1995. "Horvat Hermeshit (Ne'ot Qedumim)". *ESI* 14: 83–85.
Yron-Lubin, M. 1996a. "Horvat Hermeshit (Ne'ot Qedumim) – 1993". *ESI* 15: 66–68.
Yron-Lubin, M. 1996b. *Settlements in the Northern Shephelah in the Early Bronze Age, Iron Age II, and the Byzantine Period: Site Cachment, Carrying Capacity and Landscape Archaeology*. Unpublished M.A. Thesis, Tel Aviv University (Hebrew).
Yron-Lubin, M. 1999. "Horbat Hermeshit (Ne'ot Qedumim) – 1995". *ESI* 19: 46*-49*.
Yron-Lubin, M. In preparation. *Four Seasons of Excavation in Horvat Hermeshit* (Hebrew).
Yron-Lubin, M. and Sebah, A. 2000. "Horbat Hermeshit (Ne'ot Qedumim) – Preservation and Restoration". *ESI* 20: 69*-73*.
Zelinger, Y. 2000. "Gophna during the Second Temple Period". In: Y. Eshel (ed.). *Judea and Samaria Research Studies* 10. Ariel: 103–112 (Hebrew).
Zori, N. 1977. *Nahalat Issachar*. Jerusalem (Hebrew).

www.ingramcontent.com/pod-product-compliance
Ingram Content Group UK Ltd.
Pitfield, Milton Keynes, MK11 3LW, UK
UKHW061843041026
15591UKWH00015B/889